AF332943

Progress in Mathematics
Volume 99

Series Editors
J. Oesterlé
A. Weinstein

Symplectic Geometry and Mathematical Physics

Actes du colloque en l'honneur de
Jean-Marie Souriau

P. Donato, C. Duval
J. Elhadad, G.M. Tuynman
Editors

1991

Birkhäuser

Boston · Basel · Berlin

P. Donato
Université de Provence
C.P.T.-CNRS Luminy Case 907
13288 Marseille Cedex 9
France

J. Elhadad
Université de Provence
C.P.T.-CNRS Luminy Case 907
13288 Marseille Cedex 9
France

C. Duval
Université d'Aix-Marseille II
CPT-CNRS Luminy Case 907
F-13288 Marseille Cedex 9
France

G. M. Tuynman
U.F.R. de Mathématiques
Université de Lille I
59655 Villeneuve d'Ascq Cedex
France

Library of Congress Cataloging-in-Publication Data

Colloque de géométrie symplectique et physique mathématique (1990 :
    Aix-en-Provence, France)
      Symplectic geometry and mathematical physics : actes du colloque
    en l'honneur de Jean-Marie Souriau / edited by P. Donato . . . [et
    al.].
          p.   cm. -- (Progress in mathematics : vol. 99)
      "Based on contributions to the conference 'Colloque de géométrie
    symplectique et physique mathématique' that was held in Aix-en-
    Provence (France). June 11--15, 1990"--P. [4] of cover.
      Includes bibliographical references.
      ISBN (invalid) 0-8176-3581-5 (hard : acid-free)
      1. Geometry, Differential--Congresses. 2. Symplectic manifolds--
    Congresses. 3. Mathematical physics--Congresses. 4. Souriau, J.-M.
    (Jean-Marie), 1922-   I. Souriau, J.-M. (Jean-Marie), 1922-
    II. Donato, P. (Paul), 1949- . III. Title. IV. Series: Progress
    in mathematics (Boston, Mass.) : vol. 99.
    QC20.7.D52C65   1990   91-25454
    516.3'6--dc20   CIP

Printed on acid-free paper.

ISBN 0-8176-3581-5
ISBN 3-7643-3581-5

Camera ready text prepared in TeX and AMS TeX.
Printed and bound by Edwards Brothers, Inc., Ann Arbor, Michigan.
Printed in the U.S.A.

9 8 7 6 5 4 3 2 1

Jean-Marie Souriau
Photograph by G. Bigot

TABLE OF CONTENTS

Preface

This volume contains the proceedings of the conference "Colloque de Géométrie Symplectique et Physique Mathématique" which was held in Aix-en-Provence (France), June 11–15, 1990, in honor of Jean-Marie Souriau. The conference was one in the series of international meetings of the Séminaire Sud Rhodanien de Géométrie, an organization of geometers and mathematical physicists at the Universities of Avignon, Lyon, Marseille, and Montpellier.

The scientific interests of Souriau, one of the founders of geometric quantization, range from classical mechanics (symplectic geometry) and quantization problems to general relativity and astrophysics. The themes of this conference cover "only" the first two of these four areas.

The subjects treated in this volume could be classified in the following way: symplectic and Poisson geometry (Arms-Wilbour, Bloch-Ratiu, Brylinski-Kostant, Cushman-Sjamaar, Dufour, Lichnerowicz, Medina, Ouzilou), classical mechanics (Benenti, Holm-Marsden, Marle), particles and fields in physics (García Pérez-Muñoz Masqué, Gotay, Montgomery, Ne'eman-Sternberg, Śniatycki) and quantization (Blattner, Huebschmann, Karasev, Rawnsley, Roger, Rosso, Weinstein). However, these subjects are so interrelated that a classification by headings such as "pure differential geometry, applications of Lie groups, constrained systems in physics, etc.," would have produced a completely different clustering!

The list of authors is not quite identical to the list of speakers at the conference. M. Karasev was invited but unable to attend; C. Itzykson and M. Vergne spoke on work which is represented here only by the title of Itzykson's talk (Surfaces triangulées et intégration matricielle) and a summary of Vergne's talk.

The organizers of the conference wish to thank the following institutions for financial support: Conseil Regional Provence-Alpes-Côte d'Azur, Mairie d'Aix-en-Provence, Université de Provence, Université d'Aix-Marseille II, Centre de Physique Théorique, Centre National de la Recherche Scientifique (G.D.R. 144). They are also grateful to the Scientific Committee: D. Bennequin, P. Dazord, A. Kirillov, B. Kostant, S. Sternberg, A. Weinstein.

P. Donato
C. Duval
J. Elhadad
G.M. Tuynman
Editors

Adresse au colloque

Mon cher Jean-Marie,

Il paraît que nous sommes réunis ici pour célébrer ton départ à la retraite. Pour ma part, franchement, je n'ai pas l'impression de m'adresser à quelqu'un qui est sur le point de cesser son activité professionnelle. N'en déplaise à nos amis universitaires et dommage pour les futures générations d'étudiants d'être privées d'excellents cours (j'en parle en connaissance de cause pour le cours d'algèbre linéaire), je suis ravi de te savoir détaché de tes tâches d'enseignement car j'ose espérer que tu seras encore plus présent au Centre de Physique Théorique. Tu en as été un des membres fondateurs, tu as été de ceux qui ont mis le bateau à l'eau et ont tenu la barre pendant huit longues années. J'ai d'ailleurs eu la lourde tâche de te succéder, aussi c'est avec un grand plaisir que je te verrai revenir à plein temps parmi nous.

Tu animes depuis plus de 20 ans les "Souristes", groupe social homogène avec son langage, sa culture, ses traditions : on ne devient pas "Souriste" du jour au lendemain !! Tu as séduit les jeunes chercheurs par tes immenses qualités scientifiques, bien sûr, mais aussi par l'image que tu sais donner de la science et de la recherche, faite de dynamisme, d'exaltation, d'honnêteté. Ta réputation va bien au delà de nos frontières car en particulier tu as développé des méthodes originales en quantification géométrique, en géométrie symplectique, ouvrant de nouvelles voies pour les mathématiciens. Je peux témoigner que certains de tes cours, vieux de 30 ans, sont gardés jalousement dans des bibliothèques privées, car ils constituent encore un outil de travail irremplaçable.

Nous espérons tous que ta vision de la structure de l'Univers s'avèrera conforme à la réalité, pour te conférer la place que tu mérites parmi les plus grands. Tu as déjà été couronné par plusieurs prix de l'Académie des Sciences et en 1986 par le Grand Prix Scientifique de la ville de Paris. Ta participation à des congrès internationaux est toujours très appréciée, même si tu t'obstines à parler en V.O. en parfait accord avec les directives ministérielles recommandant l'usage du français !!

Spécialiste mondial de la relativité, tu es même sollicité par les medias pour des émissions de télévision, des conférences grand public, car tu as ce don, assez rare dans la famille des chercheurs, de savoir parler de concepts ardus à des non spécialistes.

Laisse moi te remercier aujourd'hui de m'avoir permis de parler de toi, car tu appartiens à la petite galaxie des "étoiles modestes". Si tu as accepté de laisser dévoiler ta vraie "grandeur" c'est sans doute parce que tu savais que nous ne pouvions pas utiliser un autre télescope que celui de l'amitié !!

Jacques Soffer
Directeur du CPT-CNRS Marseille
Aix en Provence, Juin 1990

Quelques mots sur Jean-Marie Souriau

Dans l'espace des mouvements d'un système dynamique, on reconnaît des mouvements à diffusion où l'on passe d'un état initial à un état final, mais aussi des orbites périodiques, ainsi que des mouvements qui ne cessent jamais de présenter des nouveautés inattendues. Bien que nous fêtions, à l'occasion de ce colloque, une transition dans la vie professionnelle de notre ami Jean-Marie Souriau, je crois pouvoir exprimer en notre nom à tous la conviction qu'il s'agit non d'un évènement de diffusion, mais seulement d'un moment, choisi plus ou moins au hasard, d'un mouvement dont le futur se révèlera aussi intéressant que le passé.

A propos de passé, il suffira de noter que Jean-Marie a été l'un des fondateurs de la géométrie symplectique dans les années 60, et qu'il a été le seul à insister sur les fondements physiques de cette géométrie. De plus nous aimons Jean-Marie pour son caractère ensoleillé, lui qui a accueilli à Aix et à Marseille, avec sa femme Christiane que nous regrettons tous, tant de symplecticiens, depuis les jeunes, inconnus, jusqu'aux moins jeunes, plus connus; lui qui a dirigé une équipe de Marseillais d'une stabilité et d'une intimité extraordinaires.

On voit dans le programme du colloque (en particulier celui de la journée de vendredi) l'influence immense que Souriau a eu pour établir des liaisons étroites entre la France et les Etats-Unis (en particulier la Californie). Je dois personnellement le remercier de m'avoir encouragé et accueilli plusieurs fois pendant près de 15 ans, à commencer par une visite de la Montagne Sainte-Victoire, où il a montré à notre fille des oeufs de dinosaures qu'il appelait "Eggs-en-Provence."

Je vous invite à vous joindre à moi pour féliciter Jean-Marie Souriau et lui adresser à cette occasion nos meilleurs voeux.

Alan Weinstein
Aix en Provence, juin 1990

Clôture du colloque

J'ai le privilège — au douteux bénéfice de l'âge — de clore ce colloque, non, comme le disent désormais nos journalistes, de le clôturer (j'ai horreur des fils de fer barbelés ou non).

Il m'appartient de dire, en votre nom à tous, notre gratitude aux membres du Comité d'organisation et aux personnalités qui les ont aidés. Cette organisation a été exemplaire. Je sais tout le temps et tous les trésors d'intelligence qu'il faut dépenser pour arriver à un tel résultat.

L'intérêt scientifique du colloque a été considérable et tous les conférenciers venus de huit ou neuf pays différents ont donné le meilleur d'eux-mêmes. Mais l'ampleur de son thème m'a paru encore un peu limité, je dirai pourquoi dans un instant.

L'une des finalités du colloque était de porter à Jean-Marie Souriau un témoignage de notre affection et de notre admiration. Parmi les scientifiques, je suis sans doute l'un des plus anciens amis de Jean-Marie. J'ai commencé à le connaître vraiment en 1948, année où j'étais redevenu parisien et où il œuvrait à l'ONERA, un lieu qui me paraissait un peu saugrenu pour lui. J'entendis de lui une remarquable conférence où il manifestait déjà son originalité. Il s'exprimait bien entendu en français, mais il utilisait des notions, des dénominations et des notations qui lui étaient propres, ce qui nuisait parfois à la clarté du message. C'est là la seule originalité dont il s'est peu à peu guéri.

Je fus ensuite son hôte à l'Institut des Hautes Etudes de Tunis, où enseignèrent à ses côtés tant de collègues de grande classe et enfin, grâce à lui et ses talents d'organisation, je connus Aix et Marseille et leur communauté scientifique. Entre temps, Jean-Marie était devenu illustre.

Ce qui me frappe en lui, c'est qu'il est le seul d'entre nous qui pourrait légitimement appartenir à quatre Unions Scientifiques Internationales. Mathématiques certes, mais aussi Mécanique, Physique et Astronomie et partout en vrai professionnel. Relativité et Astronomie m'ont paru un peu absentes du thème principal du Colloque et c'est là mon seul et léger regret.

Si l'ampleur et la profondeur de sa pensée sont patentes, c'est l'unité de cette pensée qui me frappe à travers la variété de ses domaines d'intérêt, qui vont de la cosmologie et de l'antimatière dans le cosmos, à la difféologie dont il fut récemment le créateur pour pouvoir dire, je crois, des choses sensées sur les groupes de transformations dits de dimension infinie sur une variété non compacte. Il s'est toujours efforcé de fournir des modes de pensée pour notre monde macroscopique ou microscopique.

Pour rejoindre le thème du colloque et voir l'importance de l'apport de Jean-Marie Souriau, il suffit de feuilleter le très beau livre d'Abraham

et Marsden intitulé "Foundations of Mechanics" et de regarder à Souriau
l'index des auteurs. On voit qu'il a été présent dans toutes les aventures
qui ont conduit à des concepts et à des théories devenus classiques, moment
d'un système dynamique, théorèmes de Kirillov-Souriau, variété de Moser-
Souriau, quantification géométrique etc ... et que malgré son obstination
à être francophone, sa voix a su franchir l'Atlantique et aborder aux rives
du Pacifique. Beaucoup s'enorgueillissent d'être ses émules ou ses élèves.

Cette continuité de la pensée va avec une grande culture, toute nourrie
de curiosité, avec un humour sans faille et la possibilité de rares amitiés,
toutes auréolées de son rayonnement humain et de sa générosité. Jean-
Marie Souriau aime certes la Science pour elle-même, mais, pour nous, il
est un gentilhomme de la Science.

André Lichnerowicz

Separation of Variables in the Geodesic Hamilton-Jacobi Equation

S. BENENTI

Dedicated to Jean-Marie Souriau

1. Introduction

This lecture is devoted to the basic notions of the theory of separation of variables for the Hamilton-Jacobi equation. Many of the results presented here are already available in the literature. However, we collect them in a systematic way according to a personal point of view, with a particular emphasis on their geometrical meaning.

Let Q be a differential manifold of dimension n. With each system of coordinates (q^i) on Q we associate canonical coordinates (q^i, p_j) on the cotangent bundle T^*Q. We call the coordinates (p_i) the *momenta* corresponding to the coordinates (q^i). Let H be a differentiable real function on the cotangent bundle, called the *Hamiltonian*. For each coordinate system on Q the Hamiltonian is locally represented by a real differentiable function $H(q^i, p_j)$ in the $2n$ real variables (q^i, p_j), and it gives rise to a partial differential equation, the *Hamilton-Jacobi equation*

$$H\left(q^i, \frac{\partial W}{\partial q^j}\right) = h, \tag{1.1}$$

where h is a real parameter, called the *energy*. A *complete integral* of this equation is a solution $W(q^i, a_j)$ depending on n real parameters (a_j) such that the $n \times n$ matrix

$$\left(\frac{\partial^2 W}{\partial q^i \partial a_j}\right) \tag{1.2}$$

is everywhere regular. When such a complete integral is known, then, by purely algebraic manipulations, we can find the integral curves of the Hamiltonian dynamical system generated by H. This is the classical *Jacobi method* which was originally used for finding the geodesics of an ellipsoid. However, the experience has shown that the Jacobi method can be applied with success for those cases in which it is possible to find a complete integral of the kind

$$W = W_1(q^1, a_j) + W_2(q^2, a_j) + \ldots + W_n(q^n, a_j). \tag{1.3}$$

This property is called *additive separation of variables*, and when this occurs we say that the H-J equation is integrable by *separation of variables* and that the dynamical system generated by the Hamiltonian H is *integrable in the sense of Jacobi* or *Jacobi-integrable*. The coordinates (q^i) for which a complete integral of the kind (1.3) exists are called *separable* (with respect to the Hamiltonian H).

Actually, due to the applications to Riemannian geometry and mathematical physics, we deserve the interest to Hamiltonians which are polynomials of second order in the momenta, i.e. of the kind

$$H = \frac{1}{2}\, g^{ij} p_i p_j + A^i p_i + V, \tag{1.4}$$

where g^{ij} are the contravariant components of a metric tensor g on Q, A^i are the components of a vector field on Q and V is a function on Q , called the *potential* (all fields are assumed to be smooth, i.e. of class C^∞, for simplicity).

We cannot give here a historical perspective on the theory of separation of variables. For an outline of this topic and an essential bibliography we refer to [H] [W] and to the recent book of Kalnins [K]. However, in the following two sections we will present, with proofs, classical fundamental results due to Levi-Civita (1904) and Stäckel (1893), on which we will base our discussion. We will restrict our attention to the separation of the geodesic H-J equation, which is a crucial topic in the theory of separation of variables. In Section 4 we will analyse those transformations of coordinates which preserve the separation and the complete integral. The results will be used in Section 5 for proving the existence of a normal form of the contravariant metric tensor components in separable coordinates, in which the separation is achieved in the simplest way. In Section 6 we will present some fundamental facts concerning the Killing tensors and the orthogonal separation. Section 7 will be finally dedicated to an outline of the geometrical aspects of separation. Some examples are presented for illustrating the theory, without the pretension of a reasonable completeness, as for the bibliography.

2. The theorem of Levi-Civita

Theorem 1. *The Hamilton-Jacobi equation (1.1) has a complete integral of the kind (1.3) if and only if the following equations are identically satisfied for each pair (i, j) of distinct indices:*

$$\partial^i H \partial^j H \partial_{ij} H + \partial_i H \partial_j H \partial^{ij} H - \partial^i H \partial_j H \partial_i^j H - \partial^j H \partial_i H \partial_j^i H = 0. \tag{2.1}$$

Here we have used the following short notation:

$$\partial_i H = \frac{\partial H}{\partial q^i}, \quad \partial^i H = \frac{\partial H}{\partial p_i}, \quad \text{etc...}$$

Proof. The additive separation of variables is equivalent to the condition $\partial_{ij} W = 0$ for $i \neq j$. By differentiating the H-J equation (1.1) with respect to a variable q^i, we get

$$\partial_i H + \partial^j H \partial_{ij} W = \partial_i H + \partial^i H \partial_{ii} W = 0 \quad (i \text{ n.s.}).$$

The notation "n.s." means that there is no summation over the repeated index. Thus we are led to consider the following system of partial differential equations in the unknowns (p_i),

$$\partial_i p_i = R_i, \quad \partial_i p_j = 0 \quad (i \neq j), \tag{2.2}$$

where

$$R_i = -\frac{\partial_i H}{\partial^i H}, \tag{2.3}$$

together with equation $p_i = \partial_i W$. The integrability conditions of this system are

$$\partial_i R_j + \partial^i R_j \, R_i = 0 \quad (i \neq j). \tag{2.4}$$

Due to (2.3) they are equivalent to (2.1). $\blacksquare$

The theorem of Levi-Civita [LC] provides only a criterion to decide whether a given coordinate system is separable or not. It gives no effective method to find separable coordinates for a given Hamiltonian. In fact, this is a hard problem, which has been solved only for particular Hamiltonians (like, for instance, the geodesic Hamiltonian of Riemannian manifolds with constant curvature [K]). However, Levi-Civita theorem can be considered as a starting point for developing the theory of the separation of variables.

If we substitute the polynomial Hamiltonian (1.3) in equations (2.1), then we get equations of fourth order in the momenta (p_i) which must be identically satisfied, so that the coefficients must vanish. These coefficients involve the functions g^{ij}, A^i, V, together with their first and second derivatives, so that we get finally a set of second order PDE's on these functions. For the sake of brevity we do not write them here explicitly; actually, they are very cumbersome. However, as Levi-Civita itself remarked, it can be seen that the coefficients of fourth degree coincide with those coming from the separation conditions (2.1) for the purely geodesic Hamiltonian

$$G = \frac{1}{2} g^{ij} p_i p_j, \tag{2.5}$$

4   S. Benenti

namely,

$$(\partial_{ij}g^{hk}g^{ir}g^{js} + \frac{1}{2}g^{ij}\partial_i g^{hk}\partial_j g^{rs} - \tag{2.6}$$

$$-\partial_i g^{jh}\partial_j g^{kr}g^{is} - \partial_j g^{ih}\partial_i g^{kr}g^{js})p_h p_k p_r p_s = 0.$$

Hence, coordinates which are separable with respect to the Hamiltonian (1.4) are separable with respect to the geodesic Hamiltonian (2.5), and the separation of the geodesic H-J equation becomes the fundamental argument.

3. The orthogonal separation and the theorems of Stäckel

The first step for studying the separation of the geodesic H-J equation is to consider *orthogonal* coordinates, i.e. to assume

$$g^{ij} = 0, \quad i \neq j, \tag{3.1}$$

so that $G = \frac{1}{2}g^{ii}(p_i)^2$. The separation conditions (2.6) reduce to equations

$$g^{ii}g^{jj}\partial_{ij}g^{hh} - g^{ii}\partial_i g^{jj}\partial_j g^{hh} - g^{jj}\partial_j g^{ii}\partial_i g^{hh} = 0 \quad (i \neq j,\ \text{n.s.}). \tag{3.2}$$

These equations, which are equivalent to

$$\partial_{ij}g^{hh} - \partial_i \log|g^{jj}|\,\partial_j g^{hh} - \partial_j \log|g^{ii}|\,\partial_i g^{hh} = 0 \quad (i \neq j,\ \text{n.s.}), \tag{3.2'}$$

are fundamental in the theory of the separation of variables. They are similar to the classical Lamé equations (see, for instance, [BI]).

The following two theorems due to Stäckel [ST1] [ST2] are fundamental for the theory of orthogonal separation.

Theorem 3.1. *The most general form of the metric tensor in orthogonal separable coordinates is*

$$g^{ii} = \phi^i_{(n)}, \tag{3.3}$$

where $\phi^i_{(n)}$ is a row (the last one, for instance) of the inverse of a Stäckel matrix $(\phi^{(j)}_i)$ in the coordinates (q^i). The functions

$$F_j = \frac{1}{2}\phi^i_{(j)}(p_i)^2 \tag{3.4}$$

are geodesic first integrals in involution (the last one is the geodesic Hamiltonian itself):

$$\{F_j, F_h\} = 0, \quad F_n = G. \tag{3.5}$$

Definition 3.1. A *Stäckel matrix* is a regular $n \times n$ matrix $(\phi_i^{(j)})$ of functions of the n variables (q^i) such that each element depends on the variable corresponding to the lower index only: $\phi_i^{(j)}(q^i)$. We denote by $(\phi_{(j)}^i)$ the inverse matrix.

This theorem shows in other words, that the most general functions satisfying equations (3.2) are of the kind (3.3) and, moreover, that the separation of variables is always concomitant with the existence of quadratic first integrals in involution. The very simple proof given by Stäckel can be found in the majority of the textbooks of analytical mechanics. However, we write it here explicitly for further needs.

Proof. Let us differentiate the geodesic H-J equation

$$\frac{1}{2} g^{ii} (\partial_i W)^2 = h \tag{3.6}$$

with respect to the parameters (a_j) entering in a complete integral W and set

$$\phi_i^{(j)} = \frac{\partial W}{\partial q^i} \frac{\partial^2 W}{\partial q^i \partial a_j}, \quad c^j = \frac{\partial h}{\partial a_j}.$$

We get an equation of the kind

$$g^{ii} \phi_i^{(j)} = c^j.$$

The matrix $(\phi_i^{(j)})$ is regular provided that no $p_i = \partial_i W$ is zero, and it is a Stäckel matrix if the coordinates are separable. By reversing these equations we find

$$g^{ii} = c^j \phi_{(j)}^i.$$

However, there is no loss of generality in assuming that one of the parameters coincides with the energy, say $a_n = h$. We have $c^j = \delta_n^j$, so that the metric components have the form (3.3). Once (3.3) is proved, we can write the most general separable Hamilton-Jacobi equation for the geodesics:

$$\frac{1}{2} \phi_{(n)}^i (\partial_i W)^2 = h.$$

Then the separation of variables can been achieved by writing the whole system of equations associated with the matrix $(\phi_{(j)}^i)$,

$$\frac{1}{2} \phi_{(j)}^i (\partial_i W)^2 = a_j, \tag{3.7}$$

by introducing n real parameters (a_j), with $a_n = h$. So that we get, by an inversion, the following system of separated ordinary differential equations,

$$\left(\frac{dW_i}{dq^i}\right)^2 = 2a_j \phi_i^{(j)}.$$

It follows that $W = \sum_{i=1}^n W_i(q^i, a_j)$ is a separated complete integral. Moreover, by the Jacobi theorem we know that the real parameters entering in a complete integral correspond to first integrals in involution. Hence, from (3.7) we deduce that the functions (3.4) are first integrals in involution. ∎

Let us consider a potential function V added to the geodesic Hamiltonian:

$$H = \frac{1}{2} g^{ii} (p_i)^2 + V.$$

The separability equations of Levi-Civita for this Hamiltonian give rise to the system of equations (3.2), involving the metric coefficients only, together with the following additional conditions on the potential V:

$$g^{ii} g^{jj} \partial_{ij} V - g^{ii} \partial_i g^{jj} \partial_j V - g^{jj} \partial_j g^{ii} \partial_i V = 0 \quad (i \neq j, \text{ n.s.}), \qquad (3.8)$$

which can also be written

$$\partial_{ij} V - \partial_i \log |g^{jj}| \, \partial_j V - \partial_j \log |g^{ii}| \, \partial_i V = 0 \quad (i \neq j, \text{ n.s.}). \qquad (3.8')$$

Theorem 3.2. *The most general potential compatible with orthogonal separation is of the kind*

$$V = \eta_i \, g^{ii}, \qquad (3.9)$$

where (η_i) are functions of the variable corresponding to the index only.

Proof. Since $\partial_i W$ is a function of the variable q^i only, the H-J equation

$$\frac{1}{2} g^{ii} (\partial_i W)^2 + V = h$$

shows that in separable coordinates V is a function of the kind (3.9), up to an inessential additive constant. ∎

Furthermore, following the same way of the proof of Theorem 3.1, it can be seen that the functions

$$F_j = \frac{1}{2} \phi_{(j)}^i (p_i)^2 + \eta_j$$

are first integrals in involution.

There is another way to prove all the above results which does not involve the theorem of Jacobi.

Let us consider a linear connection Γ on the manifold Q. We can think of invariant 1-forms ϕ and vector fields X and write the following two differential systems:

$$\partial_i \phi_j - \Gamma_{ij}^h \phi_h = 0, \quad \partial_i X^j + \Gamma_{ih}^j X^h = 0. \tag{3.10}$$

If ϕ and X are two solutions then

$$<X, \phi> = X^i \phi_i = \text{const.} \tag{3.11}$$

Both systems (3.10) are completely integrable if and only if the connection is locally flat. Then the general solution of the first system (3.10) has the form

$$\phi_i = c_j \phi_i^{(j)}, \quad (c_j) \in \mathbb{R}^n, \tag{3.12}$$

where $(\phi_i^{(j)}; j = 1, \dots, n)$ are independent solutions: $\det(\phi_i^{(j)}) \neq 0$. The inverse matrix $\phi_{(j)}^i$ gives the general solution of the second system (3.10):

$$X^i = c^j \phi_{(j)}^i, \quad (c^j) \in \mathbb{R}^n, \tag{3.13}$$

If the connection Γ is symmetric, then two invariant vector fields (solutions of $(3.10)_2$) commute. This is equivalent to

$$\phi_{(h)}^i \partial_i \phi_{(k)}^j - \phi_{(k)}^i \partial_i \phi_{(h)}^j = 0. \tag{3.14}$$

We say that the connection Γ is *separable* with respect to the coordinates (q^i) (or, conversely, that the coordinates (q^i) are separable with respect to the connection Γ) if the coefficients of Γ in these coordinates are such that

$$\Gamma_{ij}^h = 0, \quad \text{for } i \neq j. \tag{3.15}$$

In this case equations (3.10) become

$$\partial_i \phi_j = \delta_{ij} B_i^h \phi_h, \quad \partial_i X^j = -B_i^j X^i, \quad \text{where } B_i^j = \Gamma_{ii}^j. \tag{3.16}$$

The integrability conditions are:

$$\partial_i B_j^h - B_j^i B_i^h = 0, \quad i \neq j \text{ n.s..} \tag{3.17}$$

For every non trivial solution of equations $(3.16)_2$, we have

$$B_i^j = -\frac{1}{X^i} \partial_i X^j. \tag{3.18}$$

If we substitute this expression in the integrability conditions (3.17), then we get equations

$$X^i X^j \partial_{ij} X^h - X^i \partial_i X^j \partial_j X^h - X^j \partial_j X^i \partial_i X^h = 0 \quad (i \neq j, \text{ n.s.}). \quad (3.19)$$

On the other hand, equations $(3.16)_1$ show that a set of independent solutions form a Stäckel matrix $(\phi_i^{(k)})$, since $\partial_i \phi_j = 0$ for $i \neq j$. Hence, we have proved that

Theorem 3.3. *The most general solution of equations (3.19) is of the form (3.13) where $(\phi_{(j)}^i)$ is the inverse of a Stäckel matrix, or more simply, of the form*

$$X^i = \phi_{(n)}^i, \quad (3.20)$$

where $(\phi_{(n)}^i)$ is a line of the inverse of a Stäckel matrix.

This is in fact the Stäckel theorem 3.1, provided we set

$$X^i = g^{ii}, \quad \phi_i = (p_i)^2 = (\partial_i W)^2$$

and look at equation (3.11) as the H-J equation. Moreover, equations (3.14) are equivalent to the commutation equations (3.5).

Remark 3.1. Equations (3.19) are the Levi-Civita separation conditions (2.1) for a linear Hamiltonian $H = X^i p_i$. For a Hamiltonian of the kind $H = X^i p_i + V$ the separation conditions are still equations (3.19) together with equations

$$X^i X^j \partial_{ij} V - X^i \partial_i X^j \partial_j V - X^j \partial_j X^i \partial_i V = 0, \quad (i \neq j, \text{ n.s.}). \quad (3.21)$$

Then we can re-state Theorem 3.2 as follows:

Theorem 3.4. *The most general solution of equations (3.21), where (X^i) are solutions of equations (3.19), are of the kind $V = \eta_i X^i$, where (η_i) are functions of the variable corresponding to the index only.*

Remark 3.2. Since the components (g^{ii}) are a particular solution of system $(3.16)_2$, if we consider equations (3.18) for $X^i = g^{ii}$ we get equations

$$\frac{1}{X^i} \partial_i X^j = \frac{1}{g^{ii}} \partial_i g^{jj},$$

which can be interpreted as a differential system in the unknown functions (X^i):

$$\partial_i X^j = \frac{1}{g^{ii}} \partial_i g^{jj} X^i. \tag{3.22}$$

According to the preceding remarks, this system is completely integrable if and only if equations (3.2) hold. One solution is given by (g^{ii}). A set of independent solutions $(X^i = \phi^i_{(j)})$ gives the inverse of a Stäckel matrix. Furthermore, if we introduce the functions

$$\rho_i = \frac{X^i}{g^{ii}},$$

then system (3.22) is equivalent to

$$\partial_i \rho_j = (\rho_i - \rho_j)\partial_i \log |g^{ii}|. \tag{3.23}$$

The integrability conditions of this new differential system are still equations (3.2). We will go back to this remark in the sequel.

Remark 3.3. Transformations of the metric components of the kind

$$g^{ii} \longmapsto \zeta_i \, g^{ii},$$

where (ζ_i) are functions of the variable corresponding to the index only, preserve the separation of variables. This can be seen straightly from the H-J equation (3.6).

Remark 3.4. Assume that a Hamiltonian H satisfies the Levi-Civita separation conditions (2.1). Let us consider a "conformal" Hamiltonian fH, where f is a real smooth function of the coordinates only. This new Hamiltonian satisfies equations (2.1) if and only if, for each pair of distinct indices:

$$f^2 H \left[\partial^i H \partial^j H \partial_{ij} f - \partial^i H \partial^j_i H \partial_j f - \partial^j H \partial^i_j H \partial_i f\right] -$$
$$2 f H \, \partial^i H \partial^j H \partial_i f \partial_j f +$$
$$f^2 \, \partial^{ij} H \left[H^2 \partial_i f \partial_j f + f H (\partial_i f \partial_j H + \partial_j f \partial_i H)\right] = 0.$$

If we set $V = f^{-1}$ (where $f \neq 0$), then we get simpler equations:

$$\partial^i H \partial^j H \partial_{ij} V - \partial^i H \partial^j_i H \partial_j V - \partial^j H \partial^i_j H \partial_i V +$$
$$V^{-2} \, \partial^{ij} H \left[H \partial_i V \partial_j V + V (\partial_i V \partial_j H + \partial_j V \partial_i H)\right] = 0.$$

These are the characteristic equations of a *conformal cofactor* V preserving the separability of an Hamiltonian H. When $\partial^{ij} H = 0$ for $i \neq j$, they simplify furtherly:

$$\partial^i H \partial^j H \partial_{ij} V - \partial^i H \partial^j_i H \partial_j V - \partial^j H \partial^i_j H \partial_i V = 0,$$

and become similar to (3.21) and (3.8). This is actually the case of a linear Hamiltonian and of an orthogonal geodesic Hamiltonian. For this reason solutions of equations (3.21) or (3.8) are called *Stäckel multipliers* [KM3]: a conformal transformation of the kind $g_{ij} \longmapsto V g_{ij}$ preserves the separability of an orthogonal metric. The fact that a potential compatible with the (orthogonal) separation has the same form as a conformal cofactor preserving the (orthogonal) separation is rather a consequence of the Maupertuis principle of mechanics: the dynamical trajectories of fixed energy h of a mechanical system, whose kinetic energy and potential energy are respectively $\frac{1}{2} g_{ij} v^i v^j$ and V, are geodesics of the conformal metric $(V - h) g_{ij}$.

Remark 3.5. When for a given coordinate system the metric tensor components satisfy equations (3.2), then by the Stäckel theorem 3.1 we only conclude that a Stäckel matrix exists such that (3.3) holds and, in order to perform the separation and to construct the first integrals, we need to know explicitly such a matrix (it is not uniquely determined), or its inverse; we know only the last line of the inverse. Actually, in practice the separation can be usually achieved by a direct inspection of the H-J equation and by extracting from it equations involving single coordinates and the so called *separation constants* (see the examples in Section 5). However, the problem of finding a Stäckel matrix, when only one line of its inverse is known, can be solved through algebraic manipulations by comparison with some *canonical* form of the inverse of a Stäckel matrix. Such canonical forms depend on the dimension n. For instance, when $n = 2$ the inverse of a Stäckel matrix has the following canonical form:

$$(\phi^i_{(j)}) = \begin{pmatrix} \phi^1_{(1)} & \phi^2_{(1)} \\ \phi^1_{(2)} & \phi^2_{(2)} \end{pmatrix} = \begin{pmatrix} \dfrac{\phi_2 \psi_1}{\phi_1 + \phi_2} & \dfrac{-\phi_1 \psi_2}{\phi_1 + \phi_2} \\ \dfrac{\psi_1}{\phi_1 + \phi_2} & \dfrac{\psi_2}{\phi_1 + \phi_2} \end{pmatrix}, \tag{3.24}$$

where ϕ_i and ψ_i are functions of the variable corresponding to the index only (or constant). Hence, if we know the second row (i.e. the components of the metric) we can construct the first row (i.e. the associated first integral). The Stäckel matrix corresponding to (3.24) is

$$(\phi^{(j)}_i) = \begin{pmatrix} \dfrac{1}{\psi_1} & -\dfrac{1}{\psi_2} \\ \dfrac{\phi_1}{\psi_1} & \dfrac{\phi_2}{\psi_2} \end{pmatrix}. \tag{3.25}$$

For $n = 3$ a canonical form is

$$
\begin{cases}
\phi^i_{(1)} = \dfrac{\psi_i}{\phi}(\mu_{i+2}\nu_{i+1} - \mu_{i+1}\nu_{i+2}), \\[2ex]
\phi^i_{(2)} = \dfrac{\psi_i}{\phi}(\nu_{i+2} - \nu_{i+1}), \\[2ex]
\phi^i_{(3)} = \dfrac{\psi_i}{\phi}(\mu_{i+1} - \mu_{i+2}),
\end{cases}
\qquad (3.26)
$$

where the functions depend on the variable corresponding to the index and ϕ is the determinant

$$
\det \begin{pmatrix}
1 & 1 & 1 \\
\mu_1 & \mu_2 & \mu_3 \\
\nu_1 & \nu_2 & \nu_3
\end{pmatrix}.
$$

Canonical forms of Stäckel matrices for $n = 2, 3, 4$ are discussed in [BF1].

4. Equivalent separable coordinates

The discussion of the Levi-Civita separation conditions (2.6) for non-orthogonal variables seems to be much more difficult than the orthogonal case. In order to simplify this problem, one way is to perform suitable transformations of coordinates which preserve the separation.

First of all we remark that every coordinate transformation which involves separately each one of the coordinates (i.e. whose Jacobian is diagonal) is always allowed: it preserves the separation property. It also preserves the coordinate surfaces. We will call such a coordinate transformation a *separated transformation*. However, one can ask if these transformations are the only ones compatible with the separation. We can give an answer to this question, following again a suggestion of Levi-Civita who, in the case of a purely geodesic Hamiltonian, distinguished the separable coordinates in two *classes* [LC]. We can extend this classification to a generic Hamiltonian as follows.

Definition 4.1. Let (q^i) be a system of separable coordinates. We say that a coordinate q^i is of *first class* if the function R_i (defined in (2.3)) is a linear function in the momenta, i.e. of the kind $R_i = B^j_i(q^h)p_j$. Otherwise, we say that q^i is of *second class*. A coordinate q^i is called *ignorable* if $R_i = 0$, i.e. if $\partial_i H = 0$ (an ignorable coordinate is of first class).

It is known that a complete integral is geometrically represented by a local Lagrangian foliation of the cotangent bundle T^*Q, transversal to the fibres, such that the Hamiltonian H is constant on each leaf. With a

12 S. Benenti

given Hamiltonian we can have different separable systems of coordinates corresponding to different foliations. One can consider, as a simple example, the case of the geodesic Hamiltonian in the Euclidean plane. Cartesian coordinates and polar coordinates are both separable, but they give rise to different complete integrals. We say that they are *not* equivalent accordingly to the following

Definition 4.2. Two separable coordinate systems are said to be *equivalent* if in every domain in which they are both defined they give rise to the same complete integral, interpreted as a Lagrangian foliation of the cotangent bundle.

In order to simplify the discussion it is convenient to adopt the following convention for the indices [B1].

Convention. Latin indices $h, i, j, \ldots$ always run from 1 to n, the dimension of Q. For indices corresponding to coordinates of second class we use the first Latin letters $a, b, c \ldots$, which run from 1 to $m \leq n$. For indices corresponding to coordinates of first class we use the first Greek letters $\alpha, \beta, \gamma, \ldots$, which run from $m + 1$ to n. We set $r = n - m$, the number of coordinates of first class.

Theorem 4.1. *Two equivalent separable systems have the same number of first class coordinates (hence, the same number of second class coordinates).*

Proof. Let (q^i) and $(q^{i'})$ be two equivalent systems of separable coordinates. The corresponding momenta are related by equations

$$p_i = A_i^{i'} p_{i'}, \quad p_{i'} = A_{i'}^i p_i, \tag{4.1}$$

where

$$A_i^{i'} = \frac{\partial q^{i'}}{\partial q^i}, \quad A_{i'}^i = \frac{\partial q^i}{\partial q^{i'}}.$$

When a complete integral W is given, we have $p_i = \partial_i W$ and $p_{i'} = \partial_{i'} W$. By applying the partial derivative $\partial_{j'} = A_{j'}^j \partial_j$ to the first set of equations (4.1) we get

$$A_{j'}^j \partial_j p_i = A_{j'}^j \partial_j A_i^{i'} p_{i'} + A_i^{i'} \partial_{j'} p_{i'}.$$

On the other hand the complete integral is a solution of the following two differential systems (see (2.2)),

$$\partial_j p_i = \delta_{ji} R_i, \quad \partial_{j'} p_{i'} = \delta_{j'i'} R_{i'}.$$

Thus the following equations must be identically satisfied:

$$A^i_{j'} R_i = A^j_{j'} \partial_j A^{i'}_i \, p_{i'} + A^{j'}_i R_{j'} \quad (i, j' \text{ n.s.}).$$

Let us consider the case $i = \alpha$ (index of first class) and $j' = a'$ (index of second class):

$$A^\alpha_{j'} R_\alpha = A^j_{a'} \partial_j A^{i'}_\alpha \, p_{i'} + A^{a'}_\alpha R_{a'} \quad (\alpha, a' \text{ n.s.}).$$

By definition of coordinate of first class, the term R_α is linear in the momenta (p_j), thus it is linear in the momenta $(p_{j'})$, and this equation shows that also $R_{a'}$ is linear, against our assumption, unless $A^{a'}_\alpha = 0$. Hence, by symmetry,

$$A^{a'}_\alpha = 0, \quad A^a_{\alpha'} = 0.$$

This means that the second class coordinates (q^a) depend on the second class coordinates $(q^{a'})$ only, and viceversa. Hence, the number of second class coordinates in both systems is the same. $\blacksquare$

Theorem 4.2. *Every separable system is equivalent to a separable system in which all first class coordinates are ignorable.*

Proof. Accordingly to the distinction in two classes of the coordinates, we can split system (2.2) into the following equations:

$$\partial_\alpha p_\alpha = B^i_\alpha p_i, \quad \partial_i p_j = 0 \quad (i \neq j), \quad \partial_a p_a = R_a. \tag{4.2}$$

The integrability conditions of the whole system (2.4), with a particular choice of the indices, give

$$\partial_a B^i_\alpha p_i + B^a_\alpha R_a = 0 \quad (a \text{ n.s.}).$$

This equation shows that R_a is linear in the momenta, which is against the assumptions, unless $B^a_\alpha = 0$, so that

$$B^a_\alpha = 0, \quad \partial_a B^\beta_\alpha = 0. \tag{4.3}$$

Hence, the system (4.2) contains an autonomous sub-system in the first coordinates only,

$$\partial_\alpha p_\beta = 0 \quad (\alpha \neq \beta), \quad \partial_\alpha p_\alpha = B^\gamma_\alpha p_\gamma.$$

This system is integrable and it is linear in the momenta (p_α). Hence, locally there exist $r = n - m$ independent solutions $(\phi^{(\beta)}_\alpha)$, such that the general solution is

$$p_\alpha = c_\beta \, \phi^{(\beta)}_\alpha, \quad (c_\beta) \in \mathbb{R}^n. \tag{4.4}$$

14	S. Benenti

Let us consider the new coordinate system (x^i) defined by

$$dx^a = dq^a, \quad dx^\alpha = \phi_\beta^{(\alpha)} dq^\beta. \tag{4.5}$$

Equations (4.4) shows that the momenta $(p_{(\alpha)})$ associated with these new coordinates, evaluated on the original first integral W, are constant: $p_{(\alpha)} = c_\alpha$. Hence, the local expression of the Hamiltonian H in these coordinates cannot depend on the coordinates (x^α), or, in other words, the coordinates (x^α) are ignorable. Since the second class coordinates are unchanged, the new system (x^i) is separable. ∎

The matrix $(\phi_\alpha^{(\beta)})$ in the preceding proof is a Stäckel matrix in the r variables (q^α).

Remark 4.1. When the Hamiltonian is linear, $H = X^i p_i$, all separable coordinates are of first class, so that we can always reduce to the case in which all separable coordinates are ignorable, and the complete integral takes the form $W = a_i q^i$.

The preceding theorems hold for whatever Hamiltonian. From now on we will consider the case of a geodesic Hamiltonian (the following two statements can be extended to polynomial Hamiltonians of second degree (1.4), but with longer proofs).

Theorem 4.3. *In two equivalent separable systems the second class coordinates are related by a separated transformation (i.e. they generate the same coordinate surfaces).*

Proof. With the same procedure of the proof of Theorem 4.1, let us apply the partial derivative $\partial_{j'}$ to the second set of equations (4.1). By assuming that $j' \neq i'$, we get

$$\partial_{j'} A_{i'}^i \, p_i + A_{j'}^i A_{i'}^i \, R_i = 0. \tag{4.6}$$

Let us consider in particular two indices of first class, $a' \neq b'$:

$$\partial_{a'} A_{b'}^i \, p_i + A_{a'}^a A_{b'}^a \, R_a + A_{a'}^\alpha A_{b'}^\alpha \, R_\alpha = 0.$$

We remark that the first and the last terms in this equation are linear in the momenta. If we assume that there is no linear combination of the kind $f^a R_a$, where f^a are functions of the coordinates (q^i) only, which reduces to a linear form in the momenta (p_i), then the above formula cannot hold, unless

$$A^a_{a'} A^a_{b'} = 0 \quad (a' \neq b', a \text{ n.s.})$$

for each index a of second class. This means that for each index of second class a there exists one and only one index of second class a' such that $A^a_{a'} \neq 0$. Indeed, if two such indices (a', b') exist, the equation above cannot be satisfied; and, on the contrary, if no such indices exist, in the whole matrix $(A^i_{i'})$ the line $i = a$ is made of zeros: absurd, since this matrix is regular. The fact that only one $A^a_{a'}$ does not vanish, proves that q^a depends on one second class coordinate only. It remains to prove that no sum of the kind $f^a R_a$ can be linear in the momenta. We have

$$R_a = \frac{1}{2}\frac{\partial_a g^{ij} p_i p_j}{g^{ak} p_k} = -\frac{1}{2v^a}\partial_a g_{ij} v^i v^j,$$

being (v^i) new variables replacing the momenta:

$$v^i = g^{ij} p_j, \quad p_i = g_{ij} v^j.$$

Let us assume that there is a combination such that $f^a R_a = L_i v^i$, where (L_i) are functions of the coordinates only. With a change of notation this equation can be written as follows:

$$\sum_a \frac{1}{2v^a} P_{aij} v^i v^j = L_i v^i. \tag{4.7}$$

Let us differentiate equation (4.7) with respect to v^b:

$$-\frac{1}{2(v^b)^2} P_{bij} v^i v^j + \sum_a \frac{1}{v^a} P_{abi} v^i = L_b.$$

And again, twice with respect to v^c, with $c \neq b$:

$$-\frac{1}{(v^b)^2} P_{bci} v^i - \frac{1}{(v^c)^2} P_{cbi} v^i + \sum_a \frac{1}{v^a} P_{abc} = 0.$$

$$-\frac{1}{(v^b)^2} P_{bcc} + \frac{2}{(v^c)^3} \sum_{i \neq c,b} P_{cbi} v^i = 0.$$

Just like the preceding ones, this last equation must be identically satisfied for all admissible values of (v^i). Hence, $P_{bcc} = 0$ and $P_{cbi} = 0$ for $b \neq c$, $c \neq i$. Let us differentiate (4.7) twice with respect to v^α and v^β (indices of first class). We get:

$$\sum_a \frac{1}{v^a} P_{a\alpha\beta} = 0.$$

This proves that $P_{a\alpha\beta} = 0$. These results show that the polynomial $P_{aij}v^i v^j$ is divisible by v^a, against the assumption that a is an index of second class.

∎

Theorem 4.4. *Let $(q^i) = (q^a, q^\alpha)$ be a separable system such that all coordinates of first class (q^α) are ignorable. Then every equivalent separable system $(q^{i'}) = (q^{a'}, q^{\alpha'})$ is related to this one, modulo a separated transformation, by equations of the kind*

$$dq^a = dq^{a'}, \quad dq^\alpha = A^\alpha_{i'} \, dq^{i'}, \tag{4.8}$$

where $A^\alpha_{i'}$ are functions depending on the variable corresponding to the lower index only.

Proof. We have already seen that coordinates of second class are invariant up to a separated transformation. Since (q^α) are ignorable coordinates, we have $R_\alpha = 0$ and equation (4.6) gives $\partial_{j'} A^\alpha_{i'} = 0$ for $i' \neq j'$.

∎

5. The normal form of the metric tensor components in separable coordinates

As we have said in the last part of the preceding section, we are considering the purely geodesic Hamiltonian: $H = G = \frac{1}{2}g^{ij}p_i p_j$.

Theorem 5.1. *In a separable system $(q^i) = (q^a, q^\alpha)$, the coordinates of second class (q^a) are orthogonal: $g^{ab} = 0$ for $a \neq b$.*

Proof. The Levi-Civita separation conditions (2.2) for two indices of second class $a \neq b$ can be written as follows,

$$H^a(H^b H_{ab} - H_b H^b_a) = H_a(H^b H^a_b - H_b H^{ab}), \tag{5.1}$$

with the further simplifying convention $H^i = \partial^i H$, $H_i = \partial_i H$, etc.. Under our assumptions we have $H^i = v^i$ and $H^{ij} = g^{ij}$. By definition of second class coordinate the ratio $R_a = H_a/H^a$ is not a homogeneous first degree polynomial in the momenta, i.e. the first degree homogeneous polynomial H^a is not a divisor of the second degree homogeneous polymomial H_a. Equation (5.1) shows that it must be a divisor of the second degree polynomial $H^b H^a_b - H_b g^{ab}$, i.e. that there exists a first degree homogenous polynomial

$$L_{ba} = L^i_{ba} p_i = L_{bai} v^i$$

such that

$$H^b H^a_b - H_b g^{ab} = H^a L_{ba}. \tag{5.2}$$

From (5.1) it follows that

$$H^b H_{ab} - H_b H^b_a = H_a L_{ba}. \tag{5.3}$$

Our aim is to prove that condition $g^{ab} \neq 0$ implies $L_{ba} = 0$ so that from (5.2) H^b would be a divisor of H_b, which is against the assumption that also b is an index of second class. For this purpose, let us differentiate equation (5.2) twice with respect to p_a:

$$H^b g^{aa}_b = g^{aa} L_{ba} + H^a L^a_{ba}, \tag{5.4}$$

$$g^{ab} g^{aa}_b = 2 g^{aa} L^a_{ba}. \tag{5.5}$$

The explicit form of equation (5.4) in the variables (v^i) is

$$v^b g^{aa}_b = g^{aa} L_{bai} v^i + v^a L^a_{ba}.$$

As a polynomial identity in (v^i), this equation implies:

$$g^{aa}_b = g^{aa} L_{bab}, \tag{5.6}$$

$$g^{aa} L_{baa} + L^a_{ba} = 0, \tag{5.7}$$

$$g^{aa} L_{bai} = 0 \quad (i \neq b, a). \tag{5.8}$$

Let us differentiate equation (5.2) by q^a and equation (5.3) by p_a:

$$H^b_a H^a_b + H^b H^a_{ba} - H_{ba} g^{ab} - H_b g^{ab}_a = H^a_a L_{ba} + H^a L_{ia,a},$$

$$g^{ab} H_{ab} + H^b H^a_{ab} - H^a_b H^b_a - H_b g^{ab}_a = H_a L^a_{ba} + H^a_a L_{ba},$$

where $L_{ba,a} = \partial_a L_{ba}$. By subtracting term by term, we get equation

$$2 H^b_a H^a_b - 2 H_{ba} g^{ba} = H^a L_{ba,a} - H_a L^a_{ba}.$$

On the other hand, equations (5.2) and (5.3) give respectively

$$H^a_b = \frac{1}{H^b}(H^a L_{ba} + H_b g^{ab}), \quad H_{ab} = H_{ba} = \frac{1}{H^b}(H_a L_{ba} + H_b H^b_a),$$

so that from the last equation we get finally:

$$H^a(2 H^b_a L_{ba} - H^b L_{ba,a}) = H_a(2 g^{ba} L_{ba} - H^b L^b_{ba}).$$

This equation shows that $H^a = v^a$ must be a divisor of the polynomial

$$H^b L^a_{ba} - 2g^{ab} L_{ba} = v^b L^a_{ba} - 2\, g^{ab} L_{bai} v^i,$$

so that

$$L^a_{ba} = 2\, g^{ab} L_{bab}, \tag{5.9}$$

$$g^{ab} L_{bai} = 0 \quad (i \neq a, b). \tag{5.10}$$

Now, we make use of all equations (5.5)-(5.10). From (5.6), (5.5) and (5.9) it follows that

$$g^{ab} g^{aa} L_{bab} = g^{ab} g^{aa}_b = 2\, g^{aa} L^a_{ba} = 4 g^{ab} g^{aa} L_{bab}.$$

Hence,

$$g^{ab} g^{aa} L_{bab} = 0, \quad g^{aa} L^a_{ba} = 0. \tag{5.11}$$

Let us assume that $g^{ab} \neq 0$. We have two cases: (i) $g^{aa} \neq 0$, (ii) $g^{aa} = 0$. (i) We have: L_{bab} from (5.11), $L_{bai} = 0\,(i \neq a, b)$ from (5.8) or from (5.10), $L^a_{ba} = 0$ from (5.11), $L_{baa} = 0$ from (5.7). Thus we have proved $L_{ba} = 0$: absurd. (ii) If $g^{bb} \neq 0$ we find case (i) by symmetry. Thus let us assume $g^{aa} = g^{bb} = 0$. We have: $L_{bai} = 0\,(i \neq a, b)$ from (5.10), $L^a_{ba} = 0$ from (5.7) and $L_{bab} = 0$ from (5.9). This means that

$$L_{ba} = L_{baa} v^a = L_{baa} g^{ak} p_k. \tag{5.12}$$

Let us differentiate equation (5.2) twice with respect to p_b. Under the assumption $g^{bb} = 0$ we get $g^{ab} L^b_{ba} = 0$, thus $L^b_{ba} = 0$. However, due to equation (5.12), $L^b_{ba} = \partial^b L_{ba} = L_{baa} g^{ab}$. This shows that $L_{baa} = 0$, too. We have proved that $L_{ba} = 0$: absurd. ∎

This proof, which is taken from [B2], is inspired by a proof given by Dall'Acqua in 1912 [DA] under the assumption $g^{aa} \neq 0$, which is always fulfilled by a positive-definite metric (see also [CN]). This proof can be extended to Hamiltonians of the kind (1.4), through suitable modifications of the classification of the coordinates. This proof suggests a further distinction between the coordinates of second class.

Definition 5.1. A coordinate q^a is said to be *isotropic* if $g^{aa} = 0$.

This means that the gradient of the coordinate q^a is an *isotropic* (or *null*) vector and the coordinate surface $q^a = $ const. is a *coisotropic* submanifold (a submanifold $S \subset Q$ of a Riemannian manifold is called coisotropic if $(T_q S)^\perp \subset T_q S$ for each point $q \in S$). Hence, isotropic coordinates cannot occur in a strictly-Riemannian manifold. Let m_1 and m_2 be the number of non-isotropic and of isotropic coordinates of second class respectively:

$m_1 + m_2 = m = n - r$. Also these numbers are invariant for equivalent systems of separable coordinates, because of Theorem 4.3. Let us denote by $\tilde{a}, \tilde{b}, \ldots$ the indices of second class corresponding to non-isotropic coordinates and running from 1 to m_1, and by $\bar{a}, \bar{b}, \ldots$ the indices of second class corresponding to the isotropic coordinates, running from $m_1 + 1$ to $m = n - r$.

Now we can state the main theorem of our discussion (see [B1]).

Theorem 5.2. *In an equivalence class of separable coordinates there exists a coordinate system (q^i) such that the first class coordinates (q^α) are ignorable and such that the $n \times n$ matrix (g^{ij}) has the following form*

$$
(g^{ij}) = \begin{array}{c} \\ m_1 \\ \\ m_2 \\ \\ r \end{array}
\begin{array}{ccc} m_1 & m_2 & r \end{array}
\left(
\begin{array}{ccc}
\delta^{\tilde{a}\tilde{b}} g^{\tilde{a}\tilde{a}} & 0 & 0 \\
0 & 0 & g^{\bar{a}\beta} \\
0 & g^{\alpha\bar{a}} & g^{\alpha\beta}
\end{array}
\right) . \tag{5.13}
$$

Proof. We have already proved the existence of equivalent systems where the first class coordinates are ignorable (Theorem 4.2) and that the second class coordinates are orthogonal (Theorem 5.1). It remains to prove that we can find equivalent separable coordinates for which $g^{\tilde{a}\alpha} = 0$, for each non-isotropic second class index $\tilde{a}$. Equations (5.1) reduce to

$$
H^a(H^b H_{ab} - H_b H_a^b) = H_a H^b H_b^a,
$$

so that H^a must be a divisor of $H^b H_b^a$. Since $H^a = v^a$ is not a divisor of $H^b = v^b$, it must be a divisor of H_b^a, so that there exists a functions f^{ab} of the coordinates only such that $H_b^a = f^{ab} H^a$, i.e. $\partial_b g^{ai} = f^{ab} g^{ai}$. A straightforward discussion shows that this equation implies

$$
g^{aj} \partial_b g^{ai} = g^{ai} \partial_b g^{aj} \quad (a \neq b), \tag{5.14}
$$

for every pair of indices (i, j). By choosing $(i, j) = (a, \alpha)$ such that $g^{aa} \neq 0$, we get equation [DA]

$$
\partial_b \left(\frac{g^{a\alpha}}{g^{aa}} \right) = 0, \quad \text{for } a \neq b. \tag{5.15}
$$

This means that

$$
\theta_a^\alpha = \frac{g^{a\alpha}}{g^{aa}}
$$

20   S. Benenti

is a function of the variable q^a only (the coordinates of first class are ignorable, thus they do not appear in the metric components). The above considerations apply to all non-isotropic coordinates of second class $(q^{\tilde{a}})$: there exist functions $(\theta^{\alpha}_{\tilde{a}})$ depending on the variable corresponding to the lower index only, such that $g^{\tilde{a}\alpha} = \theta^{\alpha}_{\tilde{a}} g^{\tilde{a}a}$. Let us consider a new coordinate system $(q^{a'}, q^{\alpha'})$ defined by equations

$$dq^{a'} = dq^a, \quad dq^{\alpha'} = dq^{\alpha} - \theta^{\alpha}_{\tilde{a}} dq^{\tilde{a}}.$$

This system is equivalent to the old one (see Theorem 4.4). The coordinates of first class $(q^{\alpha'})$ are still ignorable and moreover,

$$g^{\tilde{a}'\alpha'} = A^{\tilde{a}'}_i A^{\alpha'}_j g^{ij} = g^{\tilde{a}j} A^{\alpha'}_j = g^{\tilde{a}\beta} A^{\alpha'}_\beta + g^{\tilde{a}\tilde{c}} A^{\alpha'}_{\tilde{c}} = g^{\tilde{a}\alpha} - g^{\tilde{a}\tilde{a}} \theta^{\alpha}_{\tilde{a}} = 0.$$

Apart from the notation, this is a coordinate system which we were looking for. ∎

Definition 5.2. We call (5.13) the *normal form* of a metric tensor in separable coordinates. The separable coordinates for which such a normal form holds are called *normal separable coordinates.*

Remark 5.1. For the number m_2 of second class isotropic coordinates we have the limits

$$m_2 \leq r, \quad m_2 \leq \min(p, q),$$

where (p, q) is the signature of the metric. Indeed, from the normal form (5.13) we see that $m_2 > r$ implies $\det(g^{ij}) = 0$, and the second limit follows from the fact that the maximal dimension of an isotropic subspace of a space of signature (p, q) is $\min(p, q)$, and the tangent subspaces spanned by the gradients of the coordinates $(q^{\tilde{a}})$ are isotropic.

Theorem 5.4. *The most general form of the metric tensor components in normal separable coordinates is the following:*

$$\begin{cases} g^{ab} = 0 \quad \text{for} \quad a \neq b \quad \text{and} \quad a, b = 1, \ldots, m \leq n, \\ g^{\tilde{a}\tilde{a}} = \phi^{\tilde{a}}_{(m)} \quad \text{for} \quad \tilde{a} = 1, \ldots, m_1 \leq m, \\ g^{\bar{a}\bar{a}} = 0 \quad \text{for} \quad \bar{a} = m_1, \ldots, m, \\ g^{\tilde{a}\alpha} = 0 \quad \text{for} \quad \alpha = m+1, \ldots, n \quad , \\ g^{\bar{a}\alpha} = \theta^{\alpha}_{\bar{a}} \, \phi^{\bar{a}}_{(m)} \quad (\bar{a} \text{ n.s.}), \\ g^{\alpha\beta} = \eta^{\alpha\beta}_a \phi^a_{(m)}, \end{cases} \qquad (5.16)$$

where $\phi^a_{(m)}$ is the last row of the inverse of a $m \times m$ Stäckel matrix in the second class coordinates (q^a), and $\theta^\alpha_{\bar{a}}$ and $\eta^{\alpha\beta}_a$ are functions depending on the coordinates corresponding to the lower index only.

Proof. We have the metric tensor components in the normal form (5.13). The first class coordinates are all ignorable. Let us consider equation (5.14) in the particular case

$$g^{\bar{a}\alpha}\partial_b g^{\bar{a}\beta} = g^{\bar{a}\beta}\partial_b g^{\bar{a}\alpha}.$$

For a fixed isotropic index of second class $\bar{a}$ there is at least one index of first class β such that $g^{\bar{a}\beta} \neq 0$ (otherwise the matrix (5.13) would be singular), so that the preceding equation can be written

$$\partial_b \left(\frac{g^{\bar{a}\alpha}}{g^{\bar{a}\beta}} \right) = 0.$$

This means that for each index $\bar{a}$ there is a function $X^{\bar{a}}$ and functions $(\theta^\alpha_{\bar{a}})$ depending on the variable corresponding to the lower index only, such that for each index α

$$g^{\bar{a}\alpha} = \theta^\alpha_{\bar{a}} X^{\bar{a}}. \tag{5.17}$$

Now, let us consider the Hamiltonian in normal separable coordinates,

$$H = \frac{1}{2} g^{\bar{a}\bar{a}}(p_{\bar{a}})^2 + g^{\bar{a}\alpha}p_{\bar{a}}p_\alpha + \frac{1}{2}g^{\alpha\beta}p_\alpha p_\beta.$$

Since the first class coordinates are ignorable, the corresponding momenta can be considered as constants of integration of the H-J equation. The Hamiltonian can be written

$$H = X^{\bar{a}}\phi_{\bar{a}} + X^{\bar{a}}\phi_{\bar{a}} + V,$$

by setting

$$X^{\bar{a}} = \frac{1}{2}g^{\bar{a}\bar{a}}, \quad X^{\bar{a}} = g^{\bar{a}\alpha}p_\alpha, \quad V = \frac{1}{2}g^{\alpha\beta}p_\alpha p_\beta, \quad \phi_{\bar{a}} = (p_{\bar{a}})^2, \quad \phi_{\bar{a}} = p_{\bar{a}}. \tag{5.18}$$

We can apply to this case the discussion of Section 3. The m functions $(X^a) = (X^{\bar{a}}, X^{\bar{a}})$ will be solutions of integrability conditions of the kind (3.19), and the "potential" V will be a solution of an equation of the kind (3.21). This means that there exists a $m \times m$ Stäckel matrix $(\phi^a_{(b)})$ such that $X^a = \phi^a_{(m)}$ and $V = \eta_a \phi^a_{(m)}$, with (η_a) function of the variable corresponding to the index only. Due to (5.17) and (5.18), the theorem is proved. ∎

Remark 5.2. Since we know the general form of the coordinate transformation from a separable system with ignorable first class coordinates (like a normal separable system) to a generic equivalent separable system (Theorem 4.4, formulæ (4.8)), we can perform the corresponding tensor transformation on the components (5.16). The result will be the general form of the contravariant metric tensor components in separable coordinates. By a suitable adaptation of the notation, this form turns out to be the following:

$$
\begin{cases}
g^{ab} = 0 \quad \text{for} \quad a \neq b \quad \text{and} \quad a, b = 1, \dots, m \leq n, \\[2mm]
g^{\tilde{a}\tilde{a}} = \phi^{\tilde{a}}_{(m)} \quad \text{for} \quad \tilde{a} = 1, \dots, m_1 \leq m, \\[2mm]
g^{\bar{a}\bar{a}} = 0 \quad \text{for} \quad \bar{a} = m_1, \dots, m, \\[2mm]
g^{a\alpha} = \phi^{\alpha}_{(\beta)} \theta^{\beta}_a \phi^{a}_{(m)} \quad (a \text{ n.s.}), \\[2mm]
g^{\alpha\beta} = \phi^{\alpha}_{(\gamma)} \phi^{\beta}_{(\delta)} \eta^{\gamma\delta}_a \phi^{a}_{(m)},
\end{cases}
\qquad (5.19)
$$

where $(\phi^{a}_{(m)})$ is the last row of the inverse of a $m \times m$ Stäckel matrix in the second class coordinates (q^a), $(\phi^{\alpha}_{(\beta)})$ is the inverse of a $r \times r$ Stäckel matrix in the first class coordinates (q^{α}), and the other functions depending on the coordinate corresponding to the lower index only (compare with [II] [CN]).

According to Theorem 5.4, the most general separable geodesic Hamiltonian in normal separable coordinates takes the form

$$
G = \frac{1}{2} \phi^{\tilde{a}}_{(m)} (p_{\tilde{a}})^2 + \phi^{\bar{a}}_{(m)} \theta^{\alpha}_{\bar{a}} p_{\bar{a}} p_{\alpha} + \frac{1}{2} \phi^{a}_{(m)} \eta^{\alpha\beta}_a p_{\alpha} p_{\beta}. \qquad (5.20)
$$

In the general form of the metric tensor components, with respect to normal separable coordinates (5.16), and in the expression of the most general Hamiltonian (5.20), only one line of the inverse of a Stäckel matrix appears (the last one, according to our notation). The analogous formulæ obtained by considering any other row $\phi^{a}_{(b)}$ of this matrix will give the components of a tensor $K^{ij}_{(b)}$, such that the m quadratic functions

$$
F_b = \frac{1}{2} K^{ij}_{(b)} p_i p_j = \frac{1}{2} \phi^{\tilde{a}}_{(b)} (p_{\tilde{a}})^2 + \phi^{\bar{a}}_{(b)} \theta^{\alpha}_{\bar{a}} p_{\bar{a}} p_{\alpha} + \frac{1}{2} \phi^{a}_{(b)} \eta^{\alpha\beta}_a p_{\alpha} p_{\beta}, \qquad (5.21)
$$

are first integrals in involution. This follows from a discussion similar to that of Section 3, concerning the procedure of separation: we interpret the H-J equation $G = h$ as the last equation of a system $F_b = a_b$ of m equations, which can be solved with respect to the momenta,

$$
(p_{\tilde{a}})^2 = 2 a_b \phi^{(b)}_{\tilde{a}} - \eta^{\alpha\beta}_{\tilde{a}} a_{\alpha} a_{\beta}, \qquad (a_{\alpha} \theta^{\alpha}_{\bar{a}}) p_{\bar{a}} = 2 a_b \phi^{(b)}_{\bar{a}} - \eta^{\alpha\beta}_{\bar{a}} a_{\alpha} a_{\beta}, \qquad (5.22)
$$

taking into account that, due to the ignoration of the first class coordinates, the momenta of first class are constant:

$$p_\alpha = a_\alpha.$$

Equations (5.22) are separated equations: they form a system of separated ordinary differential equations in the m unknown functions $W_a(q^a)$ such that $p_a = dW_a/dq^a$. The constants $(a_i) = (a_a, a_\alpha)$ will appear in the complete integral

$$W = \sum_{a=1}^{m} W_a(q^a, a_i) + a_\alpha q^\alpha.$$

They are the so called *separation constants*.

We conclude that the separation of variables is concomitant with the existence of r ignorable coordinates (q^α), which generate r linear first integrals (p_α), and the existence of a complementary number $m = n - r$ of quadratic first integrals (F_b), one of which is the geodesic Hamiltonian itself. All these first integrals are in involution due to the Jacobi theorem. As we will see in the next sections, in order to discuss the intrinsic characterization of the separation of variables, it is more convenient to think of these first integrals as *Killing vectors* or *Killing tensors*. The Killing vectors are the partial derivatives (∂_α) associated with the ignorable coordinates, interpreted as vector fields. The components of the Killing tensors are the coefficients (K_b^{ij}) of the quadratic forms (F_b).

When there are no isotropic second class coordinates (like in the case of a positive definite metric) the normal form (5.13) reduces to

$$(g^{ij}) = \begin{array}{c} \\ m \\ r \end{array} \overset{\begin{array}{cc} m & r \end{array}}{\begin{pmatrix} \delta^{ab} g^{aa} & 0 \\ 0 & g^{\alpha\beta} \end{pmatrix}}, \tag{5.23}$$

where

$$g^{aa} = \phi^a_{(m)}, \quad g^{\alpha\beta} = \eta^{\alpha\beta}_a g^{aa}. \tag{5.24}$$

The non vanishing components of the Killing tensors are:

$$K^{aa}_{(b)} = \phi^a_{(b)}, \quad K^{\alpha\beta}_{(b)} = \eta^{\alpha\beta}_a K^{aa}_{(b)}. \tag{5.25}$$

Several examples of separable metrics are given by exact solutions of Einstein's field equations (see for instance [CR] [W] [KSMH] [BF2]). We present here only three of them, showing how they fit the theory.

Example 5.1. The *Reissner-Nordström metric*

$$ds^2 = r^2(d\theta^2 + \sin^2\theta \, d\phi^2) + \frac{r^2}{\Delta} dr^2 - \frac{\Delta}{r^2}, \tag{5.26}$$

where $\Delta = r^2 + e^2 - 2mr$, provides a mathematical model for the exterior field of a spherically symmetric body of charge e and mass m. For $e = 0$ it reduces to the *Schwarzschild metric*. Coordinates ϕ and t are ignorable, so that, according to our notation it is convenient to set

$$q^1 = r, \quad q^2 = \theta, \quad q^3 = \phi, \quad q^4 = t. \tag{5.27}$$

These coordinates are orthogonal and the contravariant components are

$$(g^{ii}) = \left(\frac{\Delta}{r^2}, \frac{1}{r^2}, \frac{1}{r^2 \sin^2 \theta}, -\frac{r^2}{\Delta} \right). \tag{5.28}$$

We can see directly from the H-J equation that these coordinates are separable. The separation is achieved by introducing the separation constant

$$c = p_2^2 + \frac{1}{\sin^2 \theta} p_3^2,$$

so that the H-J equation reduces to

$$\Delta p_1^2 + c - \frac{r^4}{\Delta} p_4^2 = 2h\, r^2,$$

where $(p_\alpha) = (p_1, p_2)$ are constants due to the ignoration of the coordinates (q^3, q^4). However, the separation can be also recognized by comparing (5.28) with the normal form (5.16). The coordinates $(q^a) = (q^1, q^2)$ are of second class and the components

$$(g^{aa}) = \left(\frac{\Delta}{r^2}, \frac{1}{r^2} \right)$$

fit the condition $g^{aa} = \phi^a_{(2)}$ by setting in the canonical form (3.24),

$$\phi_1 = r^2, \quad \phi_2 = 0, \quad \psi_1 = \Delta, \quad \psi_2 = 1.$$

The condition $g^{\alpha\beta} = \eta^{\alpha\beta}_a g^{aa}$ is fulfilled by setting

$$\eta^{33}_1 = 0, \quad \eta^{44}_1 = -\frac{r^4}{\Delta^2}, \quad \eta^{33}_2 = \frac{1}{\sin^2 \theta}, \quad \eta^{44}_2 = 0.$$

The conclusion is that the given coordinates are normal separable coordinates. Moreover, by using formula (5.25) and (3.24) we can construct the components of the non trivial Killing tensor $K = K_2$. The result is

$$K^{11} = 0, \quad K^{22} = -1, \quad K^{33} = -\frac{1}{\sin^2 \theta}, \quad K^{44} = 0,$$

and, up to the inessential factor $-\frac{1}{2}$, the corresponding first integral coincides with the separation constant c. If instead of t we use a new coordinate u defined by equation (see [KSMH])

$$dt = du + \frac{r^2}{\Delta}dr, \tag{5.29}$$

then the metric becomes

$$ds^2 = r^2(d\theta^2 + \sin^2\theta\, d\phi^2) - 2du\, dr - \frac{\Delta}{r^2}du^2.$$

The new system (r, θ, ϕ, u) is separable and equivalent to the old one, since the transformation (5.29) fits the transformation formulæ (4.8). However, these coordinates are no more normal, since the matrix (g^{ij}) has now the form:

$$(g^{ij}) = \begin{pmatrix} 1 & 0 & 0 & -\dfrac{r^2}{\Delta} \\ 0 & \dfrac{1}{\Delta} & 0 & 0 \\ 0 & 0 & \dfrac{1}{\Delta\sin^2\theta} & 0 \\ -\dfrac{r^2}{\Delta} & 0 & 0 & 0 \end{pmatrix}.$$

Example 5.2. The *Kerr-Newmann metric*

$$ds^2 = \rho^2\left(\frac{dr^2}{\Delta} + d\theta^2\right) + \sin^2\theta\left(r^2 + a^2 - \varepsilon\frac{a^2\sin^2\theta}{\rho^2}\right)d\phi^2$$
$$- \left(1 + \frac{\varepsilon}{\rho^2}dt^2 + 2\varepsilon a\frac{\sin^2\theta}{\rho^2}\right)dtd\phi,$$

where $\Delta = r^2 + a^2 + e^2 - 2mr = r^2 + a^2 + \varepsilon$ and $\rho^2 = r^2 + a^2\cos^2\theta$, is a model for the exterior gravitational field of a rotating source of charge e, mass m and angular momentum a. For $e = 0$ it reduces to the *Kerr metric*, and for $e = 0$ and $a = 0$ to the Schwarzschild metric. As in (5.27) we set $q^1 = r$, $q^2 = \theta$, $q^3 = \phi$, $q^4 = t$. The contravariant components form the matrix

$$(g^{ij}) = \begin{pmatrix} \dfrac{\Delta}{\rho^2} & 0 & 0 & 0 \\ 0 & \dfrac{1}{\rho^2} & 0 & 0 \\ 0 & 0 & \dfrac{1}{\rho^2}\left(\dfrac{1}{\sin^2\theta} - \dfrac{a^2}{\Delta}\right) & \dfrac{a}{\rho^2}\left(1 - \dfrac{a^2+r^2}{\Delta}\right) \\ 0 & 0 & \dfrac{a}{\rho^2}\left(1 - \dfrac{a^2+r^2}{\Delta}\right) & \dfrac{1}{\rho^2}\left(a^2\sin^2\theta - \dfrac{(a^2+r^2)^2}{\Delta}\right) \end{pmatrix}$$

This is a normal form, where $(q^a) = (r, \theta)$ are second class coordinates and $(q^\alpha) = (\phi, t)$ are first class ignorable coordinates. Indeed we have

$$\phi_1 = r^2, \quad \phi_2 = a^2 \cos^2 \theta, \quad \psi_1 = \Delta, \quad \psi_2 = 1,$$

and

$$(\eta_1^{\alpha\beta}) = -\frac{1}{\Delta^2} \begin{pmatrix} a^2 & a(a^2 + r^2) \\ a(a^2 + r^2) & (a^2 + r^2)^2 \end{pmatrix},$$

$$(\eta_2^{\alpha\beta}) = \begin{pmatrix} \dfrac{1}{\sin^2 \theta} & a \\ a & a^2 \sin^2 \theta \end{pmatrix}.$$

These normal separable coordinates are not orthogonal.

Example 5.3. The *Friedmann metric*

$$ds^2 = dt^2 - R(t) \left(\frac{dr^2}{1 - kr^2} + r^2 d\theta^2 + r^2 \sin^2 \theta d\eta^2 \right),$$

where k is a constant and $R(t)$ an arbitrary smooth function without zeros. The coordinates

$$q^1 = r, \quad q^2 = t, \quad q^3 = \theta, \quad q^4 = \eta,$$

are normal separable coordinates; (r, t, θ) are of second class, and η is ignorable. Indeed, the contravariant components of the metric are

$$(g^{ii}) = \left(\frac{kr^2 - 1}{R(t)}, 1, -\frac{1}{r^2 R(t)}, -\frac{1}{r^2 R(t) \sin^2 \theta} \right).$$

The separation can be achieved through the separation constants

$$a_1 = R(\tfrac{1}{2}p_2^2 - h), \quad a_2 = p_3^2 + \frac{1}{\sin^2 \theta} p_4^2.$$

We can fit the canonical form (3.26) with

$$\psi_1 = \frac{kr^2 - 1}{1 - r^2}, \quad \psi_2 = R(t), \quad \psi_3 = 1,$$

$$\mu_1 = \frac{1}{1 - r^2}, \quad \mu_2 = 0, \quad \mu_3 = 1,$$

$$\nu_1 = 0, \quad \nu_2 = -R(t), \quad \nu_3 = 0.$$

and formula (5.25) with

$$(\eta_a^{44}) = \left(0, 0, \frac{1}{\sin^2\theta}\right).$$

Example 5.4. This is an example of normal separable coordinates with an isotropic second class coordinate (it is taken from [KM1], where further examples can be found). The three-dimensional metric

$$ds^2 = y\,dz^2 + 2dx\,dy + \frac{1}{4}\frac{z^2}{y}dy^2$$

is a locally flat Lorentzian metric. If we set

$$q^1 = z, \quad q^2 = y, \quad q^3 = x,$$

then the contravariant components are

$$(g^{ij}) = \begin{pmatrix} \dfrac{1}{y} & 0 & 0 \\ 0 & 0 & 1 \\ 0 & 1 & -\dfrac{1}{4}\dfrac{z^2}{y} \end{pmatrix}.$$

The coordinate q^3 is ignorable, q^1 is of second class and q^2 is coisotropic of second class. The normal form of the metric (5.16) and the canonical form (3.24) can be fulfilled by setting

$$\phi_1 = 0, \quad \phi_2 = y, \quad \psi_1 = 1, \quad \psi_2 = 1, \quad \theta_2^3 = y, \quad \eta_1^{33} = -\frac{1}{4}z^2, \quad \eta_2^{33} = 0.$$

6. Intrinsic characterization of orthogonal separation

With each smooth contravariant symmetric tensor field K over a manifold Q we associate a real smooth function E_K on the cotangent bundle T^*Q defined by

$$E_K = \frac{1}{k!}K^{i_1\cdots i_k}p_{i_1}\cdots p_{i_k}, \tag{6.1}$$

where k is the order of K. This fact establishes a natural identification between symmetric contavariant tensors on Q and functions on T^*Q which are homogeneous polynomial in the momenta. Due to this identification, the two natural algebraic structures on the space of these functions induce two algebraic structures on the space of contravariant symmetric tensors. The ordinary product of real functions induces a *symmetric product* denoted by $\cap$ and defined by equation:

$$E_{K \cap L} = E_K E_L. \tag{6.2}$$

This product is commutative and associative. The Poisson brackets on T^*Q, induces Lie brackets $[K, L]$ defined by equation

$$E_{[K,L]} = \{E_K, E_L\}. \tag{6.3}$$

Moreover, the Lie brackets are a bi-derivation with respect to the symmetric product, i.e.

$$[K, L \cap M] = [K, L] \cap M + [K, M] \cap L. \tag{6.4}$$

We say that symmetric contravariant tensors K and L *commute* or that they *are in involution* when $[K, L] = 0$. If Q is a Riemannian manifold, with metric tensor g, then there is a natural identification between contravariant and covariant (symmetric) tensors, and $E_g = G$ is the geodesic Hamiltonian. A *Killing tensor* is a symmetric tensor which commute with the metric: $[K, g] = 0$; i.e. such that the corresponding function E_K is a geodesic first integral: $\{E_K, G\} = 0$.

In the following discussion we will deal only with Killing tensors of order 1, which are called *Killing vectors*, and of order 2, which we will call briefly *Killing tensors*, by understanding the order 2.

A Killing vector X is characterized by the following equivalent properties. (i) The Lie derivative of the metric tensor g with respect to X is zero. (ii) The (local) action on Q generated by X is made of isometries (i.e. of *rigid motions*, according to the old terminology). (iii) In a coordinate system (q^i) such that $\partial_1 = X$, the coordinate q^1 is ignorable: $\partial_1 g_{ij} = 0$. (iv) The covariant derivative ∇X with respect to the Levi-Civita connection is skew-symmetric. (v) As a derivation over the functions, $X = X^i \partial_i$ commute with the Laplacian $\Delta = g^{ij} \nabla_i \partial_j$.

Tensor fields of order 2 can be interpreted as linear operators over the spaces of vector fields and 1-forms. Let us denote by $X \cdot K$ and by $\phi \cdot K$ the value of a 2-tensor field K on a vector field X and a 1-form ϕ. The tensor is symmetric if and only if $X \cdot K \cdot Y = Y \cdot K \cdot X$. In this sense, we can also consider fields of *eigenvectors*, *eigenforms* and *eigenvalues* of K.

It can be proved that

Theorem 6.1. *(i) [E2] If a symmetric tensor field K of order 2 is diagonalized in a coordinate system (q^i), then it is a Killing tensor if and only if the following equations are satisfied*

$$\partial_i \rho_j = (\rho_i - \rho_j)\, \partial_i \log |g^{jj}|, \tag{6.5}$$

where ρ_i is the eigenvalue corresponding to the vector fields $\partial_i = \frac{\partial}{\partial q^i}$.

(ii) If two Killing tensors and are both diagonalized in a coordinate system (q^i), then they are in involution.

We remark that equations (6.5) coincide with equations (3.23). This circumstance represents the link between Killing tensors and orthogonal separation. Indeed, as we have seen in Section 3, orthogonal separation of variables gives rise to a set of n independent quadratic first integrals. Interpreted as Killing tensors they have the following characteristic properties [E1] [W] [KM2]:

(a) they are pointwise independent and in involution;
(b) one of them is the metric tensor;
(c) they have n common orthogonal closed eigenforms.

Property (c) follows from the fact that the Killing tensors are all diagonalized in the separable coordinates. This means that the differentials (dq^i) of such coordinates are common orthogonal eigenforms (or, equivalently, that the partial derivatives (∂_i) with respect to these coordinates, interpreted as vector fields, are common orthogonal eigenvectors). It follows from Theorem 6.1 that the existence of independent Killing tensors satisfying the above requirements is equivalent to the complete integrability of system (6.5). On the other side, as we have seen in Section 3, the integrability conditions of this system coincide with the Levi-Civita separation conditions in the case of a geodesic Hamiltonian in orthogonal coordinates.

Remark 6.1. Due to equations (6.5), the characteristic equations (3.8) of a potential compatible with orthogonal separation can be written in the following equivalent form:

$$(\rho_i - \rho_j)\partial_{ij}V = \partial_i\rho_j\,\partial_j V - \partial_j\rho_i\,\partial_i V. \tag{6.1}$$

Example 6.1. *Two-dimensional manifolds.* On Riemannian manifolds of dimension 2, the separation of the H-J equation always occurs in orthogonal coordinates and without second class isotropic coordinates. Indeed, we have three cases. (i) If a separable system has only coordinates of second class, then, due to the normal form (5.13), the coordinates are necessarily orthogonal and non-isotropic. (ii) If a separable system has one coordinate of second class, and this coordinate is not isotropic, then the normal form (5.13) shows again that the coordinates are orthogonal. Hovever, the case of one isotropic coordinate of second class is excluded; indeed, in this case the normal form is

$$(g^{ij}) = \begin{pmatrix} 0 & g^{12} \\ g^{12} & g^{22} \end{pmatrix},$$

with $g^{12} \neq 0$, and $H_1 = \partial_1 g^{ij} p_i p_j$ is divisible by $H^1 = g^{1i} p_i = g^{12} p_2$, which is in contrast with the assumption that q^1 is of second class. (iii) If there are two separable coordinates of first class, then they are equivalent to ignorable coordinates. A manifold admitting a system of ignorable coordinates is locally flat, since the components of the metric tensor in these coordinates are constant. So we can always choose orthogonal (ignorable) coordinates (i.e. *orthogonal Cartesian coordinates*).

Due to this property and to the preceding discussion, equivalence classes of orthogonal separable coordinates are then in one-to-one correspondence with equivalence classes of Killing tensors, if we say that two Killing tensors are equivalent if they are related by a linear relation with constant coefficients to the metric tensor g. In other words, every Killing tensor K gives rise to an equivalence class of orthogonal separable system; a Killing tensor of the kind $aK + bg$, where $a, b \in \mathbb{R}$, gives rise to the same class. The only requisite is that the Killing tensor is not proportional to the metric tensor and that it has smooth real eigenvalues (this second requisite is automatically fulfilled if the metric is positive-definite). Hence, with every Killing tensor K we associate a *singular set* made of *singular points* in which the two eigenvalues of K coincide or are complex. The separable coordinates are then defined (locally) on the complementary set.

A straightforward discussion based on equations (6.5) shows that [BR]: (i) If the eigenvalues $(\rho_i) = (\rho_1, \rho_2)$ of K are independent functions, then they define a separable system of coordinates: $q^1 = \rho_2$, $q^2 = \rho_1$. More precisely, equations $\rho_i = \text{const.}$ define two orthogonal families of unparametrized curves which are the coordinate surfaces of equivalent orthogonal separable systems. (ii) If the eigenvalue ρ_1 is constant, then the eigenvector X_2 corresponding to the other eigenvalue is, up to a factor, a Killing vector. In case (i), the characteristic equations (6.1) of a separable potential reduce to the single equation

$$(q^1 - q^2)\partial_{12} V = \partial_1 V - \partial_2 V. \tag{6.2}$$

Example 6.2. *The Euclidean plane.* On the Euclidean plane $\mathbb{E}_2$ every Killing tensor is reducible, i.e. it is a linear combination, with constant coefficients, of symmetric product of Killing vectors. This is in fact a property of all manifolds with constant curvature [KL]. In particular it can be shown [BR] that every Killing tensor has the form

$$K = a\, R_P \cap R_Q + b\, g, \quad a, b \in \mathbb{R},$$

where R_P denotes the unitary rotation centered at the point $P \in \mathbb{E}_2$. This is a Killing vector defined by $R_P(x) = \omega \times x$, where ω is a unitary vector orthogonal to the plane $\mathbb{E}_2$ inbedded in the three-dimensional Euclidean space, and x is the position vector of a generic point of the plane. We do not exclude the case in which the point P goes to the infinity; in this case P is a direction and R_P is a unitary vector field orthogonal to this direction. As a consequence of the preceding discussion (Example 6.1), we have that every equivalence class of orthogonal separable coordinates in the plane is characterized by a Killing tensor of the form

$$K = R_P \cap R_Q.$$

The singular points of K are precisely the points P and Q, when they are "true" points (not at the infinity). Hence, we have four possibilities:

(i) The two points P and Q are both at the infinity.
(ii) Only one point (say Q) is at the infinity.
(iii) The two points P and Q are true points and $P \neq Q$.
(iv) The two points P and Q are true points and coincide.

It can also be proved that the eigenvalues of K give rise to families of confocal conics, whose focuses are precisely the points P and Q. Hence, in correspondence with the above four cases, we have the known four kinds of separable coordinates:

(i) cartesian;
(ii) parabolic (with focus P and axis PQ);
(iii) elliptic-hyperbolic (with focuses P and Q);
(iv) polar (with center $P = Q$).

Furthermore, through equations (6.1), it is possible to characterize all possible potentials compatible with the separation. We mention here, as an example, only one result (a detailed discussion is contained in [BR]).

Theorem 6.1. *The dynamical H-J equation of a material point in the Euclidean plane, submitted to two symmetrical fields centered at two distinct points P and Q, is integrable by separation of variables by means of elliptic-hyperbolic coordinates with focuses P and Q, if and only if the potential is of the kind*

$$V = a(r_P^2 + r_Q^2) + \frac{b}{r_P} + \frac{c}{r_Q} \quad (a, b, c \in \mathbb{R}),$$

where r_P denotes the distance from the point P, i.e. if and only if the two fields are a combination of Coulombian fields of any charge (in particular, Newtonian fields) and of elastic or centrifugal fields with the same constant (negative or positive).

Example 6.3. *The sphere.* As far as the Killing vectors are concerned, on the sphere $\mathbb{S}_2$ we have results similar to the Euclidean plane. We consider the unitary sphere centered at a point O of the three-dimensional Euclidean space. Separable coordinates are generated by Killing vectors of the kind

$$K = R_u \cap R_v,$$

where u and v are unitary vectors, and R_u denotes the unitary rotation around the axis (O, u) and defined by $R_u(x) = u \times x$. We have only two possibilities:

(i) the vectors u and v coincide;
(ii) the vectors u and v are distinct.

In case (i) we have two opposite singular points N and S, and the corresponding coordinates are the spherical-polar coordinates. In case (ii) we have two pairs of opposite singular points, (N_1, N_2) and (S_1, S_2), and the separable coordinates are those considered by Neumann [NE]: the coordinate curves are two families of confocal spherical conics, one family has the pair of points (N_1, N_2) (or (S_1, S_2)) as focuses, the other one has the pair (N_1, S_2) (or (S_1, N_2)), being S_i the opposite of N_i. We notice that spherical ellipses with focuses (N_1, N_2) are spherical hyperbolae with focuses (N_1, S_2).

Example 6.4. *The asymmetric ellipsoid.* This is a classical subject with a wide literature originated by Jacobi. Through the so called *Jacobian coordinates* it is possible to integrate by separation of variables the geodesic H-J equation. These coordinates are the restriction to the ellipsoid of separable orthogonal coordinates on the three-dimensional Euclidean space generated by three families of confocal quadrics (see [BI]). It can be proved that a Killing tensor generating these coordinates is

$$K = \kappa^{-\frac{3}{4}} B,$$

where B is the second fundamental form and κ is the Gaussian curvature, i.e. the product of the two eigenvalues of B (the *main curvatures*). The singular points of K are the umbilical points. Moreover, it can be proved that the separation constant corresponding to K is the *Joachimsthal constant* [BI]: for every geodesic the product of the distance of the center from the tangent plane at a point of the geodesic, times the length of the diameter parallel to the tangent to the geodesic at the same point is constant.

Example 6.5. *Manifolds with constant curvature.* Kalnins [K] has proved that on the Euclidean spaces $\mathbb{E}_n$, on the spheres $\mathbb{S}_n$ and on the pseudo-spheres $\mathbb{H}_n$ every separable system has an orthogonal equivalent. This means that on strictly-Riemannian manifolds with constant curvature the separation always occurs in orthogonal coordinates. In [B3] this property has been extended to Lorentzian manifolds (signature $(+\ldots+-)$) with non-negative curvature (for systems without isotropic second class coordinates). The proof is based on the geometrical characterization of separable coordinates in general, illustrated in the next section.

7. Intrinsic characterization of separable systems without second class isotropic coordinates

As we have seen in Section 5, the separation of variables gives rise to r linear first integrals and to a complementary number $m = n - r$ of quadratic first integrals, which are all in involution and independent. If we interpret these first integrals as Killing vectors (X_α) and Killing tensors (K_a), then, in the case in which there are not second class isotropic coordinates, by looking at formulæ (5.23), (5.24) and (5.25) (we always refer to normal separable coordinates (q^a, q^α)) we can derive for them the following properties:

(a) they are in involution; the Killing vectors are pointwise independent; the Killing tensors are pointwise independent;
(b) one of the Killing tensors is the metric tensor;
(c) the Killing tensors have m common orthogonal closed eigenforms (ϕ^a) (or m orthogonal eigenvectors (X_a) in involution: $X_a \cdot X_b = 0$, $[X_a, X_b] = 0$);
(d) The eigenforms are orthogonal to the Killing vectors: $<X_\alpha, \phi^a> = 0$ (or, equivalently, $X_\alpha \cdot X_a = 0$).
(e) The eigenforms are invariant with respect to the Killing vectors: $d_{X_\alpha} \phi^a = 0$ (or, equivalently, $[X_\alpha, X_a] = 0$).
(f) The Killing vectors generate a normal distribution.

Indeed, we recognize these conditions by setting

$$\phi^a = dq^a, \quad X_a = \partial_a, \quad X_\alpha = \partial_\alpha.$$

The symbol d_X used in (e) denotes the Lie derivative with respect to the vector field X. We remind that, for a closed 1-form ϕ, $d_X \phi = d<X, \phi>$. The last property (f) follows from the fact that the vectors (∂_a) generate the distribution orthogonal to the Killing vectors. Moreover, since the Killing vectors commute, they generate an integrable distribution whose integral foliation is made of locally flat manifolds. Hence, with a separable system we can associate two mutually orthogonal foliations: a foliations $\mathcal{E}$ made of locally flat submanifolds of dimension r, tangent to the Killing vectors and defined by equations $q^a = $ const., and an orthogonal foliation $\mathcal{S}$ defined by equations $q^\alpha = $ const., made of submanifolds of dimension m which admits orthogonal systems of separable coordinates.

We can interpret this fact in a slightly different manner. The Killing vectors generate an Abelian free action of isometries on the manifold Q, whose fibres are locally flat manifolds which form the foliation $\mathcal{E}$. Let us denote by $\bar{Q}$ the quotient manifold and by $\pi : Q \to \bar{Q}$ the quotient projection. The metric tensor g is reduced to a metric tensor $\bar{g}$ on $\bar{Q}$ by the projection π. The restriction of this projection to every manifold of the foliation $\mathcal{S}$ orthogonal to the fibres is an isometry. This is actually a geometrical picture which holds only locally, i.e. for open subset $U \subset Q$. We remark that the orthogonal separable coordinates existing on the reduced manifold $\bar{Q}$, even if they correspond to second class coordinates on the whole manifold Q, they are not all necessarily of second class. So that we can perform for $\bar{Q}$ the same procedure of reduction, and so on.

All the properties listed above can be considered as necessary conditions for the existence of separable coordinates. As sufficient conditions they are redundant (see [KM3]). However, there is not a unique or privileged minimal subset of these conditions to be taken as sufficient conditions. We do not enter in this discussion. Instead, we give an example of how the geometrical characterization of separation can be used.

Up to a linear transformation with constant coefficients, we can always reduce the Killing vectors to be orthogonal on any chosen leaf of the locally flat foliation $\mathcal{E}$. These transformations are compatible with the separation: they give rise to equivalent separable systems. However, it is not possible in general to reduce the Killing vectors to be orthogonal everywhere. When this happens, we find orthogonal separable systems. Indeed, the Killing tensors $(K_a, K_\alpha = X_\alpha \cap X_\alpha)$ satisfy the requirements listed in Section 6.

Actually, this always happens in the affine Euclidean spaces $\mathbb{E}_n$ and in the affine Minkowskian spaces $\mathbb{M}_n$. Indeed, in these spaces every system of independent commuting Killing vectors generating a normal distribution (properties (a) and (f)) can be orthogonalized, i.e. it can be transformed into an orthogonal system through a linear transformation with constant coefficients [B3]. In particular this property holds for Killing vectors which are *rotations* around a fixed point O of these spaces. Thus, if we consider the restrictions of these rotations to the *fundamental hyperquadrics* (see [E2]) centered at the point O, then we conclude that the same property holds for the Killing vectors on these manifolds. The fundamental hyperquadrics are spheres $\mathbb{S}_{n-1} \subset \mathbb{E}_n$ defined by equation $x \cdot x = 1$, hyperboloids $\mathbb{H}_{n-1} \subset \mathbb{M}_n$ defined by equation $x \cdot x = -1$, and hyperboloids $\mathbb{L}_{n-1} \subset \mathbb{M}_n$ defined by equation $x \cdot x = 1$. The first two are strictly-Riemannian manifolds with constant curvature, positive and negative respectively. The third ones are Lorentzian manifolds with constant positive curvature. The spaces and surfaces so far considered provide a local model for all strictly-Riemannian manifolds with constant curvature and all Lorentzian manifolds with constant non-negative curvature. Hence, as we said at the end of last section, on these manifolds the separation of variables always occurs in orthogonal coordinates.

REFERENCES

[B1] S. Benenti, *Separability structures on Riemannian manifolds*, in L. N. Math. **863**, 512-538 (1980).

[B2] S. Benenti, *On the orthogonality of the second class coordinates in the separation of the Hamilton-Jacobi equation*, Atti Accad. Sci. Torino **123**, 167-171 (1989).

[B3] S. Benenti, *Orthogonal separation of variables on manifolds with constant curvature*, in preparation.

[BF1] S. Benenti, M. Francaviglia, *Remarks on certain separability structures and their applications to general relativity*, GRG Journal **10**, 79-82 (1979); *Canonical forms of $S_{n-3}-$ structures in pseudo-Riemannian manifolds*, Ann. Inst. H. Poincaré **30**, 143-157 (1979); *Canonical forms for separability structures with less than five Killing tensors*, Ann. Inst. H. Poincaré **34**, 45-64 (1981).

[BF2] S. Benenti, M. Francaviglia, *The theory of separability of the Hamilton-Jacobi equation and its applications to general relativity*, in "General Relativity and Gravitation" (A. Held Ed.), Plenum, 1980.

[BI] L. Bianchi, "Lezioni di geometria differenziale", Spoerri, Pisa, 1902-1903.

[BR] S. Benenti, G. Rastelli, *Sistemi di coordinate separabili nel piano euclideo*, Memorie Accad. Sci. Torino (forthcoming).

[CR] B. Carter, *Hamilton-Jacobi and Schrödinger separable solutions of Einstein equations*, Comm. Math Phys **10**, 280-310 (1968).

[CN] F. Cantrijn, *Separation of variables in the Hamilton-Jacobi equation for non-conservative systems*, J. Phys. A: Math. Gen. **10**, 491-505 (1977).

[DA] F. A. Dall'Acqua, *Le equazioni di Hamilton-Jacobi che si integrano per separazione delle variabili*, Rend. Circ. Mat. Palermo **33**, 341-351 (1912).

[E1] L. P. Eisenhart, *Separable systems of Stäckel*, Ann. Math. **35**, 284-305 (1934).

[E2] L. P. Eisenhart, "Riemannian geometry", Princeton University Press, Princeton, 1949.

[H] A. Huaux, *Sur la séparation des variables dans l'équation aux dérivées partielles de Hamilton-Jacobi*, Ann. Mat. Pura Appl. **108**, 251-282 (1976).

[II] M. S. Iarov-Iarovoi, *Integration of the Hamilton-Jacobi equation by the methos of separation of variables*, J. Appl. Math. Mec. **27**, 1499-1520 (1963).

[K] E. G. Kalnins, "Separation of Variables for Riemannian Spaces of Constant Curvature", Pitman Monographs **28**, 1986.

[KL] H. Katzin, J. Levine, *Quadratic first integrals of the geodesics in spaces of constant curvature*, Tensor **10**, 97-104 (1965).

[KM1] E. G. Kalnins, W. Miller Jr., *Separable coordinates for three-dimensional complex Riemannian spaces*, J. Diff. Geom. **14**, 221-236 (1979).

[KM2] E. G. Kalnins, W. Miller Jr., *Killing tensors and variable separation for Hamilton-Jacobi and Helmholtz equations*, SIAM J. Math. Anal. **11**, 1011-1026 (1980).

[KM3] E. G. Kalnins, W. Miller Jr., *Killing tensors and nonorthogonal variable separation for Hamilton-Jacobi equations*, SIAM J. Math. Anal. **12**, 617-629 (1980).

[KSMH] D. Kramer, H. Stephani, M. MacCallum, E. Herlt, *Exact solutions of Einstein's field equations*, VEB-DVW, Berlin (1980).

[LC] T. Levi-Civita, *Sulla integrazione della equazione di Hamilton-Jacobi per separazione di variabili*, Math. Ann. **59**, 383-397 (1904).

[NE] C. Neumann, *De problemate quodam mechanico, quod ad primam integralium ultraellipticorum classem revocatur*, J. Reine Angew. Math. **56**, 46-63 (1859).

[ST1] P. Stäckel, *Über die Bewegung eines Punktes in einer n-fachen Mannigfaltigkeit*, Math. Ann. **42**, 537-563 (1893).

[ST2] P. Stäckel, *Über quadatizche Integrale der Differentialgleichungen der Dynamik*, Ann. Math. Pura Appl. **26**, 55-60 (1897).

[W] N. M. J. Woodhouse, *Killing tensors and the separation of the Hamilton-Jacobi equation*, Commun. Math. Phys. **44**, 9-38 (1975).

This research has been sponsored by Gruppo Nazionale di Fisica Matematica, Consiglio Nazionale delle Ricerche and by Ministero dell'Università e della Ricerca Scientifica e Tecnologica.

Istituto di Fisica Matematica "J.-L. Lagrange"
Università di Torino
Via Carlo Alberto 10
10123 Torino
ITALY

Some Remarks on Quantization

ROBERT J. BLATTNER

Dedicated to Jean-Marie Souriau

0. Introduction

Any treatment of quantum mechanics must face the problem of how one determines which quantum observables correspond to which classical observables. Dirac, in his famous book *The Principles of Quantum Mechanics* (3rd edition, 1947), discusses his method for attacking this problem, which is based on the Poisson bracket (P.B.). The P.B. is fundamental to Hamiltonian mechanics, and Dirac gives an argument to support the idea that the quantum analogue of the P.B. is given by $\{u, v\} = (i\hbar)^{-1}(uv - vu)$, where $\hbar$ is a universal constant. He then goes on to say "The strong analogy between the quantum P.B. ... and the classical P.B. ... leads us to make the quantum P.B.'s, or at any rate the simpler ones of them, have the same values as the corresponding classical P.B.'s."(p. 87). He leaves open the question of exactly how one determines which quantum P.B.'s would have the same values as their classical counterparts.

Now the spaces $\mathcal{P}$ of classical observables and $\mathcal{A}$ of quantum observables form Lie algebras under their respective P.B.'s. Geometric quantization, due to Kostant and Souriau, proposes to find a Lie algebra isomorphism between Lie subalgebras of $\mathcal{P}$ and $\mathcal{A}$ that are as large as possible. Their program is divided into two parts, prequantization and quantization, and indeed prequantization succeeds in constructing a Lie algebra isomorphism between all of $\mathcal{P}$ and all of $\mathcal{A}$.

In this review, we will look first at prequantization as a problem in constructing a functor subject to certain natural-seeming axioms derived from Dirac. Then we will go on to consider further axioms which Dirac seems to want. This will lead us close to quantization via polarizations. The cost involved in this approach is a marked shrinkage in the size of the Lie subalgebra of $\mathcal{P}$ which can be quantized. In an effort to quantize all of $\mathcal{P}$ we will find ourselves forced to consider deformation quantization, a method pioneered by Weyl, Moyal, Lichnerowicz, Flato, Sternheimer, Fronsdal and others. Finally we will talk briefly about strict deformation quantization (Rieffel) and quantum states (Souriau).

38   R. J. Blattner

1. Prequantization

We sketch here the approach to prequantization found in [B]. Let n be a positive integer. We define the *classical category* $\mathfrak{C}_n$, a typical object of which is a symplectic manifold (X^{2n}, ω) of dimension $2n$ together with the space $\mathcal{P}_X$ of $\mathbb{R}$-valued C^∞ functions on X. A map in $\mathfrak{C}_n$ is a symplecto-morphism Φ mapping (X_1, ω_1) onto an open subset of (X_2, ω_2). Then Φ maps $\mathcal{P}_{X_2}$ contravariantly into $\mathcal{P}_{X_1}$ by composition,

$$\varphi \longmapsto \varphi \circ \Phi, \qquad \varphi \in \mathcal{P}_{X_2},$$

and is a Lie algebra map with respect to P.B.'s. (X, ω) is called a *classical state space* and $\mathcal{P}_X$ is called its space of *classical observables*. Similarly, we define the *quantum category* $\mathfrak{Q}$ (independently of n), a typical object of which is the projective unit sphere $\mathcal{S}_\mathcal{H}$ of a Hilbert space $\mathcal{H}$ together with a Lie algebra of Hermitian operators $\mathcal{A}_\mathcal{H}$ on $\mathcal{H}$ with a common dense domain $\mathcal{D}$. (We finesse the question of whether these operators should be essentially self-adjoint.) A map in $\mathfrak{Q}$ is a linear isometry U from $\mathcal{H}_1$ into $\mathcal{H}_2$ modulo composition with scalar multiples λI of identity maps, $\lambda \in \mathbb{T} = \{z \in \mathbb{C} : |z| = 1\}$. Again, U should map $\mathcal{A}_{\mathcal{H}_2}$ contravariantly into $\mathcal{A}_{\mathcal{H}_1}$ by the law

$$T \longmapsto U^* T U, \qquad T \in \mathcal{A}_{\mathcal{H}_2},$$

and is a Lie algebra map with respect to quantum P.B.'s. (Note that our vagueness as to which operators are associated with $\mathcal{H}$ makes this a real assumption.) $\mathcal{B}_\mathcal{H}$ is called a *quantum state space* and $\mathcal{A}_\mathcal{H}$ is called its space of *quantum observables*.

Definition 1. A *prequantization functor* is a functor $Q : \mathfrak{C}_n \rightsquigarrow \mathfrak{Q}$ such that:

(1) On $\mathcal{D}$, $[Q(\varphi), Q(\psi)] = i\hbar Q(\{\varphi, \psi\})$ for $\varphi, \psi \in \mathcal{C}_X$ and $Q(1) = I$;
(2) $Q(\Phi_2 \circ \Phi_1) = Q(\Phi_2)Q(\Phi_1)$ up to multiplication by $\mathbb{T}$ and $Q(id_X) \in \mathbb{T}I_{Q(X)}$;
(3) $Q(\varphi \circ \Phi) = Q(\Phi)^* Q(\varphi) Q(\Phi)$ on $\mathcal{D}_1$, where $\Phi : (X_1, \omega_1) \hookrightarrow (X_2, \omega_2)$, $\varphi \in \mathcal{P}_{X_2}$.

There are problems with constructing a prequantization functor in complete generality. However, by modifying our category $\mathfrak{C}_n$, we can get a workable but restricted theory. Let the objects in $\mathfrak{C}_n^0$ be pairs (X^{2n}, α), where α is a non-degenerate 1-form (so that $(X, d\alpha)$ is symplectic), X is connected, and every closed real 1-form β is of the form $i\hbar\frac{du}{u}$, where $u \in C^\infty(X), |u| = 1$ (we say that β is $\hbar$-logarithmically exact). So every contractible (X, α) is in $\mathfrak{C}_n^0$. Q is defined as follows:

(1) $Q(X, \alpha) = \mathcal{S}_{L^2(X, \mu_\alpha)}$, where μ_α is Liouville measure for $d\alpha$.

(2) $Q(\varphi) = i\hbar\xi_\varphi + \langle \alpha, \xi_\varphi \rangle + \varphi$ defined on $\mathcal{D} = C_c^\infty(X)$ for $\varphi \in \mathcal{P}_X$.

(3) Let $\Phi : (X_1, d\alpha_1) \hookrightarrow (X_2, d\alpha_2)$ be as above. Choose $u \in C^\infty(X_1)$, $|u| = 1$ such that $i\hbar\dfrac{du}{u} = \alpha_1 - \Phi^*\alpha_2$; u is unique up multiplication by $\mathbb{T}$. Then $Q(\Phi)$ is defined by

$$(Q(\Phi)v, w) = (u \cdot v, w \circ \Phi),$$

i.e., $Q(\Phi) = W_\Phi M_u$, where $W_\Phi{}^* w = w \circ \Phi$ and M_u is multiplication by u.

Here, as usual, ξ_φ denotes the Hamiltonian vectorfield determined by φ, that is, the unique vectorfield satisfying

$$\langle d\varphi, \eta \rangle = \omega(\xi_\varphi, \eta)$$

for arbitrary vectorfields η.

Lemma 2. $Q : \mathfrak{C}_n^0 \rightsquigarrow \mathfrak{Q}$ *is a prequantization functor.*

Sketch of proof. We verify the conditions of Definition 1:

(1) This is obvious by calculation.

(2) We want $W_{\Phi_2 \circ \Phi_1} M_{u_{21}} = W_{\Phi_2} M_{u_2} W_{\Phi_1} M_{u_1}$. Now $W_{\Phi_2 \circ \Phi_1} = W_{\Phi_2} W_{\Phi_1}$ and
$M_{u_2} W_{\Phi_1} = W_{\Phi_1} M_{u_2 \circ \Phi_1}$. So we must show that u_{21} can be taken to be $(u_2 \circ \Phi_1) \cdot u_1$. But this is obvious by calculation.

(3) We need to show that if $\varphi \in \mathcal{P}_{X_2}$, then $W_\Phi M_u Q(\varphi \circ \Phi) = Q(\varphi) W_\Phi M_u$ on $C_c^\infty(X_1)$. Note that $M_u(C_c^\infty(X_1)) \subseteq C_c^\infty(X_1)$ and $W_\Phi(C_c^\infty(X_1)) \subseteq C_c^\infty(X_2)$. Using item (3) in the definition of Q, we get the desired result by calculation. $\qquad\square$

This restricted form of prequantization can be used to recover the Kostant-Souriau prequantization theory ([K], [S1]): Let (X, ω) be a symplectic manifold and let $\{U_j\}$ be a Leray covering of X. As usual, let $U_{ij} = U_i \cap U_j, U_{ijk} = U_i \cap U_j \cap U_k$, etc. On each non-empty $U_{i...m}$, choose a 1-form $\alpha_{i...m}$ so that $d\alpha_{i...m} = \omega|_{U_{i...m}}$. Let $\Phi_i : (U_{ij}, \alpha_{ij}) \hookrightarrow (U_i, \alpha_i)$ and $\Phi_j : (U_{ij}, \alpha_{ij}) \hookrightarrow (U_j, \alpha_j)$ be the canonical inclusions. Then $Q(\Phi_i) = W_{\Phi_i} M_{u_i}$ and $Q(\Phi_j) = W_{\Phi_j} M_{u_j}$ as above. Set $u_{ij} = u_i u_j{}^{-1}$. The u_{ij} satisfy $u_{ji} = u_{ij}{}^{-1}$, and

$$u_{ij} u_{jk} = b_{ijk} u_{ik}, \qquad b_{ijk} \in \mathbb{T}$$

whenever $U_{ijk} \neq \emptyset$.

The u_{ij} are obviously independent of the choice of the α_{ij} and the b_{ijk} determine a Čech 2-cocycle with values in $\mathbb{T}$ and hence an element $\gamma \in H^2(X, \mathbb{T})$, which is independent of the choice of α's and the subsequent choice of u's. If $\gamma = 1$, then it is possible to adjust the data so that the $b_{ijk} \equiv 1$. A necessary and sufficient condition that $\gamma = 1$ is that $h^{-1}[\omega]$ be an integral de Rham cohomology class. In that case it is possible to use the u_{ij} as transition functions to form an hermitian complex line bundle L over X. With this choice of the α_i on U_i and the u_{ij} on U_{ij}, we define

$$Q_i(\varphi) = i\hbar\xi_\varphi + \langle \alpha_i, \xi_\varphi \rangle + \varphi$$

on $C^\infty(U_i)$ for $\varphi \in \mathcal{P}_X$. By an argument similar to the one in part 3 of Lemma 2, the $Q_i(\varphi)$ fit together to define a first order differential operator $Q(\varphi)$ on the smooth sections of L and so on $\mathcal{D} = C_c^\infty(L)$, which is easily seen to be an hermitian operator with respect to the natural pre-Hilbert space structure on the $\mathbf{L}^2$ sections of L with respect to Liouville measure. Associated with each α_i is a connection ∇^i on $L|_{U_i}$ defined by

$$\nabla^i_\xi = \xi + (i\hbar)^{-1}\langle \alpha_i, \xi \rangle,$$

where sections of $L|_{U_i}$ are identified with $C^\infty(U_i)$ as above. As before, the ∇^i fit together to define a a connection ∇ on L, and this connection preserves the hermitian structure on L. Then

$$Q(\varphi) = i\hbar\nabla_{\xi_\varphi} + \varphi.$$

2. An Example

Let $X = \mathbb{R}^2$ with the usual symplectic structure $\omega = dp \wedge dq$. Let $\alpha = p\,dq$. Then, for any $\varphi \in C^\infty(X)$ we have

$$Q(\varphi) = i\hbar\left(\frac{\partial\varphi}{\partial q}\frac{\partial}{\partial p} - \frac{\partial\varphi}{\partial p}\frac{\partial}{\partial q}\right) - p\frac{\partial\varphi}{\partial p} + \varphi.$$

Now $\mathcal{P}_2$, the space of polynomials of degree at most 2 in p and q, is a Lie algebra of $\mathcal{P}_X$ under P.B. isomorphic to the semidirect product $\mathfrak{h} \rtimes \mathfrak{sl}(2, \mathbb{R})$, where $\mathfrak{h}$ is the 3-dimensional Heisenberg algebra.

We calculate the representation of $\mathcal{P}_2$ we get from prequantization. Let

$$\mathcal{F}(q, k) = (2\pi\hbar)^{-\frac{1}{2}} \int e^{\frac{1}{i\hbar}pk} f(q, p)\, dp,$$

a partial Fourier transform of f with respect to p. Calculate $Q'(\varphi) = \mathcal{F} \circ Q(\varphi) \circ \mathcal{F}^{-1}$ and make the change of variables

$$x = -k + q, \qquad y = k.$$

We get $Q'(q) = x, Q'(p) = -i\hbar\dfrac{\partial}{\partial x}$, and $Q'(1) = 1$. Moreover, we have

$$Q'(q^2) = x^2 - y^2 \quad , \qquad Q'(p^2) = (-i\hbar)^2\left[\frac{\partial^2}{\partial x^2} - \frac{\partial^2}{\partial y^2}\right],$$

$$Q'(pq) = -i\hbar\left[x\frac{\partial}{\partial x} + y\frac{\partial}{\partial y} + 1\right].$$

The last formula can be rewritten as

$$Q'(pq) = -\frac{i\hbar}{2}\left[\left(x\frac{\partial}{\partial x} + \frac{\partial}{\partial x}\circ x\right) + \left(y\frac{\partial}{\partial y} + \frac{\partial}{\partial y}\circ y\right)\right].$$

This representation of $\mathcal{P}_2$ operates on $\mathbf{L}^2(\mathbb{R}^2) = \mathbf{L}^2(\mathbb{R}) \otimes \mathbf{L}^2(\mathbb{R})$.

Let π be the *metaplectic representation* of $\mathcal{P}_2$ on $\mathbf{L}^2(\mathbb{R})$ given by $\pi(q) = x, \pi(p) = -i\hbar\dfrac{\partial}{\partial x}, \pi(1) = 1, \pi(q^2) = \pi(q)^2, \pi(p^2) = \pi(p)^2$, and $\pi(pq) = \frac{1}{2}\left(\pi(p)\pi(q) + \pi(q)\pi(p)\right)$. Then our representation Q is unitarily equivalent to $\pi \otimes id$ on polynomials of degree at most 1 and to $\pi \otimes \pi'$ on homogeneous polynomials of degree 2, where π' is the *anti-metaplectic representation* of $\mathfrak{sl}(2,\mathbb{R})$ defined by $\pi' = \pi \circ \mathrm{ad}\, a$, where $a = \begin{pmatrix} 1 & 0 \\ 0 & -1 \end{pmatrix}$.

These calculations show that $[\pi \otimes id] \oplus [\pi \otimes \pi']$ can be extended to all of $\mathcal{P}$ from $\mathfrak{h} \rtimes \mathfrak{sl}(2,\mathbb{R})$. It is a theorem of Van Hove ([V]) and Groenwald ([Gr]) that π itself cannot be so extended.

3. Quantization using Polarizations

In §22 of [D] Dirac indicates that he wants the quantization functor to satisfy additional conditions. In particular, at the top of p. 92, he argues that in $\mathbb{R}^{2n}$ with its usual canonical coordinates $p_1, \ldots, p_n, q_1, \ldots, q_n$ the quantized q's should form a complete set of commuting observables. This leads us to impose the following conditions on Q : Fix a maximal abelian Lie subalgebra $\mathcal{A}$ of the Lie algebra $(\mathcal{P}_X, \{\cdot, \cdot\})$. Then

(4) $\{Q(\varphi) : \varphi \in \mathcal{A}\}$ generates a maximal abelian algebra of operators;

(5) $Q(f(\varphi_1, \ldots, \varphi_m)) = f(Q(\varphi_1), \ldots, Q(\varphi_m))$ if $\varphi_1, \ldots, \varphi_m \in \mathcal{A}$ and $f \in C^\infty(\mathbb{R}^m)$.

Remark. In (5), $f(\varphi_1, \ldots, \varphi_m)$ necessarily belongs to $\mathcal{A}$.

These requirements almost imply that $\{\xi_\varphi : \varphi \in \mathcal{A}\}$ is the space of sections of some polarization F of X and that we must restrict our attention to sections s of L which satisfy $\nabla_{\xi_\varphi} s = 0$ for all $\varphi \in \mathcal{A}$. So (4) and (5) almost recover the theory of geometric quantization of polarized symplectic manifolds due to Kostant and Souriau. (For a quick review of this theory see, for example, the discussion in Section 2 of [BR].)

Let us make these statements a bit more precise. In [B] we showed that if $\mathcal{A}$ is maximal abelian and if $F_x = \mathrm{span}\{(\xi_\varphi)_x : \varphi \in \mathcal{A}\}$ for $x \in X$, then F_x is isotropic with respect to ω, but $\dim F_x$ need not be constant on X. It is even unclear whether there is $x \in X$ such that $\dim F_x = n = \frac{1}{2}\dim X$. But F *will* be a polarization providing $\mathcal{A}$ satisfies the conditions in

Definition 3. A maximal abelian subalgebra of $\mathcal{P}_X$ is *hypermaximal abelian* if $\dim F_x = n$ for all $x \in X$ and if $\mathcal{A} + \bar{\mathcal{A}}$ is a Lie subalgebra of $\mathcal{P}_X$.

Now suppose that $\mathcal{A}$ is hypermaximal abelian. Then, as shown in [B], conditions (4) and (5) taken literally imply that $\nabla_{\xi_\varphi}\nabla_{\xi_\psi} = 0$ for all $\varphi, \psi \in \mathcal{A}$. Since this is impossible to achieve as an operator equation, we regard it as a restriction on states, that is, on sections of L. So we let

$$C_{\mathcal{A}}^\infty(L) = \{s \in C^\infty(L) : \nabla_{\xi_\varphi} s = 0, \varphi \in \mathcal{A}\}$$

and we let

$$C_{\mathcal{A}(2)}^\infty(L) = \{s \in C^\infty(L) : \nabla_{\xi_\varphi}\nabla_{\xi_\psi} s = 0, \varphi, \psi \in \mathcal{A}\}.$$

It is shown on p. 232 of [B] that locally $C_{\mathcal{A}(2)}^\infty(L)$ is a multiple of $C_{\mathcal{A}}^\infty(L)$ as an $\mathcal{A}$-module under Q. Moreover, under a regularity condition, the only φ whose quantizations leave $C_{\mathcal{A}}^\infty(L)$ fixed belong to the Lie normalizer of $\mathcal{A}$ in $\mathcal{P}_X$.

However arrived at, the theory of geometric quantization using a polarization F gives us a space $C_F^\infty(L)$ of sections of L annihilated by F under ∇ and the Lie subalgebra $\mathcal{N}_F$ of $\mathcal{P}_X$ which is the inverse image of the normalizer of $C^\infty(F)$ in $C^\infty(T_{\mathbb{C}}(X))$ under the map $\varphi \mapsto \xi_\varphi$. The Lie algebra $\mathcal{N}_F$ acts on $C_F^\infty(L)$ via Q. There remains the problem of making a Hilbert space out of (a subspace of) $C_F^\infty(L)$, and this can lead to problems which can often be solved using the theory of half-densities or half-forms, and various kinds of square-integrable cohomology. (See [GS] and [BR] and the references cited therein.)

4. Deformation Quantization

The standard theory of geometric quantization is unsatisfactory in some ways: The only space of observables whose associative algebra structure is being represented is $\mathcal{A}$ and the only space of observables that is being quantized at all is the normalizer of $\mathcal{A}$. Moreover, we know that there is a representation of $\mathcal{P}_2(\mathbb{R}^2)$, viz., the metaplectic representation, which cannot be achieved from the kind of geometric quantization outlined above using any choice of $\mathcal{A}$. The method of half-forms and their pairing can be used to produce the metaplectic representation, but in the context of "plain" quantization it looks like a *deus ex machina* which gets its meaning more from the theory of pseudodifferential and Fourier integral operators and their symbols than it does from quantum mechanics.

It seems to the present author that these difficulties have a common source, requirement (1) of Definition 1, a requirement that is much too strong. We propose replacing it by

(1a) $[Q(\varphi), Q(\psi)] = i\hbar Q(\{\varphi, \psi\}) + O(\hbar^2)$.

The precise meaning of $O(\hbar^2)$ depends on the sort of theory under study.

One way of producing something satisfying (1a) may be found in the formal deformation theory of Bayen, Flato, Fronsdal, Lichnerowicz, and Sternheimer ([BS]) which is based on the theory of deformations of algebras due to Gerstenhaber ([G]). The setup is as follows: Let $\mathcal{A}$ be a complex Poisson *-subalgebra of $\mathcal{P}_X$. Define a new product on the formal power series algebra $\mathcal{A}[[\hbar]]$ by

$$\varphi *_\hbar \psi = \sum_{n=0}^{\infty} (i\hbar)^n C_n(\varphi, \psi),$$

which is extended bilinearly and continuously (with respect to the usual formal power series topology) to all of $\mathcal{A}[[\hbar]]$. Setting $[\varphi, \psi]_\hbar = \varphi *_\hbar \psi - \psi *_\hbar \varphi$, we require

$$[\varphi, \psi]_\hbar = i\hbar\{\varphi, \psi\} + O(\hbar^2)$$

and $C_0(\varphi, \psi) = \varphi\psi$.

The associativity of the $*_\hbar$ product reduces to the following equations in the complex for the Hochschild cohomology of $\mathcal{A}$:

$$(c(n)): \sum_{r+s=n; r,s>0} [C_r(C_s(\varphi, \psi), \theta) - C_r(\varphi, C_s(\psi, \theta))] = (\partial C_n)(\varphi, \psi, \theta).$$

Gerstenhaber showed that if equations $(c(j))$ hold for $j = 1, \dots, n$, then the left side of equation $(c(n{+}1))$ belongs to $Z^3(\mathcal{A}, \mathcal{A})$ and is the obstruction to solving $(c(n{+}1))$. Equation $(c(0))$ expresses the associativity of $\mathcal{A}$ and $(c(1))$ follows from the Leibniz rule for $\{\cdot, \cdot\}$ if we take, as usual, $C_1(\varphi, \psi) = \frac{1}{2}\{\varphi, \psi\}$. The existence of $*_\hbar$ on an arbitrary symplectic manifold has been shown by De Wilde and Lecomte ([DW]).

Formal deformation quantization is, of course, only formal and needs to have some analytic teeth put into it, including a natural Hilbert space on which the deformed algebra acts. The predecessors of deformation quantization, Weyl quantization and the Moyal product and bracket ([W],[M]), do give analytic deformations, but only in the case where $X = \mathbb{R}^{2n}$. What is needed is a good general definition of analytic deformation quantization. One such definition, due to Rieffel ([R]), is the following:

Definition 4. Let $(M, \{\cdot, \cdot\})$ be a Poisson manifold. Let $\mathcal{A}$ be a complex Poisson *-subalgebra of $C^\infty(M, \mathbb{C})$ such that $C_c^\infty(M) \subseteq \mathcal{A} \subseteq C_\infty(M)$, the continuous $\mathbb{C}$-valued functions vanishing at ∞ on M. A *strict deformation quantization* of $\mathcal{A}$ consists of the following data: An $\epsilon > 0$ is given, and for each $\hbar$ with $|\hbar| < \epsilon$ an associative product $*_\hbar$ on $\mathcal{A}$, an involution $\cdot^{*\hbar}$ for the algebra $(\mathcal{A}, *_\hbar)$, and a C^*-norm $|| \cdot ||_\hbar$ for $(\mathcal{A}, *_\hbar, \cdot^{*\hbar})$ are given such that:

(1) $*_0$, $\cdot^{*_0}$, and $|| \cdot ||_0$ are ordinary multiplication, complex conjugation, and sup norm of functions, respectively;

(2) $\hbar \longmapsto ||\varphi||_\hbar$ is continuous for $\varphi \in \mathcal{A}$;

(3) $\lim_{\hbar \to 0} ||(i\hbar)^{-1}[\varphi, \psi]_\hbar - \{\varphi, \psi\}||_\hbar = 0$ for $\varphi, \psi \in \mathcal{A}$.

Rieffel points out a difficulty with this definition: examples are scarce. Almost all of them make use of the Fourier transform and so must come from a setting in which the Fourier transform makes sense. For example, one obtains the Moyal product on $\mathcal{S}(\mathbb{R}^{2n})$ using Weyl's quantization formula, and that makes use of the Fourier transform. Also note that Weyl quantization does satisfy properties (4) and (5) imposed on Q in §3 if we take $\mathcal{A}$ to be all functions φ which are annihilated by a fixed translation invariant polarization F of $\mathbb{R}^{2n}$. Properties (4) and (5) are not built in to Rieffel's definition.

5. Quantum States

Finally, we quickly sketch Souriau's interesting notion of *quantum state* ([S2]). We do not have space to go into the details of the machine he builds to make this notion work. It involves a generalization of the notions of differentiable manifold and Lie group which he calls *diffeological* spaces and groups. A *quantum space* is a diffeological space Ξ fibred by circles and satisfying axioms analogous to those found in Chapter V of [S1], together with the requirements that the group $\mathcal{G}$ of quantomorphisms of Ξ act transitively and generate the diffeological structure of Ξ, and that the moment map define an injection of the base space of the circle bundle Ξ into the invariant 1-forms on G. A *dynamical group* is then a diffeological group G together with a homomorphism ρ into $\mathcal{G}$ so that G has the same transitivity, generating, and moment properties as $\mathcal{G}$. Then a quantum state is a positive definite function m on the discretized group G such that

$$\sum_{j,k} \bar{c}_j c_k m(u_j(-1) u_k(1)) \leq \sup_{\xi \in \Xi} \left| \sum_k c_k e^{i h_k(\xi)} \right|^2$$

for every choice of a finite set of mutually commuting 1-parameter subgroups u_k of G and of complex numbers $c_k, k = 1, \ldots, N$. Here h_k is the *hamiltonian* of u_k (see [S2], §§4.2, 4.3, 4.6, and 5.1). Quantum states are used to produce representations of G by the GNS construction. Such representations, when continuous, may be thought of as giving integrated versions of the commutation relations for certain observables, in much the same way that Weyl wrote down an integrated version of the Heisenberg commutation relations and considered its representations.

6. Discussion

We have surveyed several theories of quantization, including geometric quantization, deformation quantization, and Souriau's theory of quantum states. Can these theories be combined in some meaningful way? One way of combining geometric quantization and deformation quantization has been suggested by Karasëv and Maslov in [KM], which is inspired by the theory of pseudodifferential and Fourier integral operators (cf. [GS]). We would like to indicate three other directions that might be useful to investigate.

The first is an analytic question raised by Gerstenhaber's deformation theory. As stated in §4, the problem of finding a formal product $*_\hbar$ reduces to solving successively equations (c(n)). However there is no guarantee that the resultant power series will converge for any $\hbar$ other than 0. This leads to the following questions:

(1) Under what conditions is it possible to solve successively equations (c(n)) in such a way that, for some $\epsilon > 0$ and all $|\hbar| < \epsilon$, the series defining $*_\hbar$ converges in some topology such as the Schwartz topology in case $X = \mathbb{R}^{2n}$ and $\mathcal{A} \subseteq \mathcal{S}(\mathbb{R}^{2n})$ (or in similar situations)?

(2) What conditions, in addition to those in (1), will guarantee that the map $(\hbar, \varphi, \psi) \mapsto \varphi *_\hbar \psi$ is jointly continuous? (Call this *continuous* deformation quantization.)

(3) Under what conditions is it possible to find asymptotically convergent solutions to equations (c(n)) when $\mathcal{A}$ is as in question (1)? (Call this *asymptotic* deformation quantization.)

These problems lead to other natural questions:

(4) Suppose we have a continuous deformation quantization. Under what conditions is it possible to embed $(\mathcal{A}, *_\hbar)$ into a strict deformation quantization?

(5) Answer question (4) when one has an asymptotic deformation quantization.

(6) What is the best way to include conditions (4) and (5) of the definition of Q (see §3) into the setup for deformation quantization or for continuous (or asymptotic, or strict) deformation quantization.

Finally, there is the question of fitting quantum states into the deformation quantization picture. As mentioned above, quantum states give representations of the dynamical group G which, when continuous, give operators corresponding to certain classical observables, viz., those φ whose prequantizations $Q(\varphi)$ are associated with 1-parameter groups of quantomorphisms in $\rho(G)$. This leads to the following problems:

(7) Letting G be the full group of quantomorphisms, under what conditions can a quantum representation be extended to all of the functions $\varphi \in \mathcal{P}_X$ of compact support, where X is the base space of the circle bundle Ξ?

(8) Is it possible to make sense of conditions (4) and (5) of the definition of Q as a restriction on quantum states?

REFERENCES

[BS] F. Bayen, M. Flato, C. Fronsdal, A. Lichnerowicz, and D. Sternheimer, *Deformation theory and quantization.* I, II, Ann. of Physics **111** (1977), 61–151.

[B] R. Blattner, "On geometric quantization", Non-Linear Partial Differential Operators and Quantization Procedures, (S. I. Andersson and H.-D. Doebner, eds.), LNM 1037, Springer- Verlag, New York, 1983, pp. 209–241.

[BR] R. Blattner and J. Rawnsley, *Quantization of the action of* $U(k, l)$ *on* $\mathbb{R}^{2(k+l)}$, J. Funct. Anal. **50** (1983), 188–214.

[DW] M. De Wilde, *Existence de déformations formelles de l'algèbra de Poisson et star-produits sur une variété symplectique*, Physique quantique et géométrie, (D. Bernard and Y. Choquet-Bruhat, eds.), Hermann, Paris, 1988, pp. 195–203.

[D] P. A. M. Dirac, *The Principles of Quantum Mechanics*, 3rd edition, Oxford, London, 1947.

[G] M. Gerstenhaber, *On deformations of rings and algebras*, Ann. of Math. **79** (1964), 59–103.

[Gr] H. Groenwald, *On the principles of elementary quantum mechanics*, Physica **12** (1946), 405–460.

[GS] V. Guillemin and S. Sternberg, *Geometric Asymptotics*, Math. Surveys 14, Amer. Math. Soc., Providence, R. I., 1977.

[KM] M. Karasëv and V. Maslov, *Asymptotic and geometric quantization*, Russian Math. Surveys **39(6)** (1984), 133–205.

[K] B. Kostant, *Quantization and unitary representations*, Lectures in Modern Analysis and Applications III, (C. Tamm, ed.), LNM 170, Springer-Verlag, New York, 1970, pp. 87–207.

[M] J. Moyal, *Quantum mechanics as a statistical theory*, Proc. Cambridge Philos. Soc. **45** (1949), 99–124.

[R] M. Rieffel, "Deformation quantization and operator algebras", Operator Theory/Operator Algebras and Applications. I, (W. Arveson and R. Douglas, eds.), Proc. Symp. Pure Math. 51, Amer. Math. Soc., Providence, R. I., 1990, pp. 411–423.

[S1] J.-M. Souriau, *Structure des systèmes dynamiques*, Dunod, Paris, 1970.

[S2] ______ , *Quantification géométrique*, Physique quantique et géométrie, (D. Bernard and Y. Choquet-Bruhat. eds.), Herman, Paris, 1988, pp. 141–193.

[VH] L. Van Hove, *Sur certaines représentations unitaires d'un groupe infini de transformations*, Mem. de l'Acad. Roy. de Belgique (Classe de Sci.) **XXVI** (1951), 61–102.

[W] H. Weyl, *The Theory of Groups and Quantum Mechanics*, Dover, New York, 1950.

Department of Mathematics
UCLA
Los Angeles, CA 90024-1555
USA

Convexity and Integrability

A.M. BLOCH* and T.S. RATIU**

A.M. BLOCH* and T.S. RATIU**

Dedicated to J.-M. Souriau

Abstract

In this review we present the main convexity results of Atiyah, Duistermaat, Guillemin, Kirwan, Kostant, and Sternberg and their relationship to the dynamic behavior of the generalized non-periodic Toda lattice. A "symplectic" proof of one of Kostant's convexity results at group level is also sketched. This review is an expanded version of the talk one of us (T.S.R.) presented at this conference and is based upon the papers, Bloch, Brockett and Ratiu [1990 a,b], Bloch, Flaschka and Ratiu [1990], and Lu and Ratiu [1991].

§1. Introduction

In symplectic and Poisson geometry one of the main objects of study is the momentum map associated with a Hamiltonian action. This map, with values in the dual of the Lie algebra of symmetries, is at the center of many geometrical facts that are useful in a variety of fields of both pure and applied mathematics.

The history of the emergence of this map and its re-emergence seventy five years later is quite interesting, and since this volume is dedicated to J.-M. Souriau, one of its modern discoverers, we shall try to sketch, in a few lines, some chronological facts related to the momentum map. It seems that the first time the momentum map can be found in the literature is in Lie's [1890] second volume of "Theorie der Transformationsgruppen" where it appears in the context of homogeneous canonical transformations, in which case its expression is given as the contraction of the canonical one-form with the infinitesimal generator of the action. A. Weinstein [1983b] has pointed out that Lie knew several key theorems regarding Poisson and symplectic geometry as well as momentum maps. We would like to thank him for his help with the following information: On page 237, Lie defines what today is called a Poisson structure. On page 300 it is shown that the momentum map is canonical and on page 329 that it is equivariant with respect to

---

*Partially supported by NSF Grant DMS 9002136 and a Seed Grant from the Ohio State University.

**Partially supported by NSF Grant DMS 8922699.

some linear action whose generators are identified on page 331. On page 338 it is proved that if the momentum map has constant rank (a hypothesis that seems to be implicit in all of Lie's work in this area), its image is Ad*- invariant, and on page 343, actions are classified by Ad*-invariant submanifolds. The title of Chapter 19 is "The Coadjoint Group", which is explicitly identified on page 334. Chapter 17, pages 294-298, defines a linear Poisson structure on the dual of a Lie algebra, today called the Lie-Poisson structure, and "Lie's Third Theorem" is proved for the set of regular elements. On page 349, together with a remark on page 367, it is shown that the Lie-Poisson structure naturally induces a symplectic structure on each coadjoint orbit. For many years this work appears to have been forgotten.

We now present the modern history of the momentum map based on information and references provided to us by B. Kostant and J.-M. Souriau. We would like to thank them for all their help.

In the 1965 U. S.-Japan Seminar on Differential Geometry Kostant [1966] introduced the momentum map to generalize a theorem of Wang and thereby classified all homogeneous symplectic manifolds; this is called today "Kostant's coadjoint orbit covering theorem". In the same year, Souriau [1965] introduced it in his Marseille lecture notes and put it in print in Souriau [1966]. The momentum map finally got its formal definition and its name, based on its physical interpretation, in Souriau [1967a] and its properties of equivariance were studied in Souriau [1967b], where the coadjoint orbit theorem is also formulated. In 1968, the momentum map appeared as a key tool in Kostant [1968] and from then on became a standard notion. Souriau [1969] discussed it at length in his book and Kostant [1970] (page 187, Theorem 5.4.1) dealt with it in his quantization lectures. Kostant and Souriau realized its importance for linear representations, a fact apparently not foreseen by Lie (Weinstein [1983b]).

Independently, work on the momentum map and the coadjoint covering theorem was done by A. Kirillov. This is described in Kirillov [1976]. This book was first published in 1972 and states that his work on the classification theorem was done about five years earlier (page 301).

The modern formulation of the momentum map became well known in mechanics and related applied areas due to the work of Smale [1970] who used it extensively in his topological program for the planar n-body problem.

In the last few years several important properties of the momentum map have emerged. One of the most striking aspects of the momentum map are its convexity properties as formulated in the now famous theorems of Atiyah [1982], Guillemin and Sternberg [1982], [1984] and Kirwan [1984], which generalize one of the fundamental Lie theoretic results of Kostant [1973], today called "Kostant's linear convexity theorem" (see §2). We will also discuss one of his other convexity results, "Kostant's nonlinear convexity theorem," in §8. No serious relationship of these properties with the dynamics of Hamiltonian systems has been forged so far even though already a superficial familiarity with these theorems suggests links to dynamic bifurcation behavior.

A specific branch of Hamiltonian dynamics, namely the theory of completely integrable systems, seems particularly suited to a study via the convexity theorems, since the hierarchy of flows comprise the action of a cylinder (often just a torus) whose momentum map is simply the collection of integrals of motion. To the disappointment of many symplectic geometers, the theory of integrable systems, which interacted so fruitfully both with algebraic geometry and representation theory, had no impact in its recent development on its own field of origin, namely symplectic geometry itself. It may turn out that the reasons are simply in the subtlety of such an interaction. In this review we want to outline several results in this direction, some of which we came across simply by accident. We shall deal here only with one class of examples: the non-periodic generalized Toda lattices. Judging from this example the convexity properties of the momentum map have a very subtle but decisive influence upon the dynamics of completely integrable systems and this whole relationship is worthwhile exploring both in general, as well as in specific examples.

The review is organized as follows. §2 is simply a collection of standard defintions and properties from Poisson geometry and the theory of momentum maps. Its sole purpose is to fix notation and conventions for the rest of the work. §3 reviews the relevant symplectic convexity results and §4 discusses their extensions in the important case of Kähler manifolds. The different metrics on orbits are also discussed since they are needed later. §5 introduces an abstract gradient system first studied in the $su(n)$-case by R. Brockett [1988]. This gradient system lives on adjoint orbits of compact Lie algebras and the dynamic behavior of its flow is entirely determined by the momentum map on this orbit. §6 shows that for a special choice of parameters, this gradient system becomes the Toda lattice on its isospectral set. Thus the famous scattering behavior of the Toda flow, so useful in numerical analysis, is simply an inheritance from its "partner", the so-far hidden gradient system. §7 discusses this relationship on a deeper level: since the Toda hierarchy is integrable, it naturally defines a non-compact torus action on an adjoint orbit which is canonically a Kähler manifold. Then why does the isospectral set not have a convex image under the momentum map? The answer is because the action is not the diagonal one. Can one embed the isospectral set in a different way so that its image under the projection is convex? The answer is positive and discussed in §7. Finally, §8 presents a symplectic proof of one of Kostant's non-linear convexity results. The relationship with the Toda flows is not clear here but several formulas and constructions are very similar and suggestive for the standard results one finds in the non-periodic Toda-lattice theory. We hope that a more serious link can be found. Also, the result in §8 suggests several generalizations.

As mentioned already at the beginning, this paper is a review. We hereby would like to thank our coauthors R. Brockett, H. Flaschka, and J.-H. Lu without whom the results presented here would not have been possible. §5 and §6 are based on Bloch, Brockett, Ratiu [1990a,b], §7 on Bloch [1990] and Bloch, Flaschka, Ratiu [1990], and §8 on Lu, Ratiu [1991]. These papers drew inspiration from the papers of Brockett [1988, 1989]. Conversations with B. Kostant, J. Lagarias, J. Marsden, A. Reyman and A. Weinstein throughout the period the above mentioned papers were written have been of indispensable importance. We would also like to thank the referee, whose comments greatly improved parts of the exposition. Finally, T. Ratiu would like to thank the Max Planck Institute for Mathematics, Bonn, Germany, where part of this review was written, and the organizers of this conference for the opportunity to participate and enjoy both the talks and the lively discussions.

§2 Poisson manifolds and momentum maps

The only purpose of this section is to introduce notations, specify conventions, and spell out definitions of concepts used throughout this review. Anyone familiar with this material can safely skip it.

A *Poisson manifold* is a pair $(P, \{,\})$ formed by a smooth manifold P whose ring of functions $\mathcal{F}(P)$ is a Lie algebra relative to an operation $\{,\}$ called a *Poisson bracket* which is a derivation in each argument. Thus, for fixed $H \in \mathcal{F}(P)$, the formula $F \in \mathcal{F}(P) \mapsto \{F, H\} \in \mathcal{F}(P)$ determines a vector field X_H called the *Hamiltonian vector field* determined by H, i.e. $\langle dF, X_H \rangle = \{F, H\}$. There are two main examples of Poisson manifolds:

a) *Symplectic manifolds*, which are manifolds P carrying a closed non-degenerate two-form ω. Defining X_H by the requirement that $\omega(X_H, Y) = \langle dH, Y \rangle$ for any vector field Y on P and setting $\{F, H\} = \omega(X_F, X_H)$ one easily proves that (P, ω) becomes a Poisson manifold.

b) *Duals of Lie algebras.* Let $\mathfrak{g}$ be a Lie algebra and let $\mathfrak{g}^*$ be the space of all linear functionals on $\mathfrak{g}$. For $F : \mathfrak{g}^* \to \mathbb{R}$ define the *functional derivative* $\delta F/\delta\mu \in \mathfrak{g}$ by the requirement

$$\mathbf{D}F(\mu) \cdot \nu = \langle \nu, \frac{\delta F}{\delta\mu} \rangle$$

for all $\nu \in \mathfrak{g}^*$; $\mathbf{D}F$ denotes the usual Fréchet derivative of F. $\mathfrak{g}^*$ is a Poisson manifold relative to the *Lie-Poisson structure*

$$\{F, H\}_\pm(\mu) = \pm\langle \mu, \left[\frac{\delta F}{\delta\mu}, \frac{\delta H}{\delta\mu} \right] \rangle.$$

A key property satisfied by the Poisson bracket on $(P, \{,\})$ is that $F \mapsto X_F$ is a Lie algebra anti-homomorphism:

$$X_{\{F, H\}} = -[X_F, X_H]$$

for all $F, H \in \mathcal{F}(P)$. Poisson manifolds stratify into symplectic manifolds called the *symplectic leaves* of P. They are described in the following way. On P define an equivalence relation which identifies two points if they can be joined by a piecewise smooth path, each segment of which is a trajectory of a locally defined Hamiltonian vector field. The equivalence classes are connected immersed Poisson submanifolds (i.e. a Hamiltonian vector field on P at a point on such a submanifold is tangent to it) with the Poisson bracket defined by a symplectic form. The tangent space at p to the leaf containing p is $\{X_H(p) | H \in \mathcal{F}(P)\}$. We refer to Weinstein [1983a] for more information on Poisson manifolds. If P is symplectic, its symplectic leaves are its connected components. If $P = \mathfrak{g}^*$, its leaves are the connected components of the coadjoint orbits of the underlying Lie group G of $\mathfrak{g}$. The symplectic form on these orbits is the *Kirillov-Kostant-Souriau form,*

$$\omega(\mu)(ad_\xi^* \mu, \ ad_\eta^* \mu) = \pm \langle \mu, [\xi, \eta] \rangle$$

for $\xi, \eta \in \mathfrak{g}$, $\mu \in \mathfrak{g}^*$, μ on the orbit in question.

Let G be a Lie group acting on $(P, \{,\})$ in a canonical fashion, i.e., preserving the Poisson bracket. For $\xi \in \mathfrak{g}$ denote by ξ_P the infinitesimal generator of the action, i.e.

$$\xi_P(p) = \frac{d}{dt}\Big|_{t=0} (\exp t\xi) \cdot p$$

where $\exp : \mathfrak{g} \to G$ is the exponential map and dot denotes the action. Since we consider left actions, $\xi \mapsto \xi_P$ is an anti-homomorphism, i.e.

$$[\xi, \eta]_P = -[\xi_P, \eta_P]$$

for all $\xi, \eta \in \mathfrak{g}$. We shall say that the G-action on P is *Hamiltonian* if there is a Lie algebra homomorphism $\rho : \mathfrak{g} \to \mathcal{F}(P)$ factoring $\xi \mapsto \xi_P$ through $H \mapsto X_H$, i.e.

$$\xi_P = X_{\rho(\xi)}$$

for all $\xi \in \mathfrak{g}$. If P is symplectic and G is semisimple such a homomorphism always exists; this is a corollary of the two Whitehead lemmas. The map $J : P \to \mathfrak{g}^*$ given by $\langle J(p), \xi \rangle = \rho(\xi)(p)$ for all $\xi \in \mathfrak{g}$, $p \in P$ is called the *momentum map* of the action. We shall denote $\rho(\xi) = J_\xi$ and therefore we have

$$J_{[\xi,\eta]} = \{J_\xi, J_\eta\}$$

for all $\xi, \eta \in \mathfrak{g}$.

§3 A brief review of symplectic convexity results

Some of the most striking global properties of the momentum map are its convexity properties. In this section we briefly review the main results in the symplectic context.

Theorem 3.1. *(Atiyah [1982], Guillemin and Sternberg [1982]). Let P be a compact connected symplectic manifold and T a torus acting in a Hamiltonian fashion on P with momentum map $J : P \to \mathbb{R}^n$, $n = \dim T$. Let P^T denote the fixed point set of T. Then:*

(i) $J(P^T)$ is finite and each point is the image of a connected component of the set where the derivative of J vanishes;

(ii)   each fiber $J^{-1}(c)$, $c \in \mathbb{R}^n$, if non-empty, is connected;

(iii) $J(P)$ is the convex hull of $J(P^T)$

This theorem generalizes to the symplectic setting the following remarkable result of Kostant:

Corollary 3.1. *(Kostant [1973]) Let G be a connected semisimple or a compact connected Lie group, T a maximal torus. Let $\mathfrak{g}, \mathfrak{t}$ denote the Lie algebras of G and T respectively. Fix a G-invariant metric on $\mathfrak{g}$. Then the orthogonal projection of an adjoint orbit $\mathcal{O}$ onto $\mathfrak{t}$ is the convex hull of the corresponding Weyl group orbit $\mathcal{O} \cap \mathfrak{t}$.*

Indeed, by the G-invariant metric, adjoint and coadjoint orbits are identified and the orthogonal projection is just the momentum map of the T-action on $\mathcal{O}$. The fixed point set of the T-action on $\mathcal{O}$ is the Weyl group orbit $W \cap \mathcal{O}$ and the corollary is thus a direct consequence of Theorem 2.1. Kostant's result is itself a generalization of the following classical result:

Corollary 3.2. *(Schur [1923], Horn [1954]) Denote by $\{a\}$ the set of diagonals of all Hermitian $n \times n$ matrices with given eigenvalues $\lambda = (\lambda_1, \ldots, \lambda_n)$. Let the symmetric group S_n act on $\mathbb{R}^n$ by permutation of coordinates. Then $\{a\}$ is the convex hull of $S_n \cdot \boldsymbol{\lambda}$, the S_n-orbit through $\boldsymbol{\lambda}$.*

This follows from Corollary 3.1 by taking $G = U(n)$. Going in the opposite direction, Kirwan has generalized Theorem 3.1 to compact connected Lie groups.

Theorem 3.2. *(Kirwan [1984]) Let P be a compact connected symplectic manifold and let the compact connected Lie group G act in a Hamiltonian fashion on P with momentum map $J : P \to \mathfrak{g}^*$. Let T be a maximal torus in G, $\mathfrak{t}$ its Lie algebra and let $\mathfrak{t}^*_+$ denote the positive Weyl chamber relative to a fixed ordering. Then*

(i)   each fiber $J^{-1}(\mu)$, if non-empty, is connected;

(ii) $J(P) \cap \mathfrak{t}_+^*$ is convex.

Theorems 3.1 and 3.2 can be proved together by a method different from the ones used in the original papers; this is done in Condevaux, Dazord and Molino [1988]. A beautiful survey of applications of theorem 2.1 can be found in Atiyah [1983].

The question naturally arises whether the image of the momentum map characterizes the action. For so-called completely integrable toral actions the answer is positive. A torus action on a symplectic manifold P is called *completely integrable* if the action is Hamiltonian, faithful, and dim P is twice the dimension of the torus.

Theorem 3.3. *(Delzant [1988]) Let P_1, P_2 be two compact connected symplectic manifolds of dimension $2n$ and T an n-dimensional torus acting in a completely integrable fashion on both P_1 and P_2 with momentum maps J_1 and J_2 respectively. If $J_1(P_1) = J_2(P_2)$ then there is a T-equivariant symplectic diffeomorphism $\varphi : P_1 \to P_2$ such that $J_2 \circ \varphi = J_1$.*

Theorem 3.3 is a companion to the Theorem 3.1. The corresponding companion to Theorem 3.2 is not known in full generality but there are results for groups of low rank by Delzant.

In order to deal with the analog of projections of real flag manifolds as opposed to complex ones, Duistermaat has proved the following:

Theorem 3.4. *(Duistermaat [1983]) Let (P, ω) be a compact connected symplectic manifold on which the torus T acts in a Hamiltonian fashion with momentum map J. Assume that $\tau : P \to P$ is an anti-symplectic involution (i.e. $\tau \circ \tau = $ identity and $\tau^* \omega = -\omega$) which leaves J invariant (and so it will necessarily invert the toral action: $\tau \circ g \circ \tau^{-1} = g^{-1}$ for all $g \in T$). Let Q be the fixed point set of P which is assumed to be non-empty. Then*

(i) $J(Q) = J(P)$

(ii) $J(Q)$ is the convex hull of the finitely many points $J(P^T \cap Q)$, where P^T is the fixed point set of the T-action on P;

(iii) $P^T \cap Q = \{p \in Q|$ the derivative of $J|Q$ at p vanishes$\}$; each connected component of this set equals the intersection of a connected component of P^T with Q; it is mapped by $J|Q$ into one point;

(iv) all results above hold if Q is replaced by one of its connected components.

There are many other convexity results in the literature. We have deliberately avoided the infinite dimensional case and those theorems for which a symplectic analog is not yet known. The notable exception is one of the non-linear convexity results of Kostant [1973] dealing with the Iwasawa projection which will be reviewed in a more natural setting in the last section of this paper; a "symplectic proof" for it will also be sketched. It is a challenge, in view of this, to try to "symplecticize" the great variety of convexity results available. The Kähler case is special and very important. It will be dealt with in the next section.

§4 The Kähler convexity theorem and coadjoint orbits

The convexity results reviewed in §3 hold also for Kähler manifolds, since every Kähler manifold is symplectic. However, due to the extra structure there is an important refinement, particularly relevant to integrable systems, proved by Atiyah [1982]; see also Atiyah [1983]. In this section we review this result and discuss it also in the important context of coadjoint orbits of compact Lie groups. In addition, we will briefly recall the relevant facts about the possible invariant metrics on these orbits; this material is needed in §7.

On a *Kähler manifold* one has a Riemannian metric g, a symplectic form ω, and an integrable almost complex structure $\mathfrak{J}$ (i.e. $\nabla\mathfrak{J} = 0$ where ∇ is the covariant derivative of the Riemannian connection defined by g). We have $\mathfrak{J}^{-1} = \mathfrak{J}^* = -\mathfrak{J}$ the adjoint being taken relative to g. The relationship between grad f and X_f, where X_f is the Hamiltonian vector field corresponding to $f : P \to \mathbb{R}$, is the following:

$$grad\, f = \mathfrak{J}X_f.$$

This follows from the interdependence of $(g, \omega, \mathfrak{J})$ established by the formula

$$\omega(u, \mathfrak{J}v) = g(u, v)$$

for all $u, v \in T_pP, p \in P$. Any two of $(g, \omega, \mathfrak{J})$ uniquely determine the third. We have $\mathfrak{J}v = iv$ for all $v \in TP$.

Assume a torus T acts on the Kähler manifold P preserving all structures. Thus the action of T consists of holomorphic mappings. Since the automorphism group of a complex manifold is a complex Lie group, the T-action extends to a holomorphic action of the complexification $T_{\mathbb{C}}$ of T. The Lie algebra $\mathfrak{t}_{\mathbb{C}}$ of $T_{\mathbb{C}}$ is $\mathfrak{t} \oplus \mathfrak{a}$, where $\mathfrak{t}$ is the Lie algebra of T and $\mathfrak{a} = i\mathfrak{t}$. Define $A = \exp \mathfrak{a} \subset T_{\mathbb{C}}$; it is a vector Lie subgroup of $T_{\mathbb{C}}$ since exp is injective on $\mathfrak{a}$. Moreover $(t, a) \in T \times A \mapsto ta \in T_{\mathbb{C}}$ is a diffeomorphism. We phrase the following theorem both in terms of complex toral orbits and in terms of the noncompact real toral orbits. As usual, the standing assumption is that *P is compact and connected and that the T-action is Hamiltonian with momentum map $J : P \to \mathfrak{t}^*$.*

Theorem 4.1. *(Atiyah [1982]) (i) Let X be the closure in P of the complex toral orbit $T_{\mathbb{C}} \cdot p, p \in P$. The set of fixed points of the T-action in X, $P^T \cap X$, equals the intersection of X with the set of points in P*

where the derivative of J vanishes. Each connected component of this set is mapped by J to one point in $\mathfrak{t}^*$. The image $J(X)$ equals the convex hull of these points which are the vertices of a convex polytope. For each open face σ of this polytope $J^{-1}(\sigma) \subset X$ consists of a single $T_{\mathbb{C}}$-orbit. Moreover, J induces a homeomorphism of X/T onto the polytope.

(ii) Let Y be the closure in P of the A-orbit $A \cdot p$, $p \in P$. $P^T \cap Y$ equals the intersection of Y with the set of points in P where the derivative of J vanishes. Each connected component of this set is mapped by J to one point and $J(Y)$ is the convex hull of these points which are the vertices of a convex polytope. For each open face σ of this polytope, $J^{-1}(\sigma)$ consists of a single A-orbit. Moreover, J induces a homeomorphism of Y onto the polytope.

(iii) For generic X and Y as in (i) and (ii), $J(P) = J(X) = J(Y)$.

The formulation (ii) was explicitly spelled out by Duistermaat [1983] and used to generalize Kostant's theorem on the convexity of certain projections of real flag manifolds.

If G is a compact Lie group it admits a complexification $G_{\mathbb{C}}$. Any coadjoint orbit of G is of the form $G/C(T')$ where $C(T')$ is the centralizer of a subtorus T' of the maximal torus T of G. So, generically, the orbits are flag manifolds. Fix a positive Weyl chamber in $\mathfrak{t}$ and hence a Borel subgroup B of $G_{\mathbb{C}}$, $B \supset T$, and a parabolic subgroup P of $G_{\mathbb{C}}$, $P \supset C(T')$. Since $G/C(T')$ and $G_{\mathbb{C}}/P$ (generically G/T and $G_{\mathbb{C}}/B$) are diffeomorphic, one can naturally induce a complex structure on $G/C(T')$ which is homogeneous under the $G_{\mathbb{C}}$-action. Thus on each coadjoint orbit of G in $\mathfrak{g}^*$ we get a homogeneous complex structure. Together with the Kirillov-Kostant-Souriau form this makes the coadjoint orbits Kähler. Theorem 4.1 immediately implies the following:

Corollary 4.1. *(Atiyah [1982]) Let A be the non-compact part of $T_{\mathbb{C}}$, i.e. $A = \exp \mathfrak{a}$, $\mathfrak{a} = i\mathfrak{t}$, $\mathfrak{t}$ the Lie algebra of T. Relative to an invariant metric on G identify coadjoint and adjoint orbits. Let Y be the closure of an A-orbit in $\mathfrak{g}^* = \mathfrak{g}$ which lies in the (co)adjoint orbit $\mathcal{O}$ of G. If $J : \mathcal{O} \to \mathfrak{t}$ is the orthogonal projection then $J(Y)$ is the convex hull of the points in $\mathfrak{t}$ which are the images of the points in Y where the derivative of J vanishes. J is a homeomorphism of Y to this closed convex polytope. For generic Y, $J(Y) = J(\mathcal{O})$.*

For our later purposes it will be useful to review here several of the many natural metrics on coadjoint orbits of compact Lie groups. Let G be a compact connected semisimple Lie group and $\mathcal{O}_{\mu_0}$ an adjoint orbit through $\mu_0 \in \mathfrak{g}$. Let κ denote the Killing form on $\mathfrak{g}$. Most of the material below may be found in Besse [1987].

(i) $\mathfrak{g}$ is a Riemannian manifold with the constant metric given by the negative of the Killing form. Pull this metric back to $\mathcal{O}_{\mu_0}$ by the inclusion.

One gets the *induced metric* given by

$$\langle [\mu, \eta], [\mu, \xi] \rangle_i = -\kappa([\mu, \eta], [\mu, \xi])$$

for all $[\mu, \eta], [\mu, \xi] \in T_\mu O_{\mu_0}$, $\mu \in O_{\mu_0}$.

(ii) Left translate $-\kappa(\cdot, \cdot)$ from $\mathfrak{g} = T_e G$ to $T_g G$ to obtain a left-invariant metric on G. This metric is actually bi-invariant since κ is invariant under the adjoint action. The quotient of G by the stabilizer of μ_0 inherits in this way a Riemannian structure. Thus O_{μ_0} becomes a Riemannian manifold. In Besse [1987] the metric obtained in this fashion is called the *normal metric*; Atiyah [1982] calls it the *standard metric*. To write an explicit formula, let $\mu \in \mathfrak{g}$, $\mathfrak{g}_\mu = \{\xi \in \mathfrak{g} \,|\, [\xi, \mu] = 0\}$, $\mathfrak{g}^\mu = im\, ad\, \mu$, and denote for $\eta \in \mathfrak{g}$ by η^μ the $\mathfrak{g}^\mu$-component of η in the direct sum decomposition $\mathfrak{g} = \mathfrak{g}_\mu \oplus \mathfrak{g}^\mu$; the two summands are orthogonal relative to $-\kappa(\cdot, \cdot)$. Then for $[\mu, \eta], [\mu, \zeta] \in T_\mu O_{\mu_0}$ the normal metric has the expression

$$\langle (\mu, \eta], [\mu, \zeta] \rangle_n = -\kappa(\eta^\mu, \zeta^\mu).$$

(iii) Finally there is the large family of G-invariant Kähler metrics on O_{μ_0} which are in bijective correspondence with the points of the positive Weyl chamber $\mathfrak{t}_+^*$ by Borel's theorem. Among all of these we single out the one corresponding to the intersection point of O_{μ_0} with $\mathfrak{t}_+^*$. An explicit formula in the spirit of the above two expressions is not possible. Let $A(\mu) = \sqrt{-(ad\,\mu)^2}$ be the positive square root of $-(ad\,\mu)^2$. Then the Kähler metric is given by

$$\langle [\mu, \eta], [\mu, \zeta] \rangle_K = \langle A(\mu)[\mu, \eta], [\mu, \zeta] \rangle_n$$

whereas the induced metric and the normal metric are related by

$$\langle [\mu, \eta], [\mu, \zeta] \rangle_i = \langle A(\mu)^2 [\mu, \eta], [\mu, \zeta] \rangle_n$$

as can be easily checked. From the formula for the Kähler metric one sees that the root space decomposition of $\mathfrak{g}$ enters quite explicitly.

Generically, if O_{μ_0} is diffeomorphic to G/T, it is easy to describe these metrics in terms of a root space decomposition (Atiyah [1982], Besse [1987], 8.51 and 8.76). Any G-invariant metric on G/T is determined by a T-invariant inner product on $\mathfrak{g}/\mathfrak{t}$. Write $\mathfrak{g}/\mathfrak{t} = \bigoplus_\alpha V_\alpha$, where V_α are the two-dimensional real root spaces, so they are real irreducible representations for T. Their complexification coincides with $\mathfrak{g}_\alpha \oplus \mathfrak{g}_{-\alpha}$, where α is now regarded as a root of the complexified Lie algebra $\mathfrak{g}^{\mathbb{C}}$ and $\mathfrak{g}_\alpha$, $\mathfrak{g}_{-\alpha}$ are the $\pm\alpha$ root spaces which are conjugate and complex one-dimensional. Let $\mu \in O_{\mu_0}$ be the unique element in the interior of the Weyl chamber, i.e. $\alpha(\mu) > 0$ for any positive root α. On each V_α the negative of the Killing form defines an inner product. Any other inner product is given by multiplying it by a positive scalar. Then these scalars are all equal to 1 for the normal metric, equal to $\alpha(\mu)$ for the Kähler metric, and equal to $\alpha(\mu)^2$ for the induced metric; the spaces V_α and V_β for different roots α, β are orthogonal.

§5. A special gradient system for a linear function on coadjoint orbits

In this section we begin by discussing a gradient system on a compact Lie group and its induced system on a specific coadjoint orbit. We will show that this new system is also gradient relative to the restriction of a linear function to the orbit. The dynamic properties of its flow turn out to be intimately connected with the convexity results of §3. The material in this section reviews the work in Bloch, Brockett, Ratiu [1990a], [1990b] some of which was in turn influenced by the results in Brockett [1988], [1989] dealing with the $SU(n)$-case.

Let K be a compact semisimple Lie group, $\mathfrak{k}$ its Lie algebra and κ its Killing form. Fix two elements $Q, N \in \mathfrak{k}$ and define the function $F : K \to \mathbb{R}$ by

$$F(\theta) = \kappa(Q, Ad_\theta N). \tag{5.1}$$

Endow K with the bi-invariant Riemannian metric $\langle \cdot, \cdot \rangle$ whose value at the identity is minus the Killing form. Denote by $\theta \cdot P$ the left translate of $P \in \mathfrak{k}$ by $\theta \in K$ to $T_\theta K$. If $v_\theta = \theta \cdot R \in T_\theta K$, $R \in \mathfrak{k}$, then

$$\begin{aligned}
dF(\theta) \cdot v_\theta &= \frac{d}{dt}\Big|_{t=0} F(\theta \exp tR) \\
&= \frac{d}{dt}\Big|_{t=0} \kappa(Ad_{\theta^{-1}}Q, Ad_{\exp tR}N) \\
&= \kappa(Ad_{\theta^{-1}}Q, [R, N]) \\
&= \kappa([N, Ad_{\theta^{-1}}Q], R) \\
&= -\langle \theta \cdot [N, Ad_{\theta^{-1}}Q], v_\theta \rangle
\end{aligned}$$

whence

$$\bigtriangledown F(\theta) = \theta \cdot [Ad_{\theta^{-1}}Q, N] \tag{5.2}$$

and we get

Proposition 5.1. *The gradient flow on K relative to $\langle \cdot, \cdot \rangle$ and the function (5.1) is given by $\dot\theta = \theta \cdot [Ad_{\theta^{-1}}Q, N]$.*

Let $\mathcal{O}$ be the adjoint orbit containing Q. The projection of K to $\mathcal{O}$ is given by

$$\theta \mapsto Ad_{\theta^{-1}}Q. \tag{5.3}$$

Via this map, the gradient flow in proposition 5.1 transforms to

$$\dot{L} = [L, [L, N]] \tag{5.4}$$

Indeed, put $L(t) = Ad_{\theta(t)^{-1}}Q$, where $\theta(t)$ is an integral curve of $\dot{\theta} = \theta \cdot [Ad_{\theta^{-1}}Q, N]$; then

$$\begin{aligned}
\dot{L}(t) &= -[\theta^{-1}(t)\dot{\theta}(t), Ad_{\theta(t)^{-1}}Q] \\
&= [L(t), \theta(t)^{-1}\dot{\theta}(t)] = [L(t), [L(t), N]].
\end{aligned}$$

Next, decompose orthogonally $\mathfrak{k} = \mathfrak{k}_L \oplus \mathfrak{k}^L$, $\mathfrak{k}_L = \{x \in \mathfrak{k} | [L, X] = 0\}$, $\mathfrak{k}^L = im(ad\, L)$, and denote by $X = X_L + X^L$ the decomposition of $X \in \mathfrak{k}$ into its $\mathfrak{k}_L$ and $\mathfrak{k}^L$-components (see the end of §4). Endow $\mathcal{O}$ with the normal metric

$$\langle [L, X], [L, Y] \rangle_n = \langle X^L, Y^L \rangle \tag{5.5}$$

and define

$$H : L \in \mathfrak{k} \mapsto \kappa(L, N) \in \mathbb{R} \tag{5.6}$$

By definition of the gradient relative to $\langle \cdot, \cdot \rangle_n$ on $\mathcal{O}$ we have for the restriction of H to $\mathcal{O}$

$$\begin{aligned}
\langle gradH(L), [L, \delta L] \rangle_n &= dH(L) \cdot [L, \delta L] \\
&= -dH(L) \cdot [\delta L, L] \\
&= -\frac{d}{dt}\Big|_{t=0} H(\exp t\delta L \cdot L) \\
&= \frac{d}{dt}\Big|_{t=0} \langle \exp t\delta L \cdot L, N \rangle \\
&= \langle [\delta L, L], N \rangle = \langle [L, N], \delta L \rangle \\
&= \langle [L, N], (\delta L)^L \rangle.
\end{aligned}$$

Therefore, if we set $grad\, H(L) = [L, X]$, the above equality and (5.5) say that

$$\langle X^L, (\delta L)^L \rangle = \langle [L, N], (\delta L)^L \rangle$$

whence $X^L = ([L, N])^L = [L, N]$. Thus

$$grad\, H(L) = [L, [L, N]]$$

which coincides with (5.4). We summarize these results in the following:

Theorem 5.1. *The projection (5.3) of the gradient flow $\dot{\theta} = \theta \cdot [Ad_{\theta^{-1}}Q, N]$ to the orbit $\mathcal{O}$ through Q is the gradient flow $\dot{L} = [L, [L, N]]$ on $\mathcal{O}$ relative to the normal metric and the function H given by (5.6).*

The connection of the dynamics defined by this gradient system and the convexity results of §3 is given by the following:

Theorem 5.2. *On the orbit $\mathcal{O}$ of K in $\mathfrak{k}$ consider the gradient flow $\dot{L} = [L, [L, N]]$ for N a fixed regular element. Let F_t be the flow of this vector field on $\mathcal{O}$. The set of equilibria equals $\mathcal{O} \cap \mathfrak{t}$ where $\mathfrak{t}$ is the Cartan subalgebra of $\mathfrak{k}$ containing N. This set $\mathcal{O} \cap \mathfrak{t}$ consists of a single Weyl group orbit. The convex hull of these equilibria is a compact-polytope $\mathcal{P}$ which is the image of $\mathcal{O}$ under the momentum map $\pi : \mathcal{O} \to \mathfrak{t}$ (the orthogonal projection) defined by the adjoint T-action on $\mathcal{O}$, where T is the maximal torus in $\mathfrak{k}$ obtained by exponentiating $\mathfrak{t}$. $\pi(F_t(\mathcal{O}))$ lies entirely in $\mathcal{P}$.*

Proof. Indeed, by Kostant's theorem, stated in Corollary 3.1, the theorem is proved provided we show that the equilibria of $\dot{L} = [L, [L, N]]$ necessarily lie in $\mathfrak{t}$. Since this gradient field is defined on the compact manifold $\mathcal{O}$, its flow is complete, so all that needs to be shown is that $\lim_{t \to \infty} L(t) \in \mathfrak{t}$ for any integral curve $L(t)$. But

$$\frac{d}{dt}(\kappa(L, N)) = \kappa(N, [L, [L, N]]) = -\kappa([L, N], [L, N]) \geq 0$$

so that $\kappa(L(t), N)$ is increasing as a function of time. It must also be bounded since $L(t) \in \mathcal{O}$ and $\mathcal{O}$ is compact. Thus $\kappa(L(t), N)$ has a limit and its derivative vanishes if and only if $L(\infty)$ and N commute, i.e. the equilibrium $L(\infty)$ must lie in $\mathfrak{t}$. $\square$

By taking advantage of this theorem, Bloch, Brockett, Ratiu [1990b] prove the following:

Theorem 5.3. *In the hypotheses and notation of Theorem 5.2 we have:*

(i) The only stable equilibrium (a sink) of this gradient vector field is the one that lies in the Weyl chamber of $-N$.

(ii) The only source of this vector field is the unique equilibrium which lies in the same Weyl chamber as N.

(iii) The dimension of the stable manifold at any of the equilibria equals the length of the Weyl group element which maps the Weyl chamber of N to the Weyl chamber containing this equilibrium.

§6. The Toda equations as a gradient system

In this section we show the relationship between the non-periodic Toda lattice equations and the gradient system in §5. It turns out that for very special choices of N, the two systems coincide on the isospectral set. Thus, the Toda flow naturally inherits all the scattering behavior of its "brother", the gradient system discussed in §5. This property of the Toda flow has found remarkable applications in numerical analysis, especially regarding the QR-algorithm; see Deift, Nanda, Tomei [1983], Lagarias [1988], Symes [1982b]. The material in this section reviews some of the results in Bloch, Brockett, Ratiu [1990a], [1990b], Bloch, Flaschka, Ratiu [1990].

We begin with a slight generalization of the non-periodic Toda lattice associated to an arbitrary Dynkin diagram. These equations will lie on orbits of subalgebras in a *compact* Lie algebra which is the compact real form of a complex semisimple Lie algebra. For certain values of the parameters, the usual generalized non-periodic Toda lattice is recovered as we shall explicitly point out later on. This set-up is the natural one in which one links the Toda equations with the gradient system in §5 and also gives rise to several interesting questions, some of which will be addressed in §7.

We build on the Lie algebraic notations and conventions at the end of §4 which are in agreement with Humphreys [1972]. $\mathfrak{g}$ denotes a complex semi-simple Lie algebra of rank ℓ. Fix a Cartan subalgebra of $\mathfrak{g}$ and denote by $\Phi, \Phi^+, \Phi^-, \Delta$ the systems of roots, of positive roots, of negative roots, and of simple roots $\Delta = \{\alpha_1, \ldots, \alpha_\ell\}$ respectively. The Killing form on $\mathfrak{g}$ is denoted by κ, and the inner product on roots by $(\cdot, \cdot)$. The notation $\langle \alpha, \beta \rangle = (\alpha, \beta)/(\beta, \beta)$ for $\alpha, \beta \in \Phi$ will also be employed. $\mathfrak{g}_\alpha$ denotes the α-root space of $\mathfrak{g}$. Fix a Chevalley basis $\{h_i, e_\alpha | i = 1, \ldots, \ell, \alpha \in \Phi\}$ and recall that e_α is the basis vector of the one-dimensional complex vector space $\mathfrak{g}_\alpha$, $h_\alpha = [e_\alpha, e_{-\alpha}]$ for $\alpha \in \Phi^+$, $h_i = h_{\alpha_i}$, and $\kappa(h_\alpha, h) = 2\alpha(h)/(\alpha, \alpha)$ for all $h \in \mathfrak{g}$ and $\alpha \in \Phi^+$. Let $x_\alpha = e_\alpha - e_{-\alpha}$, $y_\alpha = i(e_\alpha + e_{-\alpha})$ for $\alpha \in \Phi^+$ and define:

(i)   the *compact real form* of $\mathfrak{g}$,

$$\mathfrak{k} = \{i \sum_{j=1}^{\ell} b_j h_j + \sum_{\alpha \in \Phi^+} i(c_\alpha x_\alpha + d_\alpha y_\alpha) | b_j, c_\alpha, d_\alpha \in \mathbb{R}\},$$

(ii)   the *normal real form* of $\mathfrak{g}$,

$$\mathfrak{g}_n = \{\sum_{j=1}^{\ell} b_j h_j + \sum_{\alpha \in \Phi} c_\alpha e_\alpha | b_j, c_\alpha \in \mathbb{R}\},$$

(iii)   the *compact toral subalgebra* of $\mathfrak{g}$,

$$\mathfrak{t} = \{i \sum_{j=1}^{\ell} b_j h_j | b_j \in \mathbb{R}\},$$

62   A.M. Bloch and T.S. Ratiu

(iv) the *non-compact toral subalgebra* of $\mathfrak{g}$,

$$\mathfrak{a} = i\mathfrak{t} = \{\sum_{j=1}^{\ell} b_j h_j \,|\, b_j \in \mathbb{R}\},$$

(v) $\mathfrak{b} = \mathfrak{a} \oplus \bigsqcup_{\alpha \in \Phi^+} \mathfrak{g}_\alpha, \quad \mathfrak{n}^\pm = \bigsqcup_{\alpha \in \Phi^\pm} \mathfrak{g}_\alpha.$

Think of $\mathfrak{g}$ as a *real* Lie algebra and thus $\dim_\mathbb{R} \mathfrak{g} = 2\dim_\mathbb{C} \mathfrak{g}$. We shall always denote by $\mathfrak{g}$ this real Lie algebra and we will write $\mathfrak{g}_\mathbb{C}$ whenever $\mathfrak{g}$ is thought of as a complex Lie algebra. We have $\mathfrak{g} = \mathfrak{k} \oplus \mathfrak{b}$. The adjoint groups corresponding to these Lie algebras are denoted by $G, K, G_n, T, A, B, N^\pm$ respectively. For the Iwasawa decomposition $G = KAN = KB$ we will write the factorization of $g \in G$ as $g = \mathbf{k}(g)\mathbf{b}(g)$ for $\mathbf{k}(g) \in K$, $\mathbf{b}(g) \in B$.

There are two Poisson structures on $\mathfrak{k}$. The first one is obtained by pulling back the Lie-Poisson structure of $\mathfrak{k}^*$ to $\mathfrak{k}$ via the restriction of the Killing form κ to $\mathfrak{k}$ which is negative definite. The other Poisson structure comes from a construction used by Lu and Weinstein [1990]. The non-degenerate bilinear form $Im\,\kappa$ on $\mathfrak{g}$ vanishes on $\mathfrak{k}$ since $\mathfrak{k}$ is a real form of $\mathfrak{g}_\mathbb{C}$. Also $\kappa(\mathfrak{g}_\alpha, \mathfrak{g}_\beta) = 0$ unless $\alpha + \beta = 0$ and thus $\kappa(\mathfrak{b}, \mathfrak{b}) = \kappa(\mathfrak{a}, \mathfrak{a}) \in \mathbb{R}$; thus $Im\,\kappa$ vanishes on $\mathfrak{b}$. The annihilator of $\mathfrak{k}$ relative to $Im\,\kappa$ is $\mathfrak{k}$ itself, and is isomorphic to $\mathfrak{b}^*$ as a real vector space. We summarize these observations together with the Adler-Kostant-Symes theorem in the following:

Proposition 6.1. *(i) $\mathfrak{k}$ is equipped with the usual Lie-Poisson bracket. The coadjoint orbits $\mathcal{O}$ are the symplectic leaves of this Poisson manifold. The symplectic form is given by the Kirillov-Kostant-Souriau formula (see §2). The coadjoint orbits $\mathcal{O}$ are Kähler relative to the canonical Kähler structure (see §4).*

(ii) $\mathfrak{k} \cong \mathfrak{b}^*$ has a second Lie-Poisson bracket given by

$$\{f, h\}(\xi) = -Im\,\kappa(\xi, [\pi_\mathfrak{b} \bigtriangledown \tilde{f}(\xi),\ \pi_\mathfrak{b} \bigtriangledown \tilde{h}(\xi)])$$

where $\tilde{f}, \tilde{h}$ are arbitrary extensions to $\mathfrak{g}$ of $f, h : \mathfrak{k} \to \mathbb{R}$, $\bigtriangledown$ is the gradient relative to $Im\,\kappa$, i.e.

$$d\tilde{f}(\xi) \cdot \delta\xi = Im\,\kappa(\bigtriangledown\tilde{f}(\xi), \delta\xi),$$

$\pi_\mathfrak{b}$ is the projection onto $\mathfrak{b}$ corresponding to the decomposition $\mathfrak{g} = \mathfrak{k} \oplus \mathfrak{b}$, and $\xi, \delta\xi \in \mathfrak{k}$.

(iii) If f is an invariant function on $\mathfrak{g}$, i.e. $[\bigtriangledown f(\zeta), \zeta] = 0$ for all $\zeta \in \mathfrak{g}$, then the Hamiltonian vector field relative to the Poisson structure in (ii) defined by $f|\mathfrak{k}$ is given by the Lax equation

$$X_{f|\mathfrak{k}}(\xi) = [\pi_\mathfrak{k} \bigtriangledown f(\xi), \xi], \quad \xi \in \mathfrak{k}.$$

Hamiltonian vector fields defined by invariant functions commute.

The Toda equations are a Hamiltonian system on a coadjoint orbit in $\mathfrak{b}^* = \mathfrak{k}$ of dimension 2ℓ. The following is a straightforward verification following the pattern in Kostant [1979]:

Proposition 6.2. *Let $\delta = (\delta_1, \ldots, \delta_\ell)$ be a vector of 0's and 1's and fix $(\theta_1, \ldots, \theta_\ell)$, $\theta_j \in [0, 2\pi]$. Let $L(\delta, \theta) = \sum_{j=1}^{\ell} \delta_j e^{i\theta_j} x_{\alpha_j}$. Then the coadjoint B-orbit through $L(\delta, \theta)$ is given by*

$$Jac := \{\pi_\mathfrak{k} Ad_b^G L(\delta, \theta) | b \in B\}$$

$$= \{L = \sum_{j=1}^{\ell} [b_j h_j + \delta_j (a_j e_{\alpha_j} - \bar{a}_j e_{-\alpha_j})] | i b_j \in \mathbb{R},$$

$$a_j \in \mathbb{C}\backslash\{0\}, \quad \arg a_j = \theta_j \text{ if } \delta_j \neq 0\}, \tag{6.1}$$

where Ad^G denotes the usual adjoint action of G on $\mathfrak{g}$.

If one takes $\theta_j = \pi/2$ and $\delta_j = 1$, for all $j = 1, \ldots, \ell$, then L is, up to a factor of i, a typical element of a so-called "Toda orbit" in $\mathfrak{g}_u$ (see Symes [1982a]). We will call all elements L appearing in (6.1) *Jacobi elements*.

Let $I_1, \ldots, I_\ell$ be a set of homogeneous generators of the ring of invariant polynomials on $\mathfrak{g}$ chosen such that their restrictions to $\mathfrak{k}$ are real and generate the invariant polynomials on $\mathfrak{k}$. The Hamiltonian equations

$$\dot{L} = [\pi_\mathfrak{k} \nabla I_j(L), \ L] \tag{6.2}$$

on the B-orbit Jac of Jacobi elements (5.1) will be called the *Toda hierarchy*. Fix an element Λ in the interior of the positive Weyl chamber of $\mathfrak{t}$ and let $\mathcal{O}_\Lambda$ be the K-orbit through Λ. The Toda hierarchy leaves $\mathcal{O}_\Lambda \cap Jac$ invariant. For $A_\ell, \mathcal{O}_\Lambda$ consists of matrices with fixed spectrum Λ and $\mathcal{O}_\Lambda \cap Jac$ is the isospectral manifold. After multiplying by i (and setting $\theta_j = \pi/2$, $\delta_j = 1$ for $j = 1, \ldots, \ell$) the first vector field in the Toda hierarchy for A_ℓ takes the familiar form $\dot{L} = [B, L]$, where

$$L = \begin{bmatrix} b_1 & a_1 & & & & \\ a_1 & b_2 - b_1 & a_2 & & & \\ & & \ddots & \ddots & & \ddots \\ & & & & b_\ell - b_{\ell-1} & a_\ell \\ & & & & a_\ell & -b_\ell \end{bmatrix},$$

$$B = \begin{bmatrix} 0 & a_1 & & & \\ -a_1 & 0 & a_2 & & \\ & \ddots & \ddots & \ddots & \\ & & & 0 & a_\ell \\ & & & -a_\ell & 0 \end{bmatrix}. \tag{6.3}$$

For an arbitrary Lie algebra

$$B = \sum_{j=1}^{\ell} a_j (e_{\alpha_j} - e_{-\alpha_j}) . \tag{6.4}$$

Theorem 6.1. *The Toda hierarchy is a completely integrable system and the 2ℓ integrals in involution are $I_1, \ldots, I_\ell$.*

For a proof of this and a wealth of additional material on the Toda lattice see Kostant [1979] and Symes [1980], [1982a]. We will return and sketch how a solution of the Toda lattice is found in terms of the Iwasawa decomposition in the next section. In the sequel we shall fix all $\theta_j = \pi/2$ and all $\delta_j = 1$ and set $arg\, a_j = \pi/2$.

Theorem 6.2. *If N is minus i times the sum of the simple coweights of $\mathfrak{g}$, then for*

$$L = \sum_{j=1}^{\ell} ib_j h_j + \sum_{j=1}^{\ell} ia_j(e_{\alpha_j} + e_{\alpha_j}),$$

$b_j, a_j \in \mathbb{R}$, *the equation $\dot{L} = [L, [L, N]]$ gives the generalized Toda flow on the isospectral set $Jac \cap O_\Lambda$. Explicitly, $N = \sum_{j=1}^{\ell} ix_j h_j$ where $(x_1, \ldots, x_\ell)$ is the unique solution of the system $\sum_{j=1}^{\ell} x_j \alpha_k(h_j) = -1$, $k = 1, \ldots, \ell$.*

For the proof, it is enough to require that $[L, N] = -B$ with B given by (5.4) and to observe that the solution of the system in $(x_1, \ldots, x_\ell)$ gives $N = -ih_{\check\delta}$, where $h_{\check\delta} = \sum_{r=1}^{\ell} \check\lambda_i$ since $(h_{\check\delta}, e_{\alpha_i}) = 1$, $(\, , \,)$ being the inner product. Therefore we can conclude

Corollary 6.1. *The generalized Toda lattice equations are a gradient flow on $Jac \cap O_\Lambda$ when O_Λ is equipped with the normal metric.*

The coefficients x_i in N are listed for all simple Lie algebras below

$\underline{\mathcal{G}}$	$\underline{x}$
A_ℓ	$x_i = -\frac{1}{2}i(\ell - i + 1),\ i = 1, 2, \ldots, \ell$
B_ℓ	$x_i = -\frac{1}{2}i(2\ell - i + 1),\ i = 1, 2, \ldots, \ell - 1$
	$x_\ell = -\frac{1}{4}\ell(\ell + 1)$
C_ℓ	$x_i = -\frac{1}{2}i(2\ell - i)\ i = 1, 2, \ldots, \ell$
D_ℓ	$x_i = -\frac{1}{2}i(2\ell - 1 - i),\ i = 1, 2, \ldots, \ell - 2$
	$x_{\ell-1} = x_\ell = -\frac{1}{4}\ell(\ell - 1)$
G_2	$x_1 = -3,\ x_2 = -5$
F_4	$x_1 = -11,\ x_2 = -21,\ x_3 = -15,\ x_4 = -8$
E_6	$x_1 = -8,\ x_2 = -11,\ x_3 = -15,$
	$x_4 = -21,\ x_5 = -15,\ x_6 = -8$
E_7	$x_1 = -17,\ x_2 = -\frac{49}{2},\ x_3 = -33,$
	$x_4 = -48,\ x_5 = -\frac{75}{2},\ x_6 = -26,$
	$x_7 = -\frac{27}{2}$
E_8	$x_1 = -46,\ x_2 = -68,\ x_3 = -91,$

$$x_4 = -135, \ x_5 = -110, \ x_6 = -84,$$
$$x_7 = -57, \ x_8 = -29$$

For the classical simple Lie algebras represented as in Sattinger and Weaver [1986], for example, we have

$A_\ell : h_1 = \mathrm{diag}(1, -1, 0, 0, \dots), \ h_2 = \mathrm{diag}(0, 1, -1, 0, \dots) \dots,$
$\qquad h_\ell = \mathrm{diag}(0, \dots, 0, 1 - 1)$ and hence

$$N = i \begin{bmatrix} -\frac{\ell}{2} & & & & \\ & \frac{-\ell+2}{2} & & & \\ & & \ddots & & \\ & & & \frac{\ell-2}{2} & \\ & & & & \frac{\ell}{2} \end{bmatrix},$$

$B_\ell : h_1 = \mathrm{diag}(0, 1, -1, -1, 1, 0, \dots), \ h_2 = \mathrm{diag}(0, 0, 0, 1, -1, -1, 1, \dots) \dots,$
$\qquad h_\ell = \mathrm{diag}(0, \dots, 0, 2, -2)$ and

$$N = \begin{bmatrix} 0 & & & & & & & \\ & -\ell & & & & & & \\ & & \ell & & & & & \\ & & & -\ell+1 & & & & \\ & & & & 1-\ell & & & \\ & & & & & \ddots & & \\ & & & & & & -1 & \\ & & & & & & & 1 \end{bmatrix}.$$

$C_\ell : h_1 = \mathrm{diag}(1, -1, -1, 1, 0, \dots), \ h_2 = \mathrm{diag}(0, 0, 1, -1, -1, 1, \dots) \dots,$
$\qquad h_\ell = (0, 0, \dots, 0, 1, -1)$ and

$$N = i \begin{bmatrix} \frac{1}{2}(1-2\ell) & & & & & & & \\ & -\frac{1}{2}(1-2\ell) & & & & & & \\ & & \frac{1}{2}(3-2\ell) & & & & & \\ & & & -\frac{1}{2}(3-2\ell) & & & & \\ & & & & \ddots & & & \\ & & & & & -\frac{3}{2} & & \\ & & & & & & \frac{3}{2} & \\ & & & & & & & -\frac{1}{2} \\ & & & & & & & & \frac{1}{2} \end{bmatrix}.$$

D_ℓ: h_i as for C_ℓ and

$$N = \begin{bmatrix} -\ell+1 \\ & \ell-1 \\ & & -\ell+2 \\ & & & 2-\ell \\ & & & & \ddots \\ & & & & & -1 \\ & & & & & & 1 \\ & & & & & & & 0 \\ & & & & & & & & 0 \end{bmatrix}.$$

Corollary 6.1 and Theorem 5.2 imply

Corollary 6.2. *The projection of the flow of the Toda lattice to* t *lies in the interior of the polytope which is the convex hull of* $O_\Lambda \cap$ t.

This polytope is, however, not filled by all flows of the Toda hierarchy. For A_2 the image is polygonal but lies inside the polytope. However, in general the image is not even polygonal as shown by some explict cases studied by M. Zou at the University of Arizona.

§7 A convexity theorem for Jacobi elements

This section surveys the main results of Bloch, Flaschka, Ratiu [1990]. The starting point is the observation at the end of §6 that the Toda hierarchy flows do *not* fill out a convex polytope by orthogonal projection. Return to the notations and conventions of §6; set $\delta_j = 1$, $\theta_j = \pi/2$ and so $\arg a_j = \frac{\pi}{2}$ for all $j = 1,\ldots,\ell$. The latter condition is inessential since the arguments of the a_j are always constant on *Jac*. Thus in this case

$$Jac = \{L = i\sum_{j=1}^{\ell}[b_j h_j + a_j(e_{\alpha_j} + e_{j-\alpha_j})] \mid a_j, b_j \in \mathbb{R}\}$$

and we shall fix once and for all the B-orbit in $\mathfrak{b}^* \cong \mathfrak{k}$. Define

$$\mathcal{J}_\Lambda^0 = \{L \in \mathcal{O}_\Lambda \cap Jac \mid a_j > 0 \text{ for all } j = 1,\ldots,\ell\}$$
$$\mathcal{J}_\Lambda = \{L \in \mathcal{O}_\Lambda \cap Jac \mid a_j \geq 0 \text{ for all } j = 1,\ldots,\ell\}.$$

The closure of $\mathcal{J}_\Lambda^0$ is $\mathcal{J}_\Lambda$ and the boundary of $\mathcal{J}_\Lambda$ is a disjoint union of $2^\ell - 1$ strata each stratum being given by the vanishing of some of the a_j's. It is straightforward to see (as in Kostant [1979] for example) that $\mathcal{J}_\Lambda^0$ is preserved by the flows of the Toda hierarchy. For the case of $K = SU(\ell+1)$

the Jacobi matrix L is

$$L = i \begin{bmatrix} b_1 & a_1 & & & \\ a_1 & b_2 - b_1 & a_2 & & \\ & \ddots & \ddots & & \ddots \\ & & & b_\ell - b_{\ell-1} & a_\ell \\ & & & a_\ell & -b_\ell \end{bmatrix},$$

and

$$\mathcal{J}_\Lambda^0 = \{L \mid L \text{ is conjugate to } \Lambda \text{ and } a_j > 0 \text{ for all } j\},$$

$$\mathcal{J}_\Lambda = \{L \mid L \text{ is conjugate to } \Lambda \text{ and } a_j \geq 0 \text{ for all } j\},$$

$\mathcal{J}_\Lambda$ is the closure of $\mathcal{J}_\Lambda^0$ which in this case is the *isospectral set* of the Jacobi matrices L. We shall refer to $\mathcal{J}_\Lambda$ as the *closed isospectral set* and to $Jac \cap \mathcal{O}_\Lambda$ as the *isospectral manifold*. We shall use the same names when dealing with a general Lie algebra. Since the Toda lattice is integrable and the integrals are $I_1, \ldots, I_\ell$ (see §6), $\mathcal{J}_\Lambda^0$ is an orbit of the non-compact torus given by the flows of the Hamiltonian vector fields corresponding to these functions. Thus $\mathcal{J}_\Lambda$ is the closure of a non-compact torus orbit in the Kähler manifold $\mathcal{O}_\Lambda$. *In view of Atiyah's Theorem 4.1, the map $(I_1, \ldots, I_\ell)$ to $\mathbb{R}^\ell$ must have convex image. But we just saw at the end of §6 that the orthogonal projection does not have convex image. Therefore, the question naturally arises: can one embed $\mathcal{J}_\Lambda$ in a different way in $\mathcal{O}_\Lambda$ such that its projection is convex?* The answer to this question is positive and solved in Bloch, Flaschka, Ratiu [1990]; we will describe below the main ideas.

There has been prior work in which $\mathcal{J}_\Lambda$ and $\mathcal{J}_\Lambda^0$ appear in the context of convexity properties. Moser [1975] proved that $\mathcal{J}_\Lambda^0$ is diffeomorphic to $\mathbb{R}^\ell$ and Tomei [1984] proved that $\mathcal{J}_\Lambda$ is homeomorphic to a convex closed polyhedron. The first pictures of convex polyhedra appear in van Moerbeke [1976] and then again in Deift, Nanda, Tomei [1983]. These polyhedra are used in these works as a useful tool to describe certain topological relationships. In view of the convexity results of §4, it should be expected that these polyhedra have a symplectic significance as we shall outline below. The construction employed to achieve this goal is relatively involved and we shall omit it in this review. But it is based on a mysterious relationship between diagonal non-compact toral actions and left dressing transformations. Together with the ideas of §6, this work has relationships to Fried's [1986] cohomology computations and to the work of Davis [1987] and Davis and Janusiewicz [1991] on aspherical manifolds. The work of Lu and Weinstein [1990] and Lu [1990] on dressing transformations seems particularly relevant to this circle of ideas as will also be seen in §8.

Before outlining the results, a very simple example is in order. Let $\mathfrak{k} = su(2)$ and identify $su(2)^*$ with $\mathbb{R}^3$. The Lie Poisson structure of $su(2)^*$ has the concentric spheres and the origin as symplectic leaves. Represent $\Lambda = diag(i\lambda, -i\lambda)$, $\lambda > 0$ by the point $(0, 0, \lambda)$ so that $\mathcal{O}_\Lambda$ is a sphere of radius λ centered at the origin. The symplectic form is, up to a factor of $-1/\lambda$, the area form, and the complex structure, making $\mathcal{O}_\Lambda$ into a Kähler manifold, is essentially that of the Riemann sphere. The orbit $\mathcal{O}_\Lambda$ consists in this simple example of Jacobi elements. The other Poisson structure of $\mathbb{R}^3 \cong \mathfrak{b}^*$ has leaves which are open half-planes containing the vertical axis and the points of the vertical axis, if one lets B be the set of upper triangular matrices in $SU(2)$. The compact torus $T = \{diag(e^{i\theta}, e^{-i\theta}) \mid \theta \in \mathbb{R}\}$ acts by rotating the sphere $\mathcal{O}_\Lambda$ about the vertical axis; namely

$$L \in \begin{pmatrix} ib & a \\ -\bar{a} & -ib \end{pmatrix} \in \mathcal{O}_\Lambda \subset su(2),$$

$b \in \mathbb{R}$, $a \in \mathbb{C}$, is sent to

$$\begin{pmatrix} ib & ae^{2\theta} \\ -\bar{a}e^{-2\theta} & -ib \end{pmatrix}.$$

The momentum map J of this action maps $L \in \mathcal{O}_\Lambda$ to b, i.e. it projects the point on the sphere $\mathcal{O}_\Lambda$ parallel to the horizontal plane onto the vertical axis. The images of the sphere $\mathcal{O}_\Lambda$ and of the "meridian", which is the intersection of $\mathcal{O}_\Lambda$ with a B-orbit (a "page" of the "book" $\mathfrak{b}^*$), coincide and equal the convex interval $[-\lambda, \lambda]$.

This example generalizes completely, except for the very difficult projection part, which does not give any problems here due to the low dimensionality of the example. What is true is that $\mathcal{J}_\Lambda$ (here the "meridian") can be embedded in a special way in $\mathcal{O}_\Lambda$ such that its projection is a convex polytope.

The convexity result alluded to above is built upon two main ingredients. The first one is the explicit solution of the Toda lattice equation in terms of the Iwasawa decomposition as it can be found in Kostant [1979], Reyman, Semenov-Tijan-Shanskii [1979], Symes [1980], or Goodman, Wallach [1984]. The goal is to solve

$$\dot{L}(t) = [\pi_\mathfrak{k} \nabla I_j(L(t)), L(t)], \quad L(0) = L, \tag{7.1}$$

where $I_1, \ldots, I_\ell$ are the homogeneous generators of the ring of real K-invariant polynomials on $\mathfrak{k}$. First one shows that if $\varphi : \mathfrak{k} \to \mathbb{R}$ is a homogeneous polynomial and one extends φ to $\mathfrak{g}_\mathbb{C}$ in the obvious way using homogeneity, then if $\chi = Re\,\varphi$ and the gradient is taken relative to $Im\,\kappa$, $\nabla\chi(\xi) \in i\mathfrak{k}$ for all $\xi \in \mathfrak{k}$. From here it follows that $\nabla I_j(L)$, $j = 1, \ldots, \ell$, generate $iC(L)$, where $C(L)$ is the centralizer of L. Secondly, carry out the factorization

$$\exp(t \nabla I_j(L)) = \mathbf{k}(\exp(t \nabla I_j(L)))\mathbf{b}(\exp(t \nabla I_j(L))); \tag{7.2}$$

then the solution of (7.1) is

$$L(t) = Ad_{\mathbf{k}(\exp(t\nabla I_j(L)))^{-1}}L = Ad_{\mathbf{b}(\exp(t\nabla I_j(L)))}L. \tag{7.3}$$

where $g = \mathbf{k}(g)\mathbf{b}(g)$ is the factorization of $g \in G$ defined by $G = KB$. Since $L \in \mathcal{O}_\Lambda$, we can write $L = Ad_k\Lambda$ for some $k \in K$. By our first comments, $Ad_{k^{-1}}i\nabla I_j(L) \in \mathfrak{t}$, or, equivalently, $Ad_{k^{-1}}\nabla I_j(L) \in \mathfrak{a} = i\mathfrak{t}$. Consequently, if $\mu \in \mathfrak{a}$, $X := Ad_k\mu$, $L := Ad_k\Lambda$, the curve $L(t) = Ad_{k(\exp tX)^{-1}}L$ is a solution of a linear combination of the equations (7.3). But $L(t)$ can clearly be rewritten as

$$L = Ad_k\Lambda \mapsto L(t) = Ad_{\mathbf{k}(\exp(t\mu)k^{-1})^{-1}}\Lambda, \quad \exp t\mu \in A, \qquad (7.4)$$

which defines the *left dressing action* of the non-compact real torus A on $\mathcal{O}_\Lambda$ (see Lu, Weinstein [1990], Lu [1990]). Unfortunately, the action of A defined by (7.4) is *not* the diagonal action

$$L = Ad_k\Lambda \mapsto Ad_{\mathbf{k}(hk)}\Lambda, \quad h \in T, \qquad (7.5)$$

and Atiyah's theorem stated in Corollary 4.1 does not apply. Let J denote the momentum map of the diagonal T-action, i.e. the orthogonal projection onto $\mathfrak{t}$ relative to the invariant metric coming from the Killing form on $\mathfrak{k}$. The image $J(\mathcal{J}_\Lambda)$ is neither convex, nor a polytope, as shown by examples by M. Zou. One can turn, however, (7.4) into (7.5) by inversion. Namely, define $\iota : \mathcal{T}_\Lambda \to \mathcal{O}_\Lambda$ by $\iota(k\Lambda k^{-1}) = k^{-1}\Lambda k$ and the Toda flow (7.4) becomes

$$\iota(L) = \iota(Ad_k\Lambda) \mapsto Ad_{\mathbf{k}(\exp(t\mu)k^{-1})}\Lambda$$

which is exactly the diagonal action. Therefore $J(\iota(\mathcal{J}_\Lambda))$ will be a convex polytope by Atiyah's theorem. The trouble is that the "inversion" $Ad_k\Lambda \mapsto Ad_{k^{-1}}\Lambda$ makes no sense on $\mathcal{O}_\Lambda$: when one replaces k by kh, $h \in T$, the image under ι *will* depend on h.

More precisely , given $L \in \mathcal{O}_\Lambda \cap Jac$, $L = Ad_k\Lambda$, determines k only up to the right multiplication of an element of T; remember, Λ is chosen once and for all in the interior of the positive Weyl chamber of $\mathfrak{t}$. And this leads to the second main ingredient, namely, k must be chosen in a smooth way and uniquely for each $L \in \mathcal{J}_\Lambda$. That this is possible is ultimately based on the Bruhat decomposition of K into cells and constitutes the main technical part of the proof. It was inspired by ideas that show up also in Flaschka, Haine [1990] and Ercolani, Flaschka, Haine [1990]. The upshot of this work is that *the "inversion map" ι can be extended continuously to the boundary $\mathcal{J}_\Lambda \backslash \mathcal{J}_\Lambda^0$ and that it is a diffeomorphism on $\mathcal{J}_\Lambda^0$ and a homeomorphism on $\mathcal{J}_\Lambda$.* Here is its very simple and concrete description in the case of $SU(\ell+1)$. Let $L = k\Lambda k^{-1}$. The columns of k are orthonormalized eigenvectors of L. Let now $L \in \mathcal{J}_\Lambda^0$; then the first row entries r_j of k do not vanish. Fix k by the requirement that $r_j/r_1 \in \mathbb{R}$. This is the smooth unique choice of k for a given L. The "inversion" ι is now easy to describe. The Toda hierarchy acts on $\iota(\mathcal{J}_\Lambda^0)$ as the noncompact torus A as follows: Consider the element

$$a = \exp(p_1 c_1 \Lambda t_1 + \ldots p_{\ell+1} c_{\ell+1} \Lambda^{\ell+1} t_{\ell+1}) \in A$$

where $c_k = i$ for k odd, $c_k = 1$ for k even, and $p_1, \ldots, p_{\ell+1}$ are arbitrary real constants. We have the mapping

$$a : \iota(L) \mapsto \mathbf{k}(ak^{-1})\Lambda\mathbf{k}(ak^{-1})^{-1}.$$

It is a straightforward check to see that this action preserves the normalization of the r_j's described above.

Now, several conclusions are possible.

Theorem 7.1. *The image of $\mathcal{J}_\Lambda$ under the map $J \circ \iota : \mathcal{J}_\Lambda \to \mathfrak{t}$ is the convex hull of the Weyl group orbit through Λ.*

Combined with the results of §6 and those in Duistermaat, Kolk, Varadarajan [1983] who show that the action of the one-parameter group $\exp t\mu$ for $\mu \in \mathfrak{a}$, on $\mathcal{O}_\Lambda$, is a gradient flow generated by $-\kappa(\mu, \cdot)$ relative to the Kähler metric, Theorem 7.1 now implies:

Theorem 7.2. *The Toda flows in $\mathcal{J}_\Lambda$ are gradient flows in the metric induced by pulling back the Kähler metric of $\iota(\mathcal{J}_\Lambda)$ to $\mathcal{J}_\Lambda$.*

It should be emphasized that the functions which generate the gradient flows are *not* the Toda Hamiltonians. Theorem 7.2 is complementary to Theorem 6.2, where the Toda flow is seen to be gradient relative to another metric which, geometrically, is more natural. Shades of the "inversion map" ι will appear also in §8 and there seems to be a connection between these results and the ones of next section.

We close with some comments on the relationship between the results in §6, §7 and Moser's [1975] gradient flows associated with the Toda lattice.

We recall from Sections 5 and 6 that the Toda lattice equations are gradient with respect to the normal metric on an orbit $\mathcal{O}_\Lambda$ and hence may be written in the form $\dot{L} = [L, [L, N]]$.

Now one can project the flow on $\mathcal{O}_\Lambda$ (which is diffeomorphic to K/T) to K/P, P a parabolic subgroup of K.

Consider specifically the case $K = SU(\ell+1)$ and let $P_0 = \mathrm{diag}(1, 0, \ldots, 0) - \frac{I}{\ell+1}$. Now take the K-orbit $\mathcal{O}_{iP_0}$ through iP_0 with points $k(iP_0)k^{-1}$, $k \in K$. This orbit consists of points of the form: rank 1 projection matrix minus a multiple of the identity, and may thus be identified with $\mathbb{CP}^\ell$. Now we can check that for $iP \in \mathcal{O}_{iP_0}$, $-(ad\,iP)^2([iP, \eta]) = [iP, \eta]$ and hence (see section 4) the normal and Kähler metric coincide. (This also applies to a general Grassmannian; see Bloch, Flaschka and Ratiu [1990].)

Now consider the gradient flow, with respect to the Kähler (or normal !) metric, of $\phi_k(iP) = -cTr(iP\Lambda^k)$ on $\mathbb{CP}^\ell$. Here $\Lambda = \mathrm{diag}(\lambda_1, \ldots \lambda_\ell)$ and c is 1 or i as needed to make ϕ_k real. The gradient flow is

$$i\dot{P} = [iP, [iP, c\Lambda^k]].$$

Taking $P = r \otimes \bar{r}$, $r = (r_1, \ldots r_{\ell+1})$ with $\sum |r_j|^2 = 1$ yields

$$|\dot{r}_j| = -|r_j|(\lambda_j^k - \sum_{i=1}^{\ell+1} \lambda_i^k |r_i|^2), \quad j = 1 \ldots \ell + 1$$

which is Moser's equation for the Toda lattice.

Note that this flow is just conjugate to the Hamiltonian flow $i\dot{P} = [iP, c\Lambda^k]$ – just apply the complex structure $[iP, \cdot]$!

Note also that Atiyah's theorem also applies to K/P. Ignoring the constant shift in P_0, the momentum mapping $J_P : K/P \to \mathfrak{t}$ is given by $k^{-1} \bmod P \to i\, diag(r \otimes \bar{r})$. The Jacobi matrix L is thus sent to $(i|r_1|^2, \ldots, i|r_{\ell+1}|^2)$ and the image in $\mathfrak{t} = \mathbb{R}^\ell$ is the standard simplex $\sum_{i=1}^{\ell+1} \xi_i = 1$.

§8 The non-linear convexity theorem of Kostant

This section reviews the results in Lu, Ratiu [1991]. It is not directly related to the previous ideas on the relationship between integrable systems and convexity. However, the dressing transformation that appeared so prominently in §7 and certain aspects of the "inversion map" on $\mathcal{O}_\Lambda$ can be found here too. There seems to be a much deeper connection between the ideas in this section and the Toda lattice. On the symplectic geometric side, we think, the result discussed below raises a host of other questions. For example, are there other convexity theorems that have ultimately a symplectic origin? There are several one can think of, among which are other convexity results proved in the same paper of Kostant [1973], as Kostant himself has pointed out to us. On the other hand, are there meaningful generalizations of the result below to a symplectic group setting? Finally, do the constructions below shed some light on the second Hamiltonian structure for the A_ℓ-Toda lattice (as opposed to the well-known results on the $g\ell(\ell + 1)$-Toda lattice)?

We return to §3 on the symplectic convexity results and set the stage for the statement of one of Kostant's non-linear convexity results at Lie group level.

G is a real connected semisimple Lie group with Lie algebra $\mathfrak{g}$ and $\mathfrak{k}$ is the Lie subalgebra corresponding to a maximal compact subgroup of the adjoint group of G. $\mathfrak{g} = \mathfrak{k} \oplus \mathfrak{p}$ is the *Cartan decomposition*, i.e. $\mathfrak{p}$ is the orthogonal complement of $\mathfrak{k}$ in $\mathfrak{g}$ relative to the Killing form. The elements of $\mathfrak{p}$ are all semisimple and $P = exp\,\mathfrak{p}$ is a closed submanifold of G; exp is a diffeomorphism between $\mathfrak{p}$ and P. At the group level, the map $(k,p) \in K \times P \mapsto kp$ is a diffeomorphism, where K is a connected subgroup of G with Lie algebra $\mathfrak{k}$; one writes $G = KP$ and calls this the *Cartan decomposition of G.*

By the invariance of the Killing form, $[\mathfrak{k}, \mathfrak{p}] \subset \mathfrak{p}$ and therefore K acts on $\mathfrak{p}$ by the adjoint action and also on P by conjugation; the exponential map $exp : \mathfrak{p} \to P$ is K-equivariant. Let $\mathfrak{a}$ denote a maximal abelian subalgebra of $\mathfrak{p}$. The elements of $\mathfrak{a}$ are all semisimple and thus the adjoint representation of $\mathfrak{a}$ on $\mathfrak{g}$ decomposes as $\mathfrak{g} = \mathfrak{l} \oplus \bigsqcup_{\alpha \in \Phi} \mathfrak{g}_\alpha$ where $\mathfrak{l}$ is the centralizer of $\mathfrak{a}$ in $\mathfrak{g}$, and for a linear functional α on $\mathfrak{a}$, $\mathfrak{g}_\alpha = \{\xi \in \mathfrak{g} \mid [\eta, \xi] = \alpha(\eta)\xi$ for all $\eta \in \mathfrak{a}\}$; Φ is the set of all such non-zero α's. Fix a basis in $\mathfrak{a}$ and introduce the lexicographic ordering on functionals on $\mathfrak{a}$. Put $\mathfrak{n} = \bigsqcup_{\alpha > 0} \mathfrak{g}_\alpha$ and then $\mathfrak{g} = \mathfrak{k} \oplus \mathfrak{a} \oplus \mathfrak{n}$ is the *Iwasawa decomposition of* $\mathfrak{g}$. Let A, N be the connected subgroups of G with Lie algebras $\mathfrak{a}$ and $\mathfrak{n}$ respectively. The *Iwasawa decomposition of* G states that $(k, a, n) \in K \times A \times N \mapsto kan \in G$ is a diffeomorphism; one writes $G = KAN$. By W we denote the relative Weyl group of $(K, \mathfrak{a})$, i.e. W is the quotient of the normalizer of $\mathfrak{a}$ in K by the centralizer of $\mathfrak{a}$ in K.

If G is a complex semisimple Lie group, but thought of as a real Lie group, the above choices are easier. $\mathfrak{k}$ is a compact real form of $\mathfrak{g}$, the Killing form on $\mathfrak{g}$ is the complex linear extension of that on $\mathfrak{k}$, $\mathfrak{p} = i\mathfrak{k}$, $\mathfrak{a} = i\mathfrak{t}$, where $\mathfrak{t}$ is a Cartan subalgebra of $\mathfrak{k}$. The Weyl group of $(K, \mathfrak{t})$ is that of $(K, \mathfrak{a})$. Fixing a basis of $\mathfrak{t}$ will ultimately define an Iwasawa decomposition of $\mathfrak{g}$ and G.

Now let G be a semisimple Lie group, real or complex. The Cartan and Iwasawa decompositions define a diffeomorphism between P and AN, namely, if $p = kan$, then associate to p the element an. Let ρ_A be the projection from G, P, or AN to A according to the Iwasawa decomposition.

Theorem 8.1. *(Kostant [1973]) For $a \in A$ denote by $\mathcal{O}_a$ the K-orbit of a in P. Then $\rho_A(\mathcal{O}_a)$ is the convex hull of the Weyl group orbit $W \cdot a$ in A. (The Lie group A is identified with its Lie alegbra $\mathfrak{a}$ via the exponential map, so convexity makes sense.)*

This statement is one of Kostant's nonlinear convexity theorems. The term is justified by the following remarks and the example to be discussed below. The differential (tangent map) of $\rho_A : P \to A$ gives the orthogonal projection $\rho_\mathfrak{a} : \mathfrak{p} \to \mathfrak{a}$ relative to the Killing form. Let $log : A \to \mathfrak{a}$ be the inverse of exp and define $J_1 := log \circ \rho_A \circ exp : \mathfrak{p} \to \mathfrak{a}$. We compare J_1 to $\rho_\mathfrak{a}$. J_1 is non-linear and its differential at zero is $\rho_\mathfrak{a}$. *What Kostant's theorem states is the remarkable fact that each K-orbit in $\mathfrak{p}$ has the same convex image under both J_1 and $\rho_\mathfrak{a}$ and this polytope is the convex hull of the corresponding Weyl group orbit.*

Let's sketch the simplest example: $G = SL(n, \mathbb{C})$, $K = SU(n)$, $P = \{n \times n$ positive definite Hermitian matrices with determinant 1$\}$, $N = \{n \times n$ strictly upper triangular complex matrices$\}$, $\mathfrak{g} = s\ell(n, \mathbb{C})$, $\mathfrak{k} = su(n)$, $\mathfrak{a} = \{n \times n$ real diagonal traceless matrices$\}$, $\mathfrak{p} = \{n \times n$ traceless Hermitian matrices $\}$, $\mathfrak{n} = \{n \times n$ strictly upper triangular complex matrices$\}$. The Cartan decompositions are $SL(n, \mathbb{C}) = SU(n)P$ and $s\ell(n, \mathbb{C}) = su(n) \oplus$

$\mathfrak{p}$. The Iwasawa decompositions are $SL(n,\mathbb{C}) = SU(n)AN$, $s\ell(n,\mathbb{C}) = su(n) \oplus \mathfrak{a} \oplus \mathfrak{n}$. If $X \in \mathfrak{p}$, $\rho_\mathfrak{a}(X)$ is the diagonal part of X. On the other hand

$$J_1(X) = (log \circ \rho_A \circ exp)(X)$$
$$= \frac{1}{2}\left(log\Delta_1(e^{2X}), log\frac{\Delta_2(e^{2X})}{\Delta_1(e^{2X})}, \ldots, log\frac{\Delta_n(e^{2X})}{\Delta_{n-1}(e^{2X})}\right)$$

where Δ_k is the determinant of the left upper corner $k \times k$ matrix. For $X = (x_1, \ldots, x_n) \in \mathfrak{a}$, the flag manifold $\mathcal{O}_X$ (the K-orbit in $\mathfrak{p}$ through X) is the set of all Hermitian matrices with $x_1, \ldots, x_n$ as their eigenvalues. The relative Weyl group is in this example the permutation group on n letters.

The expressions appearing in $J_1(X)$ remind one of the explicit solution of the Toda lattice equations (see Kostant [1979]); we don't believe this to be an accident but cannot explain it so far.

Let us outline below the main ideas going into the "symplectic proof" of Kostant's theorem. We begin with the complex case, i.e. G is a complex semisimple Lie group, K is a real compact form of G, and $G = KAN$ is the Iwasawa decomposition. We let $\mathfrak{b} := \mathfrak{a} \oplus \mathfrak{n}$, $B := AN$ and remark as in §7 that $\mathfrak{k}$ is isomorphic to $\mathfrak{b}^*$ via the imaginary part of the Killing form κ. Let $\rho_\mathfrak{k}$ and $\rho_\mathfrak{b}$ be the projections of $\mathfrak{g}$ on $\mathfrak{k}$ and $\mathfrak{b}$ respectively relative to the Iwasawa decomposition $\mathfrak{g} = \mathfrak{k} \oplus \mathfrak{b}$. We will denote by R_b the right translation in B by b, its differential, and all its tensorially induced maps. The following theorem is deduced form abstract considerations in Lu, Weinstein [1990], Lu [1990] and carried out "by hand" in Lu, Ratiu [1991].

Theorem 8.2. *On B define a bivector field π by*

$$(R_{b^{-1}}\pi(b))(X, Y) = (Im\,\kappa)(\rho_\mathfrak{k}(Ad_{b^{-1}}X), \rho_\mathfrak{b}(Ad_{b^{-1}}Y))$$

where $b \in B$, $X, Y \in \mathfrak{k} \cong \mathfrak{b}^*$. Then π defines a Poisson structure on $B = AN$ which is multiplicative.

The property of *multiplicativity* means that π satisfies

$$\pi(b_1 b_2) = L_{b_1}\pi(b_2) + R_{b_2}\pi(b_1)$$

where L_b is left translation by b. A *Poisson Lie group* is a Lie group endowed with a multiplicative Poisson structure. Thus theorem 8.2 gives a concrete natural formula for a Poisson Lie group structure on B.

The Iwasawa decomposition at group level $G = KB$ induces a smooth projection $\rho_B : G \to B$, $\rho_B(kb) = b$. Define the K-action on B by

$$\sigma : (k, b) \in K \times B \mapsto \rho_B(kbk^{-1}) \in B.$$

Theorem 8.3. *The symplectic leaves of the Poisson structure π in Theorem 8.1 are the K-orbits in B relative to the action σ.*

Let $\mathfrak{t} = i\mathfrak{a}$ and T be the connected subgroup of K with Lie algebra $\mathfrak{t}$. T is a maximal torus in K.

Theorem 8.4. *The restriction σ to T leaves the Poisson structure π on B invariant and the map*

$$J = \log \circ \rho_A : B = AN \to \mathfrak{a}$$
$$J(an) = \log(a)$$

is the momentum map for this T-action.

It should be noted that the Poisson structure π on B is not K-invariant, but only T-invariant. In Lu, Weinstein [1990] it is shown that there is a natural Poisson structure on K such that the map $\sigma : K \times B \to B$ is a Poisson map if one thinks of $K \times B$ as a product Poisson manifold (see Weinstein [1983a]). This Poisson structure on K alluded to above happens to vanish on T and this is why this T-action is Hamiltonian.

The Poisson structure π on B and the T-momentum map $J : B \to \mathfrak{a}$ are the first ingredient in the proof of Kostant's theorem in the complex case. The second ingredient is the convexity Theorem 3.1. P and $AN = B$ are identified via ρ_B (and both of them are diffeomorphic to $K\backslash G$). The action σ of K on B becomes the conjugation of K on P. We will still denote by π the push-forward to P of the Poisson structure on B and so the symplectic leaves are the K-orbits on P. Every K-orbit in P intersects A and the fixed point set of the T-action on the K-orbit is the corresponding Weyl group orbit. Now apply Theorem 3.1 to the equivariant momentum map J of Theorem 8.4 to immediately conclude Kostant's convexity result in Theorem 8.1.

To prove Theorem 8.1 for real Lie groups, a last ingredient is needed: namely Duistermaat's Theorem 3.4. For this, we need to identify an anti-symplectic involution on the K-orbits in P. We will do better and find an anti-Poisson involution; it will be induced by a Cartan involution. Here is the construction. Let G be a connected real semisimple Lie group with Lie algebra $\mathfrak{g}$. Without loss of generality we may assume G has trivial center so it admits a complexification $G_{\mathbb{C}}$. Then the Lie alegbra of $G_{\mathbb{C}}$ is $\mathfrak{g}_{\mathbb{C}} = \mathfrak{g} \oplus i\mathfrak{g}$, the complexification of $\mathfrak{g}$. Let $\tau(X + iY) = X - iY$ be conjugation in $\mathfrak{g}_{\mathbb{C}}$, $X, Y \in \mathfrak{g}$, and denote also by τ the unique automorphism of $G_{\mathbb{C}}$ whose differential at the identity is the conjugation. The identity component of the fixed point set of τ is G. If $\mathfrak{g} = \mathfrak{k} \oplus \mathfrak{p}$ is the Cartan decomposition of $\mathfrak{g}$, let $\mathfrak{k}_{\mathbb{C}} = \mathfrak{k} \oplus i\mathfrak{p}$, $\mathfrak{p}_{\mathbb{C}} = i\mathfrak{k} \oplus \mathfrak{p} = i\mathfrak{k}_{\mathbb{C}}$ so that both $\mathfrak{k}_{\mathbb{C}}$ and $\mathfrak{p}_{\mathbb{C}}$ are τ-invariant and the fixed point sets of $\tau \mid \mathfrak{k}_{\mathbb{C}}$, $\tau \mid \mathfrak{p}_{\mathbb{C}}$ are $\mathfrak{k}$ and $\mathfrak{p}$ respectively. $\mathfrak{k}_{\mathbb{C}}$ is a compact real form of $\mathfrak{g}_{\mathbb{C}}$ and $\mathfrak{g}_{\mathbb{C}} = \mathfrak{k}_{\mathbb{C}} \oplus \mathfrak{p}_{\mathbb{C}}$ is a Cartan decomposition of $\mathfrak{g}_{\mathbb{C}}$. Let $\mathfrak{a}$

be a maximal abelian subalgebra of $\mathfrak{p}$, $\mathfrak{a}'$ a maximal abelian subalgebra of $\mathfrak{g}$ containing $\mathfrak{a}$ and let $\mathfrak{a}_\mathbb{C} = \mathfrak{a} \oplus i(\mathfrak{a}' \cap \mathfrak{k})$. Then $\mathfrak{a}_\mathbb{C}$ is a maximal abelian subalgebra of $\mathfrak{p}_\mathbb{C}$ and $\mathfrak{a}$ is the fixed point set of $\tau \mid \mathfrak{a}_\mathbb{C}$. Bases in $\mathfrak{a}$ and $\mathfrak{a}_\mathbb{C}$ are chosen such that the basis in $\mathfrak{a}_\mathbb{C}$ is a prolongation of the basis in $\mathfrak{a}$. Relative to such a choice (and a lexicographic ordering), we have the Iwasawa decomposition

$$\mathfrak{g} = \mathfrak{k} \oplus \mathfrak{a} \oplus \mathfrak{n}, \quad \mathfrak{g}_\mathbb{C} = \mathfrak{k}_\mathbb{C} \oplus \mathfrak{a}_\mathbb{C} \oplus \mathfrak{n}_\mathbb{C}.$$

$\mathfrak{n}_\mathbb{C}$ is τ-invariant and, as before, the fixed point set of $\tau|\mathfrak{n}_\mathbb{C}$ is $\mathfrak{n}$. Now pass to the group level and use the same notational scheme.

Theorem 8.5. *Let π be the multiplicative Poisson structure on $B_\mathbb{C} = A_\mathbb{C} N_\mathbb{C}$ defined in Theorem 8.2. Then $\tau|B_\mathbb{C}$ is an anti-Poisson automorphism, i.e. $\tau^*\pi = -\pi$.*

Now Kostant's Theorem 8.1 for real Lie groups is a direct consequence of Duistermaat's Theorem 3.4 applied to a K-orbit in P viewed as the fixed point set of τ on the $K_\mathbb{C}$-orbit in $P_\mathbb{C}$ through the same point. The only missing link is the momentum map $P_\mathbb{C} \to \mathfrak{a}$, which, after a suitable identification becomes the Iwasawa projection $P_\mathbb{C} \to A$. This is constructed in the following way. Let $T_\mathbb{C}$ be the maximal torus of $K_\mathbb{C}$ with Lie algebra $i\mathfrak{a}_\mathbb{C}$. By Theorem 8.4, $T_\mathbb{C}$ leaves the Poisson structure on $P_\mathbb{C}$ invariant (recall, we identify $P_\mathbb{C}$ and $B_\mathbb{C} = A_\mathbb{C} N_\mathbb{C}$). Let T be the subtorus generated by $\mathfrak{t} = i\mathfrak{a}$. Apply theorem 8.4 to conclude that this action has equivariant momentum map

$$J_T : (k_\mathbb{C} a_\mathbb{C} n_\mathbb{C}) \in P_\mathbb{C} \mapsto proj_\mathfrak{a}(\log a_\mathbb{C})$$

where $k_\mathbb{C} \in K_\mathbb{C}$, $a_\mathbb{C} \in A_\mathbb{C}$, $n_\mathbb{C} \in N_\mathbb{C}$ and $proj_\mathfrak{a} : \mathfrak{a}_\mathbb{C} \to \mathfrak{a}$ is the orthgonal projection relative to the Killing form. Let $\mathfrak{a}_0$ be the orthogonal complement of $\mathfrak{a}$ in $\mathfrak{a}_\mathbb{C}$ and define $A_0 = \exp \mathfrak{a}_0$. Then $A_\mathbb{C} = AA_0$ and we can regard J_T as

$$J_T : (k_\mathbb{C} a a_0 n_\mathbb{C}) \in P_\mathbb{C} \mapsto a \in A$$

for $k_\mathbb{C} \in K_\mathbb{C}$, $a \in A$, $a_0 \in A_0$, $n_\mathbb{C} \in N_\mathbb{C}$. As before, this map is τ-invariant. Now restrict the T-action and the momentum map to a $K_\mathbb{C}$-orbit in $P_\mathbb{C}$. By Theorem 3.4 the image of the τ-fixed point set, which is the K-orbit in P through the same point, has J_T-image in $\mathfrak{a}$ equal to the convex hull of the image of the fixed point set of the T-action on the K-orbit. But this fixed point set is the intersection of the K-orbit with A, i.e. the corresponding Weyl group orbit in A. Finally, J_T restricts to the identity map on A. This proves Theorem 8.1 in the real case.

It is worthwhile noting that the symplectic leaves of π in $\mathfrak{p}$ and the symplectic leaves of the Lie-Poisson structure on $\mathfrak{k}^* = \mathfrak{p}$ coincide. Moreover, a theorem of Conn [1985] guarantees that locally, around zero, these two

Poisson structures are isomorphic (the key assumption in Conn's theorem, that $\mathfrak{k}$ is compact semisimple, is automatically fulfilled in our case). We suspect that these Poisson structures are globally isomorphic and hope that the isomorphism is relevant to questions regarding the multi-Hamiltonian structure of the Toda lattice equations. Duistermaat [1984] has already shown that the momentum maps for these two structures can be obtained from each other by a homotopy argument. The hoped for Poisson isomorphism between π and the Lie-Poisson structure cannot be K-invariant since π is not whereas the Lie-Poisson structure is; it must, however, be T-invariant.

REFERENCES

1. M.F. Atiyah, *Convexity and commuting Hamiltonians*, Bull. London Math. Soc. **14** (1982), 1–15.
2. M.F. Atiyah, *Angular momentum, convex polyhedra and algebraic geometry*, Proc. Edinburgh Math. Soc. **26** (1983), 121–138.
3. A. Besse, *Einstein Manifolds*, Ergebnisse der Mathematik und ihrer Grenzgebiete, 3. Folge, Springer-Verlag, New York, 1987.
4. A.M. Bloch, *The Kähler structure of the total least squares problem, Brockett's steepest descent equations, and constrained flows*, Realization and Modelling in Systems Theory (M. Kaashoek, A.C.M. Ran and J.H. van Schuppen eds.), Birkhäuser, Boston, 1990.
5. A.M. Bloch, R.W. Brockett, T.S. Ratiu, *A new formulation of the generalized Toda lattice equations and their fixed point analysis via the momentum map*, Bull. Amer. Math. Soc. **23(2)** (1990a), 477–486.
6. A.M. Bloch, R.W. Brockett, T.S. Ratiu, *Completely integrable gradient flows*, (1990b) to appear.
7. A.M. Bloch, H. Flaschka, T.S. Ratiu, *A convexity theorem for isospectral manifolds of Jacobi matrices in a compact Lie algebra*, Duke Math. Journ. **61(1)** (1990), 41–66.
8. R.W. Brockett, *Dynamical systems that sort lists and solve linear programming problems*, (Previously pulished in "Proc. 27th IEEE Conf. on Decision and Control" (1988), 799–803), Linear Algebra and its Applications **146** (1991), 79–91.
9. R.W. Brockett, *Least squares matching problems*, Lin. Alg. and Appl. **122/123/124** (1989), 761–777.
10. M. Condevaux, P. Dazord, P. Molino, *Géometrie du moment*, Publ. Dép.Math,Univ.Cl-Bernard-Lyon.
11. J. Conn, *Normal forms for smooth Poisson structures*, Ann. of Math. **121** (1985), 565–593.
12. M. Davis, *Some aspherical manifolds*, Duke Math. Journ. **55** (1987), 105–138.

13. M. Davis, T. Janusiewicz, *Convex polytopes, Coxeter polytopes and torus actions*, Duke. Math. Journ. **62(2)** (1991), 417–451.

14. P. Deift, T. Nanda, C. Tomei, *Differential equations for the symmetric eigenvalue problem*, SIAM Journ. Num. Anal. **20** (1983), 1–22.

15. T. Delzant, *Hamiltoniens périodiques et images convexes de l'application moment*, Bull. Soc. Math. France **116** (1988), 315–339.

16. J.J. Duistermaat, *Convexity and tightness for restrictions of Hamiltonian functions to fixed point sets of an antisymplectic involution*, Trans. Amer. Math. Soc. **275** (1983), 417–429.

17. J.J. Duistermaat, *On the similarity between the Iwasawa projection and the diagonal part*, Bull. Soc. Mat. France, 2e Série, Mémoire **15** (1984), 129–138.

18. J.J. Duistermaat, J.A. Kolk, V.S. Varadarajan, *Functions, flows, and oscillatory integrals on flag manifolds and conjugacy classes in real semisimple Lie groups*, Compositio Math. **49** (1983), 309–398.

19. N. Ercolani, H. Flaschka, L. Haine, *The level varieties of the complex Toda equations*, preprint (1990).

20. H. Flaschka, L. Haine, *Torus orbits in G/P*, Pac. Journ. Math. (1990), (to appear).

21. D. Fried, *The cohomology of an isospectral flow*, Proc. Amer. Math. Soc. **98** (1986), 363–368.

22. R. Goodman, N. Wallach, *Classical and quantum mechanical systems of Toda lattice type, II*, Comm. Math. Phys. **94** (1984), 177–217.

23. V. Guillemin, S. Sternberg, *Convexity properties of the moment mapping*, Invent. Math. **67** (1982), 491–513.

24. V. Guillemin, S. Sternberg, *Convexity properties of the moment mapping II*, Invent. Math. **77** (1984), 533–546.

25. A. Horn, *Doubly stochastic matrices and the diagonal of a rotation matrix*, Amer. Journ. Math. **76** (1954), 620–630.

26. J. Humphreys, *Introduction to Lie Algebras and Representation Theory*, Springer-Verlag, New York, 1972.

27. A. Kirillov, *Elements of the Theory of Representations*, Springer-Verlag, Berlin, 1976.

28. F. Kirwan, *Convexity properties of the moment mapping III*, Invent. Math. **77** (1984), 547–552.

29. B. Kostant, *Orbits, symplectic structures and representation theory*, Proc. U. S.-Japan Seminar in Differential Geometry (Kyoto 1965), p. 71, Nippon Hyoronsha, Tokyo, 1966.

30. B. Kostant, *Symplectic manifolds and unitary representations*, mimeographed notes 238, 20Kos.68, Bibliothèque Interuniversitaire Scientifique, Jussieu, Math. Res (1968).

31. B. Kostant, *Quantization and unitary representations*, Lecture Notes in Math. 170, Springer-Verlag, New York, 1970.

32. B. Kostant, *On convexity, the Weyl group, and the Iwasawa decomposition*, Ann. Sci. Ec. Norm. Sup. **6** (1973), 413–455.

33. B. Kostant, *The solution to a generalized Toda lattice and representation theory*, Adv. in Math. **34** (1979), 195–338.

34. J. Lagarias, *Monotonicity properties of the generalized Toda flow and QR flow*, SIAM Journ. Matrix Analysis and Applications (1988), (to appear).

35. S. Lie, *Theorie der Transformationsgruppen*, Bg. Teubner (1890), reprinted, Chelsea, New York, 1970.

36. J.-H. Lu, *Multiplicative and affine Poisson structures on Lie groups*, Ph. D. Thesis, U. C. Berkeley (1990).

37. J.-H. Lu, T. Ratiu, *On the nonlinear convexity theorem of Kostant*, Journ. Amer. Math. Soc. (1991), (to appear).

38. J.-H. Lu, A. Weinstein, *Poisson Lie groups, dressing transformations and Bruhat decompositions*, Journ. Diff. Geom. **31** (1990), 501–526.

39. J. Moser, *Finitely many mass points on the line under the influence of an exponential potential–an integrable system*, Springer Lecture Notes in Physics 38, 1975, pp. 467–497.

40. A. Reyman, M. Semenov-Tijan-Shanskii, *Reduction of Hamiltonian systems, affine Lie algebras, and Lax equations, I*, Invent. Math. **54** (1979), 81–100; *II*, Invent. Math. **63** (1981 423–432).

41. D.H. Sattinger and D.L. Weaver, *Lie Groups and Lie Algebras with Applications to Physics, Geometry and Mechanics*, Applied Mathematical Sciences 69, Springer-Verlag, New York, 1986.

42. I. Schur, *Über eine Klasse von Mittelbildungen mit Anwendungen auf der Determinantentheorie*, Stizungsberichte der Berliner Math. Gesellschaft **22** (1923), 9–20.

43. S. Smale, *Topology and Mechanics, I*, Invent. Math. **10** (1970), 305–311; *II*, Invent. Math. **11**, 45–64.

44. J.-M. Souriau, *Géométrie de l'espace de phases, calcul des variations et mécanique quantique*, Fac. Sc. Marseille, mimeographed notes (1965).

45. J.-M. Souriau, *Quantification géométrique*, Comm. Math. Phys. **1** (1966), 374–398.

46. J.-M. Souriau, *Quantification géométrique. Applications*, Ann. Inst. H. Poincaré **6(4)** (1967a), 311–341.

47. J.-M. Souriau, *Modèles classiques quantifiables pour les particules élémentaires II*, Comptes Rend. Acad. Sci. **265** (1967b), 165–167.

48. J.-M. Souriau, *Structure des Systèmes Dynamiques*, Dunod, Paris, 1969.

49. W.W. Symes, *Hamiltonian group actions and integrable systems*, Physica D **1** (1980), 39–376.

50. W.W. Symes, *Systems of Toda type, inverse spectral problems and representation theory*, Invent. Math. **59** (1982a), 13–51.

51. W.W. Symes, *The QR algorithm and scattering for the nonperiodic Toda lattice*, Physica D **4** (1982b), 275–280.

52. C. Tomei, *The topology of isospectral manifolds of tridiagonal matrices*, Duke Math. Journal **51** (1984), 981–996.

53. P. van Moerbeke, *The spectrum of Jacobi matrices*, Invent. Math. **37** (1976), 45–81.
54. A. Weinstein, *The local structure of Poisson manifolds*, J. Diff. Geom. **18** (1983a), 523–557.
55. A. Weinstein, *Sophus Lie and symplectic geometry*, Expo. Math. **1** (1983b), 95–96.

A.M. Bloch
Department of Mathematics
The Ohio State University
Columbus, OH 43210
U.S.A.

T.S. Ratiu
Department of Mathematics
University of California
Santa Cruz, CA 95064
U.S.A.

The Variety of all Invariant Symplectic Structures on a Homogeneous Space and Normalizers of Isotropy Subgroups

RANEE BRYLINSKI* and BERTRAM KOSTANT**

Dedicated to Jean-Marie Souriau

Introduction

A symplectic structure on a manifold, traditionally, is a real valued 2-form. However as one knows (e.g. as used in $\mathfrak{D}$-module theory) symplectic structures are analogously defined in the complex algebraic category and as such are complex valued 2-forms. One very nice aspect of the latter case is a marriage of symplectic machinery with complex algebraic geometry. In this paper we are considering smooth complex algebraic varieties and our consideration of symplectic structures will be, appropriately, in the sense of that latter case.

Let G be a complex simply-connected semisimple Lie group and let $\mathfrak{g} = \text{Lie } G$. We will identify $\mathfrak{g}$ with its dual space using the Killing form so that if M is a Hamiltonian G-space the corresponding moment map takes its values in $\mathfrak{g}$. If G operates on a smooth variety M and ω is a G-invariant symplectic structure on M then note that M is automatically a Hamiltonian G-space and there is a corresponding uniquely defined moment map

$$\mu_\omega : M \longrightarrow \mathfrak{g}$$

This is the case because the Lie algebra cohomology spaces $H^1(\mathfrak{g})$ and $H^2(\mathfrak{g})$ vanish. (See e.g. [K2] for the argument in the real category. The same proof readily applies in the complex case.)

Now assume that M is a complex homogeneous space for G. Let *Symp M* be the set of all G-invariant symplectic structures on M. Assume that *Symp M* is not empty. One of the objectives of this paper will be the determination of the structure of *Symp M*. More then that let $Q = Q(M)$ be the group of all maps $\alpha : M \longrightarrow M$ which commute with the action of G. Then *Symp M* is a Q-set and we will also be concerned with the structure of Q and its action on *Symp M*.

---

* Sloan Fellow

** Supported in part by NSF Grant No. DMS-8703278

We deal with these questions in the following way. Fix $\chi \in M$ so that we may regard $Q = N^\chi / G^\chi$ where G^χ is the G-isotropy subgroup at χ and N^χ is the normalizer of G^χ in G. Now if $\omega \in Symp\ M$ and $\mu_\omega(\chi) = x \in \mathfrak{g}$ then

$$\mu_\omega : M \longrightarrow O_x \tag{1}$$

is a G-equivariant orbit covering where $O_x \subset \mathfrak{g}$ is the adjoint orbit containing x. One notes then that $Symp\ M$ parameterizes all possible orbit covering maps (1) so that if $\sigma_\chi(\omega) = \mu_\omega(\chi)$ then

$$\sigma_\chi : Symp\ M \longrightarrow Symp_\chi M \tag{2}$$

is a Q -bijection where $Symp_\chi M = \{\mu_\omega(\chi) \mid \omega \in Symp\ M\}$. Then (2) is an isomorphism of algebraic varieties and also relegates the problem to the consideration of the Q-variety $Symp_\chi M \subset \mathfrak{g}$.

We will say that M is semisimple (resp. nilpotent) if there exists $\omega \in Symp_\chi M$ such that O_x is an orbit of semisimple (resp. nilpotent) elements. The two notions are mutually exclusive. Let $\mathfrak{g}^\chi = Lie\ G^\chi$. In case M is semisimple all is known. In that case Q is finite and is isomorphic to the "Weyl group" of $Cent\ \mathfrak{g}^\chi$. Moreover

$$Symp_\chi M = Reg\ Cent\ \mathfrak{g}^\chi \tag{3}$$

where if $\mathfrak{h}$ is a Lie algebra of a torus in G and $\mathfrak{g}^\mathfrak{h}$ is the centralizer of $\mathfrak{h}$ then $Reg\ \mathfrak{h} = \{x \in \mathfrak{h} \mid \mathfrak{g}^x = \mathfrak{g}^\mathfrak{h}\}$.

We deal with the general case but it is the case where M is nilpotent that mainly concerns us. In fact this paper is in effect the first of a series of papers dealing with symplectic structures on nilpotent orbits. For simplicity of exposition in this introduction assume that M is simply-connected (this restriction is not imposed however in the rest of the paper) but otherwise arbitrary. In the general case Q is no longer finite. We prove

Theorem A. *The identity component Q_o of Q is a solvable algebraic group. Furthermore*

$$dim\ Q = dim\ (Cent\ \mathfrak{g}^\chi)_u$$

where

$$Cent\ \mathfrak{g}^\chi = (Cent\ \mathfrak{g}^\chi)_r + (Cent\ \mathfrak{g}^\chi)_u \tag{4}$$

is the Levi decomposition of $Cent\ \mathfrak{g}^\chi$ with the subscripts r and u denoting, respectively, the reductive and nilpotent components. Moreover if M is nilpotent then $(Cent\ \mathfrak{g}^\chi)_r = 0$ so that in this case

$$dim\ Q = dim\ Cent\ \mathfrak{g}^\chi$$

As a corollary note that if e is principal nilpotent in $\mathfrak{g}$ so that $\dim\,\mathfrak{g}^e = l$ where $l = rank\,\mathfrak{g}$ and $\mathfrak{n}^e$ is the normalizer of $\mathfrak{g}^e$ then

$$\dim\,\mathfrak{n}^e = 2l$$

In the general case one may associate to M a distinguished nilpotent orbit $O \subset \mathfrak{g}$.

Theorem B. *There exists a unique nilpotent orbit $O \subset \mathfrak{g}$ — referred to hereafter as the nilpotent orbit corresponding to M — such that for any $\omega \in Symp\,M$ and any $\zeta \in M$ one has*

$$e \in O$$

where e is the nilpotent component of the Jordan decomposition $\mu_\omega(\zeta) = z + e$ of $\mu_\omega(\zeta)$. Furthermore every element in O is obtained this way.

As to the structure of $Symp\,M$ we find that it is an affine space minus a finite number of hyperplanes (generalizing a well-known fact for the case where M is semisimple).

Theorem C. *$Symp\,M$ is an affine variety. In fact there exists a polynomial function p on $Cent\,\mathfrak{g}^\chi$ whose prime factors are linear functionals such that*

$$Symp_\chi M = \{x \in Cent\,\mathfrak{g}^\chi|\ p(x) \neq 0\}$$

Furthermore in the notation of (4) if O is the nilpotent orbit corresponding to M then $O \cap (Cent\,\mathfrak{g}^\chi)_u$ is non-empty and Zariski open in $(Cent\,\mathfrak{g}^\chi)_u$ and with respect to the Levi decomposition (4) one has

$$Symp_\chi M = Reg\,(Cent\,\mathfrak{g}^\chi)_r + O \cap (Cent\,\mathfrak{g}^\chi)_u \tag{5}$$

In particular if M is nilpotent then

$$Symp_\chi M = O \cap Cent\,\mathfrak{g}^\chi$$

Finally with respect to the action of Q on $Symp\,M$ we have

Theorem D. *Let $x \in Symp_\chi M$. Then*

$$Q \cdot x = O \cap Cent\,\mathfrak{g}^\chi$$

Furthermore there exists a subgroup W_χ of the Weyl group of $(Cent\,\mathfrak{g}^\chi)_r$ such that if $x = z + e$ is the Jordan decomposition of x then one has

$z \in Reg\,(Cent\ \mathfrak{g}^\chi)_r$ and $e \in (Cent\ \mathfrak{g}^\chi)_u$ and with respect to the Levi decomposition (4)

$$Q \cdot x = W_\chi \cdot z + O \cap (Cent\ \mathfrak{g}^\chi)_u$$

In particular if M is nilpotent then

$$Q \cdot x = O \cap Cent\ \mathfrak{g}^\chi$$

so that Q is transitive on $Symp\ M$ if M is nilpotent.

1. The variety $Symp\ M$.

Let G be a connected complex semisimple Lie group and let M be a G-homogeneous algebraic variety. If $\chi \in M$ let G^χ be the isotropy subgroup at χ so that as G-homogeneous spaces one has the identification

$$G/G^\chi \xrightarrow{\sim} M$$

Now let N^χ be the normalizer of G^χ in G and let $Q(M) = N^\chi/G^\chi$ so that $Q = Q(M)$ operates on M using the symbol $\bullet$where if $n \in N^\chi$ and $\bar{n}$ is its image in Q then $\bar{n} \bullet gG^\chi = gn^{-1}G^\chi$ for any $g \in G$. We will use the same symbol to denote the induced action of Q on the (complex) tangent bundle $T(M)$ to M. The following proposition is a well-known simple formality.

Proposition 1. *The group Q operates as the group of all (necessarily variety isomorphisms) maps of M into itself which commute with the action of G. Furthermore for any $\chi \in M$ one has*

$$Q \bullet \chi = \{\eta \in M \mid G^\eta = G^\chi\} \tag{6}$$

In what follows we will consider the notion of symplectic structures in the category of complex algebraic varieties. As one knows, the standard facts about symplectic structures and moment maps have immediate complex analogues so that in particular the image of a moment map is in the complex dual space to the relevant complex Lie algebra.

Now let $Symp\ M$ be the set of all G-invariant symplectic structures on M. Henceforth we will assume that M is such that $Symp\ M$ is not empty.

We observe that the set $Symp\ M$ has naturally the structure of a quasi-affine complex algebraic variety. Indeed if $\chi \in M$ then clearly any alternating G-invariant 2-form ω on M is determined by its restriction ω_χ to the tangent space $T_\chi(M)$ to M at χ. It follows then that the set of G-invariant alternating 2-forms on M is a finite-dimensional vector space $\Omega_G^2(M)$ and the subset non-degenerate forms is a Zariski open set U in

$\Omega_G^2(M)$. But the subset of $\Omega_G^2(M)$ consisting of closed forms is a vector subspace $Z_G^2(M)$. Hence $Symp\ M = Z_G^2(M) \cap U$ is Zariski open in $Z_G^2(M)$ and thus acquires a variety structure. Furthermore, if $T_\chi^*(M)$ is the cotangent space to M at χ, then clearly the correspondence $\omega \mapsto \omega_\chi$ defines an injection

$$Symp\ M \longrightarrow (\wedge^2 T_\chi^*(M))^{G^\chi} \tag{7}$$

which is an embedding of varieties. In addition $Symp\ M$ has a Q-action where   if
$n \in N^\chi, \omega \in Symp\ M, \chi \in M$ and $u, v \in T_\chi(M)$ then

$$(\bar{n} \cdot \omega)_\chi(u, v) = \omega_{\bar{n}^{-1} \cdot \chi}(\bar{n}^{-1} \cdot u, \bar{n}^{-1} \cdot v) \tag{8}$$

Clearly Q acts morphically on $Symp\ M$ and (7) is equivariant for the action of $Q \xrightarrow{\sim} N^\chi/G^\chi$.

Now let $\mathfrak{g} = \mathrm{Lie}(G)$ be the Lie algebra of G. We identify $\mathfrak{g}$ with its (complex) dual space using the Killing form $(\ ,\ )$ on $\mathfrak{g}$ and let, for any $\omega \in Symp\ M$,

$$\mu_\omega : M \longrightarrow \mathfrak{g} \tag{9}$$

be the moment map defined by ω. We recall the definition. For any $x \in \mathfrak{g}$ let ξ^x be the (algebraic) vector field on M corresponding to x with respect to the differential of the action of G on M. But since $\mathfrak{g}$ is semisimple the first and second cohomology groups of $\mathfrak{g}$ vanish and hence there exists for each $x \in \mathfrak{g}$ a unique regular function $\varphi^x = \varphi_\omega^x$ whose corresponding hamiltonian vector field (with respect to ω) is ξ^x and which is such that for $x, y \in \mathfrak{g}$ one has (with respect to the Poisson bracket defined by ω) $[\varphi^x, \varphi^y] = \varphi^{[x,y]}$. One then has for $\omega \in Symp\ M, \chi \in M$, and $y \in \mathfrak{g}$,

$$(\mu_\omega(\chi), y) = \varphi_\omega^y(\chi) \tag{10}$$

A familiar example of a homogeneous space M where $Symp(M)$ is not empty is the case where M is an adjoint orbit $O \subset \mathfrak{g}$. As one knows there is then a canonical symplectic structure. This will be denoted by $\omega(O)$. In case $x \in \mathfrak{g}$ we will sometimes write O_x for the adjoint orbit containing x. Throughout the paper $\chi \in M$ can be replaced by $x \in \mathfrak{g}$ — it being then understood that we are taking $M = O_x$. In particular for any $x \in \mathfrak{g}$ one has the G-equivariant identification

$$G/G^x \xrightarrow{\sim} O_x \tag{11}$$

Let G_o^χ be the identity component of G^χ. In fact throughout the paper the subscript o for a Lie group will always denote its identity component. Since G^x is algebraic G_o^x has finite index in G^x. But for the fundamental group $\pi_1(O_x)$ one has the usual isomorphism

$$\pi_1(O_x) \xrightarrow{\sim} G^x/G_o^x \tag{12}$$

so that

$$\pi_1(O_x) \text{ is a finite group} \tag{13}$$

Now for any $\chi \in M$ let

$$\sigma_\chi : Symp\ M \longrightarrow \mathfrak{g} \tag{14}$$

be the map defined by putting $\sigma_\chi(\omega) = \mu_\omega(\chi)$. Our first preliminary description of $Symp\ M$ is given in the last part of

Proposition 2. *Let $\omega \in Symp\ M$. Then the image of the map μ_ω is an adjoint orbit $O \subset \mathfrak{g}$ and the restricted map*

$$\mu_\omega : M \longrightarrow O \tag{15}$$

is a G-equivariant finite covering. Furthermore one has

$$\mu_\omega{}^*(\omega(O)) = \omega \tag{16}$$

Let $\chi \in M$. Then the map

$$\sigma_\chi : Symp\ M \longrightarrow \mathfrak{g} \tag{17}$$

is injective and the image is given by

$$\sigma_\chi(Symp\ M) = \{x \in \mathfrak{g} \mid G_o^x \subset G^\chi \subset G^x\} \tag{18}$$

Proof. A proof that the moment map in the smooth category is equivariant, the image is an orbit, the restricted map is an orbit covering and one has (16) is given in [K2]. The argument easily goes over word for word in the presently considered complex algebraic category. The finiteness of (15) follows from (13). By equivariance $\sigma_\chi(\omega)$ determines μ_ω. But , by (16), μ_ω determines ω. Thus σ_χ is injective. Let $X = \{x \in \mathfrak{g} \mid G_o^x \subset G^\chi \subset G^x\}$. Since (15) is a G-equivariant covering one has $\sigma_\chi(Symp\ M) \subset X$. On the other hand if $x \in X$ then clearly there exists a G-equivariant covering $\mu : M \longrightarrow O_x$ such that $\mu(\chi) = x$. Let $\omega = \mu^*(\omega(O_x))$ so that $\omega \in Symp\ M$. But then clearly $\sigma_\chi(\omega) = x$ and hence $X = \sigma_\chi(Symp\ M)$. QED

For any subset or element S in G let $\mathfrak{g}^S$ be the subspace of $Ad\ S$ invariants in $\mathfrak{g}$. Also for any $\chi \in M$ let

$$\mathfrak{g}^\chi = Lie\ (G^\chi)$$

Obviously $x \in Cent\ \mathfrak{g}^x$ for any $x \in \mathfrak{g}$. In fact one has immediately has

Remark 2.1 If $x, y \in \mathfrak{g}$ then note that $y \in Cent\ \mathfrak{g}^x$ if and only if

$$\mathfrak{g}^x \subset \mathfrak{g}^y \tag{19}$$

Let $\chi \in M$ and put

$$(Cent\ \mathfrak{g}^\chi)^* = \{x \in \mathfrak{g} \mid \mathfrak{g}^x = \mathfrak{g}^\chi\} \tag{20}$$

Proposition 3. *Let $\chi \in M$. Then $(Cent\ \mathfrak{g}^\chi)^*$ is a non-empty Zariski open subset of $Cent\ \mathfrak{g}^\chi$. In addition*

$$(Cent\ \mathfrak{g}^\chi)^* = \{x \in \mathfrak{g} \mid Cent\ \mathfrak{g}^x = Cent\ \mathfrak{g}^\chi\} \tag{21}$$

Moreover for any $y \in Cent\ \mathfrak{g}^\chi$ one has rank ad $y \leq dim\ \mathfrak{g} - dim\ \mathfrak{g}^\chi$ and

$$(Cent\ \mathfrak{g}^\chi)^* = \{x \in Cent\ \mathfrak{g}^\chi \mid rank\ ad\ x = dim\ \mathfrak{g} - dim\ \mathfrak{g}^\chi\} \tag{22}$$

Proof. If $x \in (Cent\ \mathfrak{g}^\chi)^*$ then by definition $\mathfrak{g}^x = \mathfrak{g}^\chi$ so that $Cent\ \mathfrak{g}^x = Cent\ \mathfrak{g}^\chi$. But $x \in Cent\ \mathfrak{g}^x$ and hence $(Cent\ \mathfrak{g}^\chi)^* \subset Cent\ \mathfrak{g}^\chi$. On the other hand

$$\sigma_\chi(Symp\ M) \subset (Cent\ \mathfrak{g}^\chi)^* \tag{23}$$

by (18) so that $(Cent\ \mathfrak{g}^\chi)^*$ is not empty. But obviously *rank ad x = dim* $\mathfrak{g}$ − *dim* $\mathfrak{g}^\chi$ for $x \in (Cent\ \mathfrak{g}^\chi)^*$. However for $y \in Cent\ \mathfrak{g}^\chi$ one obviously has $\mathfrak{g}^\chi \subset Ker\ ad\ y$ so that *rank ad y ≤ dim $\mathfrak{g}$ − dim $\mathfrak{g}^\chi$* and hence $(Cent\ \mathfrak{g}^\chi)^*$ is Zariski open in $Cent\ \mathfrak{g}^\chi$. It remains only to establish (21). Assume that $x \in \mathfrak{g}$ is such that $Cent\ \mathfrak{g}^x = Cent\ \mathfrak{g}^\chi$. But then $Cent\ \mathfrak{g}^x = Cent\ \mathfrak{g}^y$ for $y \in (Cent\ \mathfrak{g}^\chi)^*$. But $\mathfrak{g}^w$ is clearly the centralizer of $Cent\ \mathfrak{g}^w$ for any $w \in \mathfrak{g}$. Thus $x \in (Cent\ \mathfrak{g}^\chi)^*$. QED

Let $\chi \in M$. Put

$$\mathfrak{z}^\chi = \mathrm{Lie}\ (Cent\ G^\chi) \tag{24}$$

Proposition 4. *Let $\chi \in M$. Then*

$$\mathfrak{z}^\chi \subset Cent\ \mathfrak{g}^\chi \tag{25}$$

and in case M is simply connected

$$\mathfrak{z}^\chi = Cent\ \mathfrak{g}^\chi \tag{26}$$

In any case

$$\mathfrak{z}^\chi = \mathfrak{g}^{G^\chi} \tag{27}$$

Proof. Clearly the identity component of $Cent\ G^\chi$ is contained in the identity component of $Cent\ G^\chi_o$ and one has equality if M is simply connected. But these statements imply (25) and (26). To establish (27) in the general case one has only to observe that if $y \in \mathfrak{g}^{G^\chi}$ then $y \in \mathfrak{g}^\chi$. So let $y \in \mathfrak{g}^{G^\chi}$. Then $y \in \mathfrak{g}^{\mathfrak{g}^\chi}$. But $\mathfrak{g}^{\mathfrak{g}^\chi} = Cent\ \mathfrak{g}^\chi$ since one may take $x \in \sigma_\chi(Symp\ M)$ and then by (18) $\mathfrak{g}^\chi = \mathfrak{g}^x$ and of course $x \in \mathfrak{g}^x$. Thus $y \in Cent\ \mathfrak{g}^\chi \subset \mathfrak{g}^\chi$. QED

Now since $Q = N^\chi/G^\chi$ it follows from (27) that $\mathfrak{z}^\chi$ has the structure (depending on the choice of χ) of a Q-module. Let

$$\mathfrak{z}^{\chi*} = (Cent\ \mathfrak{g}^\chi)^* \cap \mathfrak{z}^\chi \tag{28}$$

Recalling (20) it is clear that $\mathfrak{z}^{\chi*}$ is stable under the action of Q.

We can present a much sharper image of $Symp\ M$.

Theorem 5. *Let M be any complex homogeneous space for a simply-connected complex semisimple Lie group G. Assume that the space $Symp\ M$ of all G-invariant symplectic structures on M is not empty; then $Symp\ M$ is a quasi-affine algebraic variety. Let $\chi \in M$ and let G^χ be the isotropy subgroup of G at χ. Let N^χ be the normalizer of G^χ in G and let $Q = N^\chi/G^\chi$. Let $\mathfrak{g} = Lie\ G$ and let $\mathfrak{z}^\chi = \mathfrak{g}^{G^\chi}$. Let the map σ_χ be defined as in (14) and let $\mathfrak{z}^{\chi*} \subset \mathfrak{z}^\chi$ be defined as (28). Then $\mathfrak{z}^{\chi*}$ is a non-empty Zariski open subset of $\mathfrak{z}^\chi$ and $\sigma_\chi(Symp\ M) = \mathfrak{z}^{\chi*}$. Furthermore the restricted map*

$$\sigma_\chi : Symp\ M \longrightarrow \mathfrak{z}^{\chi*} \tag{29}$$

is a Q-equivariant isomorphism of algebraic varieties where the action of Q on $Symp\ M$ is given by (8).

Proof. Let $x \in \mathfrak{g}$. Note that the condition $G^x_o \subset G^\chi \subset G^x$ implies $\mathfrak{g}^x = \mathfrak{g}^\chi$ so that (see (20)) $x \in (Cent\ \mathfrak{g}^\chi)^*$. But also the second inclusion implies $x \in \mathfrak{z}^\chi$. Thus the condition implies that $x \in \mathfrak{z}^{\chi*}$. But the converse is also clear so that one has

$$G^x_o \subset G^\chi \subset G^x \text{ if and only if } x \in \mathfrak{z}^{\chi*} \tag{30}$$

But then $\sigma_\chi(Symp\ M) = \mathfrak{z}^{\chi*}$ by Proposition 2. In particular $\mathfrak{z}^{\chi*}$ is not empty. But then $\mathfrak{z}^{\chi*}$ is necessarily a non-empty Zariski open subset of $\mathfrak{z}^\chi$ by (25) and Proposition 3. However, since σ_χ is injective, by Proposition 2, it follows then that the map (29) is a bijection.

To see that (29) is an algebraic variety isomorphism let $\omega \in Symp\ M$. Since $\mu_\omega{}^*(\omega(O_x)) = \omega$ one has, for any $y, z \in \mathfrak{g}$, the equation

$$(\sigma_\chi(\omega), [y, z]) = \omega_\chi(\xi^z{}_\chi, \xi^y{}_\chi) \tag{31}$$

The proof of (31) in the smooth category is given in [K2]. The same argument readily applies in the complex algebraic category. Now it follows from (31) that (29) is the restriction to *Symp M* of a linear map of vector spaces

$$Z_G^2(M) \longrightarrow \mathfrak{z}^\chi \tag{32}$$

But then since *Symp M* is open dense in $Z_G^2(M)$ and (29) is a bijection then (32) must be a linear isomorphism and (29) must be a variety isomorphism.

In the following we will use the action of G on the tangent bundle $T(M)$ to M corresponding to its action on M. Let $x \mapsto w(x)$ be the inverse map to σ_χ Let $n \in N^\chi$. Then in the notation of (31) $(n \cdot x, [y, z]) = w(n \cdot x)_\chi(\xi^z{}_\chi, \xi^y{}_\chi)$. But $(n \cdot x, [y, z]) = (x, [n^{-1} \cdot y, n^{-1} \cdot z])$. Thus

$$w(n \cdot x)_\chi(\xi^z{}_\chi, \xi^y{}_\chi) = w(x)_\chi(\xi^{n^{-1} \cdot z}{}_\chi, \xi^{n^{-1} \cdot y}{}_\chi)$$

But $\xi^{n^{-1} \cdot z}{}_\chi = n^{-1} \cdot (\xi^z{}_{n \cdot \chi})$ so that

$$w(n \cdot x)_\chi(\xi^z{}_\chi, \xi^y{}_\chi) = w(x)_\chi(n^{-1} \cdot (\xi^z{}_{n \cdot \chi}), n^{-1} \cdot (\xi^y{}_{n \cdot \chi}))$$

But by the invariance of $w(x)$ under the (ordinary) action of n on M one then has

$$w(n \cdot x)_\chi(\xi^z{}_\chi, \xi^y{}_\chi) = w(x)_{n \cdot \chi}(\xi^z{}_{n \cdot \chi}, \xi^y{}_{n \cdot \chi}) \tag{33}$$

But clearly $n \cdot \chi = \bar{n}^{-1} \cdot \chi$. On the other hand since the action of Q on M commutes with the action of G one therefore has $\xi^z{}_{n \cdot \chi} = \bar{n}^{-1} \cdot (\xi^z{}_\chi)$. Consequently

$$w(n \cdot x)_\chi(\xi^z{}_\chi, \xi^y{}_\chi) = w(x)_{\bar{n}^{-1} \cdot \chi}(\bar{n}^{-1} \cdot (\xi^z{}_\chi), \bar{n}^{-1} \cdot (\xi^y{}_\chi))$$

But then $w(n \cdot x) = \bar{n} \cdot w(x)$ by (8) so that (29) is a Q-map. QED

Remark 5.1. If we ignore symplectic structures on M and just consider the question as to which adjoint orbits in $\mathfrak{g}$ are equivariantly covered by M note that Theorem 5 answers (see (30)) this question. That is, an adjoint orbit O is equivariantly covered by M if and only if $O = O_x$ for some $x \in \mathfrak{z}^{\chi *}$.

2. The action of Q

Now using Theorem 5 we can understand the action of Q on *Symp M* by considering its action on $\mathfrak{z}^{\chi *}$. If $\mu : A \longrightarrow B$ is a covering map of connected (say) smooth complex algebraic varieties then one has an embedding of fundamental groups $\pi_1(A) \longrightarrow \pi_1(B)$. The group of deck transformations — i.e. the group of variety isomorphisms of A with itself which preserve the fibers of μ — is then isomorphic to $\mathrm{norm}_{\pi_1(B)}(\pi_1(A))/\pi_1(A)$

where $norm_{\pi_1(B)}(\pi_1(A))$ is the normalizer of $\pi_1(A)$ in $\pi_1(B)$. We denote this group by the triple $Deck(\mu, A, B)$.

Theorem 6. *Let the notation be as in Theorem 5. Let $x \in \mathfrak{z}^{\chi *}$ and let Q^x be the isotropy subgroup of Q at x. Then Q^x is a finite group and one has an isomorphism*

$$Q^x \xrightarrow{\sim} Deck(\mu_{\omega(x)}, M, O_x) \tag{34}$$

using the notation of (22). In particular

$$Q^x \xrightarrow{\sim} \pi_1(O_x) \tag{35}$$

if M is simply-connected.

Furthermore $O_x \cap \mathfrak{z}^{\chi *}$ *is stable under the action of Q. Moreover there are only a finite number of Q_o orbits in $O_x \cap \mathfrak{z}^{\chi *}$ each of which has dimension equal to $\dim Q$ and these orbits are exactly the connected components of $O_x \cap \mathfrak{z}^{\chi *}$*

Proof. By (18) one has

$$G_o^x \subset G^\chi \subset N^\chi \cap G^x \subset G^x \tag{36}$$

noting that $N^\chi \cap G^x$ is the normalizer of G^χ in G^x. Clearly $Q^x = (N^\chi \cap G^x)/G^\chi$. But

$$\begin{aligned}
(N^\chi \cap G^x)/G^\chi &= ((N^\chi \cap G^x)/G_o^x)/(G^\chi/G_o^x) \\
&= norm_{\pi_1(O_x)}(\pi_1(M))/(\pi_1(M)) \\
&= Deck(\mu_{\omega(x)}, M, O_x)
\end{aligned}$$

Now obviously $O_x \cap \mathfrak{z}^{\chi *}$ is stable under the action of N^χ (and hence of Q) since both O_x and $\mathfrak{z}^{\chi *}$ are stable under the action of N^χ. Let $\mathcal{P}$ be set of all subgroups of $\pi_1(O_x)(\simeq G^x/G_o^x)$ which are isomorphic to $\pi_1(M)(\simeq G^\chi/G_o^x)$ Clearly $\pi_1(O_x)$ operates on $\mathcal{P}$ by conjugation of subgroups. Let $\mathcal{R}$ be the set of all $\pi_1(O_x)$ orbits in $\mathcal{P}$. We now define a map

$$\rho : O_x \cap \mathfrak{z}^{\chi *} \longrightarrow \mathcal{R} \tag{37}$$

Let $y \in O_x \cap \mathfrak{z}^{\chi *}$ so that (see proof of Theorem 5) $G_o^y \subset G^\chi \subset G^y$. Let $g \in G$ be such that $g \cdot y = x$. But then $G_o^x \subset gG^\chi g^{-1} \subset G^x$. But then $gG^\chi g^{-1}/G_o^x \in \mathcal{P}$. Moreover g is unique up to left multiplication by an element in G^x. Thus $gG^\chi g^{-1}/G_o^x$ defines an element $\rho(y)$ in $\mathcal{R}$ which in fact depends only on y. We now assert that if $y, z \in O_x \cap \mathfrak{z}^{\chi *}$ then

$$y \text{ and } z \text{ are } Q - \text{conjugate if and only if } \rho(y) = \rho(z) \tag{38}$$

Indeed if $\rho(y) = \rho(z)$ then clearly we can choose $g, h \in G$ so that $g \cdot y = x$ and $h \cdot z = x$ and such that $g G^\chi g^{-1} = h G^\chi h^{-1}$. But then $n g^{-1} = h^{-1}$ for some $n \in N^\chi$ so that $n \cdot y = z$. That is, y and z are Q-conjugate. The argument is clearly reversible establishing (38). Since $\mathcal{R}$ is a finite set it follows then that there are a finite number of Q-orbits in $O_x \cap \mathfrak{z}^{\chi *}$. But then of course there are a finite number of Q_o-orbits. The latter are connected and since all isotropy groups are finite the orbits all have dimension equal to $dim\, Q$. These orbits then, from elementary algebraic group theory, are necessarily the connected components of $O_x \cap \mathfrak{z}^{\chi *}$. QED

We will say that M is *semisimple* if there exists $\omega \in Symp\, M$ such that $O = \mu_\omega(M) \subset \mathfrak{g}$ is an orbit of semisimple elements. That is if $\chi \in M$ then M is semisimple if there exists $x \in \mathfrak{z}^{\chi *}$ such that x is a semisimple element. In such a case one then has the following well-known result.

Proposition 7. *Assume that M is semisimple. Then M is simply-connected. Furthermore if $\chi \in M$ then G^χ is connected and reductive. In particular $\mathfrak{g}^\chi$ is reductive and $\mathfrak{z}^\chi = Cent\, \mathfrak{g}^\chi$ is the Lie algebra of a torus in G. Moreover there exists a homogeneous polynomial function p on $\mathfrak{z}^\chi$ which is a product of linear functionals such that*

$$\mathfrak{z}^{\chi *} = \{x \in \mathfrak{z}^\chi \,|\, p(x) \neq 0\} \tag{39}$$

In particular (using the notation of (22))

$$\mu_{\omega(x)} : M \longrightarrow O_x \tag{40}$$

is an isomorphism for any $x \in \mathfrak{z}^{\chi *}$ and O_x is an orbit of semisimple elements. Finally Q is a finite group, $O_x \cap \mathfrak{z}^{\chi *}$ is a finite set for any $x \in \mathfrak{z}^{\chi *}$ and Q is simply-transitive on $O_x \cap \mathfrak{z}^{\chi *}$. In fact if $\mathfrak{h} \subset \mathfrak{g}^\chi$ is a Cartan subalgebra of $\mathfrak{g}$ so that $\mathfrak{h}_1$ is a Cartan subalgebra of $\mathfrak{g}_1$ where $\mathfrak{g}_1 = [\mathfrak{g}^\chi, \mathfrak{g}^\chi]$ and $\mathfrak{h}_1 = \mathfrak{h} \cap \mathfrak{g}_1$ then Q is isomorphic to the subgroup of the Weyl group of $(\mathfrak{h}, \mathfrak{g})$ which stabilizes a Weyl chamber of $(\mathfrak{h}_1, \mathfrak{g}_1)$.

Proof. For convenience we recall some of the main points in the proof of Proposition 7. O_x is simply-connected, and hence $M = O_x$, since as one knows G^x is connected for any semisimple $x \in \mathfrak{g}$. The set of linear functional prime factors of p is of course the set of restricted roots of $(Cent\, \mathfrak{g}^\chi, \mathfrak{g})$. The given description of Q is an immediate consequence of well-known properties of the Weyl group together with the also well-known fact that any automorphism of the semisimple Lie algebra $\mathfrak{g}_1$ is inner equivalent to one which stabilizes a Weyl chamber of $(\mathfrak{h}_1, \mathfrak{g}_1)$. The simple transitivity of Q on $O_x \cap \mathfrak{z}^{\chi *}$ follows (i) from the fact that Q^x reduces to the identity by (35) and (ii) — transitivity — from (37) and (38) since $\mathcal{R}$ has only one element. QED

We will now see that the situation changes markedly when M covers orbits $O \subset \mathfrak{g}$ which are not orbits of semisimple elements. The dissimilarity will be particularly evident when M covers nilpotent orbits. The change that we wish to focus on has to do with the nature of Q and its action on $\mathfrak{z}^{\chi *}$. Instead of being finite we will find that Q has positive dimension and that it operates transitively on $\mathfrak{z}^{\chi *}$.

We will say that M is *nilpotent* if there exists $\omega \in Symp\ M$ such that $O = \mu_\omega(M)$ is a nilpotent orbit.

Theorem 8. *If $e \in \mathfrak{g}$ is nilpotent then every element in $Cent\ \mathfrak{g}^e$ is nilpotent. In particular if M is nilpotent and $\chi \in M$ then every element in $\mathfrak{z}^\chi$ is nilpotent.*

Let $e \in \mathfrak{g}$ be nilpotent. Assume that $\omega \in Symp\ M$ and $\chi \in M$ is such that $\mu_\omega(\chi) = e$. Then

$$\mathfrak{z}^{\chi *} = O_e \cap \mathfrak{z}^\chi \tag{41}$$

In particular (taking M to be a simply-connected covering of O_e) one has

$$\{z \in \mathfrak{g} \mid \mathfrak{g}^z = \mathfrak{g}^e\} = O_e \cap Cent\ \mathfrak{g}^e \tag{42}$$

Proof. Let $e \in \mathfrak{g}$ be nilpotent. Let $x \in Cent\ \mathfrak{g}^e$ and let $x = y + z$ be the Jordan decompositon of x where y is semisimple and z is nilpotent. But then both $y, z \in Cent\ \mathfrak{g}^e$. Assume $x \in Cent\ \mathfrak{g}^e$ is not nilpotent. Then $0 \neq y$ and hence for some scalar $\lambda \neq 0$ the eigenspace $\mathfrak{g}_\lambda$ for $ad\ y$ corresponding to λ is non-zero. But obviously $\mathfrak{g}_\lambda$ is stable under $ad\ e$ and since e is nilpotent there exists $0 \neq w \in \mathfrak{g}_\lambda$ such that $[e, w] = 0$. That is $w \in \mathfrak{g}^e$. This contradicts the fact that $y \in Cent\ \mathfrak{g}^e$ since $[y, w] = \lambda w \neq 0$. Thus every element in $Cent\ \mathfrak{g}^e$ is nilpotent. If M is nilpotent and $\chi \in M$ then by definition $x = \mu_\omega(\chi)$ is nilpotent for some $\omega \in Symp\ M$. But then every element in $\mathfrak{z}^\chi$ is nilpotent by (25) since $\mathfrak{g}^x = \mathfrak{g}^\chi$.

Now to prove (41) it clearly suffices to prove (42), by definition of $\mathfrak{z}^{\chi *}$. Choose M so that M is a simply-connected covering space of O_e and let $\chi \in M$ lie in the fiber above e. Then $\mathfrak{g}^e = \mathfrak{g}^\chi$ so that $e \in (Cent\ \mathfrak{g}^\chi)^*$. Now since there are only a finite number of nilpotent orbits there exists a unique nilpotent orbit O such that $O \cap Cent\ \mathfrak{g}^e$ is a non-empty Zariski open subset of $Cent\ \mathfrak{g}^e$. But then $O \cap (Cent\ \mathfrak{g}^\chi)^*$ is not empty. But noting, by Proposition 3, that $dim\ O_w \leq dim\ O_e$ for any $w \in Cent\ \mathfrak{g}^e$ and that equality holds if and only if $w \in (Cent\ \mathfrak{g}^\chi)^*$ one has $O \cap Cent\ \mathfrak{g}^e \subset (Cent\ \mathfrak{g}^\chi)^*$. But any element $w \in (Cent\ \mathfrak{g}^\chi)^*$ is in the closure of $O \cap Cent\ \mathfrak{g}^e$. In particular w is in the closure of O. But then $w \in O$ since $dim\ O = dim\ O_w$. Thus $O \cap Cent\ \mathfrak{g}^e = (Cent\ \mathfrak{g}^\chi)^*$. QED

We saw that if M is semisimple then it covers only semisimple orbits — and an infinite number of such orbits. In the nilpotent case not only does M only cover nilpotent orbits but in fact one has

Corollary 9. *Assume that the G-homogeneous space M is nilpotent. Then it covers (equivariantly) only one — necessarily nilpotent — adjoint orbit.*

Proof. This is immediate from Theorem 8 and Remark 5.1. QED

If M is nilpotent, by Corollary 9, we can now use the terminology "its corresponding nilpotent orbit". Another striking dissimilarity with the semisimple case has to do with the nature of the "commuting group" Q. Instead of being finite and related to the Weyl group one has

Theorem 10. *Assume M is nilpotent. Let $\chi \in M$. Then*

$$dim\ Q = dim\ \mathfrak{z}^{\chi} \tag{43}$$

Furthermore Q and also its identity component Q_o operate transitively on $\mathfrak{z}^{\chi *}$ so that in particular if $e \in \mathfrak{z}^{\chi *}$ and Q_o^e is the isotropy subgroup of Q_o at e then Q_o^e is a finite group and

$$Q_o/Q_o^e \xrightarrow{\sim} \mathfrak{z}^{\chi *} \tag{44}$$

as Q_o-homogeneous spaces.

Proof. This is immediate from Theorem 6 since in the present case, being Zariski open, $O_e \cap \mathfrak{z}^{\chi *} = \mathfrak{z}^{\chi *}$ by (41). QED

Remark 10.1. Note that by Theorem 5 we may replace $\mathfrak{z}^{\chi *}$ in Theorem 10 by the space *Symp M* of all G-invariant symplectic structures on M. In particular unlike the case where M is semisimple we find that in the case where M is nilpotent any two G-invariant symplectic structures are conjugate with respect to the commuting group Q.

Now in general if $\chi \in M$ let $\mathfrak{n}^{\chi} = \mathrm{Lie}\ N^{\chi}$ so that

$$\mathfrak{n}^{\chi} \subset Norm\ \mathfrak{g}^{\chi} \tag{45}$$

where $Norm\ \mathfrak{g}^{\chi}$ is the normalizer of $\mathfrak{g}^{\chi}$. Of course one has equality in (45) in case M is simply-connected.

Lemma 10.2. *Let $\chi \in M$ and let $x \in \mathfrak{z}^{\chi *}$. Then one has*

$$\mathfrak{n}^{\chi} = \{y \in \mathfrak{g}|\ [y, x] \in \mathfrak{z}^{\chi}\} \tag{46}$$

Proof. Clearly one has $[\mathfrak{n}^{\chi}, x] \subset \mathfrak{z}^{\chi}$ since $\mathfrak{z}^{\chi}$ is stable under N^{χ}. Now assume $y \in \mathfrak{g}$ and $[y, x] \in \mathfrak{z}^{\chi}$. Let $g \in G^{\chi}$. Then clearly $[(g \cdot y) - y, x] = 0$.

But as $[y, x]$ lies in the center of $\mathfrak{g}^x$ one also has that $[y, \mathfrak{g}^x] \subset \mathfrak{g}^x$. It follows from the Campbell-Hausdorff formula that $g(exp\,y)g^{-1} = (exp\,y)h$ where $h \in G_o^x$. But then $(exp-y)g(exp\,y) \in G^x$ and hence $y \in \mathfrak{n}^x$. QED

Now in the nilpotent case the statement in Theorem 10 that the Q-orbit $(= {}_3{}^{x*})$ of e is open in ${}_3{}^x$ immediately implies

Corollary 11. *Let the notation be as in Theorem 10. Then if* $\mathfrak{n}^x =$ *Lie N^x one has*

$$[\mathfrak{n}^x, e] = {}_3{}^x \tag{47}$$

In particular

$$dim\ \mathfrak{n}^x = dim\ \mathfrak{g}^x + dim\ {}_3{}^x \tag{48}$$

This result for the case where M is taken to be simply-connected is simpler to formulate. The statement is as follows:

Corollary 12. *Let $\mathfrak{g}$ be a complex semisimple Lie algebra and let $e \in \mathfrak{g}$ be nilpotent. Let $\mathfrak{g}^e$ be the centralizer of e in $\mathfrak{g}$ and let $Norm\ \mathfrak{g}^e$ be the normalizer of $\mathfrak{g}^e$ in $\mathfrak{g}$. Then*

$$[Norm\ \mathfrak{g}^e, e] = Cent\ \mathfrak{g}^e \tag{49}$$

so that

$$dim\ Norm\ \mathfrak{g}^e = dim\ \mathfrak{g}^e + dim\ Cent\ \mathfrak{g}^e \tag{50}$$

Corollary 13. *Let the notation be as in Corollary 12. Assume however that e is principal nilpotent. Let $l = rank\ \mathfrak{g}$ so that $dim\ \mathfrak{g}^e = l$. Then*

$$dim\ Norm\ \mathfrak{g}^e = 2l \tag{51}$$

and $Norm\ \mathfrak{g}^e$ is given by

$$Norm\ \mathfrak{g}^e = \{x \in \mathfrak{g}|\ [x, e] \in \mathfrak{g}^e\} \tag{52}$$

Proof. If e is principal nilpotent then one knows that $\mathfrak{g}^e$ is commutative and hence $Cent\ \mathfrak{g}^e = \mathfrak{g}^e$ in this case. See [K1]. But then (50) implies (51). But $G^e = Cent\ G \times G_o^e$ so that $\mathfrak{g}^e = Cent\ \mathfrak{g}^e$ and $Norm\ \mathfrak{g}^e = \mathfrak{n}^e$. Thus (52) follows from (46). QED

3. Invariant vector fields

The differential of the action of Q on M is of course a map of the Lie algebra of Q into the Lie algebra of algebraic vector fields on M. The image is necessarily the set of G-invariant vector fields. Thus (since Q has positive dimension in the nilpotent case) in contrast to the semisimple case where there are no non-trivial G-invariant vector fields one finds a non-trivial Lie algebra of such vector fields in case M is nilpotent.

Let $\mathcal{I}$ denote the Lie algebra of all (necessarily algebraic) G-invariant vector fields on M. Let $\chi \in M$ and let $\mathcal{I}_\chi = \mathfrak{n}^\chi/\mathfrak{g}^\chi$. On the one hand there is clearly a natural isomorphism

$$\mathcal{I}_\chi \xrightarrow{\ \sim\ } \operatorname{Lie} Q \tag{53}$$

On the other hand if we identify $\mathfrak{g}/\mathfrak{g}^\chi$ with the tangent space $T_\chi(M)$ then as an immediate consequence of properties of the normalizer one has

$$\mathcal{I}_\chi = (T_\chi(M))^{G^\chi} \tag{54}$$

It follows easily then that evaluation at χ defines a linear isomorphism

$$\mathcal{I} \xrightarrow{\ \sim\ } \mathcal{I}_\chi \quad \eta \mapsto \eta_\chi \tag{55}$$

Now let $e \in \mathfrak{z}^{\chi\,*}$ so that $O = O_e$ is the nilpotent orbit corresponding to M. In the notation of (34) let $\mu = \mu_{\omega(e)}$ so that $\mu : M \longrightarrow O$ is the G-equivariant covering map such that $\mu(\chi) = e$. For any $\zeta \in M$ let $d\mu_\zeta$ be the differential of μ at ζ so that if we identify $T_e(\mathfrak{g})$ with $\mathfrak{g}$ in a usual manner — using directional differentiation — one has the linear isomorphism

$$d\mu_\chi : \mathfrak{g}/\mathfrak{g}^e \longrightarrow [g, e] \quad \tilde{x} \mapsto [x, e] \tag{56}$$

where, for $x \in \mathfrak{g}$, $\tilde{x}$ denotes its image in $\mathfrak{g}/\mathfrak{g}^e$. But then by (47) one has the linear isomorphism

$$d\mu_\chi : \mathcal{I}_\chi \longrightarrow \mathfrak{z}^\chi \tag{57}$$

We have proved

Proposition 14. *If $e \in \mathfrak{g}$ is nilpotent then $\operatorname{Cent} \mathfrak{g}^e$ is tangent to the adjoint orbit O_e at e. In particular if M is nilpotent, $\chi \in M$ and $e \in \mathfrak{z}^{\chi\,*}$ then $\mathfrak{z}^\chi = \mathfrak{g}^{G^\chi}$ is tangent at e to the nilpotent orbit O corresponding to M.*

Proof. We have already proved the second statement. The first statement clearly follows by choosing M to be simply-connected. QED

Remark 14.1. Note that Proposition 14 is false if M is semisimple. In that case $\mathfrak{z}^\chi$ is in fact transversal to O_x at any $x \in \mathfrak{z}^{\chi\,*}$

Now let $R = R(M)$ denote the ring of regular functions on M. We may identify R with the ring of all regular functions ϕ on G such that $\phi(gh) = \phi(g)$ for all $g \in G$ and $h \in G^\chi$. The action of $G \times Q$ on M induces the structure of a $G \times Q$- module on R and any $\phi \in R$ is $G \times Q$-finite. Solely as a G-module consider the "primary" component of R which transforms under G according to the adjoint representation of G. Quotations marks are used since we are not assuming that $\mathfrak{g}$ is simple. Nonetheless one may still consider the analogous vector space $Hom_G(\mathfrak{g}, R)$ and note that the action of Q on R induces a Q-module structure on $Hom_G(\mathfrak{g}, R)$. But now for any $w \in \mathfrak{z}^\chi$ one clearly obtains an element $\alpha_w \in Hom_G(\mathfrak{g}, R)$ by putting, for $x \in \mathfrak{g}$ and $g \in G$,

$$(\alpha_w(x))(g) = (x, g \cdot w) \tag{58}$$

and that furthermore one has a linear map

$$\Gamma : \mathfrak{z}^\chi \longrightarrow Hom_G(\mathfrak{g}, R) \quad w \mapsto \alpha_w \tag{59}$$

Obviously $\mathfrak{z}^\chi$ has the structure of a $Q = N^\chi/G^\chi$-module.

Proposition 15. *For any M (i.e. not necessarily nilpotent) the map Γ (see (59)) is a Q-isomorphism.*

Proof. If $\phi \in R$, $g \in G$ and $(a, \bar{n}) \in G \times Q$ for $n \in N^\chi$ then

$$((a, \bar{n}) \cdot \phi)(g) = \phi(a^{-1}gn) \tag{60}$$

It follows easily then from (58) that Γ is a Q-map. It is also clearly injective. On the other hand if $\alpha \in Hom_G(\mathfrak{g}, R)$ let $w \in \mathfrak{g}$ be defined by the relation $(\alpha(x))(\chi) = (x, w)$ for any $x \in \mathfrak{g}$. But then, as a function on G,

$$\begin{aligned}
(\alpha(x))(g) &= (\alpha(x))(g \cdot \chi) \\
&= (g^{-1} \cdot \alpha(x))(\chi) \\
&= (\alpha(g^{-1} \cdot x))(\chi) \\
&= (g^{-1} \cdot x, w) \\
&= (x, g \cdot w)
\end{aligned}$$

But then clearly $w \in \mathfrak{g}^{G^\chi} = \mathfrak{z}^\chi$ and $\alpha = \alpha_w$. QED

Now if M is nilpotent it follows from Proposition 15, (55) and (57) that $\mathfrak{z}^\chi$ parameterizes both $\mathcal{I}$ and $Hom_G(\mathfrak{g}, R)$ and hence there is a natural isomorphism
$\gamma : Hom_G(\mathfrak{g}, R) \longrightarrow \mathcal{I}$. More precisely let $\beta : \mathfrak{z}^\chi \longrightarrow \mathcal{I}$ be the isomorphism be defined so that for $\eta \in \mathcal{I}$ then $\beta^{-1}(\eta) = d\mu_\chi(\eta_\chi)$. Then the isomorphism γ is such that the following diagram is commutative.

$$(61)$$

We wish to be much more explicit about γ. Let $\{y_i\}$ be a basis of $\mathfrak{g}$ and let $\{\partial_i\}$ be the corresponding constant coefficient vector fields on $\mathfrak{g}$. Let $\{x_i\}$ be the basis of $\mathfrak{g}$ which is dual to the $\{y_i\}$ with respect to the Killing form. Now for any $\alpha \in Hom_G(\mathfrak{g}, R)$ let $\phi_i^\alpha = \alpha(x_i) \in R$.

Theorem 16. *Assume M is nilpotent and $\mu : M \longrightarrow O$ is a G-covering. For any $\alpha \in Hom_G(\mathfrak{g}, R)$ and $\zeta \in M$ one has that the vector $\sum_{i=1} \phi_i^\alpha(\zeta)(\partial_i)_{\mu(\zeta)}$ is tangent to O at $\mu(\zeta)$. Furthermore $\eta = \gamma(\alpha) \in \mathcal{I}$ may be given by the condition that*

$$d\mu_\zeta(\eta_\zeta) = \sum_{i=1} \phi_i^\alpha(\zeta)(\partial_i)_{\mu(\zeta)} \tag{62}$$

Proof. The map $m : M \longrightarrow T(\mathfrak{g})$ given by putting

$$m(\zeta) = \sum_{i=1} \phi_i^\alpha(\zeta)(\partial_i)_{\mu(\zeta)}$$

clearly satisfies the invariance condition

$$m(g \cdot \zeta) = g \cdot m(\zeta) \tag{63}$$

Thus to prove the first statement it suffices to prove the statement for $\zeta = \chi$. But clearly regarding x_i as a linear function of $\mathfrak{g}$ one has $(m(\chi))(x_i) = \phi_i^\alpha(\chi)$. But now $\alpha = \alpha_w$ for some $w \in \mathfrak{z}^\chi$ by Proposition 15 and hence for any $x \in \mathfrak{g}$, regarded also as a linear function on $\mathfrak{g}$, one has $(m(\chi))(x) = (\alpha_w(x))(\chi) = (x, w)$ by (58). Thus for $\alpha = \alpha_w$ one has

$$m(\chi) = w \in \mathfrak{z}^\chi \tag{64}$$

regarded as a tangent vector at $e = \mu(\chi)$. But then $m(\chi)$ is tangent to O at e by Proposition 14. But now since μ is a covering it follows from (63) that there exists a unique $\xi \in \mathcal{I}$ such that $m(\zeta) = d\mu_\zeta(\xi_\zeta)$ for any $\zeta \in M$. But then $d\mu_\chi(\xi_\chi) = w$. Thus $\xi = \beta(w)$. But this implies, by definition of γ, that $\gamma(\alpha_w) = \xi$. QED

Let $S(\mathfrak{g}^*)$ denote the symmetric algebra of polynomial functions on $\mathfrak{g}$ and let $\mathcal{J}$ denote the space of all G-invariant vector fields on $\mathfrak{g}$. For any $\tau \in Hom_G(\mathfrak{g}, S(\mathfrak{g}^*))$ one clearly obtains an element $\xi^\tau \in \mathcal{J}$ by putting

$$\xi^\tau = \sum_{i=1} \psi_i^\tau \partial_i \tag{65}$$

where using notation implicit in (62) one puts $\psi_i^\tau = \tau(x_i)$ so that $\xi^\tau(x) = \tau(x)$ for any $x \in \mathfrak{g}$. One easily sees that the map

$$Hom_G(\mathfrak{g}, S(\mathfrak{g}^*)) \longrightarrow \mathcal{J}, \quad \tau \mapsto \xi^\tau \tag{66}$$

is a linear isomorphism; see also [R].

Theorem 16.5. *Any G-invariant vector field on $\mathfrak{g} = Lie\ G$ is tangent to any nilpotent orbit in $\mathfrak{g}$. Furthermore if $e \in \mathfrak{g}$ is nilpotent and we put $\mathcal{J}_e = \{\xi_e|\ \xi \in \mathcal{J}\}$ then, using the notation of Proposition 14 where we take $M = O_e$ and let μ equal to the identity map (so that $\chi = e$), one has*

$$\mathcal{J}_e \subset \mathfrak{z}^e \tag{67}$$

and

$$\mathcal{J}_e = \mathfrak{z}^e \tag{68}$$

in case the closure $\overline{O}_e$ is a normal variety. In any case one has dim $\mathcal{J}_e \leq$ rank $\mathfrak{g}$ so that certainly dim $\mathfrak{z}^e \leq$ rank $\mathfrak{g}$ in case $\overline{O}_e$ is a normal variety.

Proof. Let $e \in \mathfrak{g}$ be nilpotent and in Theorem 16 let $M = O_e$, μ be the identity map and $\zeta = \chi = e$. Now if $\tau \in Hom_G(\mathfrak{g}, S(\mathfrak{g}^*))$ let $\alpha = \alpha^\tau \in Hom_G(\mathfrak{g}, R(O_e))$ be defined so that $\alpha(x) = \tau(x)|O_e$ for any $x \in \mathfrak{g}$. But then $\gamma(\alpha)_e = (\xi^\tau)_e$ by (62) so that $(\xi^\tau)_e$ is tangent to O_e at e by Theorem 16. Furthermore $(\xi^\tau)_e \in \mathfrak{z}^e$ by (64). This establishes (67). But now if $\overline{O}_e$ is a normal variety then $R(O_e) = S(\mathfrak{g}^*)|O_e$ so that any $\alpha \in Hom_G(\mathfrak{g}, R(O_e))$ is of the form $\alpha = \alpha^\tau$ for some $\tau \in Hom_G(\mathfrak{g}, S(\mathfrak{g}^*))$ and hence $\mathcal{J}_e = \mathfrak{z}^e$ by the commutative diagram (61) of isomorphisms. But now if $\mathcal{N} \subset \mathfrak{g}$ is the variety of all nilpotent elements in $\mathfrak{g}$ then $dim\ Hom_G(\mathfrak{g}, S(\mathfrak{g}^*)|\mathcal{N}) = rank\ \mathfrak{g}$ by [K1]. Thus $dim\ \mathcal{J}_e \leq rank\ \mathfrak{g}$. QED

4. Nilpotent homogeneous spaces

Throught this section we assume that M is a nilpotent homogeneous space, $\chi \in M$ and that $e \in \mathfrak{z}^{\chi\,*}$ is such that $\mathfrak{g}^e = \mathfrak{g}^\chi$. In addition we assume that M is not a point or equivalently that e is non-zero.

Let $h \in \mathfrak{g}$ be semisimple and $f \in \mathfrak{g}$ be nilpotent such that (h, e, f) is an S-triple so that in particular $[h, e] = 2e$. Let $\mathfrak{a} \subset \mathfrak{g}$ be the TDS spanned by h, e and f. If U is an h-stable subquotient of an $\mathfrak{a}$-module then the eigenvalues of h on U are integers. For any $k \in \mathbb{Z}$ let U_k denote the corresponding k-eigensubspace of U. Clearly $\mathfrak{g}^e$ is ad h stable and one knows $\mathfrak{g}^e_k = 0$ for $k \leq -1$ so that one has the direct sum

$$\mathfrak{g}^e = \sum_{k=0} \mathfrak{g}^e_k \tag{69}$$

One then notes that the relation $[h, e] = 2e \in \mathfrak{z}^\chi$ implies that

$$h \in \mathfrak{n}^\chi \tag{70}$$

by (46). In particular $\mathfrak{z}^\chi \subset \mathfrak{g}^e$ is ad h stable.

Proposition 17. *One has* $\mathfrak{z}^\chi_k = 0$ *for* $k \leq 1$ *so that*

$$\mathfrak{z}^\chi = \sum_{k=2} \mathfrak{z}^\chi_k \tag{71}$$

Proof. One has $\mathfrak{z}^\chi \subset Cent\ \mathfrak{g}^e$ and equality holds if M is simply-connected. Thus it suffices to assume that $\mathfrak{z}^\chi = Cent\ \mathfrak{g}^e$. But clearly $Cent\ \mathfrak{g}^e_0 \subset Cent\ \mathfrak{g}^\mathfrak{a}$ where $\mathfrak{g}^\mathfrak{a}$ is the centralizer of $\mathfrak{a}$ in $\mathfrak{g}$. But if B is the Killing form on $\mathfrak{g}$ then, since $\mathfrak{a}$ is semisimple,

$$B|\mathfrak{g}^\mathfrak{a} \text{ is non-singular} \tag{72}$$

and hence $\mathfrak{g}^\mathfrak{a}$ is reductive. But then all elements in $Cent\ \mathfrak{g}^\mathfrak{a}$ are semisimple. This implies that $Cent\ \mathfrak{g}^e_0 = 0$ by Theorem 8. Now assume that $0 \neq u \in Cent\ \mathfrak{g}^e_1$. Let $\mathfrak{r}$ be the primary component in $\mathfrak{g}$ corresponding to the irreducible 2-dimensional representation of $\mathfrak{a}$ so that $u \in \mathfrak{r}$ and $0 \neq [f, u] \in \mathfrak{r}_{-1}$. But clearly $B|\mathfrak{r}$ is non-singular and hence there exists $w \in \mathfrak{r}_1$ such that $(w, [f, u]) = 1$. But then $([u, w], f) = 1$ so that $[u, w] \neq 0$. This is a contradiction since, clearly $w \in \mathfrak{g}^e$, and by assumption $u \in Cent\ \mathfrak{g}^e$. Thus $Cent\ \mathfrak{g}^e_1 = 0$. QED

It is of course a simple fact that $\mathfrak{g}^\chi_k = 0$ for $k \leq -1$. The following result asserts, somewhat surprisingly, that the same result is true for $\mathfrak{n}^\chi$. Also one finds an explicit complement for $\mathfrak{g}^\chi$ in $\mathfrak{n}^\chi$. Let

$$\mathfrak{d} = [f, \mathfrak{z}^\chi] \tag{73}$$

Note that $\mathfrak{d}$ is stable under $ad\ h$ and that $\mathfrak{d}_k = 0$ for $k < 0$ by (71) so that

$$\mathfrak{d} = \sum_{k=0} \mathfrak{d}_k \tag{74}$$

Theorem 18. *One has* $\mathfrak{n}^\chi_k = 0$ *for negative* k *so that*

$$\mathfrak{n}^\chi = \sum_{k=0} \mathfrak{n}^\chi_k \tag{75}$$

Furthermore

$$dim\ \mathfrak{d} = dim\ \mathfrak{z}^\chi \tag{76}$$

and one has the direct sum of vector spaces

$$\mathfrak{n}^\chi = \mathfrak{d} \oplus \mathfrak{g}^\chi \qquad (77)$$

Proof. Clearly $[[f, \mathfrak{z}^\chi], e] = [h, \mathfrak{z}^\chi] \subset \mathfrak{z}^\chi$ and hence $[f, \mathfrak{z}^\chi] \subset \mathfrak{n}^\chi$ by (46). On the other hand $[f, \mathfrak{z}^\chi] \cap \mathfrak{g}^\chi = 0$ since ad e is injective on the image of ad f. But one has (76) since the spectrum of ad h is positive on $\mathfrak{z}^\chi$ by Proposition 17 but of course the spectrum of ad h on the kernel of ad f is non-positive. But then (77) follows by dimension, using (48). The first statement follows from (77) and (74). QED

In the proof above we have implicitly used that the decomposition (77) is *ad h*- stable. Explicitly we emphasize now that $\mathfrak{d}_0 = [f, \mathfrak{z}^\chi_2]$.

Proposition 19. *One has that $\mathfrak{n}^\chi_0$ is a Lie algebra of $\mathfrak{n}^\chi$. In fact*

$$\mathfrak{n}^\chi_0 = \mathfrak{d}_0 \oplus \mathfrak{g}^\mathfrak{a} \qquad (78)$$

is a direct sum of Lie algebras and $\mathfrak{d}_0$ is abelian. Furthermore $\mathfrak{d}_0$ is the B-orthocomplement of $\mathfrak{g}^\mathfrak{a}$ in $\mathfrak{n}^\chi_0$ where B is the Killing form on $\mathfrak{g}$.

Proof. One obtains (78) as a linear space direct sum by intersecting (77) with $\mathfrak{g}_0$ noting that $\mathfrak{g}^\mathfrak{a} = \mathfrak{g}^e \cap \mathfrak{g}^h$. Since f and $\mathfrak{z}^\chi$ both commute with $\mathfrak{g}^\mathfrak{a}$ it follows from (72) and the invariance of B that $\mathfrak{d}_0$ is the B-orthocomplement of $\mathfrak{g}^\mathfrak{a}$ in $\mathfrak{n}^\chi_0$. Thus $\mathfrak{d}_0$ is a Lie algebra since obviously $\mathfrak{n}^\chi_0$ is a Lie algebra and hence (78) is a Lie algebra direct sum. But it also follows that $[\mathfrak{d}_0, \mathfrak{g}^\mathfrak{a}] = 0$ which then implies that $[\mathfrak{d}_0, \mathfrak{d}_0]$ is B-orthogonal to $\mathfrak{g}^\mathfrak{a}$. Now assume $w_1, w_2 \in \mathfrak{d}_0$. Then $w_i = [f, z_i]$, $i = 1, 2$ for $z_i \in \mathfrak{z}^\chi_2$. Now $[e, [w_1, w_2]] = [e, [[f, z_1], [f, z_2]]]$. But $[e, [f, z_i]] = [h, z_i] = 2z_i$ so that $[e, [w_1, w_2]] = 2([z_1, [f, z_2]] + [[f, z_1], z_2])$ so that $[e, [w_1, w_2]] = 2[f, [z_1, z_2]] = 0$. Hence $[\mathfrak{d}_0, \mathfrak{d}_0] \subset \mathfrak{g}^e$. But $\mathfrak{d}_0 \subset \mathfrak{g}^h$ and hence $[\mathfrak{d}_0, \mathfrak{d}_0] \subset \mathfrak{g}^\mathfrak{a}$. This however implies by (72) that $[\mathfrak{d}_0, \mathfrak{d}_0] = 0$. QED

Let $A \subset G$ be the subgroup which corresponds to $\mathfrak{a}$. Also let $\kappa \in A$ be such that $\kappa \cdot h = -h$, $\kappa \cdot e = -f$ and $\kappa \cdot f = -e$ so that if $\zeta = \kappa \cdot \chi \in M$ then clearly

$$\kappa \cdot \mathfrak{z}^\chi_2 = \mathfrak{z}^\zeta_{-2} \qquad (79)$$

Now let $\mathfrak{s} \subset \mathfrak{g}$ be the $\mathfrak{a}$-(or A) submodule generated by $\mathfrak{z}^\chi_2$. Thus recalling that $\mathfrak{d}_0 = [f, \mathfrak{z}^\chi_2]$ one has that if

$$d = dim \; \mathfrak{z}^\chi_2 \qquad (80)$$

then $dim \; \mathfrak{s} = 3d$ and that, as an A-module, $\mathfrak{s}$ is equivalent to d copies of the irreducible 3-dimensional representation of $\mathfrak{a}$. Furthermore

$$\mathfrak{s}_2 = \mathfrak{z}^{\chi}_2$$
$$\mathfrak{s}_0 = \mathfrak{d}_0,$$
$$\mathfrak{s}_{-2} = \mathfrak{z}^{\varsigma}_{-2} \tag{81}$$

Remark 19.1. Note that since $e \in \mathfrak{s}_2$ one has $\mathfrak{a} \subset \mathfrak{s}$ with $h \in \mathfrak{s}_0$ and $f \in \mathfrak{s}_{-2}$.

We now find that $\mathfrak{s}$ is not only a Lie subalgebra of $\mathfrak{g}$ but in fact

Proposition 20. *One has that (1) $\mathfrak{s}$ is a semisimple Lie subalgebra of $\mathfrak{g}$ and $\mathfrak{s}$ is isomorphic to a direct sum of d copies of $\mathfrak{sl}(2, \mathbb{C})$, (2) $\mathfrak{d}_0$ is a Cartan subalgebra of $\mathfrak{s}$, and (3) relative to $(\mathfrak{d}_0, \mathfrak{s})$ the space $\mathfrak{s}_2 = \mathfrak{z}^{\chi}_2$ is the span of a choice of positive (in fact simple positive) root vectors.*

Proof. We will first prove that $\mathfrak{s}$ is a Lie subalgebra. Since $\mathfrak{z}^{\chi}_2$ lies in $Cent \; \mathfrak{g}^e$ one has $[\mathfrak{z}^{\chi}_2, \mathfrak{z}^{\chi}_2] = 0$. Applying κ one then has $[\mathfrak{z}^{\varsigma}_{-2}, \mathfrak{z}^{\varsigma}_{-2}] = 0$. We have already observed in Proposition 19 that $\mathfrak{d}_0$ is a Lie subalgebra. Since of course $\mathfrak{n}^{\chi}$ normalizes $\mathfrak{z}^{\chi}$ one has $[\mathfrak{n}^{\chi}_0, \mathfrak{z}^{\chi}_2] \subset \mathfrak{z}^{\chi}_2$. But then $[\mathfrak{d}_0, \mathfrak{z}^{\chi}_2] \subset \mathfrak{z}^{\chi}_2$ by (78). On the other hand since κ operates as -1 on a zero weight vector of a 3-dimensional representation of $\mathfrak{a}$ one has

$$\kappa \cdot \mathfrak{d}_0 = \mathfrak{d}_0 \tag{82}$$

and hence upon applying κ one also has $[\mathfrak{d}_0, \mathfrak{z}^{\varsigma}_{-2}] \subset \mathfrak{z}^{\varsigma}_{-2}$. Thus to show that $\mathfrak{s}$ is a Lie subalgebra it suffices to show that $[\mathfrak{z}^{\chi}_2, \mathfrak{z}^{\varsigma}_{-2}] \subset \mathfrak{d}_0$. Now by the representation theory of $\mathfrak{a}$ one has

$$[\mathfrak{d}_0, e] = \mathfrak{z}^{\chi}_2 \tag{83}$$

and

$$[e, \mathfrak{z}^{\varsigma}_{-2}] = \mathfrak{d}_0 \tag{84}$$

But since $\mathfrak{z}^{\varsigma}_{-2}$ is stable under $ad \; \mathfrak{d}_0$ it follows from (83), upon applying $ad \; \mathfrak{d}_0$ to (84), that

$$[\mathfrak{z}^{\chi}_2, \mathfrak{z}^{\varsigma}_{-2}] \subset \mathfrak{d}_0 \tag{85}$$

so that $\mathfrak{s}$ is a Lie subalgebra. Next we observe that $\mathfrak{s}$ is algebraic (i.e the corresponding subgroup in G is algebraic). Indeed, by its definition, $\mathfrak{z}^{\chi}_2$ is algebraic and $\mathfrak{a}$ is algebraic since $\mathfrak{a}$ is semisimple. On the other hand it is obvious, from the definition of $\mathfrak{s}$, that $\mathfrak{s}$ is generated by $\mathfrak{z}^{\chi}_2$ and $\mathfrak{a}$. It follows then that $\mathfrak{s}$ is algebraic by [B, Corollary 7.7, (9), P.193]. Let $\mathfrak{s} = \mathfrak{s}_r + \mathfrak{s}_u$ be the Levi decomposition of $\mathfrak{s}$ where $\mathfrak{s}_r$ is reductive and $\mathfrak{s}_u$ is the nilradical. Since $\mathfrak{a}$ is semisimple we may choose $\mathfrak{s}_r$ so that $\mathfrak{a} \subset \mathfrak{s}_r$. But since $\mathfrak{s}$ is a primary $\mathfrak{a}$ -module for the irreducible 3-dimensional representation of $\mathfrak{a}$ the same statement is necessarily true for $Cent\ \mathfrak{s}_r$. But this implies that $Cent\ \mathfrak{s}_r = 0$ so that $\mathfrak{s}_r$ is semisimple. To show that $\mathfrak{s}$ is semisimple it suffices then to show that $\mathfrak{s}_u = 0$. Assume that $\mathfrak{s}_u \neq 0$. But clearly $\mathfrak{s}_u$ is a primary $\mathfrak{a}$-module and hence there exists $0 \neq w \in \mathfrak{s}_u \cap \mathfrak{z}^x_2$. Note then that $w \in Cent\ \mathfrak{g}^e$. But now, as one knows, (Weisfeiler, Borel-Tits [B-T]) there exists a parabolic subalgebra of $\mathfrak{g}$ with a Levi factor which contains $\mathfrak{s}_r$ and whose nilradical contains $\mathfrak{s}_u$. That is there exists a real semisimple element $x \in \mathfrak{g}$ such that (i) $[x, \mathfrak{s}_r] = 0$ and (ii) $[x, w] \neq 0$ (since w is contained in the nilradical of the parabolic subalgebra). But $x \in \mathfrak{g}^e$ by (i) since $\mathfrak{a} \subset \mathfrak{s}_r$. But then (ii) is a contradiction since $w \in Cent\ \mathfrak{g}^e$. Thus $\mathfrak{s}$ is semisimple. But now clearly $\mathfrak{d}_0$ is the centralizer of h in $\mathfrak{s}$. Since $\mathfrak{s}$ is semisimple this implies that $\mathfrak{d}_0$ is reductive and contains a Cartan subalgebra of $\mathfrak{s}$. But $\mathfrak{d}_0$ is commutative by Proposition 9. Hence $\mathfrak{d}_0$ is a Cartan subalgebra of $\mathfrak{s}$. But $dim\ \mathfrak{d}_0 = d$ so that there are $2d$ simple (positive and negative) roots. But $dim\ \mathfrak{s} = 3d$. Thus the Dynkin diagram of $\mathfrak{s}$ is totally disconnected and hence $\mathfrak{s}$ is isomorphic to the direct sum of d copies of $\mathfrak{sl}(2, \mathbb{C})$. Obviously then $\mathfrak{z}^x_2$ is spanned by a choice of $(\mathfrak{d}_0, \mathfrak{s})$ simple root vectors. QED

Let $\mathfrak{n}^x_+ = \sum_{k>0} \mathfrak{n}^x_k$ so that $\mathfrak{n}^x_+$ is an ideal in $\mathfrak{n}^x$ and let $N^x_+ \subset N^x$ be the corresponding normal (and unipotent) subgroup. Let $\mathfrak{z}^x_{++} = \sum_{k>2} \mathfrak{z}^x_k$.

Lemma 21. *One has*

$$N^x_+ \cdot e = e + \mathfrak{z}^x_{++} \tag{86}$$

and if $n \in N^x_+$ then $n \in G^x$ if and only if $n \cdot e = e$.

Proof. By (47) one immediately has

$$[\mathfrak{n}^x_+, e] = \mathfrak{z}^x_{++} \tag{87}$$

Since $\mathfrak{z}^x_{++}$ is certainly stable under the action of N^x_+ it then follows that $N^x_+ \cdot e$ is Zariski open and dense in $e + \mathfrak{z}^x_{++}$. But $N^x_+ \cdot e$ is closed since N^x_+ is unipotent. This establishes (86).

Since $G^x \subset G^e$ then $n \in G^x$ implies $n \cdot e = e$. Conversely if $n \cdot e = e$ for $n \in N^x_+$ then $n \in G^e_o$ since n is unipotent. But $G^e_o = G^x_o$ and hence $n \in G^x$. QED

By (78) (and Remark 19.1) one clearly has

$$H \subset D_0 \subset N^x_o \tag{88}$$

where $D_0 \subset G$ is the subgroup corresponding to $\mathfrak{d}_0$ and H is the 1- dimensional torus in G corresponding to $\mathbb{C}h$. Since $\mathfrak{d}_0$ is a Cartan subalgebra of the semisimple Lie algebra $\mathfrak{s}$, by Proposition 20, note that D_0 is a torus in G and that in fact one has an isomorphism of algebraic groups

$$D_0 \xrightarrow{\sim} T^d \tag{89}$$

where T^d denotes a d-torus.

Let $(N_o^\chi)^H$ be the commuting group of H in N_o^χ

Proposition 22. *The subgroup $(N_o^\chi)^H$ of N_o^χ is a connected reductive algebraic subgroup of G. Furthermore N^χ_+ is the unipotent radical of N_o^χ and $(N_o^\chi)^H$ is a Levi factor so that*

$$N_o^\chi = (N_o^\chi)^H \ltimes N^\chi_+ \tag{90}$$

is a Levi decomposition of N_o^χ. In addition

$$Lie\ (N_o^\chi)^H = \mathfrak{n}^\chi_0 \tag{91}$$

so that

$$\mathfrak{n}^\chi = \mathfrak{n}^\chi_0 + \mathfrak{n}^\chi_+ \tag{92}$$

is the corresponding Levi decomposition of $\mathfrak{n}^\chi$.

Proof. Let K be the subgroup of G corresponding to $\mathfrak{n}^\chi_0$. Then if $G_o^\mathfrak{a}$ is the subgroup of G which corresponds to $\mathfrak{g}^\mathfrak{a}$ one has the commuting product

$$K = D_0 G_o^\mathfrak{a} \tag{93}$$

by Proposition 19. But D_0 is a torus and of course $G_o^\mathfrak{a}$ is reductive so that K is a connected algebraic reductive subgroup of G. But $N_o^\chi = K N^\chi_+$ since in any case one has the semi-direct sum (92). But obviously $K \subset (N_o^\chi)^H$. On the other hand $(N_o^\chi)^H \cap N^\chi_+ = \{1\}$ since $ad\ h$ has a positive spectrum on $\mathfrak{n}^\chi_+$. Thus

$$K = (N_o^\chi)^H \tag{94}$$

and hence one has the semi-direct product (90) which necessarily is then a Levi decomposition of N_o^χ. QED

Now recalling Proposition 20 let e_i, $i = 1, \ldots, d$, be a root vector (with respect to $(\mathfrak{d}_0, \mathfrak{s})$) basis of $\mathfrak{z}^\chi_2$. Note that the e_i may be normalized so that

$$e = e_1 + \cdots + e_d \tag{95}$$

Indeed otherwise e has a zero coefficient with respect to some e_i — readily contradicting the fact that $\mathfrak{s}$ is a primary $\mathfrak{a}$-module. Now let β_i, $i =$

$1, \ldots, d$, be the linear functionals on $\mathfrak{z}^{\chi}$ defined so that for any $y \in \mathfrak{z}^{\chi}$ one has for some $w \in \mathfrak{z}^{\chi}_{++}$

$$y = w + \sum_{i=1}^{d} \beta_i(y)e_i \tag{96}$$

For any subset $S \subset N^{\chi}$ let $\overline{S}$ denote its image in $Q = N^{\chi}/G^{\chi}$. Let $Q_+ = \overline{N^{\chi}_+}$ and let Q_o^H be the commuting group of $\overline{H}$ in Q_o.

We can now establish the following result concerning the structure of the identity component Q_o of Q.

Theorem 23. *The connected algebraic group Q_o is solvable. Furthermore Q_+ is the unipotent radical of Q_o and Q_o^H is a Levi factor so that*

$$Q_o = Q_o^H \ltimes Q_+ \tag{97}$$

is a Levi decomposition. Moreover Q_o^H is a torus. In fact

$$Q_o^H \xrightarrow{\sim} T^d \tag{98}$$

where $d = \dim \mathfrak{z}^{\chi}_2$ and T^d is a d-dimensional torus. Finally (see Theorem 10) one has $Q_o^e \subset Q_o^H$ and an isomorphism of finite groups

$$Q_o^e \xrightarrow{\sim} (\mathbb{Z}_2)^{d'} \tag{99}$$

where $d' \leq d$.

Proof. Obviously $\overline{N_o^{\chi}} = Q_o$ so that $Q_o = \overline{D_0}\,\overline{G_o^a}Q_+$ by (90),(93) and (94). But of course $G_o^a \subset G_o^e \subset G^{\chi}$ so that

$$Q_o = \overline{D_0}Q_+ \tag{100}$$

But since D_0 is a torus it follows that $\overline{D_0}$ is a torus and since N^{χ}_+ is unipotent it follows that Q_+ is unipotent so that Q_o is solvable and (100) is a Levi decomposition of Q_o. Now by Lemma 21 the isotropy subgroup of N^{χ}_+ at e is $N^{\chi}_+ \cap G^{\chi}$ so that the map

$$Q_+ \longrightarrow e + \mathfrak{z}^{\chi}_{++}, \qquad q \mapsto q \cdot e \tag{101}$$

is an isomorphism. Now clearly $\overline{D_0} \subset (Q_o)^H$. But any element in $(Q_o)^H$ necessarily stabilizes $\mathfrak{z}^{\chi}_2$ so that by (100) and (101) one has

$$\overline{D_0} = Q_o^H \tag{102}$$

Furthermore by (100) and the isomorphism (101) one that $\overline{D_0} = \{a \in Q_o | a \cdot e \in \mathfrak{z}^{\chi}_2\}$. Thus $Q_o^e \subset \overline{D_0}$.

Let F be the inverse image of Q_o^e in D_0 so that clearly $F = D_0 \cap G^e$. But now if S is the subgroup of G corresponding to $\mathfrak{s}$ then D_0 is a maximal torus in S, by Proposition 20. But then

$$F = Cent\ S \tag{103}$$

by (95). Since $\mathfrak{s}$ is isomorphic to a direct sum of d copies of $\mathfrak{sl}(2, \mathbb{C})$ this implies that $F \xrightarrow{\sim} (\mathbb{Z}_2)^k$ where $k \leq d$. But that yields (99). But now

$$\overline{D_0} = D_0/F^\chi \tag{104}$$

where $F^\chi = D_0 \cap G^\chi \subset D_0 \cap G^e = F$. Note then that one has an isomorphism

$$F^\chi \xrightarrow{\sim} (\mathbb{Z}_2)^{k-d'} \tag{105}$$

and that $\overline{D_0}$ is also a d-torus. QED

Remark 23.1. In a subsequent paper we will prove that, relative to any fixed decomposition of $\mathfrak{g}$ into simple components, d is equal to the number of those components to which e has non-zero projection.

We can now determine the structure of $\mathfrak{z}^{\chi *} (\xrightarrow{\sim} SympM)$. Not only is $\mathfrak{z}^{\chi *}$ a rational affine variety but, as in the semisimple case, $\mathfrak{z}^{\chi *}$ is an affine space minus a finite number of hyperplanes. We recall from Theorem 10 that

$$Q_o/Q_o^e \xrightarrow{\sim} Q_o \cdot e = \mathfrak{z}^{\chi *} \tag{106}$$

Theorem 24. *Let β_i, $i = 1, \ldots, d$, be the linear functionals on $\mathfrak{z}^\chi$ defined as in (96). Let p be the polynomial function on $\mathfrak{z}^\chi$ defined by putting*

$$p = \beta_1 \cdots \beta_d \tag{107}$$

Then

$$\mathfrak{z}^{\chi *} = \{w \in \mathfrak{z}^\chi |\ p(w) \neq 0\} \tag{108}$$

Proof. Let the notation be as in (95),(96), and as in the proof of Theorem 23. Then since the simple roots of $(\mathfrak{d}_0, \mathfrak{s})$ corresponding to the e_i are linearly independent one necessarily has

$$Q_o^H \cdot e = D_0 \cdot e$$
$$= \{x \in \mathfrak{z}^\chi |\ \beta_i(x) \neq 0,\ i = 1, \ldots, d\} \tag{109}$$

But (108) follows from (97) and (101). QED

5. General homogeneous spaces

Henceforth we no longer assume that M is nilpotent. That is the only assumption we are making about M is that it is a G-homogeneous space such that the set $Symp\ M$ of G-invariant symplectic structures on M is non-empty.

Let $\chi \in M$. Then $\mathfrak{z}^\chi$ is an abelian algebraic Lie subalgebra of $\mathfrak{g}$ by (25) and (27). Thus $\mathfrak{z}^\chi$ has a unique Levi decomposition

$$\mathfrak{z}^\chi = \mathfrak{z}^\chi_r + \mathfrak{z}^\chi_u \tag{110}$$

where $\mathfrak{z}^\chi_r$ corresponds to a torus $Z^\chi \subset G$ and is the set of all semisimple elements in $\mathfrak{g}$ and where $\mathfrak{z}^\chi_u$ is the set of all nilpotent elements in $\mathfrak{z}^\chi$. Thus if $x \in \mathfrak{z}^\chi$ and $z = z(x) \in \mathfrak{z}^\chi_r$ and $e = e(x) \in \mathfrak{z}^\chi_u$ are the components of x relative to (110) so that

$$x = z + e \tag{111}$$

then (111) is the Jordan decomposition of x.

We now observe that we can associate to M a distinguished nilpotent orbit in $\mathfrak{g}$. Indeed since there are only a finite number of nilpotent orbits there exists a unique nilpotent orbit $O = O(M)$ of G such that

$$O \cap \mathfrak{z}^\chi_u \text{ is non-empty and Zariski open in } \mathfrak{z}^\chi_u \tag{112}$$

Furthermore by conjugation O is clearly independent of the choice of $\chi \in M$. We will refer to O by the expression "corresponding nilpotent orbit" noting that this generalizes the use of that expression for the case where M is nilpotent (see Theorem 8 and Corollary 9).

With regard to $\mathfrak{z}^\chi_r$ we will say that $z \in \mathfrak{z}^\chi_r$ is $\mathfrak{z}^\chi_r$-*regular* in case $\mathfrak{g}^z$ equals the centralizer of $\mathfrak{z}^\chi_r$ in $\mathfrak{g}$. One notes that if $Reg\ \mathfrak{z}^\chi_r$ is the set of all $\mathfrak{z}^\chi_r$-regular elements then $Reg\ \mathfrak{z}^\chi_r$ is the non-empty Zariski open subset of $\mathfrak{z}^\chi_r$ given by

$$Reg\ \mathfrak{z}^\chi_r = \{w \in \mathfrak{z}^\chi_r|\ \varphi(w) \neq 0 \text{ where } \varphi \text{ is any (restricted) root of } (\mathfrak{z}^\chi_r, \mathfrak{g})\} \tag{113}$$

Now assume $x \in \mathfrak{z}^{\chi\,*}$ and let the notation be as in (111). By (30) one has

$$G^x_o \subset G^\chi \subset G^x \tag{114}$$

and in particular

$$\mathfrak{g}^\chi = \mathfrak{g}^x$$

But now G^z is a connected reductive subgroup of G and if $(G^z)^e$ denotes the isotropy subgroup of G^z at e then standard properties of the Jordan decomposition imply that

$$G^x = (G^z)^e \tag{115}$$

Now let $\mathfrak{k} = [\mathfrak{g}^z, \mathfrak{g}^z]$ so that $\mathfrak{k}$ is a semisimple Lie algebra and

$$e \in \mathfrak{k}$$

and also one has the direct sum

$$\mathfrak{g}^z = Cent\ \mathfrak{g}^z + \mathfrak{k} \tag{116}$$

Now let $G^{[z]}$ be the (semisimple) subgroup of G which corresponds to $\mathfrak{k}$ and let $(G^{[z]})^e$ be the isotropy subgroup of $G^{[z]}$ at e. Thus

$$G^z = (Cent\ G^z)_o G^{[z]} \tag{117}$$

But then clearly

$$G^x = (Cent\ G^z)_o (G^{[z]})^e \tag{118}$$

by (115) and one must have (since $(Cent\ G^z)_o \cap (G^{[z]})^e$ is finite)

$$G_o^x = (Cent\ G^z)_o (G^{[z]})_o^e \tag{119}$$

But then by (114) one must have

$$G^\chi = (Cent\ G^z)_o (G^{[z]})^\chi \tag{120}$$

where with regard to its action on M the group $(G^{[z]})^\chi = G^\chi \cap G^{[z]}$ is the isotropy subgroup of $G^{[z]}$ at χ. In addition one has

$$(G^{[z]})_o^e \subset (G^{[z]})^\chi \subset (G^{[z]})^e \tag{121}$$

by (114),(118), (119) and (120). But now one has the direct sum

$$\mathfrak{z}^\chi = Cent\ \mathfrak{g}^z + \mathfrak{k}^{(G^{[z]})^\chi} \tag{122}$$

by (27), (116) and (120).

Now we will apply our earlier results in the nilpotent case to the second term on the right side of (122). Let K be a simply-connected (complex semisimple) Lie group with Lie algebra $\mathfrak{k}$ so that there is a covering homomorphism

$$K \longrightarrow G^{[z]} \tag{123}$$

Obviously (123) induces an action of K on M. Let L be the K-orbit of χ. But then taking inverse images of (121) in K one readily has (even though the inverse image of $(G^{[z]})_o^e$ is not necessarily connected)

$$K_o^e \subset K^\chi \subset K^e \tag{124}$$

Thus if $O(K)_e$ is the (nilpotent) K-orbit of e in $\mathfrak{k}$ the inclusions (124) yield a K-equivariant orbit covering

$$L \longrightarrow O(K)_e \qquad (125)$$

This establishes

Proposition 25. *The set Symp L of K-invariant symplectic structures on L is not-empty. Furthermore L is nilpotent and $O(K)_e$ is the corresponding nilpotent orbit.*

We can now prove the following result about the Levi decomposition of $\mathfrak{z}^\chi$.

Theorem 26. *Let M be any G-homogeneous space such that Symp M is not empty. Let $\chi \in M$ and let*

$$\mathfrak{z}^\chi = \mathfrak{z}^\chi_{\,r} + \mathfrak{z}^\chi_{\,u}$$

be the Levi decomposition of $\mathfrak{z}^\chi = \mathfrak{g}^{G^\chi}$. Let $x \in \mathfrak{z}^{\chi\,*}$, so that M is a G-equivariant covering of the orbit $O_x \subset \mathfrak{g}$, and let $x = z + e$ be the Jordan decomposition of x. Then

$$\mathfrak{z}^\chi_{\,r} = Cent\ \mathfrak{g}^z \qquad (126)$$

so that (see 95.2) in particular

$$z \in Reg\ \mathfrak{z}^\chi_{\,r} \qquad (127)$$

Let $\mathfrak{k} = [\mathfrak{g}^z, \mathfrak{g}^z]$ so that $e \in \mathfrak{k}$ and let K be a corresponding simply-connected semisimple Lie group — which then naturally operates on M. Then Symp L is not empty where L is the K-orbit of χ. In addition L is nilpotent and $K \cdot e$ is the corresponding nilpotent orbit in $\mathfrak{k}$. Finally one has

$$\mathfrak{z}^\chi_{\,u} = \mathfrak{k}^{K^\chi} \qquad (128)$$

and, using the definitions of (27) and (28) where K and $\mathfrak{k}$ replace G and $\mathfrak{g}$, one has

$$e \in (\mathfrak{k}^{K^\chi})^* \qquad (129)$$

Proof. Recalling (122) and (123) one clearly has $\mathfrak{k}^{(G^{[z]})^\chi} = \mathfrak{k}^{K^\chi}$ so that

$$\mathfrak{z}^\chi = Cent\ \mathfrak{g}^z + \mathfrak{k}^{K^\chi} \qquad (130)$$

is a direct sum. But every element in $\mathfrak{k}^{K^\chi}$ is nilpotent by Proposition 25 and Theorem 8. Thus (130) must be the Levi decomposition of $\mathfrak{z}^\chi$. Hence

$Cent\ \mathfrak{g}^z = \mathfrak{z}^{\chi}_r$ and $\mathfrak{k}^{K^{\chi}} = \mathfrak{z}^{\chi}_u$. The theorem then follows from (124) and (125). QED

Let the notation be as in Theorem 26. Obviously

$$K \cdot e \subset O_e = G \cdot e$$

Lemma 27. O_e *is the nilpotent orbit corresponding to* M *(see (112)). Furthermore*

$$\mathfrak{z}^{\chi}_u \cap K \cdot e = \mathfrak{z}^{\chi}_u \cap O_e \tag{131}$$

Proof. Obviously $\mathfrak{z}^{\chi}_u \cap K \cdot e \subset \mathfrak{z}^{\chi}_u \cap O_e$. But $\mathfrak{z}^{\chi}_u \cap K \cdot e$ is a non-empty Zariski open subset of z^{χ}_u by (128) since

$$\mathfrak{k}^{K^{\chi}} \cap K \cdot e = (\mathfrak{k}^{K^{\chi}})^* \tag{132}$$

by (129) and (41). Thus $\mathfrak{z}^{\chi}_u \cap O_e$ is a non-empty Zariski open subset of z^{χ}_u and hence O_e is the nilpotent G-orbit corresponding to M. Now if $w \in \mathfrak{z}^{\chi}_u \cap O_e$ then, by the density of $\mathfrak{z}^{\chi}_u \cap K \cdot e$ in $\mathfrak{z}^{\chi}_u$, one has $w \in \overline{K \cdot e}$. This, however, implies that $rank\ ad_{\mathfrak{k}}w \leq rank\ ad_{\mathfrak{k}}e$ and $rank\ ad_{\mathfrak{g}/\mathfrak{k}}w \leq rank\ ad_{\mathfrak{g}/\mathfrak{k}}e$. But $rank\ ad_{\mathfrak{g}}w = rank\ ad_{\mathfrak{g}}e$ since w and e are G-conjugate. Thus the inequalities above must be equalities and hence $w \in \mathfrak{z}^{\chi}_u \cap K \cdot e$ by (23) and Proposition 3. QED.

We can give a very simple characterization of the nilpotent orbit corresponding to M.

Theorem 28. *Let M be any G-homogeneous space such that $Symp\ M$ is not empty. Then for any $\omega \in Symp\ M$ and any $\chi \in M$ one has*

$$e \in O \tag{133}$$

where $O \subset \mathfrak{g}$ is the nilpotent orbit corresponding to M, $\mu_\omega(\chi) = z + e$ is the Jordan decomposition of $\mu_\omega(\chi)$ and μ_ω is the moment map corresponding to ω. Furthermore every element in O is obtained this way.

Proof. One has (133) by (29) and Lemma 27. The last statement is immediate by conjugation. QED

We can now determine the structure of variety $Symp\ M$ of all symplectic structures on a homogeneous space M for a complex semisimple Lie group . Not only is $Symp\ M$ affine but $Symp\ M$ is isomorphic to an affine vector space minus a finite number of hyperplanes.

Theorem 29. *Let $\chi \in M$ so that $\operatorname{Symp} M$ is isomorphic to $\mathfrak{z}^{\chi\,*}$ (See Theorem 5). Let*

$$\mathfrak{z}^{\chi} = \mathfrak{z}^{\chi}_{r} + \mathfrak{z}^{\chi}_{u}$$

be the Levi decomposition of $\mathfrak{z}^{\chi}$. Then one has

$$\mathfrak{z}^{\chi\,*} = \operatorname{Reg} \mathfrak{z}^{\chi}_{r} + (O \cap \mathfrak{z}^{\chi}_{u}) \tag{134}$$

where O is the nilpotent orbit corresponding to M. Furthermore there exists linear functionals ψ_{i}, $i = 1,\ldots,k$, on $\mathfrak{z}^{\chi}$ such that if $p = \psi_{1}\cdots\psi_{k}$ then

$$\mathfrak{z}^{\chi\,*} = \{y \in \mathfrak{z}^{\chi} \mid p(y) \neq 0\} \tag{135}$$

Proof. Let $D = \operatorname{Reg} \mathfrak{z}^{\chi}_{r} + (O \cap \mathfrak{z}^{\chi}_{u})$. One has $\mathfrak{z}^{\chi\,*} \subset D$ by Theorem 26. Now let $y \in D$ so that we may write

$$y = w + b \tag{136}$$

where $w \in \operatorname{Reg} \mathfrak{z}^{\chi}_{r}$ and $b \in O \cap \mathfrak{z}^{\chi}_{u}$ noting then that (136) is the Jordan decomposition of y. Let $x \in \mathfrak{z}^{\chi\,*}$ and let the notation be as in Theorem 26. Then $\mathfrak{g}^{y} = (\mathfrak{g}^{w})^{b}$. But $\mathfrak{g}^{w} = \mathfrak{z}^{\chi}_{r} + \mathfrak{k}$ by (126) and (127). Hence $\mathfrak{g}^{y} = \mathfrak{z}^{\chi}_{r} + \mathfrak{k}^{b}$. But $b \in (\mathfrak{k}^{K^{\chi}})^{*}$ by (128), (131) and (132). Thus $\mathfrak{g}^{y} = \mathfrak{g}^{x}$. But then $y \in \mathfrak{z}^{\chi\,*}$ by (20) and (28) establishing (134).

Now we may rewrite (134) as

$$\mathfrak{z}^{\chi\,*} = \operatorname{Reg} \mathfrak{z}^{\chi}_{r} + (\mathfrak{k}^{K^{\chi}})^{*}$$

But then (135) follows immediately from (113) and Theorem 24. QED

Finally we want to make statements about the G-commuting group Q in the general case we are considering and the action of Q on $\operatorname{Symp} M$. Let $\chi \in M$ so that we can, as done so previously, regard $Q = N^{\chi}/G^{\chi}$.

We will use the notation of Theorem 26 and also of (124) so that $x \in \mathfrak{z}^{\chi\,*}$ and $x = z + e$ is the Jordan decomposition of x. Also

$$\mathfrak{g}^{z} = \mathfrak{z}^{\chi}_{r} + \mathfrak{k} \tag{137}$$

Let $\mathfrak{l}$ be a Cartan subalgebra of $\mathfrak{k}$ so that

$$\mathfrak{h} = \mathfrak{z}^{\chi}_{r} + \mathfrak{l} \tag{138}$$

is a Cartan subalgebra of $\mathfrak{g}$. Fix a Weyl chamber C in $\mathfrak{l}$ (relative to $\mathfrak{k}$) and let W_{C} be the subgroup of the Weyl group of $(\mathfrak{h}, \mathfrak{g})$ which stabilizes C. Clearly W_{C} stabilizes the decomposition (138) since C spans $\mathfrak{l}$. Now consistent with our earlier notation N^{z} is the normalizer of G^{z} in G. Obviously

N^z stabilizes the decomposition (137) and since for any automorphism a of $\mathfrak{k}$ there exists $k \in K$ such that $a\,Ad\,k$ stabilizes C one easily has an epimorphism $\gamma : N^z \longrightarrow W_C$ inducing a well known exact sequence

$$1 \longrightarrow G^z \longrightarrow N^z \xrightarrow{\gamma} W_C \longrightarrow 1 \qquad (139)$$

or if $W_z = N^z/G^z$ then one has a natural isomorphism

$$W_z \xrightarrow{\sim} W_C \qquad (140)$$

This is just a restatement of Proposition 7 noting that $W_z = Q(G/G^z)$. In addition we have observed in Proposition 7 that $Reg\ {}_3\chi_r$ is stable under the action of W_z and that for any $v \in Reg\ {}_3\chi_r$ one has

$$O_v \cap Reg\ {}_3\chi_r = W_z \cdot v \qquad (141)$$

Also W_z is simply-transitive on $O_v \cap Reg\ {}_3\chi_r$.

Remark 29.1. Even though we have indexed W_z by an element z in $Reg\ {}_3\chi_r$ it clearly depends only on ${}_3\chi_r$ and will be referred to as the Weyl group of ${}_3\chi_r$. For notational convenience we will identify W_z with W_C using the isomorphism (140).

Now recalling the K-homogeneous space $L = K/K^\chi$ of Theorem 26 let $\mathcal{L}$ be the set of all K-isotropy groups with respect to the action of K on L. That is, $\mathcal{L}$ is the set of K-conjugates of K^χ. But now N^z operates as a group of automorphisms of K since $\mathfrak{k}$ is stabilized by $Ad\,N^z$. Let $N^z(\chi) = \{n \in N^z \mid n \text{ stabilizes } \mathcal{L}\}$. Obviously $N^z(\chi)$ is a subgroup of N^z. Note, however, that if $n \in N^z$, then

$$n \in N^z(\chi) \text{ if and only if } n \cdot K^\chi \in \mathcal{L} \qquad (142)$$

Clearly $G^z \subset N^z(\chi)$ so that by (140) there exists a subgroup W_χ of W_z such that (139) induces the exact sequence

$$1 \longrightarrow G^z \longrightarrow N^z(\chi) \xrightarrow{\gamma} W_\chi \longrightarrow 1 \qquad (143)$$

Lemma 30. *One has $N^\chi \subset N^z$ and furthermore*

$$\gamma(N^\chi) = W_\chi \qquad (144)$$

Proof. Let $n \in N^\chi$. Then it folllows from (24) and (126) that n normalizes $Cent\ \mathfrak{g}^z$. But then n normalizes $\mathfrak{g}^z$ and also G^z and $G^{[z]}$. In particular $n \in N^z$ and also n normalizes $(G^{[z]})^\chi$. But then $n \cdot K^\chi = K^\chi$ by (123). Thus $N^\chi \subset N^z(\chi)$ by (142) and hence $\gamma(N^\chi) \subset W_\chi$. Now if $g \in N^z(\chi)$ then by definition (noting that any subgroup of K in $\mathcal{L}$ contains the kernel of the covering map (123)) there clearly exists $a \in G^z$ such that $ag \cdot K^\chi = K^\chi$. But then $ag \in N^\chi$ by (120). But $\gamma(g) = \gamma(ag)$. Thus $\gamma(N^z(\chi)) = \gamma(N^\chi)$. QED

Now let $N(K)^\chi$ be the normalizer of K^χ in K so that

$$Q(L) = N(K)^\chi / K^\chi \tag{145}$$

is the K-commuting group with respect to the action of K on the nilpotent homogeneous space $L = K \cdot \chi$. On the other hand if $N(G^{[z]})^\chi$ is the normalizer of $(G^{[z]})^\chi$ in $G^{[z]}$ then (123) clearly induces an isomorphism

$$Q(L) \xrightarrow{\sim} N(G^{[z]})^\chi / (G^{[z]})^\chi \tag{146}$$

On the other hand one immediately has $N^\chi \cap G^z = (Cent\ G^z)_o N(G^{[z]})^\chi$ by (117) and (120) so that the exact sequence (143) and Lemma 30 yields the exact sequence

$$1 \longrightarrow (Cent\ G^z)_o N(G^{[z]})^\chi \longrightarrow N^\chi \xrightarrow{\gamma} W_\chi \longrightarrow 1 \tag{147}$$

We can now make the following statements about the G-commuting group Q and its action on $Symp\ M$ in the general case.

Theorem 31. *Let M be any (complex) algebraic homogeneous space for a complex simply-connected semisimple Lie group G such that $Symp\ M$ is not empty. Let $\chi \in M$ so that $Q = N^\chi / G^\chi$ is isomorphic to the group self-maps of M which commute with the action of G. Let*

$$\mathfrak{z}^\chi = \mathfrak{z}^\chi{}_r + \mathfrak{z}^\chi{}_u$$

be the Levi decomposition of $\mathfrak{z}^\chi = \mathfrak{g}^{G^\chi}$ so that one has a Q-isomorphism

$$Symp\ M \xrightarrow{\sim} Reg\ \mathfrak{z}^\chi{}_r + O \cap \mathfrak{z}^\chi{}_u = \mathfrak{z}^{\chi\,*}$$

where $O \subset \mathfrak{g}$ is the nilpotent orbit corresponding to M. Then

$$dim\ Q = dim\ \mathfrak{z}^\chi{}_u \tag{148}$$

and the identity component Q_o is a solvable algebraic group. Furthermore if $x \in \mathfrak{z}^{\chi\,*}$, so that for some $\omega \in Symp\ M$ the corresponding moment map μ_ω is an orbit covering

$$\mu_\omega : M \longrightarrow O_x \tag{149}$$

and the Jordan decomposition $x = z + e$ of x is such that $z \in Reg\, {}_3\chi_r$ and $e \in O \cap {}_3\chi_u$ then

$$Q_o \cdot x = z + O \cap {}_3\chi_u \tag{150}$$

and

$$Q \cdot x = W_\chi \cdot z + O \cap {}_3\chi_u \tag{151}$$

where W_χ is the subgroup of the Weyl group of ${}_3\chi_r$ defined by (143). Moreover one has

$$Q \cdot x \subset O_x \cap {}_3\chi \tag{152}$$

and there is equality in (152) in case either M is simply connected or (149) is an isomorphism. Finally using the notation of (145) the map γ of (147) induces an exact sequence

$$1 \longrightarrow Q(L) \longrightarrow Q \longrightarrow W_\chi \longrightarrow 1 \tag{153}$$

Proof. Recalling (120) note that G^χ is contained in the first 2 non-trivial subgroups of the exact sequence (147). But then, by (146), taking the quotient of these subgroups by G^χ yields (153). Regarding $Q(L)$ as a subgroup of Q, using (153), one clearly has $Q(L)_o = Q_o$ by (153). But then Q_o is solvable and one has (148) by Theorem 26 since L is nilpotent. See Theorems 10 and 23. But also one has (150) by Theorem 10, (128), (131)and (132). In fact then

$$Q(L) \cdot x = z + O \cap z^\chi_u$$

by Theorem 10. But then (151) is an immediate consequence of (153). Furthermore since $Q \cdot x = N^\chi \cdot x$ one obtains (152).

Now let $y \in O_x \cap {}_3\chi$ and let $g \in G$ be such that $g \cdot x = y$. But then *rank ad x = rank ad y* so that $\mathfrak{g}^x = \mathfrak{g}^y$ by Proposition 3 and (20). Thus g normalizes G_o^χ. But if M is simply-connected then $G^\chi = G_o^\chi$ and hence $g \in N^\chi$. This establishes equality in (152) if M is simply-connected. In any case one has $G_o^\chi = G_o^y$. But $y \in \mathfrak{g}^{G^\chi}$. Thus

$$G_o^y \subset G^\chi \subset G^y \tag{154}$$

But now applying g^{-1} to (154) yields

$$G_o^x \subset g^{-1} \cdot G^\chi \subset G^x \tag{155}$$

But

$$G_o^x \subset G^\chi \subset G^x$$

by (30). But now if (149) is an isomorphism one has $G^\chi = G^x$ so that (155) implies that $g^{-1} \cdot G^\chi \subset G^\chi$. But then $g^{-1} \cdot G^\chi = G^\chi$ since G_o^χ has finite index in G^χ. Thus $g \in N^\chi$. Hence there is equality in (152) if (149) is an isomorphism. QED

REFERENCES

[B] A. Borel, *Linear Algebraic Groups*, W. A. Benjamin 1969.

[B-T] A. Borel and J. Tits, *Elements unipotents et sous-groupes paraboliques de groupes reductifs I*, Inv. Math. **12** (1971), 95–104.

[K1] B. Kostant, *Lie group representations on polynomial rings*, Am. J. Math. **85** (1963), 327–404.

[K2] B. Kostant, *Quantization and unitary representations*, Lectures in Modern Analysis and Applications III, Lecture Notes in Math., No. 170, Springer-Verlag, 1970, 87–208

[R] R.W. Richardson, "Derivatives of invariant polynomials on a semisimple Lie algebra," *Harmonic analysis and operator theory*, Proc. Cent. Math. Anal. Aust. Natl. Univ. **15** (1987), 228–241.

Ranee Brylinski
Department of Mathematics
The Penn State University
McAllister Hall
University Park, PA 16801

Bertram Kostant
Department of Mathematics
MIT
Bldg2-282
Cambridge, MA 02139

On Singular Reduction of Hamiltonian Spaces

RICHARD CUSHMAN and REYER SJAMAAR *

*Partially supported by a grant from the Netherlands Organization for Scientific Research (NWO)

A Jean-Marie Souriau

Le timbre de la trompette est noble et éclatant; il convient aux idées guerrières, aux cris de fureur et de vengeance, comme aux chants de triomphe, il se prête à l'expression de tous les sentiments énergiques, fiers et grandioses, à la plus part des accents tragiques. Il peut même figurer dans un morceau joyeux, pourvu que la joie y prenne un caractère d'emportement ou de grandeur pompeuse.

—H. Berlioz [4]

Introduction

Symplectic reduction of Hamiltonian spaces is a rich source of symplectic manifolds. It has been widely applied to the study of Hamiltonian systems with symmetries since the days of Jacobi (see [16] for a historical survey). Symplectic reduction also arises naturally in classical field theories, such as Yang-Mills theory, and in Guillemin's and Sternberg's work [8] on asymptotic multiplicity formulas for group representations. A general framework for symplectic reduction at regular values of a momentum map has been set up by Meyer [18] and by Marsden and Weinstein [17]. In many of the applications, however, one would like to carry out reduction at singular values of a momentum map. It is therefore of interest to devise a reduction scheme for singular levels of a momentum map and to study the singularities arising from it.

Over the past decade many results have been obtained in this direction. See [2] for a survey of and a comparison between various different approaches. Our point of view is that a reduced space of a Hamiltonian action of a compact Lie group is a *stratified symplectic space*. (Our results hold also if the group is noncompact, provided that the action is proper.) Roughly speaking, a stratified symplectic space is a topological space X with the following properties:

i. the space X decomposes into a locally finite disjoint union of symplectic manifolds;

ii. these manifolds fit together in a nice way, i.e., the decomposition is a *stratification*;

---

*Partially supported by a grant from the Netherlands Organization for Scientific Research (NWO)

iii. there is given a subclass of the set of continuous functions on X, a set of 'smooth functions', which forms a Poisson algebra;

iv. the inclusion of each stratum is a Poisson map;

v. there is a unique open stratum in X. It is open and dense in X.

This paper is an overview of joint work of the first author with J. Arms and M. Gotay [1], and of the second author with E. Lerman [13], [21]. The exposition is informal; for most of the proofs we shall refer to the original articles. Instead we shall consider a few simple examples of the theory in detail. Further examples will be described in [14]. We would like to suggest the following two problems for future research:

A. Try to generalize the results of this paper to any of the various infinite-dimensional situations arising in mathematical physics, e.g. Yang-Mills theory (cf. [3]) and the Einstein equations (cf. [11]);

B. Try to generalize the multiplicity formulas of Guillemin and Sternberg [8] to singular reduced spaces, using the concept of a stratified symplectic space.

1   A Poisson Bracket on a Reduced Phase Space

Throughout this article (M, ω) will denote a symplectic manifold and G a compact Lie group acting on M in a Hamiltonian fashion, with momentum map $J : M \to \mathfrak{g}^*$. We shall always assume that the map J is equivariant with respect to the given action on M and the coadjoint action on $\mathfrak{g}^*$. In the sequel we shall make frequent use of the following equality, which follows easily from the definition of a momentum map:

$$\operatorname{im} dJ_p = \operatorname{ann} \mathfrak{g}_p, \tag{1}$$

for all points p in M. Here $\operatorname{im} dJ_p$ denotes the image of the tangent map dJ_p, $\mathfrak{g}_p$ denotes the infinitesimal stabilizer of the point p, and $\operatorname{ann} \mathfrak{g}_p$ denotes the annihilator of $\mathfrak{g}_p$ in $\mathfrak{g}^*$.

For any point μ in $\mathfrak{g}^*$, let $\mathcal{O}_\mu$ denote the coadjoint orbit through μ. If μ is a regular value of the momentum mapping, the pre-image $J^{-1}(\mathcal{O}_\mu)$ of the coadjoint orbit is a G-invariant submanifold of M, and it follows immediately from (1) that the action of G on $J^{-1}(\mathcal{O}_\mu)$ is locally free. Marsden and Weinstein defined the reduced space M_μ to be the quotient orbifold

$$M_\mu = J^{-1}(\mathcal{O}_\mu)/G,$$

and they showed that there is a unique symplectic form ω_μ on the space M_μ such that the pullback of ω_μ to $J^{-1}(\mathcal{O}_\mu)$ equals the restriction of the form ω to $J^{-1}(\mathcal{O}_\mu)$.

If μ is not a regular value of J, then the set $J^{-1}(\mathcal{O}_\mu)$ is not in general a manifold, and one sees from (1) that the action of G on $J^{-1}(\mathcal{O}_\mu)$ is no longer locally free. However, the definition $M_\mu = J^{-1}(\mathcal{O}_\mu)/G$ still makes sense topologically. We will see that the space M_μ, which is not in general a manifold, or even an orbifold, has many nice properties.

In the first place, there is a natural concept of 'smooth functions' on M_μ. A continuous function $f : M_\mu \to \mathbf{R}$ is called *smooth* if there exists a smooth G-invariant function $F : M \to \mathbf{R}$ such that

$$F|_{J^{-1}(\mathcal{O}_\mu)} = f \circ \pi,$$

where $\pi : J^{-1}(\mathcal{O}_\mu) \to M_\mu$ is the quotient map. We denote the algebra of all smooth functions on M_μ by $C^\infty(M_\mu)$.

If the group G is finite, its 'Lie algebra' is $\{0\}$ and the 'reduced space' is just the quotient M/G. In this case, the smooth functions on M/G are the G-invariant functions on M.

Secondly, we define the bracket of two smooth functions f and h on M_μ by

$$\{f, h\}_{M_\mu} = \{F, H\}_M,$$

where F and H are smooth G-invariant functions on M such that $F|_{J^{-1}(\mathcal{O}_\mu)} = f \circ \pi$, resp. $H|_{J^{-1}(\mathcal{O}_\mu)} = h \circ \pi$. One then checks that the bracket is well-defined, i.e., it does not depend on the choice of F and H, and so the algebra $C^\infty(M_\mu)$ becomes a Poisson algebra with this bracket; see [1] for a proof. Moreover, it turns out that if μ is a regular value of the momentum map, this Poisson structure coincides with the Poisson structure defined by the Marsden-Weinstein symplectic form ω_μ.

1.1. Example. Let M be $\mathbf{C}^2$ with its standard symplectic form and let G be the circle S^1, acting on M by $e^{i\theta} \cdot (z_1, z_2) = (e^{i\theta} z_1, e^{-i\theta} z_2)$. This is the $(1, -1)$-*resonance*, the action generated by the Hamiltonian

$$J(z_1, z_2) = (|z_1|^2 - |z_2|^2)/2.$$

We want to compute the reduced space at the zero level. To this end we compute the algebra of G-invariant *real* polynomials on $\mathbf{C}^2$. It is easy to check that this algebra is generated by the following polynomials:

$$
\begin{aligned}
\sigma_1 &= |z_1|^2 & \sigma_2 &= (z_1 z_2 + \overline{z_1 z_2})/2 \\
\sigma_3 &= (z_1 z_2 - \overline{z_1 z_2})/2i & \sigma_4 &= 2J = |z_1|^2 - |z_2|^2.
\end{aligned}
$$

(We choose this particular set of generators because it is well-suited to our further computations.) The only relation among the generators is: $\sigma_1(\sigma_1 - \sigma_4) = \sigma_2^2 + \sigma_3^2$. Define the *Hilbert map* $\sigma : M \to \mathbf{R}^4$ by

$$\sigma(m) = (\sigma_1(m), \sigma_2(m), \sigma_3(m), \sigma_4(m)).$$

The image $\sigma(J^{-1}(0))$ of the zero level set is contained in the hyperplane $\{(x_1, x_2, x_3, x_4) : x_4 = 0\}$. Let us identify this hyperplane with $\mathbf{R}^3$. Then $\sigma(J^{-1}(0))$ is the quadratic half-cone in $\mathbf{R}^3$ given by

$$x_1^2 = x_2^2 + x_3^2 \quad \text{and} \quad x_1 \geq 0. \tag{2}$$

Since the Hilbert map is invariant, its restriction to $J^{-1}(0)$ descends to a map $\tilde{\sigma} : M_0 \to \mathbf{R}^3$. It is not difficult to show that this map is a homeomorphism onto its image. This describes the reduced space as a topological space.

We claim that the algebra of smooth functions on M_0 is isomorphic to the algebra $C^\infty(\mathbf{R}^3)/\mathcal{I}$, where $\mathcal{I}$ is the ideal in $C^\infty(\mathbf{R}^3)$ of all functions vanishing on the set given by (2), and where the isomorphism is given by pulling back functions via $\tilde{\sigma}$. Indeed, if h is a smooth function on $\mathbf{R}^3$, the function H given by $H(m) = h(\sigma_1(m), \sigma_2(m), \sigma_3(m))$ is a smooth invariant function on $M = \mathbf{C}^2$, and therefore descends to a smooth function on M_0. This gives us the map

$$\tilde{\sigma}^* : C^\infty(\mathbf{R}^3) \to C^\infty(M_0). \tag{3}$$

By definition, its kernel is the ideal $\mathcal{I}$. To see that the map is surjective, pick an arbitrary smooth function f on M_0. Let F be an invariant smooth function on M such that $F|_{J^{-1}(0)}$ descends to f. By a theorem of Schwarz [20], a smooth function on $\mathbf{R}^n$ invariant under a linear action of a compact Lie group K is always a smooth function of the K-invariant *polynomials* on $\mathbf{R}^n$. In our case, this means that we can find a smooth function h defined on $\mathbf{R}^4$ such that $h \circ \sigma = F$. But then $\tilde{\sigma}^*(h|_{\mathbf{R}^3}) = f$. So $\tilde{\sigma}$ is surjective.

We can describe the Poisson bracket on M_0 in terms of the generators $\bar{x}_1, \bar{x}_3, \bar{x}_3$ of the algebra $C^\infty(\mathbf{R}^3)/\mathcal{I}$. (Here $\bar{f}$ denotes the equivalence class $f \bmod \mathcal{I}$ of a function f.) It is given by the following table:

$\{\cdot, \cdot\}$	$\bar{x}_1$	$\bar{x}_2$	$\bar{x}_3$
$\bar{x}_1$	0	$4\bar{x}_3$	$4\bar{x}_2$
$\bar{x}_2$	$-4\bar{x}_3$	0	$-4\bar{x}_1$
$\bar{x}_3$	$-4\bar{x}_2$	$4\bar{x}_1$	0

1.2. Remark. In the above table, replace the $\bar{x}_i$'s by the coordinates x_i on $\mathbf{R}^3$. It is easy to check that the resulting table is the structure matrix of the Lie-Poisson bracket on $\mathfrak{sl}(2, \mathbf{R})^*$ (with respect to a suitable basis). This means that there exists a Poisson bracket on $\mathbf{R}^3$ such that the map (3) becomes a morphism of Poisson algebras and the ideal $\mathcal{I}$ a Poisson ideal.

Now let V be an arbitrary symplectic representation space for a compact Lie group, and let j be any embedding of the reduced space V_0 into a Euclidean space $\mathbf{R}^n$ constructed by means of invariant polynomials, analogous to the embedding of Example 1.1.

118 R. Cushman and R. Sjamaar

1.3. Conjecture. *There exists a Poisson bracket on $\mathbf{R}^n$ such that the embedding j of the reduced space V_0 into $\mathbf{R}^n$ becomes a Poisson map.*

A proof of this conjecture in the special case where the algebra of invariant polynomials has a basis of polynomials of degree ≤ 2 can be found in [13].

As illustrated by Example 1.1, there is a natural notion of an embedding of a reduced space into a Euclidean space. Likewise, there is a natural concept of diffeomorphisms between reduced spaces, and of a smooth action of a Lie group on a reduced space.

1.4. Example. Here is another way of computing the reduced space of Example 1.1. Consider the diagonal map $\Delta : \mathbf{C} \to M = \mathbf{C}^2$, assigning $w \mapsto (w, w)$. The image of Δ is contained in the zero level set $J^{-1}(0)$, and one readily checks that it intersects each orbit in $J^{-1}(0)$. In other words, the diagonal map induces a surjective map $\mathbf{C} \to M_0$. Under this map two points w, w' in $\mathbf{C}$ are mapped to the same orbit in the reduced space M_0 if and only if they are antipodes, $w' = -w$. So we get a bijection between orbit spaces,

$$\widetilde{\Delta} : \mathbf{C}/\mathbf{Z}_2 \to M_0$$

(where the nontrivial element in $\mathbf{Z}_2$ acts on $\mathbf{C}$ as the antipodal map). We claim that this map is an isomorphism in the sense that smooth functions on M_0 pull back to smooth functions on $\mathbf{C}/\mathbf{Z}_2$ and that the pullback map

$$\widetilde{\Delta}^* : C^\infty(M_0) \to C^\infty(\mathbf{C}/\mathbf{Z}_2)$$

is an isomorphism of Poisson algebras. In other words, the reduced space M_0 is isomorphic to the symplectic orbifold $\mathbf{C}/\mathbf{Z}_2$. That the map $\widetilde{\Delta}^*$ is well-defined follows from the fact that G-invariant functions on $M = \mathbf{C}^2$ restrict to $\mathbf{Z}_2$-invariant functions on the diagonal. That it is a morphism of Poisson algebras follows from the fact that the diagonal map Δ is symplectic. It is easy to see that $\widetilde{\Delta}^*$ is injective. Finally, we have to show that it is surjective. Consider the algebra of $\mathbf{Z}_2$-invariant real polynomials on $\mathbf{C} = \mathbf{R}^2$. It is generated by the three polynomials

$$\rho_1 = |w|^2 \qquad \rho_2 = (w^2 + \overline{w}^2)/2 \qquad \rho_3 = (w^2 - \overline{w}^2)/2i.$$

Each of these is the pullback of a G-invariant polynomial on M, namely:

$$\rho_1 = \sigma_1 \circ \Delta \qquad \rho_2 = \sigma_2 \circ \Delta \qquad \rho_3 = \sigma_3 \circ \Delta.$$

Now let h be an arbitrary $\mathbf{Z}_2$-invariant smooth function on $\mathbf{C}$. By Schwarz's theorem (loc. cit.) there exists a smooth function of three variables H such that

$$h(w) = H(\rho_1(w), \rho_2(w), \rho_3(w)).$$

Define the G-invariant smooth function F on $\mathbf{C}^2$ by $F = H \circ \sigma$. Then F descends to a smooth function f on M_0, and we have $\widetilde{\Delta}^*(f) = h$. So $\widetilde{\Delta}^*$ is surjective.

1.5. Example. Consider the symplectic manifold $M \times \mathcal{O}_\mu^-$, the symplectic product of M with the coadjoint orbit $\mathcal{O}_\mu$ through μ, endowed with the opposite of the Kirillov symplectic form. The diagonal action of G on $M \times \mathcal{O}_\mu^-$ is Hamiltonian with momentum map J_μ given by $J_\mu(m,\nu) = J(m) - \nu$. It is easy to check that the cartesian projection $\Pi : M \times \mathcal{O}_\mu^- \to M$ restricts to an equivariant bijection $J_\mu^{-1}(0) \cong J^{-1}(\mathcal{O}_\mu)$. As a result, Π descends to a bijection between reduced spaces,

$$\tilde{\Pi} : (M \times \mathcal{O}_\mu^-)_0 \xrightarrow{\sim} M_\mu.$$

1.6. Proposition (the 'shifting trick'). *The map $\tilde{\Pi}$ is an isomorphism of reduced spaces.*

Sketch of Proof. We want to show that $\tilde{\Pi}$ induces an isomorphism of Poisson algebras,

$$\tilde{\Pi}^* : C^\infty(M_\mu) \xrightarrow{\cong} C^\infty((M \times \mathcal{O}_\mu^-)_0).$$

It is not hard to show that smooth functions on $(M \times \mathcal{O}_\mu^-)_0$ pull back to smooth functions on M_μ and that $\tilde{\Pi}^*$ is injective. Moreover, the map $\tilde{\Pi}^*$ is a morphism of Poisson algebras, because the Cartesian projection $M \times \mathcal{O}_\mu^- \to M$ is a Poisson map.

The gist of the argument is showing that $\tilde{\Pi}^*$ is surjective. Pick an arbitrary smooth function ϕ on $(M \times \mathcal{O}_\mu^-)_0$. Let Φ be a smooth G-invariant function on $M \times \mathcal{O}_\mu^-$ such that the restriction of Φ to $J_\mu^{-1}(0)$ equals the pullback of ϕ to $J_\mu^{-1}(0)$. Because the group G is compact, the coadjoint orbit $\mathcal{O}_\mu$ is a *closed* subset of $\mathfrak{g}^*$. Therefore the product $M \times \mathcal{O}_\mu$ is included in $M \times \mathfrak{g}^*$ as a closed submanifold, and we can find a smooth extension $\bar{\bar{\Phi}}$ of Φ to the whole of $M \times \mathfrak{g}^*$. We may furthermore assume that $\bar{\bar{\Phi}}$ is G-invariant (if not, replace it by its average over G). Now let $j : M \to M \times \mathfrak{g}^*$ be the natural map from M onto the graph of the momentum map J, $j(m) = (m, J(m))$, and define a function F on M by putting $F = \bar{\bar{\Phi}} \circ j$. Then F is a smooth G-invariant function on M, and therefore induces a smooth function f on the reduced space M_μ. For all (m,ν) in $J_\mu^{-1}(0)$ one has

$$F \circ \Pi(m,\nu) = \bar{\bar{\Phi}}(m, J(m)) = \bar{\bar{\Phi}}(m,\nu) = \Phi(m,\nu).$$

This shows that $\tilde{\Pi}^* f = \phi$. Thus $\tilde{\Pi}^*$ is surjective. ∎

As far as we know, this proposition was first proved by Guillemin and Sternberg [8] under the assumption of regularity. It allows us to restrict our attention from now on to reduction at the zero level.

1.7. Remark. The shifting trick breaks down for Hamiltonian actions of a noncompact group. The reason is that for a noncompact group G the coadjoint orbits are not necessarily closed in $\mathfrak{g}^*$; cf. also [1, § 2, last two paragraphs]. It is not hard to show that the shifting trick works for reduction at a closed coadjoint orbit of a noncompact group G, provided the action of G on M is proper in the sense of [19]. All the other results in this paper go through for Hamiltonian actions of an arbitrary Lie group G, provided that the action is proper. (The only statement that requires a slight modification is the density theorem, Theorem 2.7; cf. the remark following that theorem.)

Many other concepts defined for smooth symplectic manifolds carry over to singular reduced phase spaces in a routine manner. For instance, one defines a *smooth curve* in M_μ as a continuous map $\gamma : [0,1] \to M_\mu$ such that the pullback $f \circ \gamma$ of a smooth function f on M_μ is a smooth function on the interval $[0,1]$. If h is a smooth function on M_μ, it makes no sense to talk about the Hamiltonian vector field of h on M_μ, since M_μ is not a manifold. One can however define the *Hamiltonian derivation* $ad\,h$ of h. It operates on a smooth function f on M_μ as follows:

$$ad\,h \cdot f = \{h, f\}_{M_\mu}. \tag{4}$$

Now an *integral curve* of the Hamiltonian h through a point m is defined as a smooth curve $\gamma(t)$ with $\gamma(0) = m$, such that for all functions $f \in C^\infty(M_\mu)$

$$\frac{d}{dt} f(\gamma(t)) = -\,ad\,h \cdot f\,(\gamma(t)). \tag{5}$$

One proves the existence and uniqueness of such integral curves by considering the flow of a lift of the Hamiltonian h to M; see [13].

If K is a Lie group acting smoothly on M_0, then for each element ξ of the Lie algebra $\mathfrak{k}$ of K one has the associated *fundamental derivation* ξ_{M_0} on the algebra $C^\infty(M_0)$. It operates on a smooth function f as follows:

$$(\xi_{M_0} \cdot f)(x) = -\,\frac{d}{dt} f(\exp t\xi \cdot x)\Big|_{t=0},$$

for x in M_0. This is well-defined since $f(\exp t\xi \cdot x)$ depends smoothly on t. We call a K-action on M_0 *Hamiltonian* if it has a momentum map, i.e., a smooth $Ad(K)$-equivariant map $I : M_0 \to \mathfrak{k}^*$ such that for each ξ in $\mathfrak{k}$ one has

$$ad\,I^\xi = -\xi_{M_0}.$$

Here I^ξ is the ξ-th component of I, defined by $I^\xi(x) = \langle I(x), \xi \rangle$, and $ad\,I^\xi$ is the Hamiltonian derivation on $C^\infty(M_0)$ corresponding to the function I^ξ. Given a Hamiltonian K-action on M_0, one can reduce once more and obtain a space with a Poisson algebra.

An important special case of a Hamiltonian action on a reduced space arises in the following situation. Assume that the group G is the product of two Lie groups G_1 and G_2. Then the momentum map for the G-action on M splits into a product of two maps, $J = J_1 \times J_2$, where $J_1 : M \to \mathfrak{g}_1^*$

and $J_2 : M \to \mathfrak{g}_2^*$ are momentum maps for the actions of G_1 and G_2, respectively. The actions of G_1 and G_2 on M commute, and it follows from the G-equivariance of J that J_1 is invariant with respect to the action of G_2. Let X_1 be the reduced space of M with respect to the G_1-action. Then the action of G_2 on M descends to an action on X_1 and the map J_2 descends to a smooth map $\bar{J}_2 : M \to \mathfrak{g}^*$. It is not difficult to show that this G_2-action is Hamiltonian with momentum map $\bar{J}_2$. Let X_{12} denote the reduced space of X_1 with respect to this action.

We might just as well carry these two reductions out in reverse order. That is, first reducing M with respect to the G_2-action we obtain a Hamiltonian G_1-space X_2 with momentum map $\bar{J}_1$, and reducing once again we obtain a space X_{21}. We claim that the answer does not depend on the order of the reductions.

1.8. Theorem (reduction in stages). *The reduced spaces X_{12} and X_{21} are isomorphic.*

Reduction in stages is a useful computational tool and it plays an important role in [13]. The proof rests on the observation that both spaces can be obtained by reducing the original manifold M with respect to the action of the product group $G = G_1 \times G_2$; see [13] for the details.

2   A Stratification of a Reduced Phase Space

For Poisson manifolds Weinstein [23] has introduced the notion of *symplectic leaves*. If P is a Poisson manifold and p a point in P, the symplectic leaf of P through p is the set of all points q in P such that q can be joined to p by means of a finite number of trajectories of Hamiltonian vector fields on P. The symplectic leaves of a Poisson manifold are always symplectic manifolds.

In the previous section we showed that the reduced space M_0 of a Hamiltonian G-space M comes equipped with a natural Poisson algebra, and that there is a suitable concept of Hamiltonian flows on M_0, despite the fact that this space may be singular. Because of this, it makes sense to talk about the symplectic leaves of M_0 (cf. also Gonçalves [7]). It will turn out that the leaves of M_0 are smooth symplectic manifolds, which fit together in a neat way.

There is an alternative definition of these symplectic pieces, which is better suited to the study of the local structure of M_0. For a closed subgroup H of G the set of points in M of *orbit type* (H) is defined as

$$M_{(H)} = \{\, m \text{ in } M : \text{stabilizer of } m \text{ is conjugate to } H \,\}.$$

It is a familiar fact from the theory of Lie group actions that $M_{(H)}$ is a locally closed submanifold of M (see e.g. Bredon [5]), possibly consisting of connected components of various dimensions. It follows from (1) that the restriction of the momentum map J to $M_{(H)}$ is of constant rank equal to the codimension of H in G. As a consequence, the intersection $J^{-1}(0) \cap M_{(H)}$ is a submanifold of M. In fact, we can say much more.

2.1. Theorem. *The quotient* $(M_0)_{(H)} := (J^{-1}(0) \cap M_{(H)})/G$ *is a manifold and there exists a unique symplectic form* $\omega_{(H)}$ *on* $(M_0)_{(H)}$ *whose pullback to* $J^{-1}(0) \cap M_{(H)}$ *equals the restriction of the symplectic form* ω *to* $J^{-1}(0) \cap M_{(H)}$.

We have thus decomposed the reduced space into a collection of symplectic manifolds,

$$M_0 = \coprod_{H < G} (M_0)_{(H)}.$$

The symplectic pieces $(M_0)_{(H)}$ are called the *strata* of M_0. It is not hard to check that the inclusion maps $(M_0)_{(H)} \to M_0$ are Poisson, i.e., smooth functions on M_0 restrict to smooth functions on $(M_0)_{(H)}$, and the restriction map $C^\infty(M_0) \to C^\infty((M_0)_{(H)})$ is a morphism of Poisson algebras. The restriction map is not surjective, unless $(M_0)_{(H)}$ is closed. It is, however, easy to see that any smooth function on $(M_0)_{(H)}$ with compact support can be extended to a smooth function on M_0. (See the argument preceding Theorem 2.5.) Therefore the image of the restriction map is dense in $C^\infty((M_0)_{(H)})$. One can also show that the Hamiltonian flow of a smooth function h on M_0 leaves the strata $(M_0)_{(H)}$ invariant. Because of this, a Hamiltonian derivation on the algebra $C^\infty(M_0)$, as defined by (4), can be regarded as a collection of Hamiltonian vector fields, one on each stratum of M_0, glued together in a smooth manner.

Sketch of Proof. Let p be a point in the zero level set of J and let H be the stabilizer of p. Let $\mathfrak{m}$ denote the tangent space at p to the orbit $G \cdot p$ through p. It is easy to see that any G-orbit in $J^{-1}(0)$ is isotropic, i.e., $\mathfrak{m} \subset \mathfrak{m}^\omega$, where $\mathfrak{m}^\omega$ denotes the skew orthogonal complement of $\mathfrak{m}$. Therefore we can define a symplectic vector space $V = \mathfrak{m}^\omega/\mathfrak{m}$, called the *symplectic slice* at p to the G-action. It is a symplectic representation space for H. Marle [15] and, independently, Guillemin and Sternberg [9] have shown how to reconstruct the Hamiltonian space M in a neighbourhood of the orbit $G \cdot p$ from the data G, H and V. The tangent space to M at p can be written (noncanonically) as a direct sum, $T_p M = \mathfrak{m} \oplus \mathfrak{m}^* \oplus V$. The topological normal bundle of the orbit $G \cdot p$ is a vector bundle,

$$Y = G \times_H (\mathfrak{m}^* \times V),$$

associated to the principal bundle $H \to G \to G/H$. The space Y can also be seen as a vector bundle with symplectic fibre V and symplectic base

$G \times_H \mathfrak{m}^* = T^*(G \cdot p)$. As a manifold, the space Y can be obtained from the Hamiltonian $G \times H$-space

$$T^*G \times V \tag{6}$$

by means of reduction with respect to the H-action. (Here G acts by left translations on T^*G, and H acts by right translations on T^*G and as the given representation on V.) This provides Y with a symplectic structure. It turns out that the left G-action on the space Y is Hamiltonian with momentum map J given by:

$$J[g, \mu, v] = (Ad^*g)(\mu + \Phi_V(v)).$$

Here $[g, \mu, v]$ denotes the class of a triple (g, μ, v) in $G \times \mathfrak{m} \times V$, and $\Phi_V : V \to \mathfrak{h}^*$ is the quadratic momentum map for the linear H-action on V, given by:

$$\langle \Phi_V(v), \eta \rangle = 1/2 \, \omega_V(\eta \cdot v, v).$$

It now follows from the G-equivariant version of Weinstein's isotropic embedding theorem [22] that a neighbourhood of the orbit $G \cdot p$ in M is G-equivariantly symplectomorphic to a neighbourhood of the zero section in the vector bundle Y over G/H. This is the local normal form of Marle-Guillemin-Sternberg.

It is easy to compute the set $(M_0)_{(H)}$ inside the local model. It is equal to the subspace V_H of fixed points in the symplectic vector space V, which is again symplectic. This proves Theorem 2.1. ∎

2.2. Remark. As a corollary of this proof we obtain a local model for the reduced space, namely the product space (6), successively reduced with respect to the H-action and the G-action. By Theorem 1.8, we may reverse the order of the two reductions. The reduced space of $T^*G \times V$ with respect to the G-action is equal to the H-space V; reducing once more we obtain the conical space $\Phi_V^{-1}(0)/H$. Consequently, any reduced space is in the neighbourhood of each point isomorphic to the reduced space of some *linear* Hamiltonian space. This observation plays an important role in the proof of the embedding theorem in [13], which says that every reduced space can be embedded in a Euclidean space in such a way that the image is a Whitney stratified space. Also, this enables one to show that the stratification of M_0 satisfies the condition of the frontier, i.e., the closure of a stratum is a union of (connected components of) strata.

2.3. Example (the 'lemon'). Let M be the product $S^2 \times S^2$ of two copies of the unit sphere in $\mathbf{R}^3$, both equipped with the symplectic form given by restriction of the standard volume form on $\mathbf{R}^3$. Consider the diagonal action of the circle on M given by rotations round a fixed axis on each factor. This space turns up in the study of perturbed Keplerian systems with axial symmetry, cf. [6]. In that article the reduced space M_0 is computed by means of invariant polynomials. Eugene Lerman showed us the following alternative computation. The Hamiltonian space M can be identified with the product of two copies of $\mathbf{C}P^1$ with a circle action defined by:

$$e^{i\theta} \cdot ((z_1 : z_2), (z_3 : z_4)) = ((e^{i\theta} z_1 : e^{-i\theta} z_2), (e^{i\theta} z_3 : e^{-i\theta} z_4)).$$

A momentum map is given by:

$$J((z_1 : z_2), (z_3 : z_4)) = \frac{1}{2} \left(\frac{|z_1|^2 - |z_2|^2}{|z_1|^2 + |z_2|^2} + \frac{|z_3|^2 - |z_4|^2}{|z_3|^2 + |z_4|^2} \right).$$

The image of J is the interval $[-1, 1]$, and its critical values are -1, 0 and 1.

There are two orbit types in M. The fixed point set consists of four points:

$$
\begin{array}{rclrcl}
P_1 & = & ((1:0), (1:0)) & P_2 & = & ((0:1), (1:0)) \\
P_3 & = & ((1:0), (0:1)) & P_4 & = & ((0:1), (0:1))
\end{array}
$$

For all other points the stabilizer is $\{ e^{i\theta} : \theta \equiv 0 \bmod \pi \} \cong \mathbf{Z}_2$. The points P_2 and P_3 are contained in the zero level set, and P_1 and P_4 are the points where J takes its maximum, respectively its minimum.

The reduced space M_{-1} is obviously a point. The Marle-Guillemin-Sternberg local model tells us that near P_4 the Hamiltonian space M can be identified with a neighbourhood of the origin in the tangent space to P_4, which is a complex plane $\mathbf{C}^2$ with its standard S^1-action, $e^{i\theta} \cdot (w_1, w_2) = (e^{i\theta} w_1, e^{i\theta} w_2)$, and momentum map

$$J(w_1, w_2) = 1/2 \left(|w_1|^2 + |w_2|^2 \right) - 1.$$

It follows that for small ε the inverse image $J^{-1}(-1 + \varepsilon)$ is a three-sphere and the reduced space $M_{-1+\varepsilon}$ is a $\mathbf{C}P^1$. But then M_ξ has to be a $\mathbf{C}P^1$ for all ξ in $(-1, 0)$. Guillemin and Sternberg [10] have shown that the space M_ε for $\varepsilon > 0$ can be obtained from $M_{-\varepsilon}$ by a succession of blowing-ups and blowing-downs. Since $M_{-\varepsilon}$ has only one degree of freedom, there can be no blowing-ups and blowing-downs, so M_ε is again a $\mathbf{C}P^1$. It also follows from Guillemin's and Sternberg's analysis that the reduced space M_0 at the critical level 0 is *homeomorphic* to $\mathbf{C}P^1$. The stratification of M_0 consists of three pieces: two single points, corresponding to the fixed points P_2 and P_3, and one open piece. Using the local model, one can easily figure out what M_0 looks like near the singular points. The symplectic slices at P_2 and P_3 are both equal to the complex plane $\mathbf{C}^2$ with the following S^1-action: $e^{i\theta} \cdot (w_1, w_2) = (e^{i\theta} w_1, e^{-i\theta} w_2)$. This is the $(1, -1)$-resonance discussed in Example 1.1. As a consequence, near both singular points the stratified symplectic space M_0 is isomorphic to the cone in $\mathbf{R}^3$ given by (2), and M_0 can be pictured as two copies of this cone glued together along the base.

From the previous examples the reader may be tempted to infer that reduced spaces are always orbifolds. The following example shows that this is not the case.

2.4. Example. The $(1, 1, -1, -1)$-*resonance* is the circle action on $\mathbf{C}^4$ generated by the Hamiltonian

$$J(z_1, z_2, z_3, z_4) = 1/2\,(|z_1|^2 + |z_2|^2 - |z_3|^2 - |z_4|^2).$$

We want to describe the topology of the reduced space at the zero level and show that it has a singularity that is worse than an orbifold singularity. The zero level set is given by:

$$|z_1|^2 + |z_2|^2 = |z_3|^2 + |z_4|^2.$$

Consider the unit seven-sphere S^7 in $\mathbf{C}^4$, given by $|z_1|^2+|z_2|^2+|z_3|^2+|z_4|^2 = 1$. Its intersection with the zero level set of J is an S^1-invariant submanifold of $\mathbf{C}^4$, namely the product of two three-spheres of radius $1/2$,

$$J^{-1}(0) \cap S^7 = S^3 \times S^3 \subset \mathbf{C}^2 \times \mathbf{C}^2.$$

The S^1-action on the first copy of S^3 is given by $e^{i\theta}\cdot(z_1, z_2) = (e^{i\theta}z_1, e^{i\theta}z_2)$, and on the second copy of S^3 it is given by $e^{i\theta}\cdot(z_3, z_4) = (e^{-i\theta}z_3, e^{-i\theta}z_4)$. The quotient of $S^3 \times S^3$ by the S^1-action is denoted by $S^3 \times_{S^1} S^3$. It is an S^3-bundle over the complex projective line $\mathbf{C}P^1$, associated to the Hopf fibration $S^3 \to \mathbf{C}P^1$. Topologically, the reduced space $(\mathbf{C}^4)_0$ can now be written as a *cone*,

$$(\mathbf{C}^4)_0 \sim C(S^3 \times_{S^1} S^3),$$

i.e., the product $(S^3 \times_{S^1} S^3) \times [0, \infty)$ with the boundary $(S^3 \times_{S^1} S^3) \times \{0\}$ collapsed to a point. It is obvious from this description that the reduced space cannot be an orbifold. For instance, an easy computation shows that the local homology in degree 3 at the singular point is nonzero,

$$H_3((\mathbf{C}^4)_0, (\mathbf{C}^4)_0 - \{\text{vertex}\}; \mathbf{Q}) = \mathbf{Q},$$

which is impossible for a six-dimensional orbifold. (All orbifolds are rational homology manifolds; see e.g. [5, Chapter III].)

One can desingularize the reduced space in the following way. Consider the S^1-equivariant map Ψ from $\mathbf{C}^4$ to itself defined by:

$$\Psi : (z_1, z_2, z_3, z_4) \mapsto ((|z_3|^2 + |z_4|^2)^{1/2}z_1, (|z_3|^2 + |z_4|^2)^{1/2}z_2, z_3, z_4).$$

It maps the subset

$$S^3 \times \mathbf{C}^2 = \{(z_1, z_2, z_3, z_4) : |z_1|^2 + |z_2|^2 = 1\}$$

of $\mathbf{C}^4$ onto $J^{-1}(0)$. Therefore it descends to a surjective map $\widetilde{\Psi}$: $S^3 \times_{S^1} \mathbf{C}^2 \to (\mathbf{C}^4)_0$. The space $S^3 \times_{S^1} \mathbf{C}^2$ is a $\mathbf{C}^2$-bundle over $\mathbf{C}P^1$. The map $\widetilde{\Psi}$ is a diffeomorphism everywhere, except on the zero section of this bundle, which is mapped to the vertex of $(\mathbf{C}^4)_0$. The reduced space $(\mathbf{C}^4)_0$ can therefore be regarded as a plane bundle over $\mathbf{C}P^1$ with the zero section collapsed to a point.

Why are the strata of a reduced space equal to its symplectic leaves, as claimed at the outset of this section? In fact, the strata may be disconnected, whereas the leaves are by definition always connected, so in general they are not the same. We claim, however, that the symplectic leaves of M_0 are precisely the connected components of the strata $(M_0)_{(H)}$. Indeed, let m be a point in $(M_0)_{(H)}$. It is not hard to show that the symplectic leaf of m has to be contained in $(M_0)_{(H)}$. It suffices therefore to prove that it is *open* in $(M_0)_{(H)}$. This amounts to showing that an arbitrary point in $(M_0)_{(H)}$ sufficiently close to p can be joined to p by means of an integral curve of a Hamiltonian that is defined *globally* on M_0. Of course, it is easy to find such a Hamiltonian, call it f, that lives on $(M_0)_{(H)}$ only. The problem is to extend f to a smooth function on the whole of M_0. We may assume that f has compact support on $(M_0)_{(H)}$. We can lift f to a compactly supported G-invariant function F on the locally closed submanifold $J^{-1}(0) \cap M_{(H)}$ of M. Because F has compact support, we can extend it to a smooth function $\check{F}$ defined on the whole of M, and we may assume that $\check{F}$ is still G-invariant. So $\check{F}$ descends to a smooth function $\check{f}$ on the reduced space, whose restriction to $(M_0)_{(H)}$ equals f, as desired. We have proved:

2.5. Theorem. *The symplectic leaves of the reduced space M_0 coincide with the connected components of its strata $(M_0)_{(H)}$, $H < G$.*

This implies that the stratification of a reduced space depends only on its Poisson algebra of smooth functions.

The manifold of orbit type (H) contains the *manifold of symmetry H*, defined by

$$M_H = \{\, m \text{ in } M : \text{stabilizer of } m \text{ is equal to } H \,\}.$$

This is a symplectic submanifold of M, which is invariant under the action of the normalizer $N_G(H)$ of H in G. Let $\mathfrak{n}$ denote the Lie algebra of $N_G(H)$. The action of $N_G(H)$ on M_H is Hamiltonian; a momentum map is given by the restriction of J to M_H, followed by the canonical projection $\mathfrak{g}^* \to \mathfrak{n}^*$. Because H iself acts trivially on M_H, the action descends to an action of the quotient group $L = N_G(H)/H$, and the momentum map descends to a momentum map for the L-action, $J_H : M_H \to \mathfrak{l}^*$. Since L acts freely on M_H, the origin in $\mathfrak{l}^*$ is a regular value of J_H.

2.6. Theorem. *The stratum $(M_0)_{(H)}$ of the reduced space of type (H) is symplectomorphic to the Marsden-Weinstein reduced space $J_H^{-1}(0)/L$.*

Because of this theorem, the procedure for lifting a Hamiltonian flow on M_0 to M is completely analogous to the regular case. Lifting an integral curve of a Hamiltonian through a point in $(M_0)_{(H)}$ amounts to solving a differential equation on the group L. Theorem 2.6 also enables one to write down a version of Smale's criterion for relative equilibria from which the assumption of regularity has been removed. See [13] for a further discussion.

We conclude by quoting one more theorem from [13]. Its proof hinges on the fact proved by Kirwan [12] that the fibres of a proper momentum map are connected.

2.7. Theorem. *Assume that the momentum map $J : M \to \mathfrak{g}^*$ is proper. Then there exists a unique open stratum in the reduced space M_0. It is connected and dense.*

2.8. Remark. If the momentum map is not proper, the reduced space is not necessarily connected. Similarly, if G is a noncompact Lie group acting properly on M, the reduced space M_0 is well-defined as a stratified symplectic space, but it needn't be connected. However, in both these cases it is still true that each component C of M_0 has a unique open stratum, which is connected and dense in C.

2.9. Remark. As an immediate corollary to Theorem 2.7 we see that the Poisson algebra $C^\infty(M_0)$ of the reduced space is nondegenerate, i.e., its centre consists of the (locally) constant functions only.

REFERENCES

[1] J. Arms, R. Cushman and M. Gotay, *A Universal Reduction Procedure for Hamiltonian Group Actions*, preprint 591, University of Utrecht, the Netherlands, 1989.

[2] J. Arms, M. Gotay and G. Jennings, *Geometric and Algebraic Reduction for Singular Momentum Maps*, Advances in Mathematics **79** (1990), 43–103.

[3] M. F. Atiyah and R. Bott, *The Yang-Mills Equations over Riemann Surfaces*, Phil. Trans. Royal Soc. **A 308** (1982), 523–615.

[4] H. Berlioz, "Grand traité d'instrumentation et d'orchestration modernes", oeuvre 10me, nouvelle édition, H. Lemoine et Cie, Paris, 1855.

[5] G. Bredon, "Introduction to Compact Transformation Groups", Academic Press, New York, 1972.

[6] R. Cushman, *A Survey of Normalization Techniques Applied to Perturbed Keplerian Systems*, preprint, University of Utrecht, 1990.

[7] J. Basto Gonçalves, *Reduction of Hamiltonian Systems with Symmetry*, preprint of the Universidade do Porto, Portugal, 1988.

[8] V. Guillemin and S. Sternberg, *Geometric Quantization and Multiplicities of Group Representations*, Invent. Math. **67** (1982), 515–538.

[9] V. Guillemin and S. Sternberg, *A Normal Form for the Moment Map*, in: "Differential Geometric Methods in Mathematical Physics", editor S. Sternberg, Reidel Publishing Company, Dordrecht, 1984.

[10] V. Guillemin and S. Sternberg, *Birational Equivalence in the Symplectic Category*, Invent. Math. **97** (1989), 485–522.

[11] J. Isenberg and J. Marsden, *A Slice Theorem for the Space of Solutions of Einstein's Equations*, Physics Reports **89** (Review Section of Physics Letters) (1982), 179–222.

[12] F. Kirwan, *Convexity Properties of the Moment Mapping III*, Invent. Math. **77** (1984), 547–552.

[13] E. Lerman and R. Sjamaar, *Stratified Symplectic Spaces and Reduction*, to appear in: Annals of Mathematics.

[14] E. Lerman, R. Montgomery and R. Sjamaar, *Examples of Singular Reduction*, preprint.

[15] C.-M. Marle, *Modèle d'action hamiltonienne d'un groupe de Lie sur une variété symplectique*, Rendiconti del Seminario Matematico **43** (1985), 227–251, Università e Politechnico, Torino.

[16] J. Marsden and T. Ratiu, "Mechanics and Symmetry", to appear.

[17] J. Marsden and A. Weinstein, *Reduction of Symplectic Manifolds with Symmetry*, Rep. Math. Phys. **5** (1974), 121–130.

[18] K. Meyer, *Symmetries and Integrals in Mathematics*, in: "Dynamical Systems", editor M. M. Peixoto, Academic Press, New York, 1973.

[19] R. S. Palais, *On the Existence of Slices for Actions of Non-Compact Lie Groups*, Annals of Math. **73** (1961), 295–323.

[20] G. W. Schwarz, *Smooth Functions Invariant under the Action of a Compact Lie Group*, Topology **14** (1975), 63–68.

[21] R. Sjamaar, "Singular Orbit Spaces in Riemannian and Symplectic Geometry", PhD thesis, University of Utrecht, 1990.

[22] A. Weinstein, "Lectures on Symplectic Manifolds", CBMS Regional Conf. Series in Math. **29**, 1977.

[23] A. Weinstein, *The Local Structure of Poisson Manifolds*, J. Diff. Geometry **18** (1983), 523–557.

Richard Cushman   Reyer Sjamaar
Mathematisch Instituut   Department of Mathematics
Rijksuniversiteit te Utrecht   Massachusetts Institute of Technology
Postbus 80.010   Cambridge, MA 02139-4307
3508 TA Utrecht   USA
The Netherlands

Formes normales de structures de Poisson

JEAN-PAUL DUFOUR

A J.-M. Souriau

Introduction

On considère une structure de Poisson $\{,\}$ sur une variété différentiable P de rang 0 en un point p_0. Si $(x_1, x_2, \ldots, x_n)$ sont des coordonnées locales au voisinage de p_0, nulles en p_0, on développe en série de Taylor les fonctions $\{x_i, x_j\}$ et on obtient

$$\{x_i, x_j\} = \sum_k c_{ij}^k x_k + o(\|x\|)$$

où les c_{ij}^k sont les constantes de structure d'une algèbre de Lie bien déterminée (à isomorphisme prés) que l'on appelle la *linéarisée* de $\{,\}$ en p_0.

Problème de linéarisation : Peut-on trouver de nouvelles coordonnées $(y_1, y_2, \ldots, y_n)$ telles que l'on ait

$$\{y_i, y_j\} = \sum_k c_{ij}^k y_k \quad ?$$

Si oui on dit que la structure de Poisson est *linéarisable* en p_0. Soit g une algèbre de Lie (réelle et de dimension finie); on dira que g est *non-dégénérée* si toute structure de Poisson de rang 0 en un point et admettant en ce point une algèbre de Lie linéarisée isomorphe à g est linéarisable.

On peut se poser le problème de la "non-dégénérescence" d'une algèbre de Lie dans différentes catégories; ainsi on peut étudier la non-dégénérescence formelle, C^∞, analytique...

Dans ce qui suit on s'intéresse essentiellement au cas C^∞.

Les résultats bien connus sont les suivants:

-l'algèbre non-triviale de dimension 2 est non-dégénérée,

-les algèbres de Lie semi-simples compactes sont non-dégénérées ([C2]) (on sait aussi que les algèbres de Lie semi-simples sont formellement et analytiquement non-dégénérées ([C1] et [W])).

Nous allons élargir un peu cette liste.

I. Cas de la dimension 3

Outre $\underline{so}(3)$ et $\underline{sl}(2)$ (la première étant non-dégénérée et la deuxième dégénérée), les algèbres de Lie de dimension 3 sont des produits semi-directs $\mathbf{R}^2 \times_A \mathbf{R}$ où $\mathbf{R}$ agit sur $\mathbf{R}^2$ par la matrice A; c'est-à-dire que, dans une base (e_1, e_2, e_3), on a

$$[e_1, e_2] = 0 \ , \ [e_1, e_3] = ae_1 + be_2 \ , \ [e_2, e_3] = ce_1 + de_2$$

avec

$$A = \begin{pmatrix} a & b \\ c & d \end{pmatrix} \quad .$$

Théorème 1 : ([D1]) *Les algébres de Lie* $\mathbf{R}^2 \times_A \mathbf{R}$ *non-dégénérées sont celles pour lesquelles A est "non-résonnant", c'est-à-dire qu'il n'existe pas de relation*

$$\lambda_i = n_1 \lambda_1 + n_2 \lambda_2 \quad ; \quad n_1 + n_2 \geq 2$$

où λ_1 et λ_2 sont les valeurs propres de A, n_1 et n_2 étant deux entiers.

Schéma de la preuve: :

1 - On choisit des coordonnées (x, y, z) nulles au point considéré avec

$$\{x, y\} = o(\|(x, y, z)\|) \quad ,$$

$$\{x, z\} = ax + by + o(\|(x, y, z)\|) \quad ,$$

$$\{y, z\} = cx + dy + o(\|(x, y, z)\|) \quad .$$

La non-résonnance implique la régularité de A; alors les équations

$$\{x, z\} = \{y, z\} = 0$$

admettent des solutions implicites

$$x = x(z) \quad , \quad y = y(z)$$

et le changement de variables

$$x\prime = x - x(z) \quad , \quad y\prime = y - y(z) \quad , \quad z\prime = z$$

nous ramène au cas où le champ de vecteurs

$$X(= X_z) = \{x, z\}\frac{\partial}{\partial x} + \{y, z\}\frac{\partial}{\partial y}$$

est nul le long de l'axe des z.

2 - On écrit le développement de Taylor de X le long de l'axe des z sous la forme

$$X = X^{(1)} + X^{(2)} + \cdots$$

où $X^{(i)}$ désigne la partie homogène de degré i en x et y. Une étude détaillée de la relation de Jacobi (et la non-résonnance) permettent de voir que $X^{(1)}$ est indépendante de z.

3 - On applique alors le théorème de linéarisation de Sternberg (à paramêtre z) pour linéariser X : pour tout z il existe un difféomorphisme local ϕ_z (dépendant différentiablement de z) avec

$$\phi_z^* X_{(z)} = X^{(1)} \quad .$$

Alors $(x, y, z) \to (\phi_z^*(x, y), z)$ est un changement de variables qui linéarise X.

4 - Le 2-vecteur Π, défini par notre structure de Poisson, a la forme

$$\Pi = a\frac{\partial}{\partial x} \wedge \frac{\partial}{\partial y} + X^{(1)} \wedge \frac{\partial}{\partial z} \quad .$$

Alors l'identité de Jacobi (et le fait que la trace de A est non nulle) permet de voir que l'on a

$$a\frac{\partial}{\partial x} \wedge \frac{\partial}{\partial y} = X^{(1)} \wedge Z$$

où Z est un champ de vecteurs de la forme

$$Z = Z_1\frac{\partial}{\partial x} + Z_2\frac{\partial}{\partial y}$$

qui commute avec $X^{(1)}$. Ainsi on a

$$\Pi = X^{(1)} \wedge (\frac{\partial}{\partial z} + Z) \quad .$$

5 - On "redresse" $\frac{\partial}{\partial z} + Z$ pour arriver à

$$\Pi = X^{(1)} \wedge \frac{\partial}{\partial z} \quad ,$$

ce qui nous donne le résultat.

Dans les deux prochains paragraphes nous donnons deux généralisations du théorème 1.

II. Systèmes complètement intégrables
sur des variétés de Poisson

Le théorème suivant peut être considéré comme une généralisation du théorème 1; c'est une légère amélioration du résultat principal de [**D2**], où l'on ne considérait que les cas avec m $\geq$ 2p + 2.

Théorème 2 : *Soit V une variété de dimension* m *munie d'une structure de Poisson de rang nul au point* p_0 *et de rang maximum* 2p. *On suppose :*

a - *Il existe* p *fonctions* $(f_1, ..., f_p)$ *indépendantes et en involution (i.e.,* $\{f_i, f_j\} = 0$ *pour tous i et j).*

b - *L'action infinitésimale* ρ *de* $\mathbf{R}^p$ *sur la sous-variété*

$$W = \cap_{i=1}^{i=p} f_i^{-1}(f_i(p_0))$$

engendrée par les champs hamiltoniens $X_{f_1}, ..., X_{f_p}$ *est "hyperbolique" en* p_0 *: on entend par là que les "poids"* $\lambda_1, ..., \lambda_{m-p}$ *de sa linéarisée en* p_0 *sont tels que*

i - leurs parties réelles sont p *à* p *indépendantes*

ii - il n'y a pas de relations de résonnance

$$\lambda_i = \sum_{j=1}^{j=m-p} n_j \lambda_j \quad ,$$

où $n_1, ..., n_{m-p}$ *sont des entiers dont la somme est au moins 2.*

Alors Π *est linéarisable, de plus, il existe des coordonnées locales* $(x_1, \ldots, x_{m-p}, u_1, \ldots, u_p)$ *avec*

$$\{u_i, u_j\} = \{x_r, x_s\} = 0 \quad , \quad \{x_r, u_i\} = \sum_k c_{ri}^k x_k \quad , \quad f_i = f_i(u_1, ..., u_p) \quad .$$

Schéma de la preuve : (Voir [**D2**] pour les cas m $\geq$ 2p + 2) On reprend le schéma de la preuve du théorème 1 (le théorème 1 se déduit du théorème 2 en prenant m = 3, p = 1, $f_1 = z...$) :

1 - On choisit des coordonnées $(x_1, \ldots, x_{m-p}, u_1, ..., u_p)$ nulles en p_0 avec $u_i = f_i$. Un premier changement de variables permet de supposer que les champs hamiltoniens $X_i = X_{f_i}$ sont tous nuls le long de la sous-variété $x_1 = \cdots = x_{m-p} = 0$.

2 - Utilisant les identités de Jacobi et l'hyperbolicité on exhibe un changement de variables, portant sur les u_i, qui nous ramène au cas où la partie linéaire en les variables x_s des X_i est indépendante des variables u.

3 - Grâce aux résultats de Chaperon ([**Ch**]) on sait alors linéariser simultanément tous les X_i. C'est là le pas essentiel de la preuve.

4 - On montre que le 2-vecteur de Poisson Π est de la forme

$$\Pi = \sum_i X_i \wedge (\frac{\partial}{\partial u_i} + Z_i) \quad ,$$

où les champs Z_i sont de la forme

$$\sum Z_i^j \frac{\partial}{\partial x_j} \quad :$$

Dans les cas m $= 2p$ ou m $= 2p + 1$ cela découle des identités de Jacobi et de l'hyperbolicité; sinon il faut utiliser le fait que le rang est au maximum $2p$ et appliquer les techniques de division de Saito ([**S**]).

5 - On achève en "redressant" tous les $\frac{\partial}{\partial u_i} + Z_i$; pour cela il nous faut encore (au moins pour m $\geq 2p + 1$) utiliser les techniques de [**Ch**].

III. Voisinage d'une courbe de singularités

On revient en dimension 3 et aux cas de singularités de type $\mathbf{R}^2 \times_A \mathbf{R}$ où A est non-résonnant. Il résulte du théorème 1 que les singularités forment une courbe Γ et le feuilletage symplectique est "localement trivial" le long de Γ . Dans le cas où Γ est une courbe fermée on peut caractériser, à isomorphisme prés, la structure de Poisson au voisinage de Γ par la forme de Jordan de A et par le "premier retour" des feuilles le long de Γ. Sans entrer dans les détails (voir le travail de B. Habch en cours de rédaction) disons que ce premier retour est défini en considérant les intersections d'une transversale à Γ avec les feuilles : c'est la façon dont on passe d'une de ces intersections à une autre en se déplaçant le long de Γ tout en restant dans une feuille. Dans cette direction on a, en particulier, le résultat suivant :

Théorème 3 : *Si la matrice A est non résonnante et a ses deux valeurs propres distinctes et de même signe (Γ étant fermée) la structure de Poisson admet, au voisinage de Γ, le modèle local suivant :*

$$\{x,y\} = axy \quad , \quad \{x,\theta\} = \lambda x \quad , \quad \{y,\theta\} = \mu y \quad ,$$

avec

$$(x,y,\theta) \in \mathbf{R}^2 \times \mathbf{S}^1 \quad ;$$

a, λ *et* μ *sont des constantes qui caractérisent le modèle à isomorphisme prés.*

La preuve de ce théorème est une amélioration de celle du théorème 1. Notons que la constante a caractérise le "premier retour" des feuilles. Remarquons enfin qu'en dimension 3 l'étude des structures de Poisson est trés voisine de celle des formes de Pfaff intégrables : on passe des unes aux autres par contraction avec une forme volume; on peut trouver des analogues des théorèmes 1 et 3 pour les formes intégrables dans l'abondante littérature qui leur a été consacrée (voir, par exemple, [CN]).

IV. Cas de la dimension 4

Théorème 4 : *Les algèbres de Lie de dimension 4 sont toutes formellement dégénérées sauf celles qui sont isomorphes à $\underline{so}(3) \times \mathbf{R}$, $\underline{sl}(2) \times \mathbf{R}$ et les produits semi-directs non triviaux de $\mathbf{R}^2$ par lui même.*

La preuve de ce théorème figure dans le travail de thèse de Ch. Molinier en cours de rédaction. Notons que les seules algèbres de dimension 4 qui pourraient être vraiment non-dégénérées (au sens C^∞) sont celles isomorphes à $\underline{so}(3) \times \mathbf{R}$ et les produits semi-directs non triviaux de $\mathbf{R}^2$ par lui même. Ce théorème se généralise (et se précise) en partie en dimension plus grande sous la forme du

Théorème : *Si $\underline{g}$ est une algèbre de Lie semi-simple alors $\underline{g} \times \mathbf{R}$ est formellement et analytiquement non dégénérée. Si $\underline{g}$ est une algèbre de Lie semi-simple compacte, alors $\underline{g} \times \mathbf{R}$ est non dégénérée.*

Ce théorème se démontre en adaptant les méthodes de Weinstein et Conn ([C1] et [C2]) pour montrer que les algèbres de Lie semi-simples (resp. compactes) sont formellement et analytiquement non-dégénérées (resp. non dégénérées) : ici il faut non seulement utiliser la nullité de la cohomologie de $\underline{g}$ d'ordre 2 mais aussi la nullité à l'ordre 1. Le "referee" nous a signalé que ce dernier résultat (bien que non publié) était connu de J. Conn.

REFERENCES

[CN] C. Camacho & A. Lins - Neto, *The topology of integrable differential forms near a singularity*, Publ. Math. I.H.E.S. **55** (1982), 5–35.

[Ch] M. Chaperon, *Géométrie différentielle et singularités de systèmes dynamiques*, Astérisque **138-139** (1986), Soc. Math. de France.

[C1] J.F. Conn, *Normal forms for analytic Poisson structures*, Ann. of Math. **119 (2)** (1984), 576–601.

[C2] ________, *Normal forms for smooth Poisson structures*, Ann. of Math. **121 (2)** (1985), 565–593.

[D1] J.-P. Dufour, *Linéarisation de certaines strutures de Poisson*, à paraitre au J. of Diff. Geometry (1990).

[D2] __________ , *Hyperbolic action of $\mathbf{R}^p$ on Poisson manifolds*, à paraitre (1990).

[S] K. Saito, *On a generalisation of De-Rham lemma*, Ann. Inst. Fourier **26, 2** (1976), 165–170.

[W] A. Weinstein, *The local structure of Poisson manifolds*, J. of Diff. Geometry **18** (1983).

Université Montpellier II
GETODIM URA 1407, Mathématiques
Place Eugène Bataillon
F-34060 MONTPELLIER CEDEX
FRANCE

Higher Order Regular Variational Problems

P.L. GARCIA PÉREZ and J. MUÑOZ MASQUÉ

A Jean-Marie Souriau

1. Introduction

The notion of regularity has been important in Mechanics and Classical Field Theory ([1], [4], [8]), basically because it allows one to establish in a natural way the equivalence between Lagrangian and Hamiltonian formulations of a variational problem.

Let us consider a first order Lagrangian density $\mathcal{L}v$ on a fibred manifold $p: Y \longrightarrow X$, where v is a volume form on X and $\mathcal{L}$ is a differentiable function on $J^1(Y)$, and let us denote by Θ the Poincaré-Cartan form associated with $\mathcal{L}v$. The extremals of the variational problem defined by $\mathcal{L}v$ are characterized as the sections s of p which satisfy the equation:

$$(j^1 s)^*(i_D d\Theta) = 0 \qquad \text{for all vector fields } D \text{ of } J^1(Y).$$

The set $\mathcal{V}$ of extremals is thus injected by means of the 1-jet prolongation $s \mapsto j^1 s$ into the set $\bar{\mathcal{V}}$ of the sections $\bar{s}$ of the canonical prolongation $p_1: J^1(Y) \longrightarrow X$ which satisfy the *Hamilton-Cartan equation:*

$$\bar{s}^*(i_D d\Theta) = 0 \qquad \text{for all vector fields } D \text{ of } J^1(Y).$$

An intrinsic definition of the notion of regularity in terms of the Poincaré- Cartan form was given in [4], by stating that the polarity $D \mapsto i_D d\Theta$ should be injective on the bundle of tangent vectors of $J^1(Y)$ vertical over Y. For such problems the previous injection $\mathcal{V} \hookrightarrow \bar{\mathcal{V}}$ is actually a bijection. In other words, every solution to the Hamilton-Cartan equation is automatically holonomic. Therefore, the extremals of such problems are the n-dimensional solutions of an exterior differential system $\mathcal{M}$ on $J^1(Y)$ which are transversal to the fibres of the projection $p_1: J^1(Y) \longrightarrow X$, where $n = \dim X$. Namely, $\mathcal{M}$ is the differential system defined for all the forms $i_D d\Theta$, where D is running over the vector fields of $J^1(Y)$. This basic fact seems to be the starting point for the extensive use of exterior differential systems in the Hamilton-Cartan formalism.

The same scheme also works in higher order Mechanics (see e.g. [6], [14]). Nevertheless, in dealing with the notion of regularity for variational problems of arbitrary order in several variables, some surprising differences appear which in the last few years have motivated a considerable number of papers on this topic, probably surpassing the initial interest in it.

Let us consider a Lagrangian density $\mathcal{L}v$ of order r on $p\colon Y \longrightarrow X$; that is, $\mathcal{L}$ is a differentiable function on $J^r(Y)$ and v a volume form on X as above. The two most outstanding differences with respect to the previous cases are the following:

Firstly, for $r > 2$ and $n > 1$ there is no canonical Poincaré-Cartan form associated with $\mathcal{L}v$. Such a form does exist, but it depends on additional parameters. Accordingly, if one wishes to give a definition of regularity in terms of the Poincaré-Cartan form, one is led to prove the independence of the definition of the particular form chosen. In this paper, we shall refer to a family of Poincaré-Cartan forms (constructed in [5], [12]) which are parametrized by the symmetric linear connections of the base manifold.

Secondly, each of the Poincaré-Cartan forms Θ associated with an r-order Lagrangian $\mathcal{L}v$ lying on a jet bundle other than that of $\mathcal{L}v$. For arbitrary Lagrangians, Θ is defined on $J^{2r-1}(Y)$. However, for a wide class of Lagrangians (precisely some of the most important ones in the applications) Θ is projectable onto some $J^k(Y)$ with $k < 2r-1$. This happens, for example, in General Relativity with the second order variational problem defined in the bundle of pseudo-Riemannian metrics by the scalar curvature, which has a Poincaré-Cartan form projecting onto the first order jet bundle. This fact also appears for first order problems, as in the case of the Lagrangian of the free Dirac field, which has a Poincaré-Cartan form projecting onto the zero jet bundle!. In this situation, we feel, unlike other authors (e.g. [9]), that different notions of regularity must be given, according to the order of the jet bundle to which the Poincaré-Cartan form —or more properly its exterior differential— can be projected in an irreducible way.

More precisely, if $d\Theta$ projects irreducibly onto $J^{2r-h}(Y)$, it seems natural to us to define the Hamilton-Cartan equation of the problem on $J^{2r-h}(Y)$, and not on $J^{2r-1}(Y)$!, that is;

$$\bar{s}^*(i_D d\Theta) = 0 \qquad \text{for all vector fields } D \text{ of } J^{2r-h}(Y),$$

where $\bar{s}$ is now a section of $p_{2r-h}\colon J^{2r-h}(Y) \longrightarrow X$, and similarly, it is natural to define the notion of regularity by means of the polarity $D \mapsto i_D d\Theta$ on the vector fields of $J^{2r-h}(Y)$.

It is true that this new point of view introduces a strong restriction to the Hamilton-Cartan equation, but in our opinion this is completely natural. It should be noted in this regard that this type of restriction has nothing to do with those leading to the so-called "Hamilton - De Donder equations", based on the Legendre transformation ([9]). We believe that these equations, probably of interest in the early development of the theory, now play a less crucial role because the Poincaré-Cartan form has become the basic geometric object of variational calculus.

It was first pointed out by P. Dedecker ([2], [3]) that for higher order problems one cannot expect, in general and even for "very good" problems,

that the solutions to the Euler-Lagrange and Hamilton-Cartan equations
will coincide, as happens for regular first order problems or for arbitrary
order regular problems in Mechanics. Moreover, the two preceding remarks
and some partial results concerning this topic ([3], [7]) allow us to formulate the main problem of regularity in higher order variational calculus as
follows:

Let $\mathcal{L}v$ be a Lagrangian density of order r on a fibred manifold
$p\colon Y \longrightarrow X$, and let us assume that the exterior differential of one of
the Poincaré-Cartan forms Θ associated with $\mathcal{L}v$ projects irreducibly onto
$J^{2r-h}(Y)$. A notion of regularity for this Lagrangian should be considered
"satisfactory" provided that from this notion the following properties can
be derived:

1) It does not depend on the particular form Θ chosen. (This obviously
   assumes that for any other form Θ' in the family $d\Theta'$ is also irreducibly
   projectable onto $J^{2r-h}(Y)$.)

2) Any solution $\bar{s}$ to the Hamilton-Cartan equation is holonomic up to
   the order $r + 1 - h$. That is, if $s = p_0^{2r-h} \circ \bar{s}$ is the projection of $\bar{s}$ onto
   Y, then $p_{r+1-h}^{2r-h} \circ \bar{s} = j^{r+1-h}(s)$, where $p_k^h\colon J^h(Y) \longrightarrow J^k(Y), h \geq k$,
   stands for the canonical projection.

3) The mapping $\bar{s} \mapsto s = p_0^{2r-h} \circ \bar{s}$ induces a fibre bundle structure
   $p_0^{2r-h}\colon \bar{\mathcal{V}} \longrightarrow \mathcal{V}$ from the set of solutions to the Hamilton-Cartan equation to the set of solutions to the Euler-Lagrange equation. This fibre
   bundle should be endowed with an affine bundle structure defined at
   each solution $s \in \mathcal{V}$ by means of the affine structure of the jet bundles.

4) For each solution $\bar{s} \in \bar{\mathcal{V}}$, the mapping p_0^{2r-h} induces a projection from
   the space $T_{\bar{s}}(\bar{\mathcal{V}})$ of the Jacobi fields of the Hamilton-Cartan equation at
   $\bar{s}$ onto the space $T_s(\mathcal{V})$ of the Jacobi fields of the Euler-Lagrange equation at $s = p_0^{2r-h} \circ \bar{s}$. Furthemore, p_0^{2r-h} projects the pre-symplectic
   metric $(\bar{\Omega}_2)_{\bar{s}}$ onto the corresponding metric $(\Omega_2)_s$, both defined by
   means of the exterior differential of the Poincaré-Cartan form.

Once the above four points have been stated as a reasonable program for the problem of regularity in higher order variational calculus,
in the present work a solution is given for the generic case (i.e., when the
Poincaré-Cartan form is irreducibly defined on the jet bundle of the highest order, $J^{2r-1}(Y)$) and further study is made of the projectable case,
with general results and a detailed analysis for second order problems. In
a forthcoming paper we shall deal with the generalization of this analysis
for r-order problems. A basic tool in our approach is constituted by r
bilinear forms $\langle\,,\,\rangle_{(k,2r+1-k)}, 1 \leq k \leq r$, canonically associated with the
Lagrangian density. These bilinear forms are described in local coordi-
nates by certain matrices in the second order partial derivatives of the
Lagrangian with respect to the r-order variables; one of them is precisely

the Hessian metric. It is also important to remark here that the vanishing of $\langle\,,\,\rangle_{(1,2r)}, \ldots, \langle\,,\,\rangle_{h-1,2r+2-h)}$ is the necessary and sufficient condition for the Poincaré-Cartan form Θ to be projectable onto $J^{2r-h}(Y)$ for $h = 2, \ldots, r$, thus connecting the notion of regularity with that of projectability of Poincaré-Cartan forms.

The need to define the notion of regularity for the different classes of variational problems with projectable Poincaré-Cartan form is now clear. Note that, unlike the generic case in which the projection $p_0^{2r-1} \colon \bar{\mathcal{V}} \longrightarrow \mathcal{V}$ is never bijective, in these classes the above projection can really be a bijection, thus stating a true Hamiltonian formalism. This is the case of the aforementioned second order variational problem defined by the Hilbert Lagrangian in General Relativity whose differential of the Poincaré-Cartan form projects irreducibly onto $J^1(Y)$. In our opinion this fact constitutes the deep geometric meaning of why General Relativity can be dealt with as a first order problem.

2. The notion of regularity: the general case

In what follows, we consider an r-order Lagrangian density $\mathcal{L}v$ on the fibred manifold $p\colon Y \longrightarrow X$.

(2.1) By using a symmetric linear connection ∇_0 on the base manifold X and a derivation law ∇ on the vertical bundle $V(Y)$, we have constructed a sequence of *structure forms* $\theta^{(k)}$ on $J^k(Y)$ in such a way that the forms $(\theta^{(1)}, \ldots, \theta^{(k)})$ are equivalent to the standard structure form on $J^k(Y)$ ([12]):

$$\theta^k = \sum_{i=1}^{m} \sum_{|\alpha|<k} \theta_\alpha^i \otimes \left(\frac{\partial}{\partial y_\alpha^i}\right), \qquad \theta_\alpha^i = dy_\alpha^i - \sum_{j=1}^{n} y_{\alpha+(j)} dx_j \qquad (2.1.1)$$

Actually, $\theta^{(k)}$ is a 1-form on $J^k(Y)$ with values in $S^{k-1}T^*(X) \otimes_{J^k} V(Y)$, and the sequence $\theta^{(1)}, \ldots, \theta^{(k)}$ in a certain sense corresponds to the decomposition of $V(J^{k-1}(Y))$ (this is the vector bundle where θ^k takes its values) into the direct sum of $S^h T^*(X) \otimes_{J^k(Y)} V(Y)$, $h = 0, \ldots, k-1$ (cf. [12], Corollary 3.2).

Furthemore, ∇_0 and ∇ allow us to construct explicitly a *Poincaré-Cartan form* Θ, depending on ∇_0 for $r > 2$, $n > 1$ but not on ∇, and a $V^*(Y)$-valued n-form $\mathcal{E}$ on $J^{2r-1}(Y)$ which is the *Euler-Lagrange form* in our approach, such that $(j^{2r-1}s)^*\mathcal{E} = 0$ is an intrinsic version of Euler-Lagrange equations. Even though $\mathcal{E}$ depends both on ∇_0 and ∇, the form obtained by pulling $\mathcal{E}$ back along any *holonomic section* no longer depends on the connections chosen ([12], Theorem 9.2).

With these preliminaries, our starting point is the following formula for the differential of the Poincaré-Cartan forms constructed by our method:

(2.1.2) ([12] Theorem 7.2) *There exist unique bilinear mappings*

$$(,)_{(k,l)} \colon (S^{k-1}T^*(X) \otimes V(Y)) \times {}_{J^{2r-1}} (S^{l-1}T^*(X) \otimes V(Y)) \to T(X)_{J^{2r-1}}$$

$$((k,l) \in I_r = \{(k,l) \in \mathbb{N} \times \mathbb{N}; 1 \le k \le l \le 2r - 1,\ k \le r,\ k + l \le 2r + 1\})$$

such that:

(i) $(,)_{(k,l)}$ *is alternating for* $k = l$,

(ii)
$$d\Theta = \theta^{(1)} \wedge \mathcal{E} + \sum_{(k,l) \in I_r} (\theta^{(k)} \underset{(k,l)}{\wedge} \theta^{(l)}) v.$$

Furthermore, the bilinear mappings $(,)_{(k,l)}$ *for* $k + l = 2r + 1$ *do not depend on* ∇_0 *and* ∇.

The notation $\eta \underset{(k,l)}{\wedge} \eta'$ stands for the exterior product of the valued forms η, η' with respect to the bilinear mapping $(,)_{(k,l)}$.

When the bilinear mapping being considered is induced by duality, this mapping is not usually included in the notation. This happens with the first term on the right hand side of the previous formula.

Let us define a further bilinear mapping

$$(,)_{(1,2r)} \colon V(Y) \times_{J^{2r-1}} \underbrace{S^{2r-1}T^*(X) \otimes V(Y)}_{\parallel} \longrightarrow T(X)_{J^{2r-1}}$$
$$T(J^{2r-1}/J^{2r-2})$$

by setting:

$$(D_0, D_1)_{(1,2r)} = \tilde{v}^{-1}((\langle i_{D_1}\mathcal{E}, D_0 \rangle)),$$

where $\langle , \rangle \colon V^*(Y) \times V(Y) \longrightarrow \mathbb{R}$ stands for the natural mapping induced by duality, and $\tilde{v} \colon T(X) \longrightarrow \bigwedge^{n-1} T^*(X)$ is the isomorphism produced by contracting with the volume form. It is also proved that:

(2.1.3) ([12], Corollary 7.4) *The bilinear mapping* $(,)_{(1,2r)}$ *does not depend on* ∇_0 *and* ∇.

The table of all these bilinear mappings looks like this:

$k\backslash l$	1	$\cdots$	$r+1$	$\cdots$	$2r-2$	$2r-1$	$2r$
1							$(,)_{(1,2r)}$
2	—					$(,)_{(2,2r-1)}$	—
3	—				$(,)_{(3,2r-2)}$	—	—
$\vdots$				$\cdots$	—	—	—
r	—	$\cdots$	$(,)_{(r,r+1)}$	—	—	—	—

In the ruled squares there are no bilinear mappings (recall that $k \leq l$ and $k + l \leq 2r + 1$), and the bilinear mappings explicitly written down are those which do not depend on the connections chosen.

Proposition 2.1.4 : *The bilinear form associated with* $(\,,\,)_{(k,2r+1-k)}$, $1 \leq k \leq r$, *on the left is totally symmetric, and hence can be considered as a bilinear form*

$$\langle\,,\,\rangle_{(k,2r+1-k)} \colon (S^k T^*(X) \otimes V(Y) \times_{J^{2r-1}} (S^{2r-k} T^*(X) \otimes V(Y))) \longrightarrow \mathbb{R}$$

Proof. The bilinear form associated with $(\,,\,)_{(k,2r+1-k)}$ on the left is the mapping

$$(\,,\,)'_{((k,2r+1-k))} \colon$$
$$(T^*(X) \otimes S^{k-1} T^*(X) \otimes V(Y)) \times_{J^{2r-1}} (S^{2r-k} T^*(X) \otimes V(Y)) \longrightarrow \mathbb{R}$$

given by
$$(\omega \otimes t, u)'_{(k,2r+1-k)} = \omega((t,u)_{(k,2r+1-k)}),$$

with $\omega \in T^*(X)$, $t \in S^{k-1} T^*(X) \otimes V(Y)$, $u \in S^{2r-k} T^*(X) \otimes V(Y)$.

From [12], (6.10), (7.6), (7.8) we conclude that the local expression of $(\,,\,)'_{(k,2r+1-k)}$ is:

$$(\,,\,)'_{(k,2r+1-k)} = \sum_{h,i,j} \sum_{\substack{|\alpha|=k-1 \\ |\beta|=2r-k}} B^{i\beta}_{jh\alpha} \left(\frac{\partial}{\partial x_j}\right) \otimes \left(\frac{\partial}{\partial x}\right)^\alpha \otimes dy_h \otimes \left(\frac{\partial}{\partial x}\right)^\beta \otimes dy_i,$$

where

$$B^{i\beta}_{jh\alpha} = \sum_{|\sigma|=r-k} (-1)^{r-k-1} \frac{1+\alpha_j}{k\binom{r}{k}} \binom{\alpha+(j)+\sigma}{\sigma} \left(\frac{\partial^2 \mathcal{L}}{\partial y^i_{\beta-\sigma} \partial y^h_{\alpha+(j)+\sigma}}\right).$$

Taking into account that $\sum_j \gamma_j (\partial/\partial x_j) \otimes (\partial/\partial x)^{\gamma-(j)} = (\partial/\partial x)^\gamma$, and setting $\gamma = \alpha + (j)$, we obtain:

$$\langle\,,\,\rangle_{(k,2r+1-k)} = (\,,\,)'_{(k,2r+1-k)} = \sum_{\substack{h,i \\ |\beta|=2r-k \\ |\gamma|=k}} B^{i\beta}_{hk} \left(\frac{\partial}{\partial x}\right)^\gamma \otimes dy_h \otimes \left(\frac{\partial}{\partial x}\right)^\beta \otimes dy_i$$

where

$$B^{i\beta}_{h\gamma} = \sum_{|\sigma|=r-k} (-1)^{r-k-1} \frac{\binom{\gamma+\sigma}{\sigma}}{k\binom{r}{k}} \left(\frac{\partial^2 \mathcal{L}}{\partial y^i_{\beta-\sigma} \partial y^h_{\gamma+\sigma}}\right).$$

Remark 2.1.5 : The bilinear forms $\langle,\rangle_{(k,2r+1-k)}$, $1 \leq k \leq r$, essentially coincide with the linear mappings introduced by I. Kolár ([10]) in a different way which, in turn, were a geometric version of the matrices obtained by Shadwick ([13]).

Remark 2.1.6 : For the sake of simplicity, we have considered all the bilinear mappings $(,)_{(k,l)}$ as defined on $J^{2r-1}(Y)$; however, a more careful analysis shows that in fact $(,)_{(k,l)}$ is defined on $J^{3r+1-k-l}(Y)$ whenever $l \geq r+1$.

Theorem 2.2 : *Let Θ be one of the Poincaré-Cartan forms associated with the r-order Lagrangian density $\mathcal{L}v$ by means of the method explained above, and let*

$$\Phi: T(J^{2r-1}(Y)) \longrightarrow \bigwedge^{n} T^*(J^{2r-1}(Y))$$

be the polarity defined by: $\Phi(D) = i_D d\Theta$. Assume $\dim X = n > 1$. Then, we have:

(i) *The maximal rank of Φ is $\rho_{m,n,r} = m\binom{n+r}{r} + m\binom{n+r-1}{r-1} - m + n$.*

(ii) *The polarity Φ takes its maximal rank if all the bilinear forms $\langle,\rangle_{(k,2r+1-k)}$ have no left radical; i.e., if the condition $\langle t,u\rangle_{(k,2r+1-k)} = 0$ for all $u \in S^{2r-k}T^*(X) \otimes V(Y)$ implies $t = 0$. In these case, $\ker \Phi \subset T(J^{2r-1}/J^r)$.*

(iii) *Assume that Φ is of maximal rank and that $\ker \Phi$ is contained in $T(J^{2r-1}/J^{r-1})$. Then the bilinear forms $\langle,\rangle_{(r,r+1)}$ and $\langle,\rangle_{(r-1,r+2)}$ have no left radical.*

Proof. According to [12], (6.6) the local expresion of Poincaré-Cartan forms is

$$\Theta = \sum_{h,j} \sum_{|\beta|\leq r-1} (-1)^{j-1}\lambda^h_{\beta j}\theta^h_\beta \wedge v_j + \mathcal{L}v,$$

where $v_j = dx_1 \wedge \ldots \wedge \widehat{dx_j} \wedge \ldots \wedge dx_n$ and the functions $\lambda^h_{\beta j}$ are determined by [12], (8.3), (8.2), (8.1), (5.3), (6.5). We have:

$$i_{\partial/\partial y^i_\alpha} d\Theta = \sum_{h,j} \sum_{|\beta|\leq r-1} (-1)^{j-1}\frac{\partial\lambda^h_{\beta j}}{\partial y^i_\alpha}\theta^h_\beta \wedge v_j \qquad (r \leq |\alpha| \leq 2r-1) \quad (2.2.1)$$

$$i_{\partial/\partial y^i_\alpha} d\Theta = \left(\frac{\partial\mathcal{L}}{\partial y^i_\alpha} - \sum_j (\lambda^i_{\alpha-(j),j} + X_j\lambda^i_{\alpha j})\right) v+ \qquad (2.2.2)$$

$$+\sum_{h,j} \sum_{|\beta|\leq r-1} (-1)^{j-1}\left(\frac{\partial\lambda^h_{\beta j}}{\partial y^i_\alpha} - \frac{\partial\lambda^i_{\alpha j}}{\partial y^h_\beta}\right)\theta^h_\beta \wedge v_j+$$

$$+\sum_{h,j}\sum_{|\beta|=r}^{2r-2}(-1)^j\frac{\partial\lambda_{\alpha j}^i}{\partial y_\beta^h}\theta_\beta^h\wedge v_j+$$

$$+\sum_{h,j}\sum_{|\beta|=2r-1}(-1)^j\frac{\partial\lambda_{\alpha j}^i}{\partial y_\beta^h}dy_\beta^h\wedge v_j\quad(|\alpha|\le r-1)$$

$$i_{X_k}d\Theta=(-1)^{k-1}\sum_h\sum_{|\beta|\le r-1}\left(\sum_j(X_j\lambda_{\beta j}^h+\lambda_{\beta-(j),j}^h)-\frac{\partial\mathcal{L}}{\partial y_\beta^h}\right)\theta_\beta^h\wedge v_k$$

$$+\sum_{h,i,j}\sum_{|\beta|\le r-1}\sum_{|\alpha|\le 2r-2}(-1)^{j-1}\frac{\partial\lambda_{\beta j}^h}{\partial y_\alpha^i}\theta_\alpha^i\wedge\theta_\beta^h\wedge(i_{\partial/\partial x_k}v_j)\quad(2.2.3)$$

$$+\sum_{h,i,j}\sum_{|\alpha|=2r-1}(-1)^{j-1}\frac{\partial\lambda_{0j}^h}{\partial y_\alpha^i}dy_\alpha^i\wedge\theta_0^h\wedge(i_{\partial/\partial x_k}v_j)\quad(1\le k\le n),$$

where

$$X_j=\frac{\partial}{\partial x_j}+\sum_h\sum_{|\beta|\le 2r-2}y_{\beta+(j)}^h\left(\frac{\partial}{\partial y_\beta^h}\right).$$

Note that $(X_j,\frac{\partial}{\partial y_\beta^h})$, $1\le j\le n$, $1\le h\le m$, $|\beta|\le 2r-1$, is a local basis of $T(J^{2r-1}(Y))$ with dual basis dx_j, θ_β^h $(|\beta|\le 2r-2)$, dy_β^h $(|\beta|=2r-1)$.

Let $\pi\colon\bigwedge^n T^*(J^{2r-1}(Y))\longrightarrow\bigwedge^n T^*(J^{2r-1}(Y))/\bigwedge^n T^*(X)_{J^{2r-1}}$ be the canonical projection and

$$\Phi'\colon V(J^{2r-1}(Y))\longrightarrow\bigwedge^n T^*(J^{2r-1}(Y))/\bigwedge^n T^*(X)_{J^{2r-1}}$$

the composition $\Phi'=\pi\circ(\Phi|_{V(J^{2r-1}(Y))})$. Let us denote by Λ the matrix of Φ' with respect to the basis induced by the previous basis. According to (2.2.1) and (2.2.2), we have:

$$\Phi'\left(\frac{\partial}{\partial y_\alpha^i}\right)=\sum_{h,j}\sum_{|\beta|\le 2r-1}\Lambda_{i\alpha}^{h\beta j}\pi(dy_\beta^h\wedge v_j)\quad(|\alpha|\le 2r-1)$$

and $\Lambda=\Lambda'-\Lambda''$, where the matrices Λ' and Λ'' are given by

$$\Lambda'=\quad\begin{array}{c}\overset{i,\,|\alpha|\le 2r-1}{\boxed{\begin{array}{c}(-1)^{j-1}\dfrac{\partial\lambda_{\beta j}^h}{\partial y_\alpha^i}\\[2ex]\hline\\ 0\end{array}}}\end{array}\quad\begin{array}{l}h,j\\[1ex]|\beta|\le r-1\\ \cdots\cdots\cdots\cdots\\ h,j\\ r\le|\beta|\le 2r-1\end{array}$$

and

$$\Lambda" = \begin{array}{cc} {\scriptstyle i} & {\scriptstyle i} \\ {\scriptstyle |\alpha| \le r-1} & {\scriptstyle r \le |\alpha| \le 2r-1} \\ \boxed{\begin{array}{c|c} (-1)^{j-1}\dfrac{\partial \lambda^i_{\alpha j}}{\partial y^h_\beta} & 0 \end{array}} & \begin{array}{c} h, j \\[1em] |\beta| \le 2r-1 \end{array} \end{array}$$

Moreover, the coefficient $\lambda^h_{\beta j}$ of the Poincaré-Cartan form satisfies the linear equations

$$\lambda^h_{\beta j} = \sum_{|\sigma|=|\beta|+1}^{r} \sigma! \mathring{a}_{\beta,\sigma-(j)} F^h_\sigma \qquad ([12],(8.3)), \qquad (*)$$

thus proving that the rank of Λ' cannot be greater than $m\sum_{k=1}^{r}\binom{n+k-1}{k} = m\binom{n+r}{r} - m$, and obviously $\operatorname{rk}\Lambda'' \le m\sum_{k=0}^{r-1}\binom{n+k-1}{k} = m\binom{n+r-1}{r-1}$. Hence $\operatorname{rk}\Lambda \le \operatorname{rk}\Lambda' + \operatorname{rk}\Lambda'' \le \rho_{m,n,r} - n$, which is equivalent to (i).

Let $B_k = (B^{i\beta}_{h\gamma})$, $|\gamma| = k$, $|\beta| = 2r-k$, be the matrix of the bilinear form $\langle\,,\,\rangle_{(k,2r+1-k)}$. It is clear that $\langle\,,\,\rangle_{(k,2r+1-k)}$ has no left radical if and only if $\operatorname{rk}B_k = m\binom{n+k-1}{k}$. Assume that this is the case for every $k = 1,\ldots,r$.

From [12], (7.8) and the local expresion for B_k obtained in the proof of (2.1.4), we deduce

$$\left(\frac{\partial \lambda^h_{\beta j}}{\partial y^i_\alpha}\right) = -(1+\beta_j)B^{i\alpha}_{h,\beta+(j)} \qquad (|\alpha| = 2r-1-|\beta|, |\beta| \le r-1) \quad (2.2.4)$$

A vector $\xi = (\xi^i_\alpha)$, $|\alpha| \le 2r-1$, belongs to the kernel of Λ if and only if:

$$\sum_{\substack{i \\ |\alpha|\le 2r-1-|\beta|}} \left(\frac{\partial \lambda^h_{\beta j}}{\partial y^i_\alpha}\right)\xi^i_\alpha = \sum_{\substack{i \\ |\alpha|\le r-1}} \left(\frac{\partial \lambda^i_{\alpha j}}{\partial y^h_\beta}\right)\xi^i_\alpha \qquad (|\beta| \le r-1) \quad (2.2.5)$$

$$\sum_{i} \sum_{|\alpha|\le 2r-1-|\beta|} \left(\frac{\partial \lambda^i_{\alpha j}}{\partial y^h_\beta}\right)\xi^i_\alpha = 0 \qquad (r \le |\beta| \le 2r-1) \quad (2.2.6)$$

Let $|\beta| = 2r-1, 2r-2,\ldots,r$ in (2.2.6); from (2.2.4) and this assumption it follows that $\xi^i_\alpha = 0$ for $|\alpha| \le r-1$; hence (2.2.5) for $|\beta| = r-1$ yields $\xi^i_\alpha = 0$ for $|\alpha| = r$, thus proving that $\ker\Lambda \subset T(J^{2r-1}/J^r)$.

We also conclude that the kernel of Λ coincides with the kernel of the submatrix $\widetilde{\Lambda}$ of Λ' given by

$$\widetilde{\Lambda} = \boxed{\ (-1)^{j-1}\frac{\partial \lambda^i_{\beta j}}{\partial y^i_\alpha}\ }\ \ \begin{array}{l} i,r \le |\alpha| \le 2r-1 \\[2pt] h,j \\[2pt] |\beta| \le r-1 \end{array}$$

Setting $\Lambda_k = (\Lambda^{h\beta j}_{i\alpha})^{|\beta|=2r-1-k}_{|\alpha|=k}, 0 \le k \le 2r-1$, we find $\Lambda^{h\beta j}_{i\alpha} = (-1)^{j-1}\partial\lambda^h_{\beta j}/\partial y^i_\alpha$ for $r \le |\alpha| \le 2r-1$, and hence $\widetilde{\Lambda}$ looks like this:

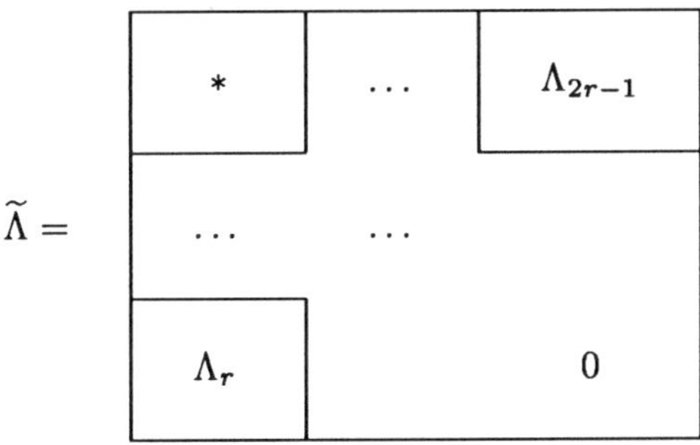

From (2.2.4) it is clear that Λ_k contains B^t_{2r-k} (the transpose of B_{2r-k}) as a submatrix, $r \le k \le 2r-1$. Therefore, the rank of $\widetilde{\Lambda}$ cannot be either less than $\operatorname{rk} B_1 + \cdots + \operatorname{rk} B_r = m\sum_{k=1}^{r}\binom{n+k-1}{k} = m\binom{n+r}{r} - m$ or greater than $m\binom{n+r}{r} - m$. Hence $\operatorname{rk}\widetilde{\Lambda} = m\binom{n+r}{r} - m$, and consequently: $\dim \ker\Lambda = \dim\ker\widetilde{\Lambda} = m\sum_{k=r}^{2r-1}\binom{n+k-1}{k} - m\binom{n+r}{r} + m$ and $\operatorname{rk}\Lambda = m\sum_{k=0}^{2r-1}\binom{n+k-1}{k} - \dim\ker\widetilde{\Lambda} = \rho_{m,n,r} - n$, thus finishing the proof of *(ii)*.

Finally, assume that the hypothesis in *(iii)* holds. Since $\ker\Lambda$ is contained in $T(J^{2r-1}/J^{r-1})$, we have:

$$\operatorname{rk}\widetilde{\Lambda} = \operatorname{rk}\left(\Lambda|_{T(J^{2r-1}/J^{r-1})}\right)$$

$$= \operatorname{rk} T(J^{2r-1}/J^{r-1}) - \operatorname{rk}\left[T(J^{2r-1}/J^{r-1})\bigcap \ker\Lambda\right]$$

$$= \operatorname{rk}\Lambda - m\binom{n+r-1}{r-1} = m\binom{n+r}{r} - m\ .$$

Let us define a submatrix $\bar{\Lambda}$ of $\widetilde{\Lambda}$ as follows:

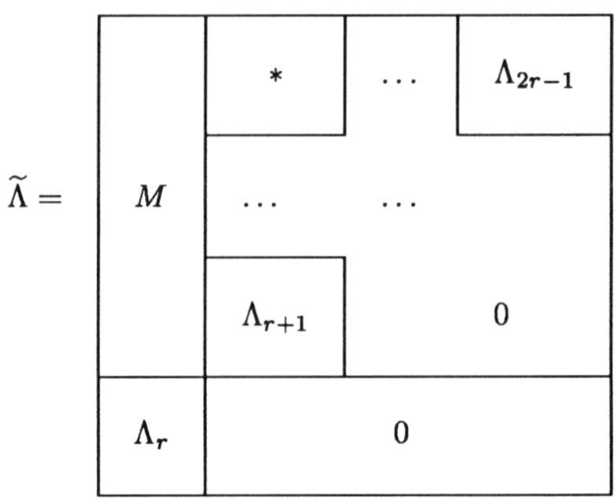

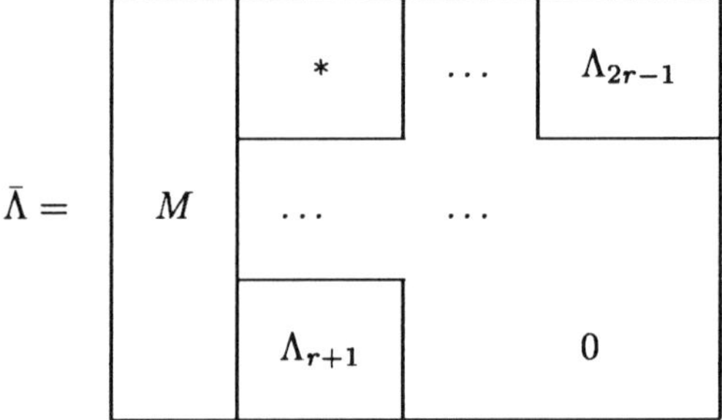

From formula $(*)$ we obtain:

$$\frac{\partial \lambda^h_{\beta j}}{\partial y^i_\alpha} = \sum_{|\sigma|=|\beta|+1}^{r-1} \sigma!\,\mathring{a}_{\beta,\sigma-(j)}\frac{\partial F^h_\sigma}{\partial y^i_\alpha} + \sum_{|\sigma|=r}\frac{\sigma!}{r!}\,\mathring{a}_{\beta,\sigma-(j)}\frac{\partial^2 \mathcal{L}}{\partial y^h_\sigma \partial y^i_\alpha}$$

Thus, subtracting a suitable linear combination of the rows of Λ_r from M, we can assume that the rows of $\bar{\Lambda}$ are

$$\sum_{|\sigma|=|\beta|+1}^{r-1} \sigma!\,\mathring{a}_{\beta,\sigma-(j)}(\partial F^h_\sigma/\partial y^i_\alpha) \quad (|\beta| \leq r-2, |\alpha| \leq r)$$

Hence $\mathrm{rk}\,\bar{\Lambda} \leq m\binom{n+r-1}{r-1} - m$, and since $\mathrm{rk}\,\Lambda_r = \mathrm{rk}\,B_r \leq m\binom{n+r-1}{r}$, $\mathrm{rk}\,\widetilde{\Lambda} = m\binom{n+r}{r} - m$, necessarily: $\mathrm{rk}\,B_r = m\binom{n+r-1}{r}$ and $\mathrm{rk}\,\bar{\Lambda} = m\binom{n+r-1}{r-1} - m$. It thus follows that the rows of Λ_r span $\mathbb{R}^{m\binom{n+r-1}{r}}$. Hence, by subtracting again a suitable linear combination we can assume $M = 0$, and behaving as in the preceding case we conclude that $\mathrm{rk}\,\Lambda_{r+1} = \mathrm{rk}\,B_{r-1} = m\binom{n+r-2}{r-1}$; but we cannot continue any further because the rows of Λ_{r+1} do not span $\mathbb{R}^{m\binom{n+r}{r+1}}$.

Definition 2.3 : *An r-order Lagrangian density is said to be regular if all its associated bilinear forms $\langle\,,\,\rangle_{(k,2r+1-k)}$, $1 \le k \le r$, have no left radical.*

Theorem 2.4 : *Let Θ be one of the Poincaré-Cartan forms associated with an r-order regular Lagrangian density $\mathcal{L}v$, and let $\bar{s}$ be a solution to the Hamilton-Cartan equation corresponding to Θ. We have:*

(i) $\bar{s}$ is holonomic up to order r; that is, if $s = p_0^{2r-1} \circ \bar{s}$ is the projection of $\bar{s}$ onto Y, then $p_r^{2r-1} \circ \bar{s} = j^r s$.

(ii) The section $s = p_0^{2r-1} \circ \bar{s}$ is a solution to the Euler-Lagrange equation. Hence, the projection p_0^{2r-1} induces a mapping

$$p_0^{2r-1} \colon \bar{\mathcal{V}} \to \mathcal{V}$$

from the set of solutions to the Hamilton-Cartan equation onto the set of solutions to the Euler-Lagrange equation.

Proof. We set $\bar{s}_\alpha^i = y_\alpha^i \circ \bar{s}, 0 \le |\alpha| \le 2r - 1$. Note that $\bar{s}_0^i = s_i = y_i \circ s$. From formulas (2.2.1) we obtain

$$\bar{s}^* \left(i_{\partial/\partial y_\alpha^i} d\Theta \right) = \sum_{h,j} \sum_{|\beta| \le r-1} \frac{\partial \lambda_{\beta j}^h}{\partial y_\alpha^i} \bar{\theta}_\beta^h \wedge v_j = 0 \quad (r \le |\alpha| \le 2r - 1),$$

where $\bar{\theta}_\beta^h = \bar{s}^* \theta_\beta^h$. We write $\bar{\theta}_\beta^h \wedge v_j = f_{\beta j}^h v$ for some function $f_{\beta j}^h$ of X. From (2.2.4) we obtain $\sum_{h,j} B_{h(j)}^{i\alpha} f_{0j}^h = 0$ for $|\alpha| = 2r - 1$, and by virtue of the hypothesis $f_{0j}^h = 0$; i.e., $\bar{s}_{(j)}^h = \partial s_h / \partial x_j$.

By recurrence on $k = 0, \ldots, r - 1$ we prove that $\bar{s}_\beta^h = D^\beta s_h$ for $|\beta| \le k + 1$. Actually,

$$\sum_{h,j} \sum_{|\beta| \le k+1} \left(\frac{\partial \lambda_{\beta j}^h}{\partial y_\alpha^i} \right) f_{\beta j}^h = 0 \qquad \text{for } |\alpha| = 2r - 2 - k,$$

and from (2.2.4) and the recurrence hypothesis we derive

$$\sum_h \sum_{|\gamma|=k+2} B_{h\gamma}^{i\alpha} \left(\sum_j \gamma_j f_{\gamma-(j),j}^h \right) = 0,$$

which yields $\sum_j \gamma_j f_{\gamma-(j),j}^h = 0$ for $|\gamma| = k + 2$; or equivalently,

$$\bar{s}_\gamma^h = \frac{1}{k+2} \sum_j \gamma_j \frac{\partial \bar{s}_{\gamma-(j)}^h}{\partial x_j} = \frac{1}{k+2} \sum_j \gamma_j \frac{\partial D^{\gamma-(j)} s_h}{\partial x_j} = D^\gamma s_h.$$

In order to prove *(ii)*, it will be sufficient to state that

$$\lambda^h_{\beta j} \circ \bar{s} = \lambda^h_{\beta j} \circ j^{2r-1}s \quad (0 \le |\beta| \le r-1), \tag{2.4.1}$$

where $\lambda^h_{\beta j}$ are, as usual, the coefficients of Θ, because in this case:

$$\bar{s}^*(i_{\partial/\partial y_i} d\Theta) = (j^{2r-1}s)^*(i_{\partial/\partial y_i} d\Theta) = 0,$$

and from the formula for $d\Theta$ in (2.1.2) we thus conclude:

$$(j^{2r-1}s)^*(i_{\partial/\partial y_i} d\Theta) = (j^{2r-1}s)^*(\partial/\partial y_i \circ \mathcal{E}) = 0;$$

hence $(j^{2r-1}s)^*\mathcal{E} = 0$.

We prove (2.4.1) by descending recurrence on $k = |\beta|$. For $k = r-1$, this is trivial because of *(i)*. From the Hamilton-Cartan equation and the recurrence hypothesis we obtain

$$\left(\sum_j \lambda^h_{\beta-(j),j}\right) \circ \bar{s} = \frac{\partial \mathcal{L}}{\partial y^h_\beta} \circ j^r s - \sum_j (D_j \lambda^h_{\beta j}) \circ j^{2r-1}s.$$

Hence,

$$\left(\sum_j \lambda^h_{\beta-(j),j}\right) \circ \bar{s} = \left(\sum_j \lambda^h_{\beta-(j),j}\right) \circ j^{2r-1}s.$$

Next, note that it follows from the formula $(*)$ in the proof of (2.2) that (2.4.1) for $k \le |\beta| \le r-1$ is equivalent to: $F^h_\sigma \circ \bar{s} = F^h_\sigma \circ j^{2r-1}s$ for $k+1 \le |\sigma| \le r$, and by again using $(*)$ we have:

$$\sum_j \lambda^h_{\beta-(j),j} = |\beta|! F^h_\beta + \sum_j \sum_{|\sigma|=|\beta|+1}^{r} \sigma! \mathring{a}_{\beta-(j),\sigma-(j)} F^h_\sigma.$$

Hence $F^h_\beta \circ \bar{s} = F^h_\beta \circ j^{2r-1}s$, and accordingly $\lambda^h_{\beta-(j),j} \circ \bar{s} = \lambda^h_{\beta-(j),j} \circ j^{2r-1}s$, thus finishing the proof.

Remark 2.5 : By using the local expressions obtained in the proof of (2.1.4) it is possible to prove that (2.4.1) for $|\beta| = r-2$ is equivalent to saying that the unique tensor $t_{r+1} \in s^*\left(S^{r+1}T^*(X) \otimes V(Y)\right)$ such that $t_{r+1} + j^{r+1}s = p^{2r-1}_{r+1} \circ \bar{s}$ belongs to the *right radical* of $\langle\,,\,\rangle_{(r-1,r+2)}$. This fact has allowed us to describe the projection $p^{2r-1}_0 : \bar{\mathcal{V}} \to \mathcal{V}$ in the particular case of second order Lagrangians as an affine fibre bundle (cf. [7], Théorème 2). It would be interesting to obtain a natural generalization of this result for problems of order $r \ge 3$.

Definition 2.6 : *Let $\bar{s}$ be a solution to the Hamilton-Cartan equation with respect to one of the Poincaré-Cartan forms Θ associated with an r-order Lagrangian density. A vector field $\bar{D}$ of $J^{2r-1}(Y)$ vertical over X defined along $\bar{s}$ is said to be a Jacobi field if and only if:*

$$\bar{s}^*(i_{D'}L_{\bar{D}}d\Theta) = 0 \quad \text{for all vector fields } D' \text{ of } J^{2r-1}(Y).$$

We denote by $T_{\bar{s}}(\bar{\mathcal{V}})$ the set of all Jacobi vector fields defined along $\bar{s}$, and we think of this space as the tangent space at a point $\bar{s} \in \bar{\mathcal{V}}$ of the "manifold of solutions" to the Hamilton-Cartan equation. In the general case, of course, the notion of a Jacobi field depends on the Poincaré-Cartan form chosen. Evidently, if $\bar{\mathcal{V}}$ depends on Θ, the same should be expected of its tangent space. We have, however, the following result:

Proposition 2.7 : *Let D be a vector field of Y vertical over X defined along a solution s to the Euler-Lagrange equation of an r-order variational problem, and let $D_{(2r-1)}$ be the infinitesimal contact transformation of order $2r - 1$ associated with D defined along $j^{2r-1}s$. If $D_{(2r-1)}$ is a Jacobi field along $j^{2r-1}s$ with respect to a Poincaré-Cartan form associated with the problem, the same holds for any other Poincaré-Cartan form.*

Proof. According to [12], Corollary 7.6, for any two Poincaré-Cartan forms Θ, Θ' there exists a valued $(n-1)$-form $\bar{\eta}$ such that $d\Theta' - d\Theta = \theta^r \wedge (\theta^{2r-1} \wedge \bar{\eta})$. Hence, if $D_{(2r-1)}$ is a Jacobi field with respect to Θ, for every vector field D' of J^{2r-1}, we have:

$$(j^{2r-1}s)^*(i_{D'}L_{D_{(2r-1)}}d\Theta') = (j^{2r-1}s)^* \left[i_{D'}L_{D_{(2r-1)}}(\theta^r \wedge (\theta^{2r-1} \wedge \bar{\eta})) \right].$$

According to the definition of an infinitesimal contact transformation, there exist endomorphisms A and B of $(j^{r-1}s)^*V(J^{r-1}(Y))$ and $(j^{2r-2}s)^*V(J^{2r-2}(Y))$ respectively, such that $L_{D_{(2r-1)}}\theta^r = A \circ \theta^r$ and $L_{D_{(2r-1)}} = B \circ \theta^{2r-1}$. As $(j^{2r-1}s)^*\theta^r = 0, (j^{2r-1}s)^*\theta^{2r-1} = 0$, the previous formula proves that $D_{(2r-1)}$ is also a Jacobi field with respect to Θ'.

We denote by $T_s(\mathcal{V})$ the set of vertical vector fields D along s whose infinitesimal contact transformation $D_{(2r-1)}$ is a Jacobi field along $j^{2r-1}s$, with respect to any Poincaré-Cartan form. According to the above proposition, for every Poincaré-Cartan form, the mapping $D \mapsto D_{(2r-1)}$ establishes an injection

$$T_s(\mathcal{V}) \hookrightarrow T_{j^{2r-1}s}(\bar{\mathcal{V}}).$$

Theorem 2.8 : *Let $\bar{s}$ be a solution to the Hamilton-Cartan equation with respect to a Poincaré-Cartan form Θ associated with an r-order regular variational problem, and let $\bar{D}$ be a Jacobi field along $\bar{s}$ with respect to Θ. Let D be the vertical vector field along $s = p_0^{2r-1} \circ \bar{s}$ defined by $D = (p_0^{2r-1})_* \bar{D}$. Then, we have:*

(i) $\bar{D}$ is holonomic up to order r; that is, $(p_r^{2r-1})_* \bar{D} = D_{(r)}$.

(ii) $D \in T_s(\mathcal{V})$; in other words, $D_{(2r-1)}$ is a Jacobi vector field along $j^{2r-1}s$ for every Poincaré-Cartan form.

Proof. Let $\bar{D} = \sum_i \sum_{|\alpha| \le 2r-1} \bar{u}_\alpha^i (\partial/\partial y_\alpha^i)$ the local expression of $\bar{D}$, and let us consider a local extension $\tilde{\neq} \sum_i \sum_{|\alpha| \le 2r-1} \tilde{u}_\alpha^i (\partial/\partial y_\alpha^i)$, such that $\tilde{u}_\alpha^i \circ \bar{s} = \bar{u}_\alpha^i, |\alpha| \le 2r - 1$. From the Jacobi equation for $r \le |\alpha| \le 2r - 1$, and taking into account (2.4)-(i) and (2.4.1), we obtain

$$\bar{s}^* \left(i_{\partial/\partial y_\alpha^i} L_{\bar{d}\Theta} \right) = \sum_{h,j} \sum_{|\beta| \le r-1} \left(\frac{\partial \lambda_{\beta j}^h}{\partial y_\alpha^i} \circ \bar{s} \right) \left(\frac{\partial \bar{u}_\beta^h}{\partial x_j} - \bar{u}_{\beta+(j)}^h \right) = 0,$$

and by proceeding as in the proof of (2.4), we have $\bar{u}_{\beta+(j)}^h = \partial \bar{u}_\beta^h/\partial x_j$, for $|\beta| \le r-1$. Hence $\bar{u}_\alpha^h = D^\alpha u_h$ ($|\alpha| \le r$), where $u_h = \bar{u}_0^h$ are the coefficients of D, i.e., $D = \sum u_h (\partial/\partial y_h)$.

In order to prove *(ii)*, it will be sufficient to state that

$$\bar{D}\lambda_{\beta j}^h = D_{(2r-1)}\lambda_{\beta j}^h \qquad (0 \le |\beta| \le r - 1), \tag{2.8.1}$$

because in this case we have:

$$\bar{s}^* \left(i_{\partial/\partial y_i} L_{\bar{D}} d\Theta \right) = (j^{2r-1}s)^* \left(i_{\partial/\partial y_i} L_{D_{(2r-1)}} d\Theta \right) = 0,$$

and from the formula for $d\Theta$ in (2.1.2) we thus conclude:

$$(j^{2r-1}s)^* \left(i_{\partial/\partial y_i} L_{D_{(2r-1)}} d\Theta \right) = (j^{2r-1})^* \left(\partial/\partial y_i \circ L_{D_{(2r-1)}} \mathcal{E} \right) = 0.$$

Hence $(j^{2r-1}s)^* \left(L_{D_{(2r-1)}} \mathcal{E} \right) = 0$, and this is precisely one of the characteristic equations for Jacobi vector fields ([11]). Finally, the proof of (2.8.1) is by descending recurrence, behaving as in the proof of (2.4.1).

Remark 2.8.1 : The mapping $T_{\bar{s}}(\bar{\mathcal{V}}) \longrightarrow T_s(\mathcal{V})$, $\bar{D} \mapsto D = (p_0^{2r-1})_* \bar{D}$, can be considered as the tangent linear mapping associated with the bundle mapping $p_0^{2r-1} : \bar{\mathcal{V}} \to \mathcal{V}$ at the point $\bar{s}$ in the sense that if $\bar{s}_t$ is a curve in $\bar{\mathcal{V}}$, then the above mapping transforms the "tangent vector" at $t = 0$ into its image $s_t = p_0^{2r-1} \circ \bar{s}_t$.

Moreover, the vector spaces $T_{\bar{s}}(\bar{\mathcal{V}})$ and $T_s(\mathcal{V})$ are endowed with presymplectic metrics $(\bar{\Omega}_2)_{\bar{s}}$ and $(\Omega_2)_s$ respectively, taking values in the space of closed $(n-1)$-forms of X ([4], [11]). We accomplish our initial program by proving the following:

Corollary 2.9 *With the above hypothesis and notations, we have*

$$(\bar{\Omega}_2)_{\bar{s}} = (p_0^{2r-1})^*(\Omega_2)_s.$$

This result proves that, although the solutions to the Hamilton-Cartan and Euler-Lagrange equations are not equivalent for higher order problems, the dynamical structures that they define are the same in the regular case.

Proof. According to the definitions of Ω_2 and $\bar{\Omega}_2$, we have:

$$(p_0^{2r-1})^*(\Omega_2)_s(\bar{D}, \bar{D}') = i_{D'_{(2r-1)}} i_{D_{(2r-1)}} d\Theta, \quad (\bar{\Omega}_2)_{\bar{s}}(\bar{D}, \bar{D}') = i_{\bar{D}'} i_{\bar{D}} d\Theta.$$

Hence, it will suffice to prove that the right hand sides in the above equations coincide. This follows from (2.8.1) taking into account that

$$i_{D'_{(2r-1)}} i_{D_{(2r-1)}} d\Theta = \sum_{h,j} \sum_{|\beta| \leq r-1} (-1)^{j-1} \Big[(D_{(2r-1)}\lambda^h_{\beta j})(D'_{(2r-1)}y^i_\beta) - (D'_{(2r-1)}\lambda^h_{\beta j})(D_{(2r-1)}y^h_\beta) \Big] v_j,$$

and similarly for $i_{\bar{D}'} i_{\bar{D}} d\Theta$.

3. Projectability of Poincaré-Cartan forms and regularity

(3.1) One of the most outstanding features of the theory of regularity in the higher order setting is that the same bilinear forms involved in the definition of regularity determine the projectability of Poincaré-Cartan forms onto a lower order jet bundle $J^k(Y), k < 2r - 1$. In fact, according to [12], Corollary 7.7 and bearing in mind Proposition (2.1.4), we can state:

(3.1.1) *The necessary and sufficient conditions for the Poincaré-Cartan forms associated with an r-order Lagrangian density to be projectable onto $J^{2r-h}(Y), 2 \leq h \leq r$ are that the bilinear forms $\langle , \rangle_{(k,2r+1-k)}$ vanish for $1 \leq k \leq h - 1$.*

Note that these conditions only involve the first $r - 1$ bilinear forms $\langle , \rangle_{(1,2r)}, \ldots, \langle , \rangle_{(r-1,r+1)}$. No mention is made of the last one, $\langle , \rangle_{(r,r+1)}$, the Hessian metric. Nevertheless, if the Hessian metric vanishes, all the other bilinear forms will do so as well. Actually, the vanishing of the Hessian metric is equivalent to saying that the Lagrangian is an affine function on $J^{r-1}(Y)$. In higher order Mechanics all the bilinear forms $\langle , \rangle_{(k,2r+1-k)}$ coincide with the Hessian metric, except for one sign. Hence, the Poincaré-Cartan form of a variational problem of arbitrary order in one variable is projectable onto a lower order jet bundle if and only if its Hessian metric vanishes. It is interesting to note that in second order Mechanics the Poincaré-Cartan form associated with an affine Lagrangian is automatically projectable onto $J^1(Y)$, as we shall see below.

According to these results, it is obvious that the class of variational problems for which the Poincaré-Cartan forms project onto the same $J^k(Y)$ must have its own notion of regularity, according to the program proposed in the introduction. We have almost completely accomplished this program for second order problems and our results are explained below. In a forthcoming paper we shall extend them to problems of order $r \geq 3$.

3.2 Second order variational problems whose Poincaré-Cartan form projects onto $J^2(Y)$.

In what follows we shall consider a second order problem defined by a Lagrangian density $\mathcal{L}v$ whose Poincaré-Cartan form Θ is projectable onto $J^2(Y)$. In this case, the polarity associated with Θ is defined on $J^2(Y)$; i.e., we have

$$\Phi: T(J^2(Y)) \longrightarrow \bigwedge^n T^*(J^2(Y)), \qquad \Phi(D) = i_D d\Theta, \qquad (3.2.1)$$

and the first result that we obtain is the following:

Proposition 3.2.2 : *With the above hypothesis we have:*
(i) If $n = 1$, then Θ is projectable onto $J^1(Y)$.
(ii) If $n > 1$, then Φ takes its maximal rank if and only if it is injective. If the Hessian metric is not singular, every solution to the Hamilton-Cartan equation which is holonomic up to the first order is also holonomic up to the second order; i.e., it is a solution to the Euler-Lagrange equation.

Proof. By using the standard notations in Mechanics, according to the hypothesis we have $\mathcal{L} = \mathcal{L}_0 + \sum_h \mathcal{L}\ddot{q}_h$ for some functions $\mathcal{L}_0, \mathcal{L}_h$ on $J^1(Y)$. Hence,

$$\Theta = \sum_h \left[\frac{\partial \mathcal{L}_0}{\partial \dot{q}_h} - \frac{\partial \mathcal{L}_h}{\partial t} - \sum_i \dot{q}_i \frac{\partial \mathcal{L}}{\partial q_i} \right] (dq_h - \dot{q}_h dt) + \sum_h \mathcal{L}_h d\dot{q}_h + \mathcal{L}_0 dt,$$

which is a differential form on $J^1(Y)$. As for *(ii)*, if $\bar{s}$ is a 1-holonomic section to the Hamilton-Cartan equation, we have:

$$\bar{s}^* \left(i_{\partial/\partial y^h_{(kl)}} d\Theta \right) = \sum_{i,j,q} (-1)^{j-1} \frac{\partial \lambda^i_{(q)j}}{\partial y^h_{(kl)}} (\bar{s}^* \theta^i_{(q)}) \wedge v_j = 0,$$

and since $\left| \partial^2 \mathcal{L}/\partial y^h_\alpha \partial y^i_\beta \right| \neq 0$ $(|\alpha| = |\beta| = 2)$, we conclude $\bar{s}^* \theta^i_{(q)} = 0$.

Remark 3.2.3 : Even for Lagrangians (with the Poincaré-Cartan form projecting to J^2) whose polarity Φ is injective we cannot ensure that the solutions to the Hamilton-Cartan equation are 1-holonomic. Furthermore, solutions $\bar{s}$ may exist to the Hamilton-Cartan equation whose projection $p_0^2 \circ \bar{s}$ is not a solution to the Euler-Lagrange equation. Examples can be taken from the family of Lagrangians in two variables ($n = 2$): $\mathcal{L} = A y_{(2,0)} + B y_{(0,2)} + C + F(y_{(2,0)} y_{(0,2)} - y_{(1,1)}^2)$, where A, B, C and F are functions on $J^1(Y)$, which represent all the Lagrangians whose Poincaré-Cartan form lies on $J^2(Y)$. Therefore, for this class of Lagrangians we cannot expect a result similar to Theorem (2.4). We shall see, however, that the situation is more satisfactory for the affine Lagrangians.

From now onwards we shall assume that $n > 1$.

Let $\mathcal{L}$ be a second order affine Lagrangian. Its fiber differential $d_{J^2/J^1}\mathcal{L}$ is a symmetric section of the bundle $\bigotimes^2 T(X) \otimes_{J^2} V^*(Y)$. In fact, $d_{J^2/J^1}\mathcal{L}$ is defined on $J^1(Y)$, because from its local expression $\mathcal{L} = \mathcal{L}_0 + \sum_{i,|\alpha|=2} \mathcal{L}_\alpha^i y_\alpha^i$ ($\mathcal{L}_0, \mathcal{L}_\alpha^i \in \mathcal{C}^\infty(J^1(Y))$) we obtain:

$$d_{J^2/J^1}\mathcal{L} = \sum_i \sum_{|\alpha|=2} \mathcal{L}_\alpha^i \left(\frac{\partial}{\partial x}\right)^\alpha \otimes dy_i.$$

Note that $d_{J^2/J^1}\mathcal{L}$ coincides with the linear form associated with the affine function $\mathcal{L}$. Hence, $d_{J^2/J^1}\mathcal{L}$ can be viewed as a mapping from $J^1(Y)$ into $\bigotimes^2 T(X) \otimes_Y V^*(Y)$, and consequently its fiber differential (over Y) determines a bilinear form

$$d_{J^1/Y}(d_{J^2/J^1}\mathcal{L}) \colon (T^*(X) \otimes V(Y)) \times_{J^1} \left(\overset{2}{\bigotimes} T^*(X) \otimes V(Y)\right) \longrightarrow \mathbb{R},$$

locally given by:

$$d_{J^1/Y}(d_{J^2/J^1}\mathcal{L})(dx_j \otimes \partial/\partial y_i, dx_k \otimes dx_l \otimes \partial/\partial y_h) = \frac{1}{2}(kl)! \frac{\partial \mathcal{L}_{(kl)}^h}{\partial y_{(j)}^i}.$$

Definition 3.2.4 : *Let $\mathcal{L}$ be a second order affine Lagrangian. We define the bilinear forms*

$$\mathbb{B}_1 \colon (T^*(X) \otimes V(Y)) \times_{J^2} (S^2 T^*(X) \otimes V(Y)) \longrightarrow \mathbb{R}$$

$$\mathbb{B}_2 \colon (T^*(X) \otimes V(Y)) \times_{J^1} \left(\overset{2}{\bigotimes} T^*(X) \otimes V(Y)\right) \longrightarrow \mathbb{R}$$

as follows:

$$\mathbb{B}_1(\omega \otimes D, \bar{D}) = \omega(\tilde{v}^{-1}(i_D i_{\bar{D}} d\Theta)),$$

where Θ is the Poicaré-Cartan form associated with $\mathcal{L}v$, and $\omega \in T^*(X)$, $D \in V(Y)$, $\bar{D} \in T(J^2/J^1) = S^2 T^*(X) \otimes_{J^2} V(Y)$.

$$\mathbb{B}_2(\omega \otimes D, \omega_1 \otimes \omega_2 \otimes D') = d_{J^1/Y}(d_{J^2/J^1}\mathcal{L})(\omega \otimes D, \omega_2 \otimes \omega_1 \otimes D')$$
$$- d_{J^1/Y}(d_{J^2/J^1}\mathcal{L})(\omega_1 \otimes D', \omega_2 \otimes \omega \otimes D),$$

where $\omega, \omega_1, \omega_2 \in T^*(X), D, D' \in V(Y)$. We also denote by

$$\mathbb{B}_2' : (T^*(X) \otimes V(Y)) \times_{J^1} (S^2 T^*(X) \otimes V(Y)) \longrightarrow \mathbb{R}$$

the restriction of $\mathbb{B}_2$.

From the local expression of Θ we obtain:

$$\mathbb{B}_1(dx_j \otimes \partial/\partial y_h, \partial/\partial y^i_{(kl)}) = \frac{\partial \lambda^h_{0j}}{\partial y^i_{(kl)}}$$

$$\mathbb{B}_2(dx_k \otimes \partial/\partial y_h, dx_q \otimes dx_j \otimes \partial/\partial y_i) = \frac{\partial \lambda^i_{(q)j}}{\partial y^h_{(k)}} - \frac{\partial \lambda^h_{(k)j}}{\partial y^i_{(q)}}$$

Theorem 3.2.5 : Let $\mathcal{L}$ be a second order affine Lagrangian, and let $\Phi \colon T(J^2(Y)) \to \bigwedge^n T^*(J^2(Y))$ be the polarity associated with the Poincaré-Cartan form defined by $\mathcal{L}v$. We have:

(i) $\operatorname{rk}\Phi \leq m + n + 2mn$. (Recall that $n > 1$.)

(ii) The polarity Φ takes its maximal rank if and only if the bilinear forms $\mathbb{B}_1$ and $\mathbb{B}_2$ have no left radical. In this case, the kernel of Φ is contained in $T(J^2/J^1)$. In fact, $\ker\Phi$ coincides with the right radical of $\mathbb{B}_1$.
Assume that Φ takes its maximal rank. Then,

(iii) Every solution to the Hamilton-Cartan equation is holonomic up to the first order and its projection onto J^0 is a solution to the Euler-Lagrange equation.

(iv) The induced mapping $p_0^2 \colon \bar{\mathcal{V}} \to \mathcal{V}$ is a fibre bundle whose structure is as follows.
The fibre of p_0^2 at a point $s \in \mathcal{V}$ is the set $t + j^2 s$, where t is an arbitrary section of $(j^2 s)^* R$ and $R \to J^2$ is the right radical of $\mathbb{B}_2'$, the sum being taken with respect to the affine structure of $J^2(Y)$ over $J^1(Y)$.

Proof. Parts *(i)* and *(ii)* follow from the local expressions for $\mathbb{B}_1$ and $\mathbb{B}_2$ obtained from above and the formulas (2.2.1) and (2.2.2), taking into account the particular case we are considering. As for *(iii)*, let $\bar{s}$ be a solution to the Hamilton-Cartan equation. The condition $\bar{s}^*(i_{\partial/\partial y^h_{(kl)}} d\Theta) = 0$ gives rise to

$$\sum_{i,j}(-1)^{j-1} \frac{\partial \lambda^i_{0j}}{\partial y^h_{(kl)}}(\bar{s}^* \theta^i_0) \wedge v_j = 0,$$

and since $\mathbb{B}_1$ has no left radical, it follows that $(\bar{s}^*\theta_0^i) \wedge v_j = 0$, or equivalently $\bar{s}^*\theta^1 = 0$, thus proving that $\bar{s}$ is holonomic up to first order. We set $s = p_0^2 \circ \bar{s}$. Then:

$$\bar{s}^* \left(i_{\partial/\partial y_h} d\Theta\right) = \left(\frac{\partial \mathcal{L}}{\partial y_h} \circ \bar{s}\right) v - \sum_j (-1)^{j-1} d(\lambda_{0j}^h \circ \bar{s}) \wedge v_j +$$

$$+ \sum_{i,j,k} (-1)^{j-1} \left(\frac{\partial \lambda_{(k)j}^i}{\partial y_h} \circ \bar{s}\right) (\bar{s}^*\theta_{(k)}^i) \wedge v_j = 0$$

$$(j^2 s)^* \left(i_{\partial/\partial y_h} d\Theta\right) = \left(\frac{\partial \mathcal{L}}{\partial y_h} \circ j^2 s\right) v - \sum_j (-1)^{j-1} d(\lambda_{0j}^h \circ j^2 s) \wedge v_j.$$

As $\frac{\partial \mathcal{L}}{\partial y_h} \circ j^2 s - \frac{\partial \mathcal{L}}{\partial y_h} \circ \bar{s} = \sum_i \sum_{|\alpha|=2} \left(\frac{\partial \mathcal{L}_\alpha^i}{\partial y_h} \circ j^1 s\right) (D^\alpha s_i - \bar{s}_\alpha^i)$ and

$$\sum_{i,j,k} (-1)^{j-1} \left(\frac{\partial \mathcal{L}_{(k)j}^i}{\partial y_h} \circ \bar{s}\right) (\bar{s}^*\theta_{(k)}^i) \wedge v_j$$

$$= \sum_i \sum_{|\alpha|=2} \left(\frac{\partial \mathcal{L}_\alpha^i}{\partial y_h} \circ j^1 s\right) (D^\alpha s_i - \bar{s}_\alpha^i) v,$$

we obtain:

$$(j^2 s)^* \left(i_{\partial/\partial y_h} d\Theta\right) = \sum_j (-1)^{j-1} [d(\lambda_{0j}^h \circ \bar{s}) - d(\lambda_{0j}^h \circ j^2 s)] \wedge v_j.$$

Developing the right hand side, we finally conclude that $(j^2 s)^* \left(i_{\partial/\partial y_h} d\Theta\right) = 0$, thus proving that s is a solution to the Euler-Lagrange equation. (Repeat the argument in the proof of (2.4)-(i).)

Moreover, we have:

$$0 = \sum_{i,j,q} (-1)^{j-1} \left[\left(\frac{\partial \lambda_{(q)j}^i}{\partial y_{(k)}^h} \circ \bar{s}\right) - \left(\frac{\partial \lambda_{(k)j}^h}{\partial y_{(q)}^i} \circ \bar{s}\right)\right] (\bar{s}^*\theta_{(q)}^i) \wedge v_j$$

$$= \bar{s}^* \left(i_{\partial/\partial y_{(k)}^h} d\Theta\right) .$$

That is,

$$\sum_{i,j,q} \left[\left(\frac{\partial \lambda_{(q)j}^i}{\partial y_{(k)}^h} \circ j^1 s\right) - \left(\frac{\partial \lambda_{(k)j}^h}{\partial y_{(q)}^i} \circ j^1 s\right)\right] \left(\frac{\partial^2 s_i}{\partial x_j \partial x_q} - \bar{s}_{(jq)}^i\right) = 0,$$

thus proving *(iv)* and finishing the proof of the theorem.

3.3 Second order variational problems
whose Poincaré-Cartan form projects onto $J^1(Y)$.

The proof of the following proposition is straightforward and is thus omitted.

Proposition 3.3.1 : *The necessary and sufficient conditions for the Poincaré-Cartan form associated with a second order Lagrangian density $\mathcal{L}v$ to be projectable onto $J^1(Y)$ are the following:*
(i) $\mathcal{L}$ is an affine function of $J^2(Y)$ over $J^1(Y)$,
(ii) The bilinear form $\mathbb{B}_1$ in (3.2.4) vanishes identically.

In what follows, we shall consider a second order Lagrangian density $\mathcal{L}v$ whose Poincaré-Cartan form is projectable onto $J^1(Y)$. For such a Lagrangian, the polarity associated with Θ is defined on $J^1(Y)$; i.e.,

$$\Phi: T(J^1(Y)) \longrightarrow \overset{n}{\bigwedge} T^*(J^1(Y)), \quad \Phi(D) = i_D d\Theta.$$

Theorem 3.3.2 : *Let $\mathcal{L}v$ be a second order Lagrangian density whose Poincaré-Cartan form projects onto $J^1(Y)$. Moreover, let us assume that $\mathcal{L}v$ satisfies the following two conditions:*
(a) The polarity $\Phi: T(J^1(Y)) \to \bigwedge^n T^*(J^1(Y))$is injective.
(b) The bilinear form $\mathbb{B}_2$ in (3.2.4) vanishes identically.
Under these conditions, we have:
(i) Every solution $\bar{s}$ to the Hamilton-Cartan equation is holonomic (i.e. $\bar{s} = j^1 s$) and the mapping $s \mapsto j^1 s$ establishes a bijection from the set of solutions to Euler-Lagrange equation onto the set of solutions to the Hamilton-Cartan equation.
(ii) Locally, there exist functions $\mathcal{L}_1, \ldots, \mathcal{L}_n$ on $J^1(Y)$ such that $\partial \mathcal{L}_j / \partial y^i_{(k)} = (-1)^{j-1} \lambda^i_{(k)j}$. By setting $p^i_j = (-1)^{j-1} \lambda^i_{0j} - \partial \mathcal{L}_j / \partial y_i$, the functions (x_j, y_i, p^i_j) constitute a system of local coordinates.
(iii) In this system, we have:

$$d\Theta = \sum_{i,j} dp^i_j \wedge dy_i \wedge v_j + dH \wedge v,$$

where $H = H_0 + \sum_j (-1)^j (\partial \mathcal{L}_j / \partial x_j), \quad H_0 = \mathcal{L}_0 - \sum_{i,j} y^i_{(j)} \lambda^i_{0j}$.
(iv) A section $\bar{s}$ is a solution to the Hamilton-Cartan equation if and only if $\bar{s}$ satisfies the following system of equations:

$$(-1)^{j-1} \frac{\partial \bar{s}_i}{\partial x_j} + \left(\frac{\partial H}{\partial p^i_j} \circ \bar{s} \right) = 0 \qquad \sum_j (-1)^j \frac{\partial \bar{s}^i_j}{\partial x_j} + \left(\frac{\partial H}{\partial y_i} \circ \bar{s} \right) = 0.$$

Proof. Condition *(b)* means $\partial\lambda^i_{(q)j}/\partial y^h_{(k)} = \partial\lambda^h_{(k)j}/\partial y^i_{(q)}$. Hence,

$$\bar{s}^*\left(i_{\partial/\partial y^h_{(k)}} d\Theta\right) = \sum_{i,j}(-1)^{j-1}\left(\frac{\partial\lambda^i_{0j}}{\partial y^h_{(k)}} - \frac{\partial\lambda^h_{(k)j}}{\partial y_i}\right)(\bar{s}^*\theta^i_0)\wedge v_j,$$

and according to *(a)* the matrix $\left(\dfrac{\partial\lambda^i_{0j}}{\partial y^h_{(k)}} - \dfrac{\partial\lambda^h_{(k)j}}{\partial y_i}\right)^{i,j}_{h,k}$ is not singular. From the equation $\bar{s}^*\left(i_{\partial/\partial y^h_{(k)}} d\Theta\right) = 0$ it thus follows that $(\bar{s}^*\theta^i_0)\wedge v_j$. Therefore, the section $\bar{s}$ is holonomic, and *(i)* is proved.

Let η_j be the differential form of $T^*(J^1/Y)$ given by:

$$\eta_j = (-1)^{j-1}\sum_{i,k}\lambda^i_{(k)j}d_{J^1/Y}y^i_{(k)}.$$

By virtue of the condition *(b)*, we obtain $d_{J^1/Y}\eta_j = 0$. Hence, functions $\mathcal{L}_j$ exist locally such that $\eta_j = d_{J^1/Y}\mathcal{L}_j$, or equivalently, $\partial\mathcal{L}_j/\partial y^i_{(k)} = (-1)^{j-1}\lambda^i_{(k)j}$. Moreover, we have:

$$\frac{\partial p^i_j}{\partial y^h_{(k)}} = (-1)^{j-1}\frac{\partial\lambda^i_{0j}}{\partial y^h_k} - \frac{\partial^2\mathcal{L}_j}{\partial y^h_{(k)}\partial y_i} = (-1)^{j-1}\left(\frac{\partial\lambda^i_{0j}}{\partial y^h_{(k)}} - \frac{\partial\lambda^h_{(k)j}}{\partial y_i}\right)$$

and hence according to *(a)* we obtain:

$$\frac{\partial(x_j, y_i, p^i_j)}{\partial(x_j, y_i, y^i_{(j)})} \neq 0,$$

thus proving *(ii)*.

Starting with the standard local expression for Θ, we have:

$$\Theta = \sum_{i,j}\sum_{|\alpha|=0}^{1}(-1)^{j-1}\lambda^i_{\alpha j}\theta^i_\alpha\wedge v_j + \mathcal{L}v$$

$$= \sum_{i,j}\sum_{|\alpha|=0}^{1}(-1)^{j-1}\lambda^i_{\alpha j}dy^i_\alpha\wedge v_j + H_0 v$$

$$= \sum_{i,j}p^i_j dy_i\wedge v_j + \sum_j d\mathcal{L}_j\wedge v_j + \left(H_0 + \sum_j(-1)^j\frac{\partial\mathcal{L}_j}{\partial x_j}\right)v,$$

and taking the exterior derivative we obtain *(iii)*. Finally, *(iv)* directly follows from *(iii)*.

It can be verified that the second order Lagrangian defined on the bundle of pseudo-Riemannian metrics by the scalar curvature satisfies conditions *(a)* and *(b)* above, such that all the results in (3.3.2) can be applied to this Lagrangian. We shall, however, consider below another example to illustrate the theory.

Example 3.3.3 : Let us consider the variational problem defined on the scalar field $p: X \times \mathbb{R} \longrightarrow X$ by the 2nd order Lagrangian density $\mathcal{L}v$, where:

$$\mathcal{L}(j_x^2 f) = \frac{1}{2}\|df\|^2 + \triangle f \qquad (f \in \mathcal{C}^\infty(X)),$$

and the norm $\|\cdot\|$, the Laplacian and the volume v are taken with respect to a Riemannian metric g on X. Note that we add to a first order Lagrangian its own Euler-Lagrange equation.

As direct calculation shows, the Poincaré-Cartan form is projectable onto $J^1(Y)$ and conditions *(a)* and *(b)* are satisfied. By using the Hamilton equations obtained in *(iv)*, we conclude that the solutions to the Euler-Lagrange equation are the harmonic functions!. We thus have a second order problem whose solutions coincide with those of a first order problem. This happens because there is a first order Lagrangian density $\mathcal{L}'v$ whose Poincaré-Cartan form Θ' satisfies $d\Theta = d\Theta'$. Actually, we can take $\mathcal{L}'(j_x^1 f) = \frac{1}{2}\|df\|^2$.

REFERENCES

[1] R. Abraham, J. Marsden. "Foundations of Mechanics". Benjamin-Cummings, Menlo Park, Calif., (1978).

[2] P. Dedecker. *Existe-t-il, en calcul des variations, un formalisme de Hamilton-Jacobi-E. Cartan pour les intégrales multiples d'ordre supérieur?* C.R.Acad.Sc.Paris **298**, Série I, (1984), 397–400.

[3] P. Dedecker. *Sur le formalisme de Hamilton-Jacobi pour une intégrale multiple d'ordre supérieur.* C.R.Acad.Sc.Paris, **299**, Série I (1984), 363–366.

[4] P.L. García. *The Poincaré-Cartan Invariant in the Calculus of Variations.* Symp. Mathematica **14**, Academic Press, London (1974), 219–246.

[5] P.L. García, J. Muñoz Masqué. *On the Geometrical Structure of Higher Order Variational Calculus.* Proc. IUTAM-ISIMM Symp. on Modern Developments in Analyt. Mech. Turin, June 1982, Vol. I-Geometrical Dynamics, Tecnoprint, Bologna (1983), 127–147.

[6] P.L. García, J. Muñoz Masqué. *Higher order analytical dynamics.* Proc. VII ELAM on Dynamical Systems and Partial Differential Equations. Caracas, July 1984, Universidad Simón Bolívar, Ed. Equinoccio, Caracas (1986), 19–47.

[7] P.L. García, J. Muñoz Masqué.*Le problème de la régularité dans le calcul des variations du second ordre.* C.R.Acad.Sc.Paris **301**, Série I, nº 12 (1985), 639–642.

[8] H. Goldschmidt, S. Sternberg. *The Hamilton-Cartan Formalism in the Calculus of Variations.* Ann.Inst.Fourier, **23** (1973), 203–267.

[9] M.J. Gotay. *A Multisymplectic Framework for Classical Field Theory and the Calculus of Variations, I. Covariant Hamiltonian Formalism.* To appear in "Mechanics, Analysis and Geometry: 200 Years After Lagrange", M. Francaviglia and D.D. Holm, Eds. (North Holland, Amsterdam, 1990).

[10] I. Kolár. *A geometrical version of the higher order Hamilton formalism in fibred manifolds.* Journal of Geom. and Physics **1** (1984),126–137.

[11] J. Muñoz Masqué. *Pre-symplectic structure for higher order variational problems.* Proc. of the Conference on Dif. Geometry and its Applications, Nové Mésto na Moravé, Sept. 1983, Vol II-Applications, Univerzita Karlova, Praga (1984), 191–205.

[12] J. Muñoz Masqué. Poincaré-Cartan forms in higher order Variational Calculus on fibred manifolds. Revista Matemática Iberoamericana **1** (1985), 85–126.

[13] W. Shadwick. *The Hamiltonian formulation of regular r-th order Lagrangian field theories.* Letters in Mathematical Physics **6** (1982), 409–416.

[14] S. Sternberg. *Some Preliminary Remarks on the Formal Variational Calculus of Gel'fand and Dikii.* (Dif.Geom.Methods in Math. Physics, II, Proceedings, 1977), Lect. Notes in Math. **676**, Springer-Verlag, Berlin (1978), 399-407.

P.L. García Pérez
Departamento de Matemáticas
Universidad de Salamanca
Plaza de la Merced 1-4
37008 Salamanca
Spain

J. Muñoz Masqué
Unidad de Matemáticas
C.S.I.C.
C. Serrano 123
28006 Madrid
Spain

An Exterior Differential Systems
Approach to the Cartan Form

MARK J. GOTAY

Dedicated to Jean-Marie Souriau

Abstract

The notion of a "Lepagean equivalent" of a given variational problem is
defined, and the basic properties of these objects are sketched. Using some
ideas of Bryant, Dedecker and Griffiths, I show that every constant rank
variational problem has a canonical Lepagean equivalent and that, as a
consequence, to every such variational problem there is assigned a canonical
"generalized Cartan form." These observations rely crucially upon the
exterior differential systems approach to the calculus of variations. Then
I prove that this generalized Cartan form is *universal* in the sense that
every Cartan form for a classical variational problem can be obtained from
it by pullback upon sectioning a certain bundle. These results lead to a
simple new proof of the existence of Cartan forms for classical variational
problems, and explain in intrinsic terms why and to what extent classical
Cartan forms are (typically) not uniquely determined by a Lagrangian.

1. Introduction

The "Cartan form" is the basic geometric object in the calculus of variations [13, 14, 16, 17, 19].[1] It generalizes the Poincaré-Cartan 1-form

$$\Theta = L\,dt + \frac{\partial L}{\partial \dot{q}^i}\left(dq^i - \dot{q}^i\,dt\right)$$

familiar from mechanics. However, in field theory this form is not uniquely
defined, even for first order Lagrangians. Indeed, consider a Lagrangian
density $\mathcal{L} = L\omega$ on the first jet bundle $J^1 Y$ of a fibration $Y \to X$. (Notation
and terminology are explained in the Appendix.) The "standard" Cartan
form corresponding to $\mathcal{L}$ [14, 16] is

$$\Theta = L\omega + \frac{\partial L}{\partial y_\nu^A}\psi^A \wedge \omega_\nu. \tag{1.1}$$

---

[1]There is substantial variance in terminology regarding Cartan forms in the
literature. For the moment, I will use the term loosely. More precise nomenclature
will be introduced later.

But if L is nowhere zero, then

$$\Theta = \frac{1}{L^{n-1}}\eta^1 \wedge \ldots \wedge \eta^n, \qquad (1.2)$$

where $\eta^\mu = L\,dx^\mu + \frac{\partial L}{\partial y_\mu^A}\psi^A$ and $n = \dim X$, is another Cartan form [6, 35]. Yet a third is [2, 26, 35]:

$$\Theta = L\omega + \sum_{s=1}^n \frac{1}{(s!)^2}\left(\frac{\partial^s L}{\partial y_{\nu_1}^{A_1}\ldots\partial y_{\nu_s}^{A_s}}\right)\psi^{A_1}\wedge\ldots\wedge\psi^{A_s}\wedge\omega_{\nu_1\ldots\nu_s}. \qquad (1.3)$$

These Cartan forms are designed with specific purposes in mind. The first one seems well suited to studying the Hamiltonian aspects of a theory and the initial value problem [18]. Carathéodory's form (1.2) has the property of being invariant under contact transformations of the Lagrangian $\mathcal{L}$, as opposed to (1.1), which is invariant only under (prolongations of) bundle automorphisms [4]. The form (1.3) has the virtue that it is closed iff $\mathcal{L}$ is variationally trivial. See [1, 3, 11, 23] for further discussion.

These three examples do not exhaust all the possibilities. In fact, they are merely special instances of the Cartan form

$$\Theta = L\omega + \frac{\partial L}{\partial y_\nu^A}\psi^A \wedge \omega_\nu + \sum_{s=2}^n \lambda_{A_1\ldots A_s}^{\nu_1\ldots\nu_s}\psi^{A_1}\wedge\ldots\wedge\psi^{A_s}\wedge\omega_{\nu_1\ldots\nu_s}, \qquad (1.4)$$

where the coefficients $\lambda_{A_1\ldots A_s}^{\nu_1\ldots\nu_s}$ are arbitrarily specifiable functions on J^1Y. This is the most general Cartan form for a first order theory which is semi-basic with respect to the projection $J^1Y \to Y$ [30, 31]. The significance of such forms is in general unclear.

The situation is correspondingly more complex for higher order theories [1, 25, 27].

This peculiarity has been extensively studied in recent years [1, 2, 8, 12, 15, 20, 21, 24-29, 32, 33, 35-38]. As a result, one now knows that Cartan forms can be associated with any "classical" variational problem – that is, one defined on an appropriate jet bundle of some fibration – and the Cartan forms corresponding to a given Lagrangian have to a large extent been characterized. Several explicit methods of constructing Cartan forms have also been developed. Thus the situation is somewhat under control, at least from a practical standpoint.

But there is more to be done. In particular, it is not altogether clear – in intrinsic, global terms – exactly *why* the Cartan form is nonunique or, more precisely, *why* there is no preferred choice in general. Part of the reason is that the techniques used to construct Cartan forms are themselves

ambiguous (in that they rely upon the specification of elements extrinsic
to the variational problem under consideration, such as connections and
partitions of unity). Put another way, it is difficult to see what the geometry
behind these constructions is. Related aspects of the problem are that it is
not obvious how to choose a Cartan form in specific cases, nor is it apparent
what the consequences of different such choices are.

Anderson has studied these issues using the variational bicomplex [1].
Here I present a different technique for constructing Cartan forms which
also sheds some light upon these questions. It is based on the exterior
differential systems approach to the calculus of variations [7, 10, 19]. In
this framework, I define the general notion of a "Lepagean equivalent" of a
given variational problem and sketch some of the basic properties of these
objects. Then, following ideas of Bryant, Dedecker and Griffiths [5, 10, 11,
19], I show that every constant rank variational problem has a canonical
Lepagean equivalent and that, as a consequence, to every such variational
problem there is *canonically* associated a "generalized Cartan form." This
is not, however, a Cartan form in the usual sense of the classical calculus
of variations, as it is not defined directly on a jet bundle. Nonetheless,
this canonical object can be used to induce "classical" Cartan forms on the
jets. In fact I show that this generalized Cartan form is *universal* inasmuch
as every Cartan form for a classical variational problem can be obtained
from it by pullback upon sectioning a certain bundle. These results lead
to a new proof of the existence of Cartan forms for classical variational
problems, and enable one to pinpoint exactly what the nonuniqueness in
the classical Cartan form is and how it arises. The method is relatively
simple and algorithmic, and so is possibly less "mysterious" than some
other approaches.

The constructions in this paper also yield a new and interesting connection between the Hamiltonian and Lagrangian formalisms in the classical
calculus of variations. I will show that one can "interpolate" between these
two pictures by means of a series of Lepagean equivalents, and that the
Hamiltonian formalism is closely related to the canonical Lepagean equivalent. In this way one can recover Dedecker's theory [11] as well as Griffiths'
formalism [19] as special cases.

The paper is laid out as follows. In §2 I introduce and discuss Lepagean
equivalents. §3 is devoted to demonstrating the existence of the canonical Lepagean equivalent of any constant rank variational problem, and to
determining the relationship between their extremals. The key result here
states that every extremal of the equivalent problem projects to an extremal of the original problem. In the next section I study the canonical
Lepagean equivalent of a classical variational problem, and prove the converse of the preceding result in this context. Then in §5 I review the theory
of classical Cartan forms on jet bundles following [25, 27]. §6 adapts the

proof of the main theorem in §4 to prove the universality of the canonical generalized Cartan form in the sense explained above. I conclude with some speculations and open problems.

2. Lepagean Equivalents

The basic data in the calculus of variations consists of a fibration $\pi :$ $M \to X^n$, a differential ideal $\mathcal{I}$ in the exterior algebra $C^\infty(\wedge^* M)$ and an n-form $\mathcal{L}$ on M. Usually the Lagrangian density $\mathcal{L}$ is required to be semi-basic, but this is not essential. In this context an *integral* of $\mathcal{I}$ is a section $\phi \in \Gamma(\pi)$ such that $\phi^* \mathcal{I} = 0$. The space of all sections of $\pi : M \to X$ which are integrals of $\mathcal{I}$ is denoted $\Gamma(\pi, \mathcal{I})$.

The *variational problem* $(M \xrightarrow{\pi} X, \mathcal{I}, \mathcal{L})$ is to extremize the action functional

$$\mathcal{A}_\mathcal{L}(\phi) = \int_X \phi^* \mathcal{L} \tag{2.1}$$

over all $\phi \in \Gamma(\pi, \mathcal{I})$.[2] This can be viewed as a "constrained" variational problem, inasmuch as one is not free to extremize over all sections of π, but rather only those which are integrals of $\mathcal{I}$. It is in principle simpler to consider "free" problems, in which one is allowed to extremize over arbitrary sections of a fibration.[3] To accomplish this is the *raison d'être* of "Lepagean equivalents."

Definition 1. A *Lepagean equivalent* of a variational problem $(M \xrightarrow{\pi} X, \mathcal{I}, \mathcal{L})$ is another variational problem $(W \xrightarrow{\rho} X, \{0\}, \Theta)$, together with a surjective submersion $\nu : W \to M$, such that: (i) $\rho = \pi \circ \nu$, and (ii) if $\gamma \in \Gamma(\rho)$ satisfies $\nu \circ \gamma \in \Gamma(\pi, \mathcal{I})$, then

$$\gamma^* \Theta = (\nu \circ \gamma)^* \mathcal{L}. \tag{2.2}$$

The problem now is to extremize

$$\mathcal{A}_\Theta(\gamma) = \int_X \gamma^* \Theta \tag{2.3}$$

over *arbitrary* sections γ of $W \to X$. Thus one replaces the constrained variational problem on M with a free problem on W. In effect the *generalized*

---

[2]Of course $\mathcal{A}_\mathcal{L}$ may not be everywhere defined. Throughout this paper I ignore such technicalities and work formally. In particular I suppose that $\partial X = \emptyset$ and neglect surface terms.

[3]In the literature "free" classical variational problems are often referred to as "zeroth order" problems.

Cartan form Θ functions as a new Lagrangian density. By a slight abuse of terminology, I sometimes refer to such a form as a "Lepagean equivalent" of the Lagrangian density. In particular, Θ is a Lepagean equivalent of $\mathcal{L}$ if

$$\Theta \equiv \nu^* \mathcal{L} \bmod \nu^* \mathcal{I};$$

thus (2.2) generalizes the classical "first congruence of Lepage" [30, 31]. A standard argument shows that the variational equations corresponding to $(W \xrightarrow{\rho} X, \{0\}, \Theta)$ take the form: $\gamma \in \Gamma(\rho)$ is extremal iff

$$\gamma^* \left(i_\zeta d\Theta \right) = 0 \tag{2.4}$$

for all vector fields ζ on W. I refer to these as the *Cartan equations*.

The relation (2.2) guarantees that the actions $\mathcal{A}_\mathcal{L}$ and $\mathcal{A}_\Theta$ are the same. But it is important to observe that (2.2) does *not* imply any relationship between the extremals of the variational problems $(M \xrightarrow{\pi} X, \mathcal{I}, \mathcal{L})$ and $(W \xrightarrow{\rho} X, \{0\}, \Theta)$. For instance, $(M \xrightarrow{\pi} X, \{0\}, \mathcal{L})$ is a Lepagean equivalent of $(M \xrightarrow{\pi} X, \mathcal{I}, \mathcal{L})$ according to the definition, but there is no *a priori* correspondence between their extremals. This trivial example indicates that some Lepagean equivalents are less useful than others. On the other hand, it would not do to impose such a relation *ab initio*. For one thing, as I shall demonstrate later, when a relation between extremals does exist, it may vary depending upon the "type" of Lepagean equivalent being considered (even of the same original variational problem)! Another point is that requiring a variational problem and its Lepagean equivalents to have the *same* extremals is too strong a condition to make the notion of Lepagean equivalent useful, since this is often true only subject to a great deal of regularity (§7). Moreover, such a requirement entails solving the Euler-Lagrange equations, at least to the extent of reducing the problem of computing the extremals of $(M \xrightarrow{\pi} X, \mathcal{I}, \mathcal{L})$ to that of computing the integral sections of the Cartan system of $d\Theta$ on W, *cf.* Remark 3 below and §1.e of [19]. But the idea behind the introduction of Lepagean equivalents is merely to *simplify* the original variational problem.[4]

In view of these comments it is useful to introduce some terminology. Let $\mathcal{E}(\mathcal{L})$ denote the set of all extremals of $(M \xrightarrow{\pi} X, \mathcal{I}, \mathcal{L})$, etc. Then say that $(W \xrightarrow{\rho} X, \{0\}, \Theta)$ is a *covariant* Lepagean equivalent if the projection $\nu : W \to M$ induces a map $\mathcal{E}(\Theta) \to \mathcal{E}(\mathcal{L})$. Similarly, it is *contravariant* if every extremal of $\mathcal{L}$ is the projection of some extremal of Θ. As the preceding discussion shows, a given Lepagean equivalent need be neither of these

---

[4]Thus the appelation Lepagean *equivalent* is somewhat misleading (but is nonetheless prevalent).

types; on the other hand, it might simultaneously be both ("bivariant"). In the latter eventuality it does not follow that $\mathcal{E}(\Theta) \approx \mathcal{E}(\mathcal{L})$, but rather only that $\mathcal{E}(\Theta) \to \mathcal{E}(\mathcal{L})$ is surjective.

Remark 1. The classical notion of a "Lepagean equivalent" of a given Lagrangian is due to Lepage [30, 31] and was later reinterpreted and extended by Dedecker [9-11] and Krupka [27]. Especially relevant is Dedecker's 1957 memoir [10], which contains a number of ideas and results intimately related to those presented here. In particular, the concepts of covariant and bivariant Lepagean equivalent are akin to his *"semi-relèvement d'une structure variationnelle"* and *"relèvement,"* respectively. Dedecker also proves, under suitable assumptions, that one can construct a (global) *semi-relèvement* (*cf.* §3 and Theorem 3 *ff.*) And in the classical case, he shows that the *semi-relèvement* so constructed is in fact a *relèvement* (*cf.* Theorem 4 *ff.*).

Remark 2. There are several conceivable alternatives to this definition of Lepagean equivalent. Firstly, if one only demands that the *actions* (2.1) and (2.3) be the same, then (2.2) need only hold modulo a divergence. Such considerations might be important when considering the divergence equivalence of variational problems [5]. Secondly, as I have defined it, a Lepagean equivalent is "larger" than the original problem in the sense that $\nu : W \to M$ is a projection. On occasion – and this happens in general relativity [27] – one encounters Lepagean equivalent variational problems which are strictly "smaller"; that is to say, there is a submersion $M \to W$. I do not consider either of these possibilities here.

Remark 3. It is possible to give a somewhat different description of Lepagean equivalent as follows. Given a form α on a manifold N, define the *Cartan system* $\mathcal{C}(\alpha)$ to be the differential ideal generated by the forms $i_\xi \alpha$ for $\xi \in C^\infty(TN)$. Applying this construction to $d\Theta$ on W, it follows from (2.4) that $(W \xrightarrow{\rho} X, \mathcal{C}(d\Theta), 0)$ and $(W \xrightarrow{\rho} X, \{0\}, \Theta)$ have the same extremals. Thus the extremals of the latter can be computed "simply" by integrating $\mathcal{C}(d\Theta)$. This implies that one could equally well define a Lepagean equivalent by trivializing the Lagrangian rather than the exterior ideal.

3. The Canonical Lepagean Equivalent

It is not obvious that every variational problem $(M \xrightarrow{\pi} X^n, \mathcal{I}, \mathcal{L})$ has a nontrivial Lepagean equivalent. But Dedeckeer [9-11], and subsequently Griffiths [19] and Bryant [5], have shown that this is the case; in fact, they have shown that every variational problem has a *canonical* Lepagean equivalent,[5] provided only that the differential ideal $\mathcal{I}$ has "constant rank" (which I shall assume henceforth). By this I mean that it is the differential ideal generated by $C^\infty(I)$ for some graded subbundle $I \subseteq \wedge^* M$.

The motivation for their construction comes from the Lagrange multiplier trick. Suppose for the moment that $n = 1$ and $\mathcal{I}$ is a Pfaffian ideal, generated by a collection $\{\theta^a\}$ of 1-forms on M. One can rephrase the variational problem $(M \xrightarrow{\pi} X, \mathcal{I}, \mathcal{L})$ as wanting to extremize the action (2.1) subject to the "side conditions" $\theta^a = 0$. The Lagrange multiplier trick states that one can accomplish this by adjoining these side conditions to the Lagrangian $\mathcal{L}$ with arbitrary multipliers λ_a, obtaining

$$\Theta = \mathcal{L} + \lambda_a \theta^a.$$

Let W be the space M enlarged by considering the λ_a as new variables. Then $(W \to X, \{0\}, \Theta)$ is a free problem.

Here is how this construction is performed intrinsically. Let $I^p = I \cap \wedge^p M$ for each $p \geq 0$ and build the affine subbundle W of $\wedge^n M$ whose fiber above $m \in M$ is

$$W_m = \{\mathcal{L}(m) + \beta_m \mid \beta_m \in I^n_m\}.$$

I denote this by

$$W = \mathcal{L} + I^n.$$

Since $\wedge^n M$ is a bundle of n-forms, it carries a canonical n-form defined in the usual manner. Let $\Theta_\mathcal{L}$ be the pullback of this form to W, let ν be the restriction of the projection $\wedge^n M \to M$ to W and set $\rho = \pi \circ \nu$.

Proposition 2. $(W \xrightarrow{\rho} X, \{0\}, \Theta_\mathcal{L})$ *so defined is a Lepagean equivalent of* $(M \xrightarrow{\pi} X, \mathcal{I}, \mathcal{L})$.

Proof. Suppose $\gamma \in \Gamma(\rho)$ with $\phi = \nu \circ \gamma \in \Gamma(\pi, \mathcal{I})$. Then there exists $w \in \Gamma(\nu)$ such that $\gamma = w \circ \phi$. By the universal property of $\Theta_\mathcal{L}$ (*viz.* $w^* \Theta_\mathcal{L} = w$ for all $w \in \Gamma(\nu)$),

$$\gamma^* \Theta_\mathcal{L} = (w \circ \phi)^* \Theta_\mathcal{L} = \phi^* w^* \Theta_\mathcal{L} = \phi^* w = \phi^* \mathcal{L} = (\nu \circ \gamma)^* \mathcal{L}$$

where in the second to last equality I have used the definition of W and the fact that ϕ is an integral of $\mathcal{I}$. ∎

Thus one is justified in calling $(W \xrightarrow{\rho} X, \{0\}, \Theta_\mathcal{L})$ the *canonical Lepagean equivalent* of $(M \xrightarrow{\pi} X, \mathcal{I}, \mathcal{L})$. It is obvious that the assignment

---

[5]The rudiments of this idea can be traced (at least) as far back as 1941, to a remark in a paper by Pâquet [34].

$(M \xrightarrow{\pi} X, \mathcal{I}, \mathcal{L}) \rightsquigarrow (W \xrightarrow{\rho} X, \{0\}, \Theta_{\mathcal{L}})$ is an (affine) functor from the category of constrained variational problems to the category of free ones. This is a welcome feature of the canonical Lepagean equivalent, as classical Lepagean equivalents are not functorial in general.

Next is the fundamental relation between the extremals of a variational problem and its canonical Lepagean equivalent; to some extent it reflects the fact that W, being a bundle of forms, is a covariant object. Although I will not use this result I include it for completeness.

Theorem 3. *The canonical Lepagean equivalent is covariant.*

Proof. (Bryant [5]). Let $\gamma \in \mathcal{E}(\Theta_{\mathcal{L}})$ and set $\phi = \nu \circ \gamma$. There are two things to prove: (i) $\phi^* \mathcal{I} = 0$ and (ii) $\mathcal{A}_{\mathcal{L}}(\phi)$ is a critical value.

For (i), let $\beta \in C^\infty(I^n)$. Then the translations $\alpha_m \rightsquigarrow \alpha_m + t\beta(m)$ define a flow on W whose generating vector field X_β satisfies

$$i\left(X_\beta\right) d\Theta_{\mathcal{L}} = \nu^* \beta. \tag{3.1}$$

(This works exactly as in the cotangent bundle case.) But as γ is an extremal,

$$\phi^* \beta = \gamma^* \nu^* \beta = \gamma^* \left(i\left(X_\beta\right) d\Theta_{\mathcal{L}}\right) = 0.$$

Thus $\phi^* C^\infty(I^n) = 0$. Of course $\phi^* C^\infty(I^p) = 0$ for $p > n$. Finally, suppose $\alpha \in C^\infty(I^p)$ for $p < n$. Choose any $(n - p)$-form η; then $\alpha \wedge \eta \in C^\infty(I^n)$ and so $\phi^*(\alpha \wedge \eta) = \phi^* \alpha \wedge \theta^* \eta = 0$. Since this is true for all η, $\phi^* \alpha = 0$. Thus (i) follows.

For (ii), let ϕ_t be any variation of $\phi = \phi_0$ through integrals of $\mathcal{I}$. Let γ_t be any lift of ϕ_t to W with $\gamma_0 = \gamma$. Then by (2.2) $\mathcal{A}_{\mathcal{L}}(\phi_t) = \mathcal{A}_{\Theta_{\mathcal{L}}}(\gamma_t)$. Hence

$$\frac{d}{dt}\mathcal{A}_{\mathcal{L}}\left(\phi_t\right)|_{t=0} = \frac{d}{dt}\mathcal{A}_{\Theta_{\mathcal{L}}}\left(\gamma_t\right)|_{t=0} = 0$$

as γ_0 extremizes $\mathcal{A}_{\Theta_{\mathcal{L}}}$. ∎

It is unclear to what extent the canonical Lepagean equivalent is also contravariant. In complete generality one has formal recourse to the Lagrange multiplier theorem (see, *e.g.*, §43.9 of [40]), but the assumptions there are rather severe. For single integral problems Hsu [22] has shown that this is so, provided $\mathcal{I}$ is Pfaffian and has no completely integrable subsystems (modulo some nontrivial technicalities concerning the existence of compactly supported variations). On the other hand, Dedecker [11] has shown that the canonical Lepagean equivalent of the classical variational problem $(J^1 Y \xrightarrow{\pi^1} X, \mathcal{J}_1, \mathcal{L})$ is contravariant. He has also proven a number of results implying that the same is true for higher order classical problems in a semi-holonomic setting [10]. In the next section I will extend these last results to fully holonomic classical problems of any order.

I want to dwell for a moment on the case when $\mathcal{I}$ is Pfaffian, so that $C^\infty(I)$ is locally generated by a collection $\{\theta^a\}$ of 1-forms. Then every element w of W can be locally written

$$
\begin{aligned}
w &= \mathcal{L} + \lambda_a^i \theta^a \wedge \tau_i + \lambda_{ab}^{ij} \theta^a \wedge \theta^b \wedge \tau_{ij} + \cdots \\
&= \mathcal{L} + \sum_r \lambda_{a_1\ldots a_r}^{i_1\ldots i_r} \theta^{a_1} \wedge \ldots \wedge \theta^{a_r} \wedge \tau_{i_1\ldots i_r},
\end{aligned}
\tag{3.2}
$$

where the forms $\{\tau_{i_1\ldots i_r}\}$ constitute a basis for a complement of I^{n-r} in $\wedge^{n-r} M$. Thus the fibers of $W \to M$ are parametrized by the coefficients $\lambda_{a_1\ldots a_r}^{i_1\ldots i_r}$. Since $\Theta_\mathcal{L}$ is a tautological form, its coordinate expression at w is the same as that of w itself, *viz.*,

$$
\Theta_\mathcal{L} = \mathcal{L} + \sum_r \lambda_{a_1\ldots a_r}^{i_1\ldots i_r} \theta^{a_1} \wedge \cdots \wedge \theta^{a_r} \wedge \tau_{i_1\ldots i_r}.
\tag{3.3}
$$

But when $\mathcal{I}$ is Pfaffian there are other constructions which also give rise to canonical Lepagean equivalents. For instance, let K^n be the subspace of I^n consisting of forms which are of degree one in the generators $\{\theta^a\}$, and set

$$
V = \mathcal{L} + K^n.
$$

In a similar manner one obtains a canonical n-form on V, etc. (I use the same notation as before.) But now $\Theta_\mathcal{L}$ looks like

$$
\Theta_\mathcal{L} = \mathcal{L} + \lambda_a^i \theta^a \wedge \tau_i.
$$

Clearly one can perform a similar construction using forms of degree at most r, $1 \le r \le n$, in the $\{\theta^a\}$. It is straightforward to check that all of these possibilities define canonical Lepagean equivalents with the same properties as the one based on I^n. Later in §6 I will show that W is intimately related to "classical Lepagean equivalents" in the sense of Krupka [27], while V corresponds to the special case of "classical Cartan forms" as defined in §5.

4. Classical Variational Problems

Let $\pi : Y \to X^n$ be a fibration and let $\mathcal{L} = L\omega$ be a kth-order Lagrangian density, now presumed to be a semi-basic n-form on $J^k Y$. The corresponding *classical variational problem* is $(J^k Y \xrightarrow{\pi^k} X, \mathcal{J}_k, \mathcal{L})$ where

$\mathcal{J}_k$ is the contact differential ideal on J^kY. As is well-known, a section ϕ of π^k is an extremal iff $\phi = j^k\varphi$ where φ satisfies the Euler-Lagrange equations

$$\frac{\delta L}{\delta y^A}\left(j^{2k}\varphi\right) = \sum_{|\underline{\mu}|=0}^{k}(-1)^{|\underline{\mu}|}\partial_{\underline{\mu}}\left(\frac{\partial L}{\partial y_{\underline{\mu}}^A}\left(j^k\varphi\right)\right) = 0.$$

Let $(W \xrightarrow{\rho} X, \{0\}, \Theta_{\mathcal{L}})$ be the canonical Lepagean equivalent of $(J^kY \xrightarrow{\pi^k} X, \mathcal{J}_k, \mathcal{L})$, where $W = \mathcal{L} + J_k^n$. Theorem 3 states that it is covariant. This section will be devoted to proving:

Theorem 4. *The canonical Lepagean equivalent of a classical variational problem is contravariant.*

Let $\phi \in \mathcal{E}(\mathcal{L})$. The plan is to show that ϕ can be lifted to a section γ of ρ satisfying (2.4).

The problem can be simplified as follows. Let γ be any lift of ϕ to W; I claim that $\gamma^*(i_\zeta d\Theta_{\mathcal{L}}) = 0$ for all ν-vertical vector fields ζ on W. Indeed, referring to part (i) of the proof of Theorem 3, one sees that the translational vector fields X_β span $C^\infty(V\nu)$, where $\beta \in C^\infty(J_k^n)$. Since $\phi \in \Gamma(\pi^k, \mathcal{J}_k)$, (3.1) yields

$$\gamma^*\left(i\left(X_\beta\right)d\Theta_{\mathcal{L}}\right) = \gamma^*\nu^*\beta = \phi^*\beta = 0.$$

Now γ can be decomposed as $\gamma = w \circ \phi$ for some $w \in \Gamma(\nu)$. Since

$$TW|w\left(J^kY\right) = Tw\left(TJ^kY\right) \oplus V\nu,$$

by the above one may suppose in (2.4) that $\zeta|w\left(J^kY\right) = Tw \cdot \xi$ for some $\xi \in C^\infty(TJ^kY)$. Then γ is an extremal iff

$$0 = \gamma^*\left(i_\zeta d\Theta_{\mathcal{L}}\right) = \phi^* w^*\left(i_{Tw\cdot\xi}d\Theta_{\mathcal{L}}\right) = \phi^*\left(i_\xi dw^*\Theta_{\mathcal{L}}\right) = \phi^*\left(i_\xi dw\right).$$

Thus to construct the extremal γ it suffices to find a section $w : J^kY \to W$ satisfying

$$\phi^*\left(i_\xi dw\right) = 0 \tag{4.1}$$

for all vector fields ξ on J^kY. Actually, slightly less will do: w and dw need only be defined along $\phi(X)$ in J^kY. Observe also that in (4.1) one has $\phi = j^k\varphi$.

What must such a form w on $J^k Y$ look like? Since J_k is Pfaffian and $\mathcal{L}$ is semi-basic, (3.2) gives

$$w = L\omega + \sum_{|\underline{\mu}|=0}^{k-1} \lambda_A^{\underline{\mu}\nu}\, \psi_{\underline{\mu}}^A \wedge \omega_\nu + \mathcal{O}\left(J_k \wedge dy_{\mu_1 \cdots \mu_k}^A\right) + \mathcal{O}\left(J_k \wedge J_k\right).$$

The presence of terms involving the top differentials $dy_{\mu_1 \cdots \mu_k}^A$ is computationally awkward. However, it suffices to assume that w is (π_{k-1}^k)-horizontal, in which case

$$w = L\omega + \sum_{|\underline{\mu}|=0}^{k-1} \lambda_A^{\underline{\mu}\nu}\, \psi_{\underline{\mu}}^A \wedge \omega_\nu + \mathcal{O}\left(J_k \wedge J_k\right). \tag{4.2}$$

The case of general w can then be handled by pulling w back to $J^{k+1}Y$ and arguing as below.

Now from (4.2)

$$\begin{aligned}
dw = {} & \sum_{|\underline{\mu}|=0}^{k} \frac{\partial L}{\partial y_{\underline{\mu}}^A}\, dy_{\underline{\mu}}^A \wedge \omega + \sum_{|\underline{\mu}|=0}^{k-1} d\lambda_A^{\underline{\mu}\nu} \wedge \psi_{\underline{\mu}}^A \wedge \omega_\nu \\
& - \sum_{|\underline{\mu}|=0}^{k-1} \lambda_A^{\underline{\mu}\nu}\, dy_{\underline{\mu}\nu}^A \wedge \omega + \mathcal{O}\left(J_k \wedge \mathcal{J}_k\right).
\end{aligned} \tag{4.3}$$

To investigate the conditions that (4.1) places on w as ξ ranges over $C^\infty(TJ^k Y)$, set $\xi = \frac{\partial}{\partial y_{\underline{\mu}}^A}$ with $|\underline{\mu}| = k, \ldots, 1$ in turn. Then with $\phi = j^k\varphi$, (4.3) yields:

$$\lambda_A^{(\underline{\mu})} \circ j^k\varphi = \begin{cases} \dfrac{\partial L}{\partial y_{\underline{\mu}}^A} \circ j^k\varphi, \ |\underline{\mu}| = k \\[3mm] \dfrac{\partial L}{\partial y_{\underline{\mu}}^A} \circ j^k\varphi - \partial_\nu\left(\lambda_A^{\underline{\mu}\nu} \circ j^k\varphi\right), \ 1 \leq |\underline{\mu}| < k, \end{cases} \tag{4.4}$$

where the parentheses denotes symmetrization. Setting $\xi = \frac{\partial}{\partial y^A}$ in (4.1) and using (4.4) recursively, one computes

$$\left(j^k\varphi\right)^* \left(i_{\partial/\partial y^A}\, dw\right) = \frac{\delta L}{\delta y^A}\left(j^{2k}\varphi\right)\omega = 0$$

as $j^k\varphi = \phi$ is an Euler-Lagrange extremal. Finally, setting $\xi = \frac{\partial}{\partial x^\mu}$ in (4.1) gives nothing new. Note that (4.1) places no restrictions whatsoever on the higher degree contact terms in w; only the "principal part" of w is relevant for contravariance.

To construct such a form w I use a standard argument [27]. Introduce a cover $\{U_i\}$ of $J^k Y$ by π^k-bundle charts, and let $\{\sigma_i\}$ be a partition of unity subordinate to this cover. Set $\tilde{U}_i = (\pi_k^{2k-1})^{-1}(U_i)$ and $\tilde{\sigma}_i = (\pi_k^{2k-1})^* \sigma_i$. Define forms

$$\tilde{w}_i = \tilde{\sigma}_i L\omega + \sum_{|\underline{\mu}|=0}^{k-1} \tilde{\lambda}^{\mu\nu}_{\underline{A},i} \psi^A_{\underline{\mu}} \wedge \omega_\nu$$

on $J^{2k-1}Y$ with $supp\ \tilde{w}_i \subset \tilde{U}_i$ by setting

$$\tilde{\lambda}^{\mu}_{\underline{A},i} = \begin{cases} \dfrac{\partial\,(\tilde{\sigma}_i L)}{\partial y^A_{\underline{\mu}}}, & |\underline{\mu}| = k \\[3mm] \dfrac{\partial\,(\tilde{\sigma}_i L)}{\partial y^A_{\underline{\mu}}} - D_\nu \tilde{\lambda}^{\mu\nu}_{\underline{A},i}, & 1 \le |\underline{\mu}| < k \end{cases} \tag{4.5}$$

where D_ν is the total derivative. Let $\tilde{w} = \Sigma_i \tilde{w}_i$; then $\tilde{w}$ is a globally defined form on $J^{2k-1}Y$ whose components satisfy (4.4) in each chart $\tilde{U}_i$. But then the equation $w|_{j^k\varphi(X)} = \tilde{w}|_{j^{2k-1}\varphi(X)}$ implicitly defines a form w along the image of $j^k\varphi$ whose differential, by construction, satisfies (4.1). This finishes the proof of Theorem 4.

Remark 4. The form $\tilde{w}$ is defined on $J^{2k-1}Y$, *not* $J^k Y$, because of the total divergence in the second of equations (4.5). These terms force the order of each successive coefficient $\tilde{\lambda}^{\mu}_{A,i}$, $|\underline{\mu}| = k,\ldots,1$, to increase by one (so that $\tilde{\lambda}^{\mu}_{A,i}$ is actually a locally defined function on $J^{2k-|\underline{\mu}|}Y$). All in all $J^k Y$ must be prolonged $k-1$ times to $J^{2k-1}Y$. This observation will be exploited in §6.

Remark 5. The construction of the form $\tilde{w}$ is somewhat subtle in one regard. In general (*i.e.*, $k > 2$) there is no globally defined form on $J^{2k-1}Y$ whose first degree contact components $\tilde{\lambda}^{\mu}_{A}$ satisfy

$$\tilde{\lambda}^{\mu}_{A} = \begin{cases} \dfrac{\partial L}{\partial y^A_{\underline{\mu}}}, & |\underline{\mu}| = k \\[3mm] \dfrac{\partial L}{\partial y^A_{\underline{\mu}}} - D_\nu \tilde{\lambda}^{\mu\nu}_{A}, & 1 \le |\underline{\mu}| < k \end{cases} \tag{4.6}$$

in *every* bundle chart, as one may verify directly from the transformation properties of such a form. (In other words, the vanishing of the $\tilde{\lambda}^{[\mu]}_A$ is not an invariant condition.) But (4.6), with $\tilde{\lambda}^{(\mu)}_A$ on the left hand side instead of $\tilde{\lambda}_A^\mu$, *is* an invariant condition, and this explains why $\tilde{w}$ as defined above satisfies (4.4). See [27] for an extensive discussion of this point.

Although this proof basically consists of a straightforward calculation, it should be noted that, at least *formally*, Theorem 4 is a direct consequence of the Lagrange multiplier theorem, *cf.* Proposition 43.21 in [40]. The reason I have proceeded in this fashion is that the proof itself is interesting, insofar as a slight reinterpretation of it leads to the main construction and results of the paper in §6.

5. Classical Lepagean Equivalents

In this section I reprise the classical theory of Lepagean equivalents as developed by Krupka [27] and Kolář [24]. Along the way I will make the connection with the theory of §2. Throughout I am concerned solely with the classical variational problem $(J^k Y \xrightarrow{\pi^k} X, \mathcal{J}_k, \mathcal{L})$. Here is the technical definition.

Definition 5. An n-form Θ on $J^{2k-1}Y$ is a *classical Lepagean equivalent* of $\mathcal{L}$ provided:

(LE1) $\Theta - \left(\pi_k^{2k-1}\right)^* \mathcal{L}$ is contact, and

(LE2) $i_\xi d\Theta$ is contact for all $\xi \in V\pi_0^{2k-1}$.

These are generalizations of the "congruences of Lepage" extended to the higher order case. Every Lagrangian density has a classical Lepagean equivalent on $J^{2k-1}Y$, but the latter is not uniquely determined unless $n = 1$ [1, 25, 27, 32]. Classical Lepagean equivalents may exist on jet bundles of lower order in particular instances, but this cannot be guaranteed in general.[6] (The reason will become apparent in the next section.) In jet charts, any classical Lepagean equivalent Θ of $\mathcal{L}$ can be expressed as follows:

$$\left(\pi_{2k-1}^{2k}\right)^* \Theta = L\omega + \sum_{|\underline{\mu}|=0}^{2k-1} \lambda_A^{\underline{\mu}\nu} \psi_{\underline{\mu}}^A \wedge \omega_\nu + \chi,$$

where χ is at least quadratic in the generators of $\mathcal{J}_{2k}$ and the coefficients

---

[6]On occasion one sees classical Lepagean equivalents defined on the semi-holonomic jet bundle $\bar{J}^{2k-1}Y \subset J^1(J^{2k-2}Y)$. This approach is useful when discussing the so-called "Poincaré-Cartan forms" [37].

$\lambda_A^{\underline{\mu}}$ are given according to

$$\lambda_A^{\underline{\mu}} = \begin{cases} c_A^{\underline{\mu}}, & |\underline{\mu}| = 2k \\ -D_\nu \lambda_A^{\underline{\mu}\nu} + c_A^{\underline{\mu}}, & k < |\underline{\mu}| < 2k \\ \dfrac{\partial L}{\partial y_{\underline{\mu}}^A} - D_\nu \lambda_A^{\underline{\mu}\nu} + c_A^{\underline{\mu}}, & 1 \le |\underline{\mu}| \le k. \end{cases} \tag{5.1}$$

Here the functions $c_A^{\underline{\mu}}$, of order at most $2k - 2$, satisfy $c_A^{(\underline{\mu})} = 0$ and $c_A^{\underline{\mu}1} = 0$ but are otherwise arbitrary. The nonuniqueness of Θ is reflected in the arbitrariness of both the $c_A^{\underline{\mu}}$ and the higher degree contact term χ.

However, *not* every classical Lepagean equivalent so defined is a Lepagean equivalent in the sense of §2. The discrepancy stems from a slight difference between (LE1) and (2.2): the former can be rewritten

$$\left(j^{2k-1}\varphi\right)^* \Theta = \left(j^k \varphi\right)^* \mathcal{L}, \tag{5.2}$$

while the latter reduces to

$$\gamma^* \Theta = \left(j^k \varphi\right)^* \mathcal{L}. \tag{5.3}$$

But γ in (5.3) need not be $(2k - 1)$-holonomic, as is required in (5.2).

This indicates that the above definition of classical Lepagean equivalent is too broad. Moreover, one sees from (5.1) that the top components $\lambda_A^{\underline{\mu}}$, $k < |\underline{\mu}| \le 2k$, carry no essential information. It is therefore useful to restrict the arbitrariness in Θ by requiring it to satisfy certain additional invariant conditions.

Definition 6. A classical Lepagean equivalent Θ is *strict* if it is (π_{k-1}^{2k-1})-horizontal.

This condition kills the above-mentioned components. Combined with (LE1) it implies that, as forms on $J^{2k-1}Y$,

$$\Theta \equiv \mathcal{L} \bmod J_k. \tag{5.4}$$

This is Lepage's first congruence for a higher order variational problem. It follows from (5.3) and (5.4) that a strict classical Lepagean equivalent is a genuine Lepagean equivalent according to Definition 1.

Every strict classical Lepagean equivalent can be locally written

$$\Theta = L\omega + \sum_{|\underline{\mu}|=0}^{k-1} \lambda_A^{\underline{\mu}\nu} \psi_{\underline{\mu}}^A \wedge \omega_\nu + \chi, \tag{5.5}$$

where now χ is at least quadratic in the generators of J_k and

$$
\lambda_A^{\underline{\mu}} = \begin{cases} \dfrac{\partial L}{\partial y_{\underline{\mu}}^A} + c_A^{\underline{\mu}}, & |\underline{\mu}| = k \\[2ex] \dfrac{\partial L}{\partial y_{\underline{\mu}}^A} - D_\nu \lambda_A^{\underline{\mu}\nu} + c_A^{\underline{\mu}}, & 1 \le |\underline{\mu}| < k. \end{cases}
\tag{5.6}
$$

Strict classical Lepagean equivalents always exist (*cf.* [27]; I will give another proof in §6); they are unique iff $n = 1$ or $k = 1$ and $N = 1$, where N is the fiber dimension of Y.

Remark 6. On the other hand, it is clear that *every* classical Lepagean equivalent, strict or not, is a Lepagean equivalent in the sense of Definition 1 for the prolonged system $(J^{2k-1}Y \xrightarrow{\pi^{2k-1}} X, \mathcal{J}_{2k-1}, (\pi_k^{2k-1})^*\mathcal{L})$. So nothing is really lost.

Remark 7. To my knowledge, every classical Lepagean equivalent which arises in specific applications is strict. In this connection note that the form (1.4) – originally considered by Lepage – is the most general strict classical Lepagean equivalent for a first order field theory.

As Definitions 5 and 6 do not fix the higher degree contact term χ, it is useful to specialize further.

Definition 7. A strict classical Lepagean equivalent such that

$$
\text{(C)} \quad i_\xi i_\eta \Theta = 0 \quad \text{for all} \quad \xi, \eta \in V\pi^{2k-1}
$$

will be called a *classical Cartan form*.

Condition (C) forces χ in (5.5) to vanish. One may therefore think of a classical Cartan form as being the (invariantly defined) "principal part" of a strict classical Lepagean equivalent. As with the latter, the former always exist [12, 21, 36] and are typically nonunique (but not so much so; they are uniquely determined when $n = 1$ or $k = 1$).

Since jets are contravariant objects, one might expect that classical Lepagean equivalents are contravariant in the sense of §2. Indeed this is the case:

Proposition 8. *(Strict) classical Lepagean equivalents are contravariant.*

Proof. Let φ satisfy the Euler-Lagrange equations; I will show that

$$\left(j^{2k-1}\varphi\right)^{*}\left(i_{\zeta}d\Theta\right) = 0 \tag{5.7}$$

for all vector fields ζ on $J^{2k-1}Y$, so that every extremal $j^{k}\varphi$ of $\mathcal{L}$ is the projection $\pi_{k}^{2k-1} \circ j^{2k-1}\varphi$ of an extremal of Θ.

By (LE2) this is true for all $\zeta \in C^{\infty}(V\pi_{0}^{2k-1})$. Next, just as in the proof of Theorem 4, one finds that (5.6) in conjunction with the fact that $j^{k}\varphi$ is an Euler-Lagrange extremal imply that (5.7) is true for $\zeta = \frac{\partial}{\partial y^{A}}$ as well. Finally, also as before, (5.7) is then automatically satisfied for $\zeta = \frac{\partial}{\partial x^{\mu}}$. ∎

Now recall that the *canonical* Lepagean equivalent of a classical variational problem is both covariant *and* contravariant. Unhappily, this is not so for *classical* Lepagean equivalents, at least not without additional regularity assumptions.[7] The problem is that $\gamma \in \mathcal{E}(\Theta)$ need not be k-holonomic. Moreover, even under appropriate regularity conditions, $\mathcal{E}(\Theta) \not\approx \mathcal{E}(\mathcal{L})$ unless n or $k = 1$. This is because a Cartan extremal γ need not be $(2k-1)$-holonomic, even when regularity guarantees that it is k-holonomic. I will return to this point in §7.

Thus the canonical Lepagean equivalent of a classical variational problem is really rather special in that one gets both covariance and contravariance without any regularity whatsoever.

6. Universality

I now establish the main result of this paper, *viz.*, the canonical Lepagean equivalent of a classical variational problem is universal in the sense that every strict classical Lepagean equivalent can be obtained from it by "pullback."

To this end let $\overline{W}$ be the subbundle of $W = \mathcal{L} + J_{k}^{n}$ which consists of (π_{k-1}^{k})-horizontal forms; $\overline{W}$ inherits a canonical form $\bar{\Theta}_{\mathcal{L}}$ from W. According to (3.3) this form has the local expression

$$\bar{\Theta}_{\mathcal{L}} = L\omega + \sum_{|\underline{\mu}|=0}^{k-1} \lambda_{A}^{\underline{\mu}\nu} \psi_{\underline{\mu}}^{A} \wedge \omega_{\nu} + \mathcal{O}\left(J_{k} \wedge J_{k}\right). \tag{6.1}$$

---

[7]See [17] for a review and detailed references.

Comparing the forms (6.1) on $\overline{W}$ and (5.5) on $J^{2k-1}Y$, it is not surprising that one can recover the latter from the former; the problem is to elucidate the precise mechanism which accomplishes this.

The key is contained in the proof of Theorem 4, which I now reprise from a slightly different angle. The basic idea is to regard (4.1) as a first order differential condition on w – now viewed as a section of $\overline{W} \to J^kY$ – and to describe the corresponding differential relation in $J^1(\overline{W})$. Set $\overline{W}^r = (\pi_k^r)^\# \overline{W}$.

Now for each $s = 1, \ldots, k$, the first s equations of (4.4), starting with $|\underline{\mu}| = k$ and continuing down, can be rewritten:

$$
\lambda_{\underline{A}}^{(\underline{\mu})} =
\begin{cases}
\dfrac{\partial L}{\partial y_{\underline{\mu}}^A}, & |\underline{\mu}| = k \\[2ex]
\dfrac{\partial L}{\partial y_{\underline{\mu}}^A} - D_\nu \lambda_{\underline{A}}^{\underline{\mu}\nu}, & k - s + 1 \le |\underline{\mu}| < k.
\end{cases}
\tag{6.2}
$$

In this expression I regard the $\lambda_{\underline{A}}^{\underline{\mu}}$ as coordinates along the fibers of $\overline{W}^{k+s-1} \to J^{k+s-1}Y$ and use the abbreviation

$$
D_\nu \lambda_{\underline{A}}^{\underline{\mu}\nu} := \left(\lambda_{\underline{A}}^{\underline{\mu}\nu} \right)_\nu + \sum_{|\underline{\rho}|=0}^{2k-|\underline{\mu}|-1} \left(\lambda_{\underline{A}}^{\underline{\mu}\nu} \right)_{\underline{B}}^{\underline{\rho}} y_{\underline{\rho}\nu}^B,
\tag{6.3}
$$

where $(\lambda_{\underline{A}}^{\underline{\mu}\nu})_\sigma$ and $(\lambda_{\underline{A}}^{\underline{\mu}\nu})_{\underline{B}}^{\underline{\rho}}$ are jet coordinates along the fibers of $J^1(\overline{W}^{k+s-1}) \to \overline{W}^{k+s-1}$.

Equations (6.2) thus specify the $\lambda_{\underline{A}}^{(\underline{\mu})}$ in terms of the derivatives of L and the jet coordinates $(\lambda_{\underline{A}}^{\underline{\mu}\nu})_\sigma$ and $(\lambda_{\underline{A}}^{\underline{\mu}\nu})_{\underline{B}}^{\underline{\rho}}$. Consequently for each s these conditions define an affine subbundle $R^{k+s-1} \subset J^1(\overline{W}^{k+s-1})$.[8] Two points are worth noting (*cf.* Remark 4): (i) Each time s increases by one, it is necessary to prolong the base once. This shows up explicitly in (6.3), since $|\underline{\mu}|$ decreases as s increases. (ii) The process terminates after k steps, so that $R^{2k-1} \to J^{2k-1}Y$ encodes the totality of equations (4.4). This construction may be summarized by the diagram:

---

[8]Secretly lurking here is the fact that the derived flag of the contact system J_k has constant rank.

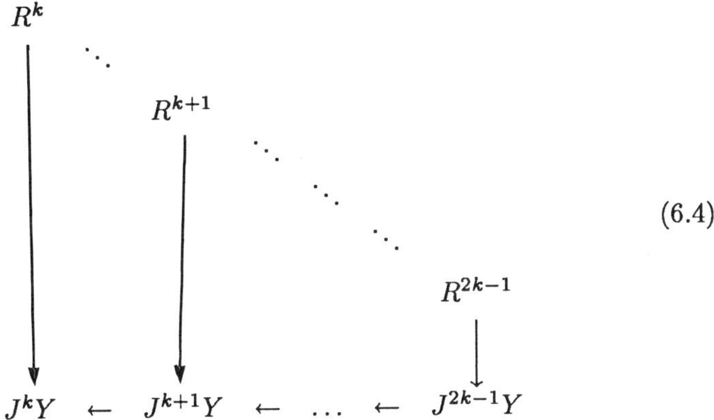

$$(6.4)$$

The goal is to induce classical Lepagean equivalents on $J^{2k-1}Y$, so consider the bundle $R^{2k-1} \to J^{2k-1}Y$. Observe that the canonical form $\bar{\Theta}_{\mathcal{L}}$ on $\overline{W}$ can be pulled back to $\overline{W}^{2k-1}$, then to $J^1(\overline{W}^{2k-1})$, and then finally to R^{2k-1}. I use the same symbol for this induced form; note that its coordinate expression remains (6.1). To obtain classical Lepagean equivalents on the jets, then, it suffices to section this bundle.

Theorem 9. (i) *Let r be any section of $R^{2k-1} \to J^{2k-1}Y$. Then $r^*\bar{\Theta}_{\mathcal{L}}$ is a Lepagean equivalent of $\mathcal{L}$ in the sense of §2.*

(ii) *If r is holonomic, then $r^*\bar{\Theta}_{\mathcal{L}}$ is a strict classical Lepagean equivalent of $\mathcal{L}$ in the sense of §5.*

(iii) *Every strict classical Lepagean equivalent on $J^{2k-1}Y$ is $r^*\bar{\Theta}_{\mathcal{L}}$ for some holonomic section r of $R^{2k-1} \to J^{2k-1}Y$.*

Proof. (i) Suppose $\gamma \in \Gamma(\pi^{2k-1})$ is such that $\pi_k^{2k-1} \circ \gamma = j^k\varphi$ for some $\varphi \in \Gamma(\pi)$. It must be shown that $\gamma^*(r^*\bar{\Theta}_{\mathcal{L}}) = (j^k\phi)^*\mathcal{L}$. But this follows immediately from (6.1).

(ii) It is necessary to verify that $r^*\bar{\Theta}_{\mathcal{L}}$ satisfies (LE1), (LE2) and Definition 6. Now (LE1) follows from part (i). Next, $r^*\bar{\Theta}_{\mathcal{L}}$ satisfies (LE2) by virtue of the construction of R^{2k-1} and the assumption that r is holonomic. Finally, Definition 6 is a consequence of the very definition of $\overline{W}$.

(iii) Consider a strict classical Lepagean equivalent Θ on $J^{2k-1}Y$. Then Θ is (π_{k-1}^{2k-1})-horizontal, and hence from (5.6) it follows that Θ defines a section of $\overline{W}^{2k-1}$ with the property that $j^1\Theta$ maps into R^{2k-1}. But as $\bar{\Theta}_{\mathcal{L}}$ is a tautological form, $(j^1\Theta)^*\bar{\Theta}_{\mathcal{L}} = \Theta$. ∎

This construction provides an alternate proof of the existence of strict classical Lepagean equivalents as well as a characterization of the ambiguity therein. In particular the proof of Theorem 4 shows that there exists a global section, call it λ, of $\overline{W}^{2k-1} \to J^{2k-1}Y$ with the property that $j^1\lambda$ maps into R^{2k-1}. Combined with part (ii) of Theorem 9, this yields:

Corollary 10. *Strict classical Lepagean equivalents always exist.*

Regarding uniqueness, Theorem 9 and a comparison of (5.5) and (5.6) with (6.1) and (6.2), respectively, immediately give:

Corollary 11. *The set of strict classical Lepagean equivalents on $J^{2k-1}Y$ is in one-to-one correspondence with the set of holonomic sections of $R^{2k-1} \to J^{2k-1}Y$.*

The algorithm (6.4) is entirely canonical except for the last step, where an extrinsic element – the section r – enters. *This* is why (strict) classical Lepagean equivalents are not uniquely determined in general; the ambiguity therein is a reflection of the freedom in the choice of holonomic section of $R^{2k-1} \to J^{2k-1}Y$. When $n = 1$ or $k = 1$ and $N = 1$ (where N is the fiber dimension of Y), this bundle "collapses" in the sense that it then has only one holonomic section. Thus in particular there is a unique strict classical Lepagean equivalent in mechanics (of any order); it is [15, 28, 38]:

$$\Theta = L\,dt + \sum_{s=0}^{k-1}\left[\sum_{r=s}^{k-1}\left[(-1)^{r-s}\frac{d^{r-s}}{dt^{r-s}}\left(\frac{\partial L}{\partial q^A_{(r+1)}}\right)\right]\right]\psi^A_{(s)}. \qquad (6.5)$$

Here I have set $t = x^1$, $q^A_{(s)} = y^A_{\mu_1\cdots\mu_1}$ (s times), and $D_t = \frac{d}{dt}$.

Remark 8. The results of this section also explain why (strict) classical Lepagean equivalents usually cannot be defined on jet bundles of order less than $2k - 1$: it is because generically the algorithm for determining the λ^μ_A only terminates with R^{2k-1}. Put another way, Theorem 9 is not necessarily valid for $R^q \to J^qY$ with $q < 2k - 1$.

Remark 9. When constructing a strict classical Lepagean equivalent by directly sectioning $R^{2k-1} \to J^{2k-1}Y$, one arbitrarily specifies the $c^\mu_A = \lambda^{[\mu]}_A$ as functions on $J^{2k-2}Y$. But if instead one constructs such a section in stages; *i.e.*, by sectioning each bundle $R^{k+s-1} \longrightarrow J^{k+s-1}Y$, $s = 1, \ldots, k$ in turn, one actually obtains the c^μ_A as functions on $J^{2k-|\mu|}Y$, *cf.* [21].

The construction presented here can be both specialized and extended. For instance, one can recover all classical Cartan forms in much the same way as all strict classical Lepagean equivalents. For this one need only replace the bundle $\overline{W}$ in the above by its subbundle $\overline{V}$ which consists of all (π^k_{k-1})-horizontal forms on $J^k Y$ which in addition are of at most degree one in the generators of J_k (*cf.* §3). One obtains analogues of Theorem 9 and Corollaries 10 and 11 in this context, where the classical Cartan form is unique provided only that $n = 1$ or $k = 1$. (The condition on the fiber dimension of Y is no longer necessary, since now $\chi = 0$). This unique classical Cartan form for a first order field theory is just (1.1). It is straightforward to specialize even further to the so-called "Poincaré-Cartan forms" defined in [12, 17, 21, 24].

Likewise, it is possible to consider the entire bundle W on $J^k Y$ as the starting point. Then the analogues of parts (i) and (ii) of Theorem 9 are still valid, but the classical Lepagean equivalents $r^* \Theta_{\mathcal{L}}$ so obtained have no special designation. In fact all classical Lepagean equivalents on $J^{2k-1} Y$ which satisfy $\Theta \equiv \mathcal{L}$ mod J_{2k-1} can be gotten by this general procedure provided, in view of Remark 6, one now uses the canonical Lepagean equivalent of the prolonged system $(J^{2k-1} Y \xrightarrow{\pi^{2k-1}} X, \mathcal{J}_{2k-1}, (\pi^{2k-1}_k)^* \mathcal{L})$. Here no prolongation of the base is required as in (6.4), as it is "built in" from the beginning.

Finally, I point out that there are Lepagean equivalents of $\mathcal{L}$ on $J^{2k-1} Y$ which are not classical in the sense of §5. For instance, if $c_A^{\mu_1 \cdots \mu_k+1}(x)$ are arbitrary functions on X antisymmetric in their last two indices, the form $\Theta = c_A^{\mu_1 \cdots \mu_k \nu}(x)\, dy^A_{\mu_1 \cdots \mu_k} \wedge \omega_\nu$ is contact and hence is a (local) Lepagean equivalent of the zero Lagrangian. But this Θ does not satisfy (5.1) and hence is not classical. This example also shows that not every Lepagean equivalent on $J^{2k-1} Y$ is the pullback of $\Theta_{\mathcal{L}}$ on W. Thus, roughly speaking, the *canonical* Lepagean equivalent is universal for *classical* Lepagean equivalents, but not for all Lepagean equivalents.

Remark 10. Muñoz has defined a "universal Poincaré-Cartan form" associated to a Lagrangian density [33]. It functions in much the same way that the universal Cartan form $\bar{\Theta}_{\mathcal{L}}$ on $\overline{V}$ does in that any classical Poincaré-Cartan form can be obtained from his universal one by sectioning a certain bundle. But Muñoz's approach differs from mine in two respects: his universal form is not itself a Lepagean equivalent as is $\bar{\Theta}_{\mathcal{L}}$, and it is unclear if his construction extends to classical Lepagean equivalents more general than Poincaré-Cartan forms.

7. Discussion

For a given variational problem $(M \xrightarrow{\pi} X, \mathcal{I}, \mathcal{L})$, it is obviously of some interest to construct a bivariant Lepagean equivalent $(W \xrightarrow{\rho} X, \{0\}, \Theta)$.

Under what circumstances this can be accomplished is problematical. In the classical case, as I have shown, the canonical Lepagean equivalent *is* bivariant, due to the fact that the contact ideal is so "nice" in many respects. But this makes it difficult to isolate those features of a classical variational problem which may be relevant for bivariance in a more general setting. A straightforward way to attack the problem would be to extend the algorithm of §6 – which in §4 (implicitly) appeared as a set of necessary and sufficient conditions for contravariance – to nonclassical situations. In §1.e of [19] Griffiths describes a similar procedure in the general single integral case. To carry this through, however, one might have to require that the derived systems of I have constant rank (but see [22]).

Naturally, the "ultimate" goal is to construct a Lepagean equivalent which has the stronger property that

$$\mathcal{E}(\Theta) \approx \mathcal{E}(\mathcal{L}). \tag{7.1}$$

This is certainly too ambitious, even classically, except in special instances. In mechanics of any order k, the canonical Lepagean equivalent satisfies (7.1) by virtue of the fact that any extremal of $\mathcal{L}$ can be *uniquely* lifted to an extremal of $\Theta_{\mathcal{L}}.^9$ The same is true for the (unique) classical Lepagean equivalent (6.5) provided L is regular in the sense that

$$\det\left(\frac{\partial^2 L}{\partial q_{(k)}^i \partial q_{(k)}^j}\right) \neq 0.$$

(This condition forces every extremal of Θ on $J^{2k-1}Y$ to actually be $(2k-1)$-holonomic, not just k-holonomic, *cf.* [21, 28].) Likewise, for first order field theories, the generalized Cartan form on $\overline{V}$ satisfies (7.1), as does the classical Cartan form (1.1) provided

$$\det\left(\frac{\partial^2 L}{\partial y_\mu^A \partial y_\nu^B}\right) \neq 0, \tag{7.2}$$

as is well-known. But the situation is more complicated than these results would indicate. For instance, there is no such isomorphism for the canonical Lepagean equivalent on $\overline{W}$ since now the lifting $\mathcal{E}(\mathcal{L}) \to \mathcal{E}(\bar{\Theta}_{\mathcal{L}})$ constructed in the proof of Theorem 4 is not uniquely determined. This is because nothing in the construction fixes the higher degree contact terms in (4.2). On the other hand, for more general classical Lepagean equivalents such as (1.4), and in particular (1.2) and (1.3), (7.2) is no longer the appropriate

---

[9]The proof is identical to that which gives the unicity of the classical Lepagean equivalent in this context.

regularity condition. In fact, a sufficient condition for "regularity" in this context is [17, 23]

$$\det \left(\frac{\partial^2 L}{\partial y_\mu^A \partial y_\nu^B} - \lambda_{AB}^{\mu\nu} - \frac{\partial \lambda_{BC}^{\nu\rho}}{\partial y_\mu^A} y_\rho^C + \dots \right) \neq 0.$$

But this is not sufficient to guarantee that an extremal of such a Θ is holonomic; global considerations intervene. In both the classical and canonical cases, then, difficulties arise from the presence of higher degree contact terms.

A more tractable problem this would be to determine conditions under which *every* lift of an extremal of $(M \xrightarrow{\pi} X, \mathcal{I}, \mathcal{L})$ is an extremal of $(W \xrightarrow{\rho} X, \{0\}, \Theta)$. (This is related to Dedecker's problem [10] of constructing a *"relèvement libre et complet."*) The reader is invited to study the above examples in this light. Further discussion of these matters in a variety of contexts may be found in [10, 11, 19, 22].

Regarding the classical theory, it is natural to ask if there is (something like) a functor which assigns to each classical variational problem on $J^k Y$ a classical Lepagean equivalent problem on $J^{2k-1} Y$. Unfortunately, for $k > 2$ Anderson has shown that no such global correspondence can exist, at least if it is to be *linear* in $\mathcal{L}$ (*cf.* Proposition 5.53 in [1]). The basic problem is that there is no "natural" way to fix the arbitrary terms c_A^μ in (5.1).[10] Very likely a similar result is true even if the correspondence is to be nonlinear, as is the case for Carathéodory's theory, *cf.* (1.2). Even so, one can ask for less. Consider a strict classical Lepagean equivalent Θ of the form (5.5). The algorithm of §6 fixes – insofar as is possible – the coefficients $\lambda_A^{\mu\nu}$ of the first degree contact terms in Θ via (5.6). But as observed above it says nothing at all about the higher degree contact term χ. Is there any procedure analogous to that presented here which can be used to specify this term, other than to force it to vanish à la (C)? If such procedures exist, what is their significance? For example, in the first order case Betounes [2] iteratively determined the coefficients $\lambda_{A_1 \dots A_s}^{\mu_1 \dots \mu_s}$ in (1.4) and thereby obtained (1.3) by requiring that $d\Theta = 0$ iff $\mathcal{L}$ is a divergence. His procedure extends to higher order field theories producing local classical Lepagean equivalents which have this property. But it does not work globally unless $k = 1$ [3].

The algorithm of §6 leads to some intriguing speculations on the relation between the Hamiltonian and Lagrangian formalisms in the classical calculus of variations. As one moves left to right in the diagram (6.4), one

---

[10]That is, without introducing some element extrinsic to the problem, such as a connection on X.

may think of the formalism as becoming "more and more Lagrangian": the process starts on the "nonclassical" bundle $\overline{W} \to J^k Y$ and ends on the "nearly classical" bundle $R^{2k-1} \to J^{2k-1}Y$. Once the latter bundle is sectioned, one recovers the usual classical Lagrangian formalism on $J^{2k-1}Y$, which one can think of as the covariant analogue of the tangent bundle. So it might be expected that as one goes toward the *left*, the formalism becomes progressively "more Hamiltonian." This is indeed the case, as I now explain.

First recall the setup for strict classical Lepagean equivalents, which is based upon the subbundle $\overline{W} \subset W \subset \wedge^n(J^k Y)$ defined in §6. Each element $\overline{w}$ of $\overline{W}$ can be written

$$\overline{w} = L\omega + \sum_{|\underline{\mu}|=0}^{k-1} \lambda_A^{\underline{\mu}\nu} \psi_{\underline{\mu}}^A \wedge \omega_\nu + \mathcal{O}\left(J_k \wedge J_k\right). \tag{7.3}$$

As just observed, the algorithm of §6 successively determines the coefficients of the first degree contact terms in such a form. But the horizontal term – *viz.* $L\omega$ – was fixed *ab initio* by the very definition of the bundle $\overline{W}$. To incorporate this into the above general scheme, consider the bundle $\wedge^n(J^{k-1}Y)$. Every element $z \in \wedge^n(J^{k-1}Y)$ takes the form

$$z = p\omega + \sum_{|\underline{\mu}|=0}^{k-1} p_A^{\underline{\mu}\nu}\, dy_{\underline{\mu}}^A \wedge \omega_\nu + \mathcal{O}\left(dy_{\underline{\mu}}^A \wedge dy_{\underline{\nu}}^B\right).$$

The reason why $\wedge^n(J^{k-1}Y)$ and not $\wedge^n(J^k Y)$ appears here is explained by a comparison of (7.3) with the pullback of z to $J^k Y$, which can be written

$$\left(\pi_{k-1}^k\right)^* z = \lambda\omega + \sum_{|\underline{\mu}|=0}^{k-1} \lambda_A^{\underline{\mu}\nu} \psi_{\underline{\mu}}^A \wedge \omega_\nu + \mathcal{O}\left(J_k \wedge J_k\right)$$

for some coefficients λ and $\lambda_A^{\underline{\mu}\nu}$.

From these expressions one sees that the "zeroth" step in the algorithm really should be to fix $\lambda = L$. As the Lagrangian is k^{th}-order, this requires prolonging the base once, from $J^{k-1}Y$ to $J^k Y$. Thus $\overline{W}$ appears naturally as a subbundle of $(\pi_{k-1}^k)^\#(\wedge^n(J^{k-1}Y))$, and the diagram (6.4) can be extended as follows:

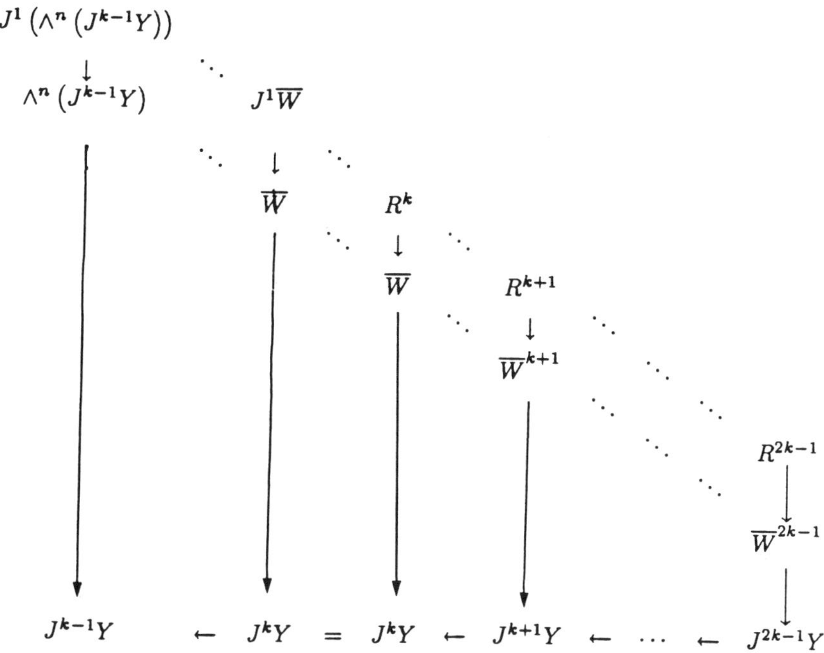

Now the point of all this is that according to [17] the space $\wedge^n(J^{k-1}Y)$ is exactly the covariant phase space corresponding to the Lagrangian system $(J^{2k-1}Y \xrightarrow{\pi^{2k-1}} X, \{0\}, \Theta)$ with Θ a strict classical Lepagean equivalent. In other words, $\wedge^n(J^{k-1}Y)$ is the covariant analogue of the cotangent bundle in this context, and this is where the Hamiltonian formalism resides![11] Naturally it carries a canonical n-form – the exterior differential of which is the *multisymplectic form* – which, when pulled back to $\overline{W}$ in the manner of §6, is just the canonical Lepagean equivalent $\bar{\Theta}_{\mathcal{L}}$. Thus the left hand column of the above diagram is purely Hamiltonian, and by invoking the algorithm it is possible to "derive" (in some sense) the Lagrangian formalism from the Hamiltonian. To close the circle, one of course has the covariant Legendre transformation which maps $J^{2k-1}Y \to \wedge^n(J^{k-1}Y)$. Thus the Legendre transformation is the "inverse" of the algorithm of §6. For further details on the covariant Hamiltonian formalism see [17].

All this might be viewed as the "Lepagean approach" to the classical calculus of variations. Turn next to the more familiar "De Donder-Weyl

---

[11]Some authors view the Hamiltonian formalism as occurring elsewhere, *e.g.*, on $J^{2k-1}Y$ [27] or R^k [19], but this conflicts with common practice.

approach," in which Θ is taken to be a classical Cartan form. The situation is then essentially similar to that above, except that $\overline{W}$ is replaced by

$$V = \left\{ \overline{w} \in \overline{W} \mid i_\xi i_\eta \overline{w} = 0 \ \forall \ \xi, \eta \in V\pi^{k-1} \right\},$$

etc., and so $\wedge^n(J^{k-1}Y)$ must be replaced by its subbundle

$$Z^{k-1} = \left\{ z \in \wedge^n \left(J^{k-1}Y \right) \mid i_\xi i_\eta z = 0 \ \forall \ \xi, \eta \in V\pi^{k-1} \right\}.$$

The resulting Hamiltonian formalism now lives on Z^{k-1} which is the multisymplectic manifold appropriate in this context [17].

An interesting corollary of this discussion is that the covariant Hamiltonian formalism depends crucially upon the specific type of classical Lepagean equivalent under consideration, and not just the original variational problem $(J^kY \xrightarrow{\pi^k} X, \mathcal{J}_k, \mathcal{L})$. Moreover, as illustrated above, the notion of regularity shares this property.

These observations are tantalizing, and deserve further investigation. They may also have analogues in the general nonclassical case, but it is not so clear how this might go. When I is Pfaffian, the analogue of the covariant Hamiltonian formalism should in some sense live on a subbundle of $\wedge^n M$, whereas the rest of the algorithm should live on bundles over prolongations of $(M \xrightarrow{\pi} X, \mathcal{I}, \mathcal{L})$.

Finally, note that besides the Hamiltonian formalism on the far left of the diagram and the classical Lagrangian formalism on the far right, there are "intermediate" formalisms based on the bundles $R^q \to J^qY$, where $q = k, \ldots, 2k - 1$. Each of these bundles carries a canonical form induced from $\overline{\Theta}_{\mathcal{L}}$ on $\overline{W}$ (or, equivalently, from the canonical form on $\wedge^n(J^{k-1}Y)$) which makes it into a Lepagean equivalent of the original problem. These Lepagean equivalents are distinctly "nonclassical"; nonetheless, people have looked at variational problems on (what amounts to) R^k. Dedecker [11] has pursued this for first order field theories, and his work was extended to the higher order case by Szapiro [39]. Particularly interesting is the fact that they obtained the multisymplectic manifold $\wedge^n(J^{k-1}Y)$ in the Hamiltonian versions of their theories. Griffiths [19] has studied both classical and general single integral problems in an analogous context.

Appendix: Notation and Terminology

If X is an n-dimensional manifold with coordinates x^μ, define $\omega = dx^1 \wedge \ldots \wedge dx^n$, $\omega_\nu = i_{\partial_\nu}\omega$, and $\omega_{\mu\nu} = i_{\partial_\nu}\omega_\mu$ etc. On occasion I will employ multi-index notation. If $\mu = (\mu_1, \ldots, \mu_r)$ is a multi-index, set $|\mu| = r$. By convention $|\mu| = 0$ means there is no multi-index. When ν is an ordinary index, the combination $\underline{\mu}\nu$ stands for the "augmented" multi-index $(\mu_1, \ldots, \mu_r, \nu)$. Parentheses about a multi-index denotes symmetrization in its last two slots, e.g., $\lambda_A^{(\mu)} = \lambda_A^{\mu_1 \cdots (\mu_{r-1}\mu_r)}$, while brackets denote antisymmetrization. Set $\partial_{\underline{\mu}} = \frac{\partial^r}{\partial x^{\mu_1} \ldots \partial x^{\mu_r}}$. Throughout I use the summation convention, except with multi-indices where the summation is made explicit.

Let $\pi : Y \to X$ be a fibration. Adapted coordinates on Y are y^A along the fibers and x^μ along the base. The space of sections of π will be denoted $\Gamma(\pi)$. If Y is a tensor bundle then I will often write $C^\infty(Y)$ for $\Gamma(\pi)$. The vertical bundle $V\pi$ is the subbundle $\ker T\pi$ of TY. A form α on Y is *semi-basic* or *π-horizontal* if $i(\ker T\pi)\alpha = 0$.

The kth jet bundle of Y is $J^k Y$ with coordinates $x^\mu, y^A, y^A_{\mu_1}, \ldots, y^A_{\mu_1 \ldots \mu_k}$. Set $J^0 Y = Y$, $\pi^0 = \pi$ and $y^A_{\mu_0} = y^A$. The various projections are:

$$
\begin{array}{ccc}
J^r Y & \xrightarrow{\;\;\pi^r_s\;\;} & J^s Y \\[2mm]
{\scriptstyle \pi^r_0} \downarrow & & \downarrow {\scriptstyle \pi^s} \\[2mm]
Y & \xrightarrow[\;\;\pi\;\;]{} & X
\end{array}
$$

for $r \geq s$. The kth jet prolongation is denoted j^k. A section ϕ of $J^r Y \to X$ is said to be *s-holonomic* if $\pi^r_s \circ \phi = j^s(\pi^r_0 \circ \phi)$.

The contact system on $J^k Y$ is denoted J_k and is locally spanned by the 1-forms

$$
\psi^A_{\underline{\mu}} = dy^A_{\underline{\mu}} - y^A_{\underline{\mu}\nu} dx^\nu,
$$

where $0 \leq |\underline{\mu}| \leq k - 1$. The differential ideal generated by J_k is denoted $\mathcal{J}_k$. A form α on $J^k Y$ is *contact* provided $(j^k \varphi)^* \alpha = 0$ for all $\varphi \in \Gamma(\pi)$. Equivalently $(\pi^{k+1}_k)^* \alpha \equiv 0 \bmod \mathcal{J}_{k+1}$.

For a kth-order function $f \in C^\infty(J^k Y, \mathbb{R})$, the *formal* (or *total*) partial derivative $D_\mu f \in C^\infty(J^{k+1} Y, \mathbb{R})$ of f in the direction x^μ is defined by $(j^{k+1}\varphi)^*(D_\mu f) = \partial_\mu(f \circ j^k \varphi)$ for all $\varphi \in \Gamma(\pi)$. In jet charts

$$
D_\mu f = \partial_\mu f + \sum_{|\underline{\nu}|=0}^{k} \frac{\partial f}{\partial y^A_{\underline{\nu}}} y^A_{\underline{\nu}\mu}.
$$

The *variational derivative* of f is the function on $J^{2k} Y$ given by

$$
\frac{\delta f}{\delta y^A} = \sum_{s=0}^{k} (-1)^s D_{\mu_1} \cdots D_{\mu_s} \left(\frac{\partial f}{\partial y^A_{\mu_1 \ldots \mu_s}} \right).
$$

Acknowledgments. I would especially like to thank Bill Shadwick for many conversations during the course of this work. In particular, a number of the ideas and results appearing in this paper were jointly developed with him. Appreciation is also due Ian Anderson and Peter Olver for discussing the intricacies of Cartan forms with me, and Paul Dedecker for his most helpful comments.

This work was supported in part by NSF grant DMS-8805699, various ONR/NARC grants and a fellowship from the Ford Foundation. I am also grateful to the Mathematical Sciences Research Institute for its hospitality while part of this research was carried out.

REFERENCES

1. I. M. Anderson, *The variational bicomplex*, Utah State University preprint, (1989).

2. D. E. Betounes, *Extension of the classical Cartan form*, Phys. Rev. D **29** (1984), 599–606.

3. ___, *Differential geometric aspects of the Cartan form: Symmetry theory*, J. Math. Phys. **28** (1987), 2347–2353.

4. A. C. Bodnar, *The Cartan equivalence problem for first order Lagrangians in m independent and N dependent variables*, M. Sc. thesis, University of Waterloo (1986).

5. R. L. Bryant, *On notions of equivalence of variational problems with one independent variable*, Contemp. Math. **68** (1987), 65–76.

6. C. Carathéodory, *Über die Variationsrechnung B Über die Variationsrechnung bei mehrfachen Integralen*, Acta Szeged **4** (1929), 193–216.

7. É. Cartan, *Les systèmes différentiels extérieurs et leurs applications géométriques*, (Hermann, Paris, 1945).

8. M. de León and P. R. Rodrigues, *A contribution to the global formulation of the higher-order Poincaré-Cartan form*, Lett. Math. Phys. **14** (1987), 353–362.

9. P. Dedecker, *Calcul des variations, formes différentielles et champs géodésiques*, Colloq. Int. de Géométrie Différentielle (Strasbourg), (C.N.R.S., Paris, 1953), 17–34.

10. ___, *Calcul des variations et topologie algébrique*, Mém. Soc. Roy. Sci. Liège **19** (1957), 1–216.

11. ___, *On the generalization of symplectic geometry to multiple integrals in the calculus of variations*, Lect. Notes in Math. **570** (1977), 395–456.

12. M. Ferraris, *Fibered connections and global Poincaré-Cartan forms in higher-order calculus of variations*, Geometrical Methods in Physics, D. Krupka, ed., (J. E. Purkyně University, Brno, Cz., 1984), 61–91.

13. M. Ferraris and M. Francaviglia, *Applications of the Poincaré-Cartan form in higher order field theories*, Differential Geometry and Its Applications, D. Krupka and A. Švec, eds. (Reidel, Boston, 1987), 31–52.

14. P. L. García, *The Poincaré-Cartan invariant in the calculus of variations*, Symp. Math. **XIV** (1974), 219–246.

15. P. L. García and J. Muñoz, *Higher order analytical dynamics*, Dynamical Systems and Partial Differential Equations, L. Lara-Carrero and J. Lewowicz, eds., (Equinoccio, Caracas, 1984), 19-47.

16. H. Goldschmidt and S. Sternberg, *The Hamilton-Cartan formalism in the calculus of variations*, Ann. Inst. Fourier **23** (1973), 203–267.

17. M. J. Gotay, *A multisymplectic framework for classical field theory and the calculus of variations. I. Covariant Hamiltonian formalism*, Mechanics, Analysis and Geometry: 200 Years after Lagrange, M. Francaviglia ed., (North Holland, Amsterdam, 1991), pp. 203–235.

18. M. J. Gotay, J. A. Isenberg, J. E. Marsden, and R. Montgomery, *Momentum mappings and the Hamiltonian structure of classical field theories with constraints*, to appear (1992).

19. P. A. Griffiths, *Exterior Differential Systems and the Calculus of Variations*, Prog. in Math. **25** (Birkhäuser, Boston, 1983).

20. T. J. Harding and F. J. Bloore, *A fresh approach to the Poincaré-Cartan form for a linear P. D. E. and a map between cohomologies*, Differential Goemetry and Its Applications, J. Janyška and D. Krupka, eds., (World Scientific, Singapore, 1990), 220–229.

21. M. Horák and I. Kolář, *On the higher order Poincaré-Cartan forms*, Cz. Math. J. **33** (1983), 467–475.

22. L. Hsu, *Calculus of variations via the Griffiths formalism*, Duke University preprint (1990).

23. H. A. Kastrup, *Canonical theories of Lagrangian dynamical systems in physics*, Phys. Rep. **101** (1983), 1–167.

24. I. Kolář, *A geometric version of the higher order Hamilton formalism in fibered manifolds*, J. Geom. Phys. **1** (1984), 127–137.

25. ___, *Some geometric aspects of the higher order variational calculus*, Geometrical Methods in Physics, D. Krupka, ed., (J. E. Purkyně University, Brno, Cz., 1984), pp. 155–166.

26. D. Krupka, *A map associated to the Lepagean forms of the calculus of variations in fibered manifolds*, Cz. Math. J. **27** (1977), 121–133.

27. ___, *Geometry of Lagrangean structures. 3*, Supp. Rend. Circ. Mat. (Palermo), ser. II, no. 14 (1987), 187–224.

28. D. Krupka and J. Musilová, *Hamilton extremals in higher order mechanics*, Arch. Math. (Brno) **20** (1984), 21–30.

29. B. Kupershmidt, *Geometry of jet bundles and the structure of Lagrangian and Hamiltonian formalisms*, Lect. Notes in Math. **775** (1980), 162–217.

30. T. H. J. Lepage, *Sur les champs géodésiques du calcul des variations I, II*, Bull. Acad. Roy. Belg., Classe des Sciences **22** (1936), 716–739, 1036–1046.

31. ___, *Champs stationnaires, champs géodésiques et formes intégrables,* I, II, Bull. Acad. Roy. Belg., Classe des Sciences **28** (1942), 73–92, 247–268.

32. M. Marvan, *On global Lepagean equivalents,* Geometrical Methods in Physics, D. Krupka, ed., (J. E. Purkyně University, Brno, Cz. 1984), 185–190.

33. J. Muñoz, *Poincaré-Cartan forms in higher order variational calculus on fibered manifolds,* Rev. Mat. Iberoamericana **1** (1985), 85–126.

34. P. V. Pâquet, *Les formes différentielles extérieures Ω_n dans le calcul des variations,* Bull. Acad. Roy. Belg., Classe des Sciences **27** (1941), 65–84.

35. H. Rund, *A Cartan form for the field theory of Carathéodory in the calculus of variations of multiple integrals,* Lect. Notes in Pure and Appl. Math. **100** (1985), 455–469.

36. D. J. Saunders, *An alternative approach to the Cartan form in Lagrangian field theories,* J. Phys. A: Math. Gen. **20** (1987), 339–349.

37. D. J. Saunders and M. Crampin, *On the Legendre map in higher-order field theories,* J. Phys. A: Math. Gen. **23** (1990), 3169–3182.

38. S. Sternberg, *Some preliminary remarks on the formal variational calculus of Gel'fand and Dikki,* Lect. Notes in Math. **676** (1977), 399–407.

39. T. Szapiro, *Geodesic fields in the calculus of variations of multiple integrals depending on derivatives of higher order,* Lect. Notes in Math. **836** (1980), 504–511.

40. E. Zeidler, *Nonlinear Functional Analysis and Its Applications.* III. *Variational Methods and Optimization,* (Springer, New York, 1985).

Mathematics Department
United States Naval Academy
Annapolis MD 21402-5002
USA

The Rotor and the Pendulum

DARRYL D. HOLM and JERROLD E. MARSDEN

In Honor of J.-M. Souriau

Abstract

We show that Euler's equations for a free rigid body, and for a rigid body with a controlled feedback torque each reduce to the classical simple pendulum equation under an explicit cylindrical coordinate change of variables. These examples illustrate several ideas in Hamiltonian mechanics: Lie-Poisson reduction, cotangent bundle reduction, singular Lie-Poisson maps, deformations of Lie algebras, brackets on $\mathbb{R}^3$, simplifications obtained by utilizing the representation-dependence of Lie-Poisson reduction, and controlling instability by inducing global bifurcations among a set of equilibria using a control parameter.

1. Introduction

Even though the free rigid body is a classical and well understood system, some new and interesting features are still being uncovered. Notable amongst these is Montgomery's [1990] formula for the change in the geometric phase for the attitude of the body when the body angular momentum vector executes one period of its motion. In this paper we present a number of other results that also seem to be new. Perhaps the most interesting of these is the fact that the rigid body system in body angular momentum space (identified with $\mathbb{R}^3$) is filled with invariant elliptical cylinders on each of which the dynamics is, in elliptical cylindrical coordinates, *exactly* the dynamics of a standard simple pendulum.

Related to this is the variety of ways the rigid body equations can be written in Hamiltonian form as a Lie-Poisson system associated to a Lie algebra structure on $\mathbb{R}^3$. The standard choice is to use the Lie algebra $SO(3)$, but one can also use the Euclidean Lie algebra $SE(2)$ or the Lie algebra $SO(2,1)$. The deformation through these algebras, discussed abstractly in Weinstein [1983], is achieved explicitly in the rigid body simply by defining new Hamiltonians and Casimirs using linear combinations of the standard ones.

We make a similar analysis of the rigid body with the stabilizing torque feedback law introduced by Bloch and Marsden [1990] (see also Bloch, Krishnaprasad, Marsden, and Sanchez de Alvarez [1990]). In particular, this sort of analysis enables one to see how the stabilization is achieved from a geometric viewpoint. Some interesting global bifurcations accompany this stabilization. Combined with the ideas about geometric phases, this stabilization process may be useful for attitude control of rigid bodies.

2. The Free Rigid Body

Euler's classical equations for the rotational dynamics of a freely spinning rigid body are

$$
\begin{aligned}
I_1\dot{\Omega}_1 &= (I_2 - I_3)\Omega_2\Omega_3 \\
I_2\dot{\Omega}_2 &= (I_3 - I_1)\Omega_3\Omega_1 \\
I_3\dot{\Omega}_3 &= (I_1 - I_2)\Omega_1\Omega_2
\end{aligned}
\tag{$EE\Omega$}
$$

where $\vec{\Omega} = (\Omega_1, \Omega_2, \Omega_3)$ is the *body angular velocity vector*, an overdot denotes time derivative, and $I_1 < I_2 < I_3$ denote the principal moments of inertia of the body. We can rewrite $(EE\Omega)$ as

$$
\frac{d}{dt}\begin{bmatrix} \Pi_1 \\ \Pi_2 \\ \Pi_3 \end{bmatrix} = \begin{bmatrix} \left(\dfrac{1}{I_2} - \dfrac{1}{I_3}\right)\Pi_2\Pi_3 \\[2mm] -\left(\dfrac{1}{I_1} - \dfrac{1}{I_3}\right)\Pi_1\Pi_3 \\[2mm] \left(\dfrac{1}{I_1} - \dfrac{1}{I_2}\right)\Pi_1\Pi_2 \end{bmatrix},
\tag{EE}
$$

where $(\Pi_1, \Pi_2, \Pi_3) = \vec{\Pi}$ denotes the *body angular momentum vector* given by $\Pi_i = I_i\Omega_i$, $i = 1, 2, 3$. Euler's equations are expressible in vector form as

$$
\frac{d}{dt}\vec{\Pi} = \nabla H \times \nabla L,
$$

where H is the energy,

$$
H = \frac{\Pi_1^2}{2I_1} + \frac{\Pi_2^2}{2I_2} + \frac{\Pi_3^2}{2I_3},
$$

$$
\nabla H = \left(\frac{\partial H}{\partial \Pi_1}, \frac{\partial H}{\partial \Pi_2}, \frac{\partial H}{\partial \Pi_3}\right) = \left(\frac{\Pi_1}{I_1}, \frac{\Pi_2}{I_2}, \frac{\Pi_3}{I_3}\right)
$$

is the gradient of H and L is the square of the body angular momentum,

$$
L = \frac{1}{2}\left(\Pi_1^2 + \Pi_2^2 + \Pi_3^2\right).
$$

Hence, both H and L are conserved, and the rigid body motion itself takes place along the intersections of the level surfaces of the energy (ellipsoids) and the angular momentum (spheres) in $\mathbb{R}^3$. The centers of the energy ellipsoids and the angular momentum spheres coincide. This, along with the $(\mathbb{Z}_2)^3$ symmetry of the energy ellipsoids, implies that the two sets of level surfaces in $\mathbb{R}^3$ develop collinear gradients (*e.g.*, tangencies) at pairs of points which are diametrically opposite on an angular momentum sphere. At these points, collinearity of the gradients of H and L implies stationary rotations, *i.e.*, equilibria.

Euler's equations for the rigid body may also be written as

$$\frac{d}{dt}\vec{\Pi} = \nabla K \times \nabla N, \qquad\qquad (EE')$$

where K and N are linear combinations of energy and angular momentum of the form

$$\begin{pmatrix} K \\ N \end{pmatrix} = \begin{bmatrix} a & b \\ c & d \end{bmatrix} \begin{pmatrix} H \\ L \end{pmatrix}, \qquad\qquad (SL2R)$$

with real constants $a, b, c,$ and d satisfying the unit determinant condition $ad - bc = 1$. Thus, the equations of rigid body motion are unchanged (so the trajectories of the motion in $\mathbb{R}^3$ remain unchanged) when the energy H and angular momentum L are replaced by the $SL(2, \mathbb{R})$-linear combinations K and N. Notice that K will be a quadratic form and that it can occur with *any signature except* $(0, 0, 0)$ *and, if the moments of inertia are distinct,* $(*, 0, 0)$ by suitably choosing a and b.

For example, one may choose to eliminate one of the terms in each of H and L, by choosing the linear combination given by

$$\begin{bmatrix} a & b \\ c & d \end{bmatrix} = \begin{bmatrix} 1 & \frac{-1}{I_3} \\ -c & \frac{c}{I_1} \end{bmatrix}, \quad \text{with } c = \frac{1}{\left(\frac{1}{I_1} - \frac{1}{I_3}\right)},$$

so that

$$K = \frac{1}{2}\left(\frac{1}{I_1} - \frac{1}{I_3}\right)\Pi_1^2 + \frac{1}{2}\left(\frac{1}{I_2} - \frac{1}{I_3}\right)\Pi_2^2,$$

and

$$N = \frac{c}{2}\left(\frac{1}{I_1} - \frac{1}{I_2}\right)\Pi_2^2 + \frac{1}{2}\Pi_3^2.$$

With this choice, the orbits for Euler's equations for rigid body dynamics are realized as motion along the intersections of two, orthogonally-oriented, *elliptic cylinders*, one elliptic cylinder a level surface of K, with its translation axis along Π_3 (where $K = 0$), and the other a level surface of N, with its translation axis along Π_1 (where $N = 0$).

For a general choice of K and N, equilibria occur at points where the gradients of K and N are collinear. This can occur at points where the level sets are tangent (and the gradients both are non-zero), or at points where one of the gradients vanishes. In the elliptic cylinder case above, these two cases are points where the elliptic cylinders are tangent, and at points where the axis of one cylinder punctures normally through the surface of the other. The elliptic cylinders are tangent at one $\mathbb{Z}_2$-symmetric pair of points along the Π_2 axis, and the elliptic cylinders have normal axial punctures at two other $\mathbb{Z}_2$-symmetric pairs of points along the Π_2 and Π_3 axes.

The stability of the equilibria at the points of tangency is determined geometrically by whether the curvatures of the two surfaces have opposite sign (stable), or the same sign (unstable). When the two surfaces have curvatures of opposite sign near the tangent point, their intersections describe nested ellipses on each surface—this is the stable case. When the two surfaces have curvatures of the same sign near the tangent point, their intersections describe hyperbolas on each surface—this is the unstable case. In the elliptic cylinder example, the curvatures of the elliptic cylinders have the same sign at the points of tangency, so the pair of equilibria at the tangent points along the Π_2 axis are unstable. Equilibria at puncture points are stable; so, the two pairs of diametrically opposite equilibria at the puncture points along the Π_2 and Π_3 axes are stable. This geometric picture thus recovers the known equilibria and the stability properties of these equilibria for the rigid body, in a new parameterization in terms of intersections of two, orthogonally-oriented, elliptic cylinders in $\mathbb{R}^3$. The situation is shown in Figure 1.

Let us pursue the elliptic cylinders example further. We now change variables in the rigid body equations within a level surface of K. To simplify notation, we first define the three positive constants k_i^2, $i = 1, 2, 3$, by setting

$$K = \frac{\Pi_1^2}{2k_1^2} + \frac{\Pi_2^2}{2k_2^2} \quad \text{and} \quad N = \frac{\Pi_2^2}{2k_3^2} + \frac{1}{2}\Pi_3^2$$

and referring term by term to the formulae

$$K = \frac{1}{2}\left(\frac{1}{I_1} - \frac{1}{I_3}\right)\Pi_1^2 + \frac{1}{2}\left(\frac{1}{I_2} - \frac{1}{I_3}\right)\Pi_2^2$$

and
$$N = \frac{c}{2}\left(\frac{1}{I_1} - \frac{1}{I_2}\right)\Pi_2^2 + \frac{1}{2}\Pi_3^2.$$

On the surface $K = $ constant, and setting $r = \sqrt{2K} = $ constant, define new variables θ and p by

$$\Pi_1 = k_1 r \cos\theta, \quad \Pi_2 = k_2 r \sin\theta, \quad \Pi_3 = p.$$

In terms of these variables, the constants of the motion become

$$K = \frac{1}{2}r^2 \quad \text{and} \quad N = \frac{1}{2}p^2 + \left(\frac{k_2^2}{2k_3^2}r^2\right)\sin^2\theta.$$

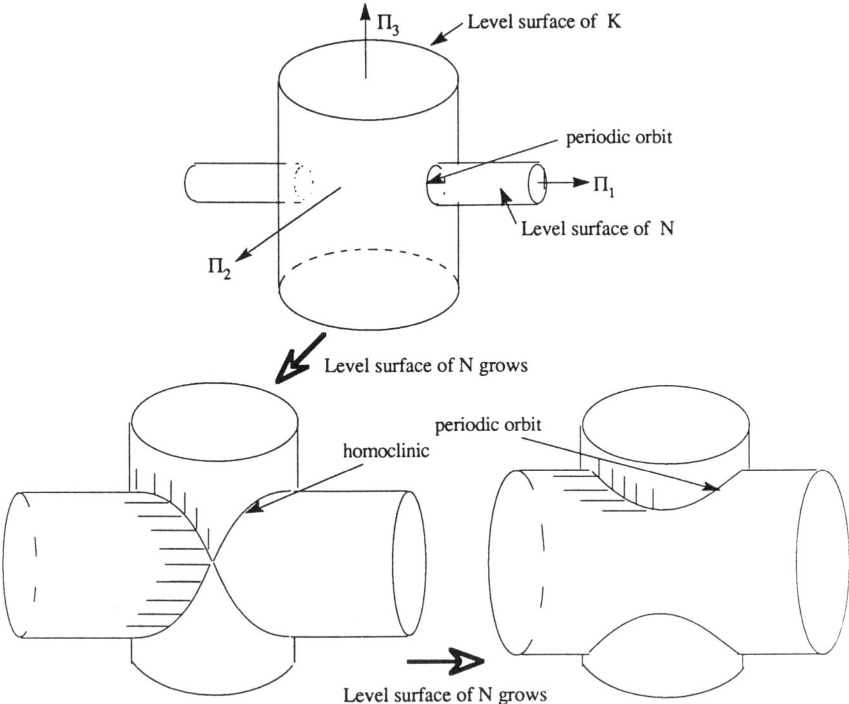

Figure 1. Intersection of the elliptic cylinders—level surfaces of K and N; orbits change from "bound" to "running" as the level surface of N passes through the critical value at which homoclinic orbits occur.

As we shall show in section **3**, using a Poisson structure relevant to the equations of motion in the form $\frac{d}{dt}\vec{\Pi} = \nabla K \times \nabla N$, the variables θ and p are, up to a scale factor, canonically conjugate, $i.e.$, the Poisson bracket of two functions of θ and p are given in standard canonical form (up to a scale factor) as follows:

$$\{F, G\}_{\text{EllipCyl}} = \frac{1}{k_1 k_2}\left(\frac{\partial F}{\partial p}\frac{\partial G}{\partial \theta} - \frac{\partial F}{\partial \theta}\frac{\partial G}{\partial p}\right).$$

In particular,

$$\{p, \theta\}_{\text{EllipCyl}} = \frac{1}{k_1 k_2}.$$

The quantity N is the Hamiltonian in these variables—note that N has the form of kinetic plus potential energy—and the equations of motion express themselves in Hamiltonian form in terms of the canonical Poisson bracket. Namely,

$$\frac{d}{dt}\theta = \{N, \theta\}_{\text{EllipCyl}} = \frac{1}{k_1 k_2}\frac{\partial N}{\partial p} = \frac{1}{k_1 k_2}p,$$

$$\frac{d}{dt}p = \{N, p\}_{\text{EllipCyl}} = \frac{-1}{k_1 k_2} \frac{\partial N}{\partial \theta} = \frac{-1}{k_1 k_2} \frac{k_2^2}{k_3^2} r^2 \sin\theta \cos\theta.$$

Combining these equations of motion gives

$$\frac{d^2}{dt^2}\theta = \frac{-r^2}{2k_1^2 k_3^2} \sin 2\theta,$$

or, in terms of the original rigid body parameters,

$$\frac{d^2}{dt^2}\theta = -K\left(\frac{1}{I_1} - \frac{1}{I_2}\right)\left(\frac{1}{I_1} - \frac{1}{I_3}\right) \sin 2\theta.$$

Thus, we have proved that

Theorem 1. *Rigid body motion reduces to pendulum motion on level surfaces of K.*

Another way of saying this is as follows: regard rigid body angular momentum space as the union of the level surfaces of K, so the dynamics of the rigid body is recovered by looking at the dynamics on each of these level surfaces. On each level surface, the dynamics is equivalent to a simple pendulum. In this sense, we have proved that:

Corollary. *The dynamics of a rigid body in three dimensional body angular momentum space is a union of two dimensional simple pendula phase portraits.*

Remarks. By restricting to a nonzero level surface of K, the pair of rigid body equilibria along the Π_3 axis are excluded. (This pair of equilibria can be included by permuting the indices of the moments of inertia.) The other two pairs of equilibria, along the Π_1 and Π_2 axes, lie in the $p = 0$ plane at $\theta = 0$, $\pi/2$, π, and $3\pi/2$. Since K is positive, the stability of each equilibrium point is determined by the relative sizes of the principle moments of inertia, which affect the overall sign of the right-hand-side of the pendulum equation. The well-known results about stability of equilibrium rotations along the least and greatest principle axes, and instability around the intermediate axis, are immediately recovered from this overall sign, combined with the stability properties of the pendulum equilibria. For $K > 0$ and $I_1 < I_2 < I_3$, this overall sign is negative, so the equilibria at $\theta = 0$ and π (along the Π_1 axis) are stable, while those at $\theta = \pi/2$ and $3\pi/2$ (along the Π_2 axis) are unstable. The factor of 2 in the argument of the sine in the pendulum equation is explained by the $\mathbb{Z}_2$ symmetry of the level surfaces of K (or, just as well, by their invariance under $\theta \mapsto \theta + \pi$). Under this discrete symmetry operation, the equilibria at $\theta = 0$ and $\pi/2$ exchange with their counterparts at $\theta = \pi$ and $3\pi/2$, respectively, while the elliptical level surface of K is left invariant. By construction, the Hamiltonian N in the reduced variables θ and p is also invariant under this discrete symmetry.

3. Alternative Poisson Structures for the Rigid Body

The standard **rigid body Poisson bracket** on two functions F_1 and F_2 of $\vec{\Pi}$ is given by the *minus Lie Poisson bracket* for $\mathfrak{so}(3)^*$:

$$\{F_1, F_2\} = -\vec{\Pi} \cdot (\nabla F_1 \times \nabla F_2). \qquad (RBB)$$

In case Euler's equations are rewritten as

$$\frac{d}{dt}\vec{\Pi} = \nabla K \times \nabla N, \qquad (EE')$$

where K and N are given as above by an $SL(2, \mathbb{R})$ matrix

$$\begin{pmatrix} K \\ N \end{pmatrix} = \begin{bmatrix} a & b \\ c & d \end{bmatrix} \begin{pmatrix} H \\ L \end{pmatrix}, \qquad (SL2R)$$

one checks that the equations (EE') are Hamiltonian with energy N and the Poisson bracket

$$\{F_1, F_2\} = -\nabla K \cdot (\nabla F_1 \times \nabla F_2). \qquad (PBK)$$

One verifies also that the bracket (PBK)—essentially a "Nambu" bracket (see Nambu [1973]), indeed defines a Poisson structure on $\mathbb{R}^3$. Clearly the function K is a Casimir for this bracket; *i.e.*, $\{K, F\} = 0$ for any function F. One can now directly verify the formula $\{F, G\}_{\text{EllipCyl}}$ for the Poisson bracket on level sets of the function K in the elliptic cylinder case by a straightforward calculation.

We shall see shortly *that the bracket (PBK) is in fact a Lie Poisson bracket.* Let $\mathbf{K}$ be the symmetric 3×3 matrix associated with the quadratic form K; *i.e.*, $K(\mathbf{v}) = \frac{1}{2}\mathbf{v}^t \cdot \mathbf{K} \cdot \mathbf{v}$, where ^{t} denotes transpose. Thus, the gradient of K is $\nabla K(\vec{\Pi}) \cdot \mathbf{v} = \mathbf{d}K(\vec{\Pi}) \cdot \mathbf{v} = \vec{\Pi} \cdot \mathbf{K}\mathbf{v}$, and so (PBK) may be written

$$\{F_1, F_2\}_K(\vec{\Pi}) = -\vec{\Pi} \cdot \mathbf{K}(\nabla F_1 \times \nabla F_2). \qquad (PBK')$$

Note that these formulas are valid even if the matrix $\mathbf{K}$ is singular.

Define the following bracket on $\mathbb{R}^3$:

$$[\mathbf{u}, \mathbf{v}]_K = \mathbf{K} \cdot (\mathbf{u} \times \mathbf{v}). \qquad (LAK)$$

This defines a Lie algebra structure on $\mathbb{R}^3$, as is to verify.

For $\mathbf{K}$ nonsingular, the Lie algebra structure (LAK) can be explicitly identified with that of the orthogonal group of $\mathbf{K}$ as follows. The orthogonal group $O(K)$ is the group of linear transformations of $\mathbb{R}^3$ that leave the quadratic form K invariant. It consists of the set of 3×3 matrices $\mathbf{A}$ that satisfy the condition $\mathbf{A}^t \mathbf{K} \mathbf{A} = \mathbf{K}$. The corresponding Lie algebra $\mathfrak{o}(K)$ is the Lie algebra of 3×3 matrices $\mathbf{S}$ that satisfy the condition $\mathbf{S}^t \mathbf{K} + \mathbf{K} \mathbf{S} = \mathbf{0}$, *i.e.*, $\mathbf{K} \mathbf{S}$ is a skew matrix, and with the standard commutator bracket of matrices. Since $\mathbf{K}$ is nonsingular, the equation

$$\mathbf{S} \mathbf{v} = \mathbf{s} \times \mathbf{K} \mathbf{v} \qquad (LAI)$$

defines an isomorphism between $\mathbf{S} \in \mathfrak{o}(\mathbf{K})$ and $\mathbf{s} \in \mathbb{R}^3$. The following is a straightforward verification:

Lemma 1.
1. *If $\mathbf{K}$ is nonsingular, the isomorphism (LAI) is a Lie algebra isomorphism between the Lie algebra $\mathfrak{o}(K)$ with the commutator bracket and $\mathbb{R}^3$ with the bracket (LAK); i.e., $[\mathbf{S_1}, \mathbf{S_2}] = \mathbf{K}(\mathbf{s_1} \times \mathbf{s_2})$.*
2. *If $\mathbf{K}$ is singular, then the Lie algebra structure is that of the euclidean Lie algebra $\mathfrak{se}(2)$ if $\mathbf{K}$ has signature $(+, +, 0)$, and that of the Heisenberg algebra if $\mathbf{K}$ has signature $(+, 0, 0)$.*

Lemma 2. *The Poisson bracket $\{F_1, F_2\}_K(\vec{\Pi}) = -\vec{\Pi} \cdot \mathbf{K}(\nabla F_1 \times \nabla F_2)$ on $\mathbb{R}^3$ is the minus Lie Poisson bracket for the Lie algebra $\mathbb{R}^3_K$, defined to be $\mathbb{R}^3$ with the Lie algebra bracket $[\mathbf{u}, \mathbf{v}]_K = \mathbf{K} \cdot (\mathbf{u} \times \mathbf{v})$.*

To see this, recall that the general minus Lie-Poisson bracket formula (see, for example, Marsden and Weinstein [1983]) is given on $\mathfrak{g}^*$, the dual of a general Lie algebra $\mathfrak{g}$ by

$$\{F_1, F_2\} = -\left\langle \mu, \left[\frac{\delta F_1}{\delta \mu}, \frac{\delta F_2}{\delta \mu}\right] \right\rangle.$$

In our case, the Lie algebra is $\mathbb{R}^3_K$ and the dual space is identified with $\mathbb{R}^3$ via the standard dot product. The functional derivative $\frac{\delta F}{\delta \mu}$ is thus the ordinary gradient, and μ gets replaced by $\vec{\Pi}$, so Lemma 2 follows.

We can gain insight into why K is a Casimir for the bracket $\{F_1, F_2\}_K$ by noting that the coadjoint action of the Lie algebra $\mathbb{R}^3_K$ on the dual space is given by taking the dual of the adjoint action:

$$\langle ad^*_\mathbf{s} \vec{\Pi}, \mathbf{r}\rangle = \langle \vec{\Pi}, ad_\mathbf{s} \mathbf{r}\rangle = \langle \vec{\Pi}, \mathbf{K} \cdot (\mathbf{s} \times \mathbf{r})\rangle = \langle \mathbf{K}\vec{\Pi} \times \mathbf{s}, \mathbf{r}\rangle$$

and so
$$ad^*_\mathbf{s} \vec{\Pi} = (\mathbf{K}\vec{\Pi}) \times \mathbf{s}.$$

Vectors of this form are tangent to the coadjoint orbit through $\vec{\Pi}$. Notice that the differential of K vanishes on vectors of this form; so K is a Casimir and therefore K is constant on (connected components of) orbits. (This corresponds to the fact, observed earlier, that K is a Casimir for the bracket $\{F, H\}_K$.) In the nonsingular case, *the coadjoint orbit through $\vec{\Pi}$ is, up to connected components, exactly the level set of K through $\vec{\Pi}$.* The orbit is given algebraically as follows: write $\vec{\Pi} = \mathbf{K}\mathbf{v} \times \mathbf{w}$; then the orbit through $\vec{\Pi}$ consists of vectors of the form $Ad^*_{\mathbf{A}}\vec{\Pi} = \mathbf{A}^t\mathbf{K}\mathbf{v} \times \mathbf{A}^t\mathbf{w}$ as $\mathbf{A}$ ranges over $O(K)$. In the case of the euclidean Lie algebra, which corresponds to the case of the elliptic cylinders, the regular coadjoint orbits are given by cotangent bundles to circles, with the canonical symplectic structure (up to a factor, depending on the radius of the circle). Again, we see by a dimension count that the orbits are the level sets of K. We have proved the following:

Lemma 3. *The (connected components of the) coadjoint orbits for $\mathbb{R}^3_K \cong \mathbb{R}^3$ are the level sets of K if $\mathbf{K}$ is nonsingular, or if $\mathbf{K}$ has signature $(+, +, 0)$. Tangent vectors to the coadjoint orbit through $\vec{\Pi}$ are given by vectors in $\mathbb{R}^3$ of the form $ad^*_{\mathbf{s}}\vec{\Pi} = (\mathbf{K}\vec{\Pi}) \times \mathbf{s}$ and the symplectic structure on the orbit is given by $\Omega\big((\mathbf{K}\vec{\Pi}) \times \mathbf{s}_1, (\mathbf{K}\vec{\Pi}) \times \mathbf{s}_2\big) = -\mathbf{K} \cdot (\mathbf{s_1} \times \mathbf{s_2})$.*

The rigid body can, correspondingly, be regarded as a left invariant system on the group $O(K)$ or $SE(2)$. The special case of $SE(2)$ is the one in which the orbits are cotangent bundles. The fact that one gets a cotangent bundle in this situation is a special case of the cotangent bundle reduction theorem using the semidirect product reduction theorem; see Marsden, Ratiu, and Weinstein [1984]. For the Euclidean group it says that the coadjoint orbits of the Euclidean group of the plane are given by reducing the cotangent bundle of the rotation group of the plane by the trivial group, giving the cotangent bundle of a circle with its canonical symplectic structure up to a factor. This is the abstract explanation of why, in the elliptic cylinder case above, the variables θ and p were, up to a factor, canonically conjugate. This general theory is also consistent with the fact that the Hamiltonian N is of the form kinetic plus potential energy. In fact, in the cotangent bundle reduction theorem, one always gets a Hamiltonian of this form, with the potential being changed by the addition of an amendment to give the *amended potential.* In the case of the pendulum equation, the original Hamiltonian is purely kinetic energy and so the potential term in N, namely $\left(\frac{k_2^2}{2k_3^2}r^2\right)\sin^2\theta$, is entirely amendment. See Abraham and Marsden [1978] for the general theory.

We summarize some of our findings as follows:

Theorem 2. *Euler's equations for a free rigid body are Lie Poisson with the Hamiltonian N for the Lie algebra $\mathbb{R}^3_K$ where the underlying Lie* *group is the orthogonal group of K if the quadratic form is nondegenerate, and is the Euclidean group of the plane if K has signature $(+,+,0)$. In particular, all the groups $SO(3)$, $SO(2,1)$, and $SE(2)$ occur as the parameters a, b, c, and d are varied. (If the body is a Lagrange body (with two moments of inertia equal), then the Heisenberg group occurs as well.)*

Remark. The same richness of Hamiltonian structure was found in the Maxwell-Bloch system in David and Holm [1990]. As in the case of the rigid body, the $\mathbb{R}^3$ motion for the Maxwell-Bloch system may also be realized as motion along the intersections of two orthogonally oriented cylinders. However, in this case, one cylinder is parabolic in cross section, while the other is circular. Upon passing to parabolic cylindrical coordinates, the Maxwell-Bloch system reduces to the ideal Duffing equation, while in circular cylindrical coordinates, the pendulum equation results. The $SL(2,R)$ matrix transformation $(SL2R)$ in the Maxwell-Bloch case provides a parametrized array of (offset) ellipsoids, hyperboloids, and cylinders, along whose intersections the $\mathbb{R}^3$ motion takes place.

4. Rigid Body with Controlled Feedback Torque

In this section we illustrate how to control the stability properties of equilibria for a dynamical system by using a control parameter that induces global bifurcations. The example is the free rigid body with a single torque about its major axis introduced by Bloch and Marsden [1990]. A similar analysis for other systems, such as that in Bloch, Krishnaprasad, Marsden, and Sánchez de Alvarez [1990] coming from a rigid body system with internal rotors is possible as well.

Euler's equations for the free rigid body with a single torque, u, about its major axis are given by

$$\frac{d}{dt}\begin{bmatrix}\Pi_1 \\ \Pi_2 \\ \Pi_3\end{bmatrix} = \begin{bmatrix}\left(\dfrac{1}{I_2} - \dfrac{1}{I_3}\right)\Pi_2\Pi_3 \\ -\left(\dfrac{1}{I_1} - \dfrac{1}{I_3}\right)\Pi_1\Pi_3 \\ \left(\dfrac{1}{I_1} - \dfrac{1}{I_2}\right)\Pi_1\Pi_2 + u\end{bmatrix}.$$

Following Bloch and Marsden [1990], we employ the feedback rule

$$u = -k\Pi_1\Pi_2,$$

where k is the feedback gain parameter. We refer to the system with this feedback as the *controlled system*. The equations of the controlled system are expressible in vector form as

$$\frac{d}{dt}\vec{\Pi} = \nabla H' \times \nabla L' \qquad\qquad (CEE)$$

where H' is the controlled-system energy,

$$H' = \frac{\Pi_1^2}{2I_1} + \frac{\Pi_2^2}{2I_2} + \frac{\Pi_3^2}{2gI_3},$$

L' is the square of the controlled-system angular momentum,

$$L' = \frac{1}{2}(g\Pi_1^2 + g\Pi_2^2 + \Pi_3^2),$$

and the parameter g is defined by

$$g = 1 - \frac{k}{\left(\frac{1}{I_1} - \frac{1}{I_2}\right)}.$$

Note that the parameter g in these equations contains elements of both the rigid body mass distribution, and its internal feedback torque along the 3-axis.

The vector-cross-product form of the controlled system (CEE) ensures that both of the quantities H' and L' are conserved, and that the motion of the controlled system takes place in $\mathbb{R}^3$ along the intersections of the level surfaces of the two conserved quantities. The level surfaces H' and L' are clearly of either elliptic type, or hyperbolic type, depending upon the sign of g. (The case $g = 0$ is degenerate and will be discussed later. The rigid body case treated earlier is recovered for $g = 1$.) Regardless of the sign of g, the centers of the level surfaces of the quadratic functions H' and L' coincide at the origin of coordinates in $\mathbb{R}^3$, so points on these level surfaces where the gradients of H' and L' are collinear (*i.e.*, equilibrium points) will occur in opposing ($\mathbb{Z}_2$-symmetric, $\mathbf{x} \mapsto -\mathbf{x}$) pairs. In particular, tangencies for which the H' and L' level surfaces have curvatures of the same sign produce unstable equilibria, while tangencies of H' and L' having opposite sign curvatures produce stable equilibria. From the expressions for H' and L' above, it is clear that the sign of g will play a key role in the stability properties of the equilibria.

Euler's equations for the controlled system may be re-expressed as

$$\frac{d}{dt}\vec{\Pi} = \nabla K' \times \nabla N',$$

where K' and N' are linear combinations of the controlled-system energy and angular momentum given by

$$\begin{pmatrix} K' \\ N' \end{pmatrix} = \begin{bmatrix} a & b \\ c & d \end{bmatrix} \begin{pmatrix} H' \\ L' \end{pmatrix},$$

with real constants a, b, c, and d satisfying the unit determinant condition $ad - bc = 1$. The controlled system is unchanged (and so the trajectories of its motion in $\mathbb{R}^3$ remain unchanged) when its energy H' and angular momentum L' are replaced by the $SL(2,\mathbb{R})$ linear combinations K' and N'. By analogy to the rigid body discussed in the previous sections, one may choose to eliminate one term in each of H' and L' by choosing the linear combination given by

$$\begin{bmatrix} a & b \\ c & d \end{bmatrix} = \begin{bmatrix} 1 & \frac{-1}{gI_3} \\ -c & \frac{c}{gI_1} \end{bmatrix}, \quad \text{with } c = \frac{g}{\left(\frac{1}{I_1} - \frac{1}{I_3}\right)},$$

so that

$$K' = \frac{1}{2}\left(\frac{1}{I_1} - \frac{1}{I_3}\right)\Pi_1^2 + \frac{1}{2}\left(\frac{1}{I_2} - \frac{1}{I_3}\right)\Pi_2^2,$$

and

$$N' = \frac{c}{2}\left(\frac{1}{I_1} - \frac{1}{I_2}\right)\Pi_2^2 + \frac{1}{2}\Pi_3^2.$$

With this choice, the controlled system may be regarded as describing motion taking place along the intersection of a level surface of K', an elliptic cylinder with its translation axis along Π_3 (where $K' = 0$), and a level surface of N', which (depending on the sign of g) is either an elliptic cylinder ($g > 0$), or a hyperbolic cylinder ($g < 0$); in either case with its translation axis along Π_1 (where $N' = 0$). For $g = 0$, the level surfaces of N' are horizontal planes, $\Pi_3^2 = \text{constant}$.

We restrict the controlled system to a level surface of K' by the same change of variables as for the rigid body. In terms of the three positive constants k_i^2, $i = 1, 2, 3$, defined in the rigid-body case, we have

$$K' = \frac{\Pi_1^2}{2k_1^2} + \frac{\Pi_2^2}{2k_2^2} \quad \text{and} \quad N' = \frac{g\Pi_2^2}{2k_3^2} + \frac{1}{2}\Pi_3^2.$$

In terms of the elliptic polar coordinates defined by

$$\Pi_1 = k_1 r \cos\theta, \ \Pi_2 = k_2 r \sin\theta, \ \Pi_3 = p,$$

and with $r = \sqrt{2K'} = \text{constant}$, the conserved quantities become

$$K' = \frac{1}{2}r^2, \quad \text{and} \quad N' = \frac{1}{2}p^2 + g\left(\frac{k_2^2}{k_3^2}r^2\right)\sin^2\theta.$$

Transforming to the canonically conjugate variables θ and p now gives

$$\frac{d}{dt}\theta = \{N', \theta\} = \frac{1}{k_1 k_2}\frac{\partial N'}{\partial p} = \frac{1}{k_1 k_2}p,$$

$$\frac{d}{dt}p = \{N', p\} = \frac{-1}{k_1 k_2}\frac{\partial N'}{\partial \theta} = \frac{-g}{k_1 k_2}\frac{k_2^2}{k_3^2}r^2 \sin\theta \cos\theta.$$

Combining these equations of motion gives

$$\frac{d^2}{dt^2}\theta = \frac{-gr^2}{2k_1^2 k_3^2}\,\sin 2\theta,$$

or, in terms of the original rigid body parameters,

$$\frac{d^2}{dt^2}\theta = -gK'\left(\frac{1}{I_1}-\frac{1}{I_2}\right)\left(\frac{1}{I_1}-\frac{1}{I_3}\right)\sin 2\theta.$$

Thus, the controlled system also reduces to the pendulum equation on level surfaces of K'. The stability of the equilibria of the controlled system now depends upon the sign of g, as well as the relative sizes of the principle moments of inertia of the body. In particular, negative g stabilizes the equilibrium along the intermediate axis, while *destabilizing* the other two equilibria. This occurs because the introduction of the control parameter g causes a global bifurcation of the rigid body system in which the equilibria exchange stability.

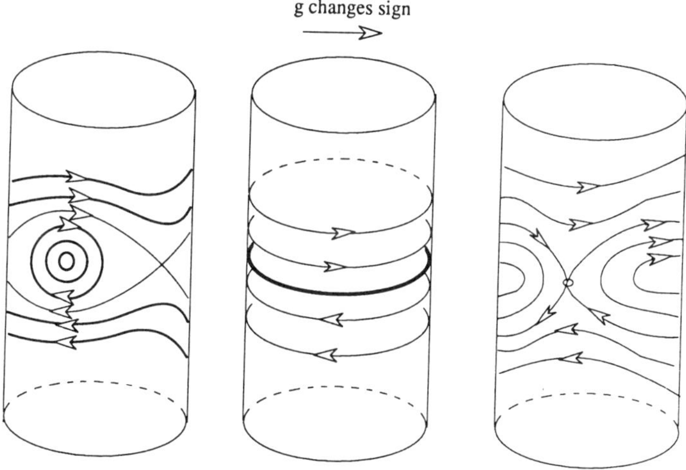

Figure 2. The change in phase portrait of the controlled rigid body as the parameter g passes through zero.

The general theory of reduction discussed in section **3** of course applies just as well to this example, and puts into a different light the fact observed by Bloch and Marsden [1990], that the controlled equations are Lie Poisson for the group $SO(2,1)$ in the stabilized situation.

5. Conclusions

In this paper we have shown how the Euler equations for a rigid body (even possibly including a stabilizing control torque) is Hamiltonian simultaneously with respect to a variety of Lie Poisson structures on $\mathbb{R}^3$. These Lie-Poisson structures are those for the three dimensional groups $SO(3)$, $SO(2,1)$, and the Euclidean group of the plane. For the euclidean group of the plane, the coadjoint orbits are elliptic cylinders with a canonical cotangent structure. In these canonical variables, which are given explicitly in elliptic cylindrical coordinates, the rigid body equations become transformed to the equations of a simple planar pendulum. This is analogous to the situation of David and Holm [1990] in which both the pendulum and the Duffing equation occur.

The geometry of the intersections of the level surfaces can be used to understand the stability of the system dynamics and how the feedback parameter affects the controlled system. There is a bifurcation and a change of stability as this parameter passes through zero (see Figure 2).

One potential application of these ideas is for reorienting satellites by controlled switching from one stable equilibrium to the opposite one (rotation by π, to the $\mathbb{Z}_2$-partner equilibrium) by momentarily destabilizing it, then restabilizing the partner equilibrium using the g control parameter. The angle of switch is a geometric phase and can be calculated using the phase method of Montgomery; see Marsden, Montgomery, and Ratiu [1990]. The switch itself is accomplished by passing from an equilibrium to its $\mathbb{Z}_2$ partner—it is a *particular* geometrical phase associated to the $\mathbb{Z}_2$ symmetry of the level surfaces of the Casimir, K'. The rate at which the switching process takes place scales with the magnitude of the control parameter g, thereby allowing fine precision control through the adjustment of g near zero.

REFERENCES

1. R. Abraham and J. Marsden, *Foundations of Mechanics*, Second Edition, Addison-Wesley Publishing Co., Reading, Mass. (1978).
2. A.M. Bloch, P.S. Krishnaprasad, J.E. Marsden, and G. Sánchez de Alvarez, *Stabilization of rigid body dynamics by internal and external torques*, preprint (1990), to appear in Automatica.
3. A.M. Bloch and J.E. Marsden, *Stabilization of rigid body dynamics by the energy-Casimir method*, Systems and Control Letters, 14, (1990) 341–346.
4. D. David and D. Holm, *Multiple Lie-Poisson structures, reductions, and geometric phases for the Maxwell-Bloch travelling wave equations*, preprint (1990).

5. J.E. Mardsen, R. Montgomery and T. Ratiu, *Reduction, symmetry, and phases in mechanics*, Memoirs AMS **436** (1990).

6. J.E. Marsden, T. Ratiu and A. Weinstein, *Semi-direct products and reduction in mechanics*, Trans. Am. Math. Soc. **281**, (1984) 147–177.

7. J.E. Marsden and A. Weinstein, *Coadjoint orbits, vortices and Clebsch variables for incompressible fluids*, Physics **7D**, (1983) 305–323.

9. Y. Nambu [1973] *Generalized Hamiltonian dynamics*, Phys. Rev. **D7**, 2405–2412

10. A. Weinstein, *The local structure of Poisson manifolds*, J. Diff. Geom. **18** (1983), 523–557.

Darryl Holm
Theoretical Division
& Center for Nonlinear Studies
Los Alamos Natl. Lab. MS B284
Los Alamos, NM 87545, USA

Jerrold Marsden
Department of Mathematics
University of California
Berkeley, CA 94720
USA

On the Quantization of Poisson Algebras

JOHANNES HUEBSCHMANN*

En hommage à J.-M. Souriau

Introduction

A *Poisson structure* on a commutative algebra A over a commutative ring R is a Lie bracket $\{\cdot,\cdot\}: A \otimes A \to A$ on A, viewed as an R-module, satisfying the Leibniz rule

$$\{ab,c\} = a\{b,c\} + b\{a,c\}, \quad a,b,c \in A. \tag{0.1}$$

An algebra A together with a Poisson structure $\{\cdot,\cdot\}$ is called a *Poisson algebra*. A *Poisson structure* on a smooth manifold is a Poisson structure on its ring of smooth functions; a smooth manifold with a Poisson structure is called a *Poisson manifold*. For a symplectic manifold (N,σ), the rule $\{f,g\} = \sigma(X_f, X_g)$, where X_f is the Hamiltonian vector field corresponding to f, defines a Poisson structure on N. However, this is not the only way in which a Poisson structure on a manifold arises, see e. g. WEINSTEIN [53].

The significance of Poisson structures in physics is classical, see LIE [38] and DIRAC [9], [10]. In fact, under nice circumstances, the classical phase space of a physical system with finitely many degrees of freedom may be described as a symplectic manifold, and the corresponding hamiltonian systems can concisely be handled in terms of the associated Poisson structure. However, there are systems that do *not* arise from a symplectic manifold, even systems which do not admit a good underlying space but still a Poisson algebra structure, see e. g. WEINSTEIN [53] and ŚNIATYCKI-WEINSTEIN [47].

The concept of a Poisson structure and in particular that of a Poisson manifold is currently of much interest, see [18] and the literature there. Moreover, it is perhaps tempting to work merely over the reals $\mathbf{R}$ as ground ring; however, in view of [52] it may be worthwhile admitting an arbitrary commutative ring R and we shall do so whenever possible.

According to DIRAC [9], [10], a *quantization* of a classical system described by a real Poisson algebra $(A, \{\cdot,\cdot\})$ as above is a representation $a \mapsto \hat{a}$ of a certain sub Lie algebra $B \subseteq A$, A and B being viewed merely as Lie algebras, by linear operators $\hat{a}$ on a complex linear space $\mathcal{H}$ such that

---

*Expanded version of the lecture given at the meeting

  — the Dirac condition

$$i\,[\hat{a},\hat{b}] = \widehat{\{a,b\}} \tag{0.2}$$

  holds;
  — for a constant c, the operator $\hat{c}$ is given by $\hat{c} = c\,\mathrm{Id}$;
  — a certain irreducibility condition is satisfied.

The irreducibility condition has something to do with phase transitions, see e. g. WOODHOUSE [57]. It has become common to refer to a procedure furnishing a representation that satisfies only (1) and (2) above as *prequantization*. Under suitable circumstances, over a smooth symplectic manifold, the *geometric quantization scheme*, due to KIRILLOV [32], KOSTANT [33], SOURIAU [48], and I. SEGAL [46], furnishes such a quantization. In our paper [18] the geometric quantization scheme has been extended to a quantization procedure for arbitrary Poisson algebras, and in a follow up paper [24] to graded Poisson algebras. The purpose of the present paper is to expose the ideas behind this quantization procedure.

The two new ideas that come into play are

  — to investigate Poisson algebras by means of their associated modules of formal differentials, in the graded sense if necessary, and
  — to develop an algebraic theory of quantization explaining the usual geometric quantization scheme in such a way that it can be extended to (graded) Poisson algebras.

In Section 1 below we reproduce some material from the theory of Lie-Rinehart algebras. In Section 2 we shall concentrate on the ungraded case, while in Section 3 we sketch the extension to the graded case. In Section 4 we give an example. Using our methods, we hope to extend elsewhere the solution of the KIRILLOV conjecture due to GUILLEMIN-STERNBERG [15] to the case where 0 is a no longer a regular value of the corresponding momentum mapping.

1. Poisson algebras and Lie-Rinehart algebras

We have already indicated that a quantization of a real Poisson algebra is a suitable representation of a sub Lie algebra of its underlying Lie algebra. In this Section we show how to reduce the requisite representation theory of the underlying Lie algebra to that of a suitable associated *Lie-Rinehart* algebra. A *Lie-Rinehart algebra* is a pair (A, L), where A is a commutative algebra, L a Lie algebra over the ground ring, where L comes with a structure of an A-module and A with that of an L-module, and the two structures satisfy two compatibility conditions modeled on the properties of the pair (A, L), A being the ring of smooth functions $C^\infty(N)$ on a smooth finite dimensional manifold N and L the Lie algebra $\mathrm{Vect}(N)$ of smooth vector fields on N. Explicitly, these compatibility conditions read

$$a(\alpha(b)) = (a\alpha)b \tag{1.1}$$

$$[\alpha, a\beta] = \alpha(a)\beta + a[\alpha, \beta], \tag{1.2}$$

where $a, b \in A$ and $\alpha, \beta \in L$. Another example of a Lie-Rinehart algebra is the pair $(A, \mathrm{Der}(A))$, with the obvious structures; here A is an arbitrary commutative algebra over R and $\mathrm{Der}(A)$ refers to the derivations of A. Further, there is an obvious notion of morphism of Lie-Rinehart algebras, and with this notion of morphism the Lie-Rinehart algebras constitute a category. More details may be found in RINEHART [44] and in Section 1 of our paper [18].

Here is the Lie-Rinehart algebra associated with a given Poisson algebra $(A, \{\cdot, \cdot\})$: Consider the A-module D_A of formal differentials of A. We remind the reader that, for a commutative algebra A, its A-module of *formal differentials* or *Kähler differentials* is the pair (D_A, d) consisting of an A-module D_A and a derivation $d: A \to D_A$ which is *universal* in the sense that, given an A-module M and a derivation $\delta: A \to M$, there is a unique morphism $\Phi_\delta: D_A \to M$ of A-modules so that the diagram

$$\begin{array}{ccc} A & \xrightarrow{\ d\ } & D_A \\ {\scriptstyle \mathrm{Id}}\Big\downarrow & & \Big\downarrow{\scriptstyle \Phi_\delta} \\ A & \xrightarrow[\ \delta\ ]{} & M \end{array} \tag{1.3}$$

is commutative. As an A-module, D_A may be taken to be generated by elements da, $a \in A$, subject to the relations

$$d(ab) = (da)b + adb, \ dr = 0,$$

where $a, b \in A$ and $r \in R$. Given a Poisson algebra $(A, \{\cdot, \cdot\})$, one defines at first the 2-form

$$\pi = \pi_{\{\cdot, \cdot\}}: D_A \otimes_A D_A \longrightarrow A \tag{1.4}$$

on D_A with values in A, the *Poisson 2-form* of $(A, \{\cdot, \cdot\})$, by

$$\pi(du, dv) = \{u, v\}. \tag{1.5}$$

Next, the adjoint

$$\pi^\sharp: D_A \longrightarrow \mathrm{Der}(A) = \mathrm{Hom}_A(D_A, A) \tag{1.6}$$

of π is a morphism of A-modules, and one defines a Lie bracket

$$[\cdot, \cdot]: D_A \otimes D_A \longrightarrow D_A$$

on D_A, viewed as an R-module, by

$$[adu, bdv] = a\{u, b\}dv + b\{a, v\}du + abd\{u, v\}. \tag{1.7}$$

More details about (1.4) – (1.7) may be found in Section 3 of [18]. Here is the first result.

Theorem 1.8. *Given a Poisson algebra* $(A, \{\cdot, \cdot\})$ *over a commutative ring* R, *the* A-*module structure on* D_A, *the bracket* $[\cdot, \cdot]$, *and the morphism* $\pi^\sharp$ *of* A-*modules endow the pair* (A, D_A) *with a structure of a Lie-Rinehart algebra in such a way that* $\pi^\sharp$ *is a morphism of Lie-Rinehart algebras.*

Proof. See (3.8) in [18].

We write $D_{\{\cdot,\cdot\}} = (D_A, [\cdot, \cdot], \pi^\sharp)$ and, for $\alpha \in D_A$, we write

$$\alpha^\sharp = \pi^\sharp(\alpha) \in \mathrm{Der}(A).$$

The 2-form $\pi_{\{\cdot,\cdot\}}$ is always defined, whatever Poisson algebra. It generalizes the concept of a symplectic structure; see Section 3 of [18] for details.

For quantization, the following observation is crucial: The Poisson 2-form gives rise to an *extension*

$$0 \to A \to \overline{L}_{\{\cdot,\cdot\}} \to D_{\{\cdot,\cdot\}} \to 0 \tag{1.9}$$

of Lie-Rinehart algebras. Here

$$\overline{L}_{\{\cdot,\cdot\}} = A \oplus D_{\{\cdot,\cdot\}} \tag{1.10}$$

as A-modules, while the Lie bracket is given by

$$[(a, du), (b, dv)] = (\{u, b\} + \{a, v\} - \{u, v\}, d\{u, v\}). \tag{1.11}$$

In general, given a commutative R-algebra A and (R, A)-Lie algebras L', L, L'', an *extension* of (R, A)-Lie algebras is a short exact sequence

$$\mathbf{e} : 0 \to L' \to L \xrightarrow{p} L'' \to 0 \tag{1.12}$$

in the category of (R, A)-Lie algebras; notice in particular that the Lie algebra L' necessarily acts trivially on A and hence is a Lie algebra in the category of A-modules in the ordinary sense. The extension theory of Lie-Rinehart algebras is studied in our papers [28], [29], [30]. Returning to the present situation, we have the following.

Proposition 1.13 [18 (4.2)]. *For an arbitrary Poisson algebra* $(A, \{\cdot, \cdot\})$, *the morphism*

$$\iota_{\{\cdot,\cdot\}} = (\mathrm{Id}, d) : A \longrightarrow \overline{L}_{\{\cdot,\cdot\}} \tag{1.13.1}$$

yields a commutative diagram

$$
\begin{array}{ccccccccc}
0 & \longrightarrow & R & \longrightarrow & A & \longrightarrow & A/R & \longrightarrow & 0 \\
& & \downarrow & & \scriptstyle{\iota_{\{\cdot,\cdot\}}}\downarrow & & \downarrow & & \\
0 & \longrightarrow & A & \longrightarrow & \overline{L}_{\{\cdot,\cdot\}} & \longrightarrow & D_{\{\cdot,\cdot\}} & \longrightarrow & 0
\end{array}
\tag{1.13.2}
$$

in the category of Lie algebras over the ground ring R.

Here the crucial observation is that this diagram reduces the representation theory of $(A, \{\cdot, \cdot\})$, viewed merely as a Lie algebra over the ground ring R, to that of $(A, \overline{L}_{\{\cdot, \cdot\}})$ as a Lie-Rinehart algebra. In classical situations, cf. the example below, as an A-module, $\overline{L}_{\{\cdot, \cdot\}}$ is finitely generated, whereas $(A, \{\cdot, \cdot\})$, viewed as a Lie algebra, is usually *not* finitely generated.

Example 1.14. Let (N, σ) be a symplectic manifold, let $\xi \colon P \longrightarrow N$ be a principal circle bundle having $[-\sigma] \in \mathrm{H}^2(N, \mathbf{R})$ as characteristic class, and let

$$0 \to \mathrm{ad}(\xi) \to E(\xi) \to \tau_N \to 0 \tag{1.14.1}$$

be its ATIYAH sequence [4]; for a complete account to Atiyah sequences see App. A in MACKENZIE [39]. The corresponding sequence

$$0 \to A \to \Gamma(E(\xi)) \to \mathrm{Vect}(N) \to 0 \tag{1.14.2}$$

of sections is an extension of Lie-Rinehart algebras. The role of the sequence (1.14.2) in geometric quantization is well known; versions of it appear crucially in any of the approaches to geometric quantization over symplectic manifolds. The extension (1.9) of Lie-Rinehart algebras is the exact algebraic analogue of (1.14.2) and hence furnishes a generalization of the concept of Atiyah sequence to arbitrary Poisson algebras.

In the extension theory of Lie-Rinehart algebras suitable notions of connection and curvature come into play. It seems worthwhile to say a few words about these notions: Let $\mathbf{e}$ be an extension of (R, A)-Lie algebras of the kind (1.12), and assume it splits in the category of A-modules; this will e. g. hold if L'' is projective as an A-module. Then the extension $\mathbf{e}$ may be represented by a 2-cocycle. More precisely, let $\omega \colon L'' \to L$ be a section of A-modules for the projection $p \colon L \to L''$. Since ω generalizes the notion of a *principal* connection, we refer it as an $\mathbf{e}$-*connection*. Likewise, given an $\mathbf{e}$-connection, the corresponding $(\mathbf{e}\text{-})$*curvature* $\Omega \colon L'' \otimes_A L'' \to L'$ is the morphism Ω of A-modules satisfying

$$[\omega(\alpha), \omega(\beta)] = \omega[\alpha, \beta] + \Omega(\alpha, \beta) \tag{1.15}$$

for every $\alpha, \beta \in L''$; a little thought reveals that Ω is indeed well defined as an alternating A-bilinear 2-form on L'' with values in L'. When (A, L) is the Lie-Rinehart algebra $(C^\infty(N), \mathrm{Vect}(N))$, for a smooth finite dimensional manifold N, and when the extension of Lie-Rinehart algebras under consideration is one of the kind (1.14.2), these notions of connection and curvature boil down precisely to the descriptions of a principal connection and curvature given by ATIYAH in his fundamental paper [4].

Using some homological algebra, we introduced a certain cohomology theory

$$\mathrm{H}^*_{\mathrm{Poisson}}(A, \{\cdot, \cdot\}; A),$$

christened *Poisson cohomology*, for an arbitrary Poisson algebra $(A, \{\cdot, \cdot\})$. We give some hints as to its definition: Given an arbitrary (R, A)-Lie algebra L, its *universal object* $(U(A, L), \iota_L, \iota_A)$ is an R-algebra $U(A, L)$ together with a morphism $\iota_A: A \longrightarrow U(A, L)$ of R-algebras and a morphism $\iota_L: L \longrightarrow U(A, L)$ of Lie algebras over R having the properties

$$\iota_A(a)\iota_L(\alpha) = \iota_L(a\,\alpha), \quad \iota_L(\alpha)\iota_A(a) - \iota_A(a)\iota_L(\alpha) = \iota_A(\alpha(a)),$$

and $(U(A, L), \iota_L, \iota_A)$ is *universal* among triples (B, ϕ_L, ϕ_A) having these properties. More precisely:

1.16. *Given*

- (i)   *another R-algebra B, viewed at the same time as a Lie algebra over R,*
- (ii)   *a morphism $\phi_L: L \longrightarrow B$ of Lie algebras over R, and*
- (iii)   *a morphism $\phi_A: A \longrightarrow B$ of R-algebras,*

so that, for $\alpha \in L, a \in A$,

$$\phi_A(a)\phi_L(\alpha) = \phi_L(a\,\alpha), \tag{1.16.1}$$

$$\phi_L(\alpha)\phi_A(a) - \phi_A(a)\phi_L(\alpha) = \phi_A(\alpha(a)), \tag{1.16.2}$$

there is a unique morphism $\Phi: U(A, L) \longrightarrow B$ of R-algebras so that $\Phi\,\iota_A = \phi_B$ and $\Phi\,\iota_L = \phi_L$.

In particular, the universal property entails that the obvious (left) (A, L)-module structure on A induces that of a left $U(A, L)$-module on A.

For example, when A is the algebra of smooth functions on a smooth manifold N and L the Lie algebra of smooth vector fields on N, then $U(A, L)$ is the *algebra of (globally defined) differential operators on N*. Moreover, when A is just the ground ring R and L a Lie algebra g in the category of R-modules in the ordinary sense, the algebra $U(A, L)$ is just the usual universal algebra for g denoted more commonly by $U(g)$.

The universal property of $U(A, L)$ for a general Lie-Rinehart algebra is not spelled out in RINEHART [44] while an explicit construction for $U(A, L)$ is given there. In Section 1 of our paper [18] an alternate construction for $U(A, L)$ is given which employs the MASSEY-PETERSON [42] algebra.

Given an (R, A)-Lie algebra L and an R-module M having the structures of a left A-module and that of a left L-module, the object M is said to be an (A, L)-*module*, provided the actions are compatible in the sense that, for $\alpha \in L, a \in A, m \in M$,

$$(a\,\alpha)(m) = a(\alpha(m)), \tag{1.17.1}$$

$$\alpha(a\,m) = a\,\alpha(m) + \alpha(a)\,m. \tag{1.17.2}$$

For example, let A be the algebra of smooth functions on a smooth finite dimensional manifold N, let L be the Lie algebra of smooth vector fields on N, let $\xi\colon E \to N$ be a smooth vector bundle on N, and let M be the A-module of smooth sections of ξ; then an (A, L)-module structure on M is precisely a flat (linear) connection on ξ.

Keeping in mind that, for a Lie algebra g over a field or more generally for a Lie algebra g over a commutative ring R so that g is projective as an R-module, the corresponding Lie algebra cohomology with values in an arbitrary left g-module M is defined by

$$\mathrm{H}^*(g, M) = \mathrm{Ext}^*_{U(g)}(R, M),$$

given an arbitrary (R, A)-Lie algebra L and a left (A, L)-module M, one defines

$$\mathrm{H}^*_A(L, M) = \mathrm{Ext}^*_{U(A,L)}(A, M).$$

When L is projective as an A-module, this cohomology can be computed by means of the complex $\mathrm{Alt}_A(L, M)$ of A-multilinear alternating forms with a differential d generalizing the classical Cartan-Chevalley-Eilenberg differential. It is given by the formula

$$
\begin{aligned}
(df)&(\alpha_1, \ldots \alpha_n)\\
&= (-1)^n \sum_{i=1}^{n} (-1)^{(i-1)} \alpha_i(f(\alpha_1, \ldots, \widehat{\alpha_i}, \ldots, \alpha_n))\\
&+ (-1)^n \sum_{j<k} (-1)^{(j+k)} f([\alpha_j, \alpha_k], \alpha_1, \ldots, \widehat{\alpha_j}, \ldots, \widehat{\alpha_k}, \ldots, \alpha_n);
\end{aligned}
\tag{1.18}
$$

see RINEHART [44] or our paper [18] for details. The reason is that the usual Koszul complex computing Lie algebra cohomology carries over to the present situation; further, in view of the Birkhoff-Witt type theorem proved by RINEHART [44], when L is projective as an A-module, the corresponding Koszul complex yields a projective resolution of A as an $U(A, L)$-module [44], [18]. In particular, when (A, L) is the Lie-Rinehart algebra $(C^\infty(N), \mathrm{Vect}(N))$, for a smooth finite dimensional manifold N, and when $M = A = C^\infty(N)$ or more generally the space of sections of a smooth vector bundle ζ over N, the complex $\mathrm{Alt}_A(L, M)$ coincides with the de Rham complex of forms on N with values in $\mathbf{R}$ or ζ as appropriate. In particular this subsumes de Rham cohomology under standard homological algebra.

For a general Lie-Rinehart algebra (A, L), the complex $\mathrm{Alt}_A(L, M)$ of A-multilinear alternating forms with the differential (1.18) makes sense also when L is no longer projective as an A-module. It then computes a kind of relative Ext-groups; we intend to give a precise description of the resulting relative homological algebra elsewhere.

We now return to an arbitrary Poisson algebra $(A, \{\cdot, \cdot\})$ over a commutative ring R. Consider the universal algebra $U(A, D_{\{\cdot,\cdot\}})$ associated with the corresponding (R, A)-Lie algebra $D_{\{\cdot,\cdot\}}$ and the corresponding complex $\mathrm{Alt}_A(D_{\{\cdot,\cdot\}}, A)$ of A-multilinear alternating forms with the generalized Cartan-Chevalley-Eilenberg differential (1.18). We *define* the *Poisson cohomology* $\mathrm{H}^*_{\mathrm{Poisson}}(A, \{\cdot, \cdot\}; A)$ of $(A, \{\cdot, \cdot\})$ to be the homology of the complex $\mathrm{Alt}_A(D_{\{\cdot,\cdot\}}, A)$. In particular, when the module D_A of formal differentials is projective as an A-module, Poisson cohomology coincides with $\mathrm{Ext}^*_{U(A, D_{\{\cdot,\cdot\}})}(A, A)$. More details may be found in Section 3 of our paper [18].

A straightforward calculation shows that the 2-form $\pi_{\{\cdot,\cdot\}}$ given by (1.5) is a 2-cocycle in the complex $\mathrm{Alt}_A(D_{\{\cdot,\cdot\}}, A)$ computing Poisson cohomology and hence yields a class

$$[\pi_{\{\cdot,\cdot\}}] \in \mathrm{H}^2_{\mathrm{Poisson}}(A, \{\cdot, \cdot\}; A),$$

the *Poisson class* of $(A, \{\cdot, \cdot\})$. The extension (1.9) represents the negative

$$-[\pi_{\{\cdot,\cdot\}}] \in \mathrm{H}^2_{\mathrm{Poisson}}(A, \{\cdot, \cdot\}; A)$$

of this class in Poisson cohomology. When the Poisson class is trivial, we refer to a 1-form $\vartheta \colon D_A \to A$ so that $d\vartheta = \pi_{\{\cdot,\cdot\}}$ as a *Poisson potential*. Explicitly, for $\alpha, \beta \in D_A$, we then have

$$
\begin{aligned}
\pi_{\{\cdot,\cdot\}}(\alpha, \beta) &= d\vartheta(\alpha, \beta) \\
&= \alpha^\sharp(\vartheta(\beta)) - \beta^\sharp(\vartheta(\alpha)) - \vartheta([\alpha, \beta]).
\end{aligned}
\tag{1.19}
$$

Poisson cohomology generalizes the de Rham cohomology of a symplectic manifold N, and, accordingly, the Poisson class $[\pi_{\{\cdot,\cdot\}}]$ generalizes the class $[\sigma] \in \mathrm{H}^2_{\mathrm{deRham}}(N, \mathbf{R})$ in de Rham cohomology of the symplectic structure σ. Indeed, it is shown in Section 3 of our paper [18] that when $(A, \{\cdot, \cdot\})$ is the Poisson algebra of smooth functions on a smooth symplectic manifold, the Poisson structure being induced from the symplectic one, Poisson cohomology boils down to de Rham cohomology and the Poisson class to the class in de Rham cohomology represented by the symplectic structure. Moreover, the notion of Poisson potential generalizes that of a symplectic potential. In Section 3 of our paper [18] we also introduced a notion of Poisson homology. Both theories extend and unify earlier theories given in LICHNEROWICZ [37], KOSZUL [36], and BRYLINSKI [6] on a somewhat ad hoc basis. In particular, the above discussion shows that, in the framework of Lie-Rinehart algebras, these theories are subsumed under standard homological algebra as Ext and Tor groups over suitable algebras of differential operators. See Section 3 in [18] for details. To our knowledge, the first ones to play such a game with de Rham theory were HOCHSCHILD-KOSTANT-ROSENBERG [17]: they subsumed the theory of Kähler differentials over a regular affine variety under standard homological algebra.

To simplify the exposition, for a smooth manifold N, it will henceforth be convenient not to distinguish between a formal differential $dh \in D_{C^\infty(N)}$ and its image in the space $\Omega^1(N)$ of 1-forms under the obvious projection map

$$D_{C^\infty(N)} \longrightarrow \Omega^1(N).$$

Of course $D_{C^\infty(N)}$ is *much* bigger than $\Omega^1(N)$ but that causes no problem here. Actually, there are different versions of the theory: For example, for an algebra A of the kind $C^\infty(N)$, the A-module $\Omega^1(N)$ with the corresponding $(\mathbf{R}, A)$-Lie algebra structure yields a smooth version of Poisson cohomology. Likewise, there is an analytic version, and an algebro-geometric one. The distinction will not be important here. More details about the smooth version may be found in Section 3 of our paper [18].

2. Quantization

We explain briefly an extension of the usual geometric quantization construction to general real Poisson algebras. We denote the complex numbers by $\mathbf{C}$. Given an $(A \otimes \mathbf{C})$-module M, keeping in mind that, cf. (1.10) above, by construction, $\overline{L}_{\{\cdot,\cdot\}} = A \oplus D_{\{\cdot,\cdot\}}$, we say that a structure

$$\chi \colon \overline{L}_{\{\cdot,\cdot\}} \longrightarrow \mathrm{End}(M) \tag{2.1}$$

of an $(A, \overline{L}_{\{\cdot,\cdot\}})$-module on M is that of a *prequantum module for* $(A, \{\cdot,\cdot\})$ provided for $a \in A$ we have

$$\chi(a, 0) = i\, a\, \mathrm{Id}_M. \tag{2.2}$$

A structure of a prequantum module can be described in terms of the notions of connection and curvature mentioned earlier. Indeed, the following holds, see also Section 4 of [18] and Section 2 of [20].

Lemma 2.3. *Given a real Poisson algebra* $(A, \{\cdot,\cdot\})$ *and an* $(A \otimes \mathbf{C})$-*module* M, *the relationship*

$$\nabla(\alpha) = \chi(0, \alpha), \quad \alpha \in D_{\{\cdot,\cdot\}}, \tag{2.3.1}$$

yields a bijective correspondence between prequantum module structures

$$\chi \colon \overline{L}_{\{\cdot,\cdot\}} \longrightarrow \mathrm{End}_{\mathbf{R}}(M)$$

and $D_{\{\cdot,\cdot\}}$-*connections*

$$\nabla \colon D_{\{\cdot,\cdot\}} \longrightarrow \mathrm{End}_{\mathbf{R}}(M)$$

having curvature

$$K_\nabla : D_{\{\cdot,\cdot\}} \otimes_A D_{\{\cdot,\cdot\}} \longrightarrow \mathrm{End}_A(M)$$

given by

$$K_\nabla(dx, dy) = -i\{x, y\}\,\mathrm{Id}_M : M \to M, \quad x, y \in A. \tag{2.3.2}$$

Proof. For illustration we sketch the argument: For any connection ∇ of the kind considered, given $x, y \in A$, we have, in view of the definition (1.7) of the bracket on D_A,

$$\nabla_{dx}\nabla_{dy} - \nabla_{dy}\nabla_{dx} = \nabla_{d\{x,y\}} + K_\nabla(dx, dy) \in \mathrm{End}_\mathbf{R}(M).$$

On the other hand, cf. (1.11), the Lie structure in $\overline{L}_{\{\cdot,\cdot\}}$ is determined by

$$[(0, dx), (0, dy)] = (-\{x, y\}, d\{x, y\}),$$

together with the axioms for an $(\mathbf{R}, A)$-Lie algebra. Hence, given a prequantum module structure

$$\chi : \overline{L}_{\{\cdot,\cdot\}} \longrightarrow \mathrm{End}_\mathbf{R}(M),$$

when we define a $D_{\{\cdot,\cdot\}}$-connection

$$\nabla : D_{\{\cdot,\cdot\}} \longrightarrow \mathrm{End}_\mathbf{R}(M)$$

by (2.3.1), we must have

$$
\begin{aligned}
\nabla_{dx}\nabla_{dy} - \nabla_{dx}\nabla_{dy} &= \chi(0, dx)\chi(0, dy) - \chi(0, dy)\chi(0, dx) \\
&= \chi([(0, dx), (0, dy)]) \\
&= \chi(-\{x, y\}, d\{x, y\}) \\
&= -\chi(\{x, y\}, 0) + \chi(0, d\{x, y\}) \\
&= -i\{x, y\}\,\mathrm{Id}_M + \nabla_{d\{x,y\}} \\
&= -i\{x, y\}\,\mathrm{Id}_M + \nabla_{[dx,dy]}.
\end{aligned}
$$

Thus the curvature of this connection is then given by (2.3.2).

The same kind of argument shows that, given a $D_{\{\cdot,\cdot\}}$-connection having curvature given by (2.3.2), the formula (2.3.1) yields a structure of a prequantum module. $\qquad\square$

The following is now immediate:

Proposition 2.4. *Given a real Poisson algebra $(A, \{\cdot, \cdot\})$ and a pre-quantum module (M, χ) for $(A, \{\cdot, \cdot\})$, the composition of $\iota_{\{\cdot, \cdot\}} \colon A \longrightarrow \overline{L}_{\{\cdot, \cdot\}}$ (cf. (1.13.1)) with $-i\chi$ yields a representation $a \mapsto \hat{a}$ of $(A, \{\cdot, \cdot\})$, viewed as a real Lie algebra, on M by $\mathbf{R}$-linear operators in such a way that,*

(1)   *the constants act by multiplication, and*

(2)   *for $a, b \in A$, $\widehat{\{a, b\}} = i\,[\hat{a}, \hat{b}]$.*

Explicitly, we have, for $a \in A$,

$$\hat{a}x = \tfrac{1}{i}\,\chi(0, da)(x) + ax, \quad x \in M. \tag{2.4.3}$$

Remark 2.5. In the classical case, the module of sections of a pre-quantum bundle is a prequantum module in our sense. Our concept of prequantum module thus generalizes the notion of prequantum bundle. See e. g. WOODHOUSE [57] for the classical notion of prequantum bundle.

Illustration 2.6. We consider $\mathbf{R}^{2n}$ with coordinates

$$(\mathbf{x}, \mathbf{p}) = (x_1, \ldots, x_n, p_1, \ldots, p_n)$$

and the usual Poisson structure $\{x_j, p_k\} = \delta_{j,k}$ etc. on its algebra $A = C^{\infty}(\mathbf{R}^{2n})$ of smooth functions. Actually, this is a kind of toy example but yet it may well serve to illustrate the ideas that come into play: Inspection shows that the 1-form $\vartheta \colon D_A \to A$ given by

$$\vartheta(dp_j) = p_j, \ 1 \le j \le n, \quad \vartheta(dx_k) = 0, \ 1 \le k \le n, \tag{2.6.1}$$

is a Poisson potential; we refer to it as the *vertical Poisson potential*. Under the obvious identification of $\mathbf{R}^{2n}$ with $\mathrm{T}^*(\mathbf{R}^n)$, the x_k and p_j being viewed as "position" and "momentum" coordinates, the Poisson potential ϑ corresponds to the usual symplectic potential for a cotangent bundle. As prequantum module for our Poisson algebra $(A, \{\cdot, \cdot\})$ we may take

$$M = A \otimes \mathbf{C} <1>, \tag{2.6.2}$$

that is, the algebra $A \otimes \mathbf{C}$ of smooth complex valued functions in the variables $(x_1, \ldots, x_n, p_1, \ldots, p_n)$, viewed as the free $A \otimes \mathbf{C}$-module with a single basis element 1, with structure map

$$\chi \colon \overline{L}_{\{\cdot, \cdot\}} \longrightarrow \mathrm{End}_{\mathbf{R}}(M) \tag{2.6.3}$$

given by

$$\chi(a, \alpha)(1) = -i\,\vartheta(\alpha) + i\,a, \quad a \in A, \ \alpha \in D_A; \tag{2.6.4}$$

notice that, since ϑ is a Poisson potential, in view of (1.19), the formula

$$\nabla_\alpha(1) = -i\,\vartheta(\alpha) \tag{2.6.5}$$

yields a $D_{\{\cdot,\cdot\}}$-connection on M having curvature K_∇ satisfying

$$K_\nabla(dx, dy) = -i\,\{x, y\}\,\mathrm{Id}_M \colon M \to M, \tag{2.6.6}$$

whence indeed χ is a prequantum module structure, cf. Lemma 2.3. In view of (2.4.3), the corresponding representation $a \mapsto \hat{a}$ of $(A, \{\cdot, \cdot\})$, viewed as a real Lie algebra, on M by $\mathbf{R}$-linear operators is now given by

$$\hat{f}(h) = -i\,\chi(f, df)(h) = -i\,\{f, h\} + (f - \vartheta(df))h\,, \tag{2.6.7}$$

$f \in A$, $h \in A \otimes \mathbf{C}$. In particular, by identifying b with 1, this yields the usual formulas

$$\hat{x}_j(h) = -i\,\frac{\partial h}{\partial p_j} + x_j h$$
$$\hat{p}_k(h) = i\,\frac{\partial h}{\partial x_j} \tag{2.6.8}$$

of prequantization of position and momentum on the space of smooth complex valued functions h in the variables $(x_1, \ldots, x_n, p_1, \ldots, p_n)$. Likewise, for $1 \leq j, k \leq n$, with $j \neq k$, the component $(\mathbf{x} \wedge \mathbf{p})_{j,k} = x_j p_k - x_k p_j$ of angular momentum satisfies

$$\vartheta d(x_j p_k - x_k p_j) = x_j \vartheta dp_k - x_k \vartheta dp_j = x_j p_k - x_k p_j,$$

whence the operator $\widehat{(\mathbf{x} \wedge \mathbf{p})}_{j,k}$ is given by

$$\widehat{(\mathbf{x} \wedge \mathbf{p})}_{j,k} = -i\,\{x_j p_k - x_k p_j, \,\cdot\,\}. \tag{2.6.9}$$

Alternatively, we may complexify and consider the 1-form

$$\vartheta^{\mathrm{hol}} \colon D_A \otimes \mathbf{C} \to A \otimes \mathbf{C}$$

given by

$$\vartheta^{\mathrm{hol}}(dp_j) = p_j,\ 1 \leq j \leq n, \quad \vartheta^{\mathrm{hol}}(dx_k) = -i\,p_k,\ 1 \leq k \leq n; \tag{2.6.10}$$

we refer to it as the *holomorphic Poisson potential*. The formula (2.6.7), with ϑ^{hol} instead of ϑ, then furnishes again a representation $a \mapsto \hat{a}$ of $(A, \{\cdot, \cdot\})$, viewed as a real Lie algebra, on M by $\mathbf{R}$-linear operators; in particular, it yields the formulas

$$\hat{x}_j^{\mathrm{hol}}(h) = -i\,\frac{\partial h}{\partial p_j} + (x_j - ip_j)h$$
$$\hat{p}_k^{\mathrm{hol}}(h) = i\,\frac{\partial h}{\partial x_j} \tag{2.6.11}$$

on the space of smooth complex valued functions h in the variables $(x_1, \ldots, x_n, p_1, \ldots, p_n)$; further, for $1 \leq j, k \leq n$, with $j \neq k$, the component

$$(\mathbf{x} \wedge \mathbf{p})_{j,k} = x_j p_k - x_k p_j$$

of angular momentum satisfies

$$\begin{aligned}
\vartheta^{\mathrm{hol}} d(x_j p_k &- x_k p_j) \\
&= x_j \vartheta^{\mathrm{hol}}(dp_k) - x_k \vartheta^{\mathrm{hol}}(dp_j) + p_k \vartheta^{\mathrm{hol}}(dx_j) - p_j \vartheta^{\mathrm{hol}}(dx_k) \\
&= x_j p_k - x_k p_j - i\,(p_k p_j - p_j p_k) = (\mathbf{x} \wedge \mathbf{p})_{j,k}
\end{aligned}$$

whence the operator $\widehat{(\mathbf{x} \wedge \mathbf{p})}^{\,\mathrm{hol}}_{j,k}$ is still given by

$$\widehat{(\mathbf{x} \wedge \mathbf{p})}^{\,\mathrm{hol}}_{j,k} = -i\,\{x_j p_k - x_k p_j, \,\cdot\,\}. \tag{2.6.12}$$

The above procedure for real Poisson algebras goes only as far as *prequantization*; indeed, up to now nothing has been said about any kind of irreducibility of the resulting representation and about Hilbert space structures. Within the geometric quantization scheme, the irreducibility problem is usually taken care of by means of a polarization, and in our framework it is straightforward to extend the usual notion of polarization: For a real Poisson algebra $(A, \{\cdot, \cdot\})$, we say that an $(\mathbf{R}, A)$-sub Lie algebra $P \subseteq D_{\{\cdot, \cdot\}}$ which is maximally coisotropic with respect to the corresponding Poisson 2-form (1.4) is a *real polarization*. Likewise, we refer to a $(\mathbf{C}, A \otimes \mathbf{C})$-sub Lie algebra $P \subseteq D_{\{\cdot, \cdot\}} \otimes \mathbf{C}$ which is maximally coisotropic with respect to the corresponding Poisson 2-form (1.4), tensored with $\mathbf{C}$, as a *complex polarization* for $(A, \{\cdot, \cdot\})$. Once such a polarization has been chosen, the composition

$$P \xrightarrow{\rho} D_{\{\cdot, \cdot\}} \xrightarrow{\kappa} \overline{L}_{\{\cdot, \cdot\}} \quad \text{or} \quad P \xrightarrow{\rho} D_{\{\cdot, \cdot\}} \otimes \mathbf{C} \xrightarrow{\kappa} \overline{L}_{\{\cdot, \cdot\}} \otimes \mathbf{C}$$

of the obvious injection map ρ with the obvious section κ for the corresponding extension (1.9), complexified if need be, is a morphism of Lie-Rinehart algebras, since by hypothesis the Poisson 2-form vanishes on P. Consequently, given a prequantum module M for $(A, \{\cdot, \cdot\})$, with structure map

$$\chi \colon \overline{L}_{\{\cdot, \cdot\}} \longrightarrow \mathrm{End}_{\mathbf{R}}(M),$$

as a Lie algebra, the object P acts on M through the composite

$$\chi \kappa \rho \colon P \to \mathrm{End}_{\mathbf{R}}(M), \tag{2.6.13}$$

the morphisms κ and χ being complexified if need be, and we may take the P-invariants M^P; these are the corresponding *polarized* elements in our prequantum module M. We then refer to M^P as the corresponding *quantum module*. Moreover, we shall say that an element $a \in A$ is *compatible with the polarization* P provided we have

$$[(a, da), (0, \alpha)] \in \kappa(P) \quad \text{for every} \quad \alpha \in P, \qquad (2.6.14)$$

where as before κ refers to the obvious section for the corresponding extension (1.9), complexified if need be. Inspection shows that, for an element $a \in A$ compatible with the polarization, the operator $\hat{a}$, cf. (2.4.3), maps polarized elements to polarized ones. This corresponds to the usual restriction in geometric quantization that only those classical observables are directly quantized that preserve the polarization. The following is a useful observation.

Lemma 2.6.15. *Let $\Phi \subseteq A$ or $\Phi \subseteq A \otimes \mathbf{C}$ (if need be) be a subset, and suppose that the polarization P is as an A-module or as an $A \otimes \mathbf{C}$-module generated by the differentials $d\phi$, $\phi \in \Phi$. The every $a \in A$ so that $\{a, \phi\}$ lies in the subalgebra generated by Φ is compatible with the polarization P.*

Proof. Indeed, let $\alpha = bd\phi$, where $b, \phi \in A$. Then

$$[(a, da), (0, bd\phi)] = (b\{a, \phi\}, 0) + (0, \{a, b\}d\phi) + (0, bd\{a, \phi\}) - (b\{a, \phi\}, 0)$$
$$= (0, \{a, b\}d\phi + bd\{a, \phi\})$$

lies in $\kappa(P)$ since $d\phi \in P$ and $d\{a, \phi\} \in P$. Since as a module over the ground ring, the module of differentials is generated by differentials of the kind $\alpha = bd\phi$, this proves the claim. $\qquad \square$

When $(A, \{\cdot, \cdot\})$ is the Poisson algebra arising from a smooth symplectic manifold, suitably interpreted, the above notions of polarization come down to the usual ones. We illustrate this by means of the above example:

Illustration 2.6. (continued). Consider the $(\mathbf{R}, A)$-sub Lie algebra $P \subseteq D_{\{\cdot, \cdot\}}$ generated by the differentials dx_j; we refer to it as the *vertical* polarization. We note that, extending a terminology used on p. 102 of WOODHOUSE [57], the vertical Poisson potential (2.6.1) is *adapted to P* in the sense that it vanishes on $P\left(\subseteq D_{\{\cdot, \cdot\}}\right)$. The corresponding quantum module M^P, that is, the polarized elements or P-invariants in our prequantum module M, are the smooth complex valued functions h in the variables $(x_1, \ldots, x_n, p_1, \ldots, p_n)$ satisfying

$$0 = (\chi(0, dx_j))(h) = \{x_j, h\} = \frac{\partial h}{\partial p_j}, \quad 1 \leq j \leq n,$$

that is to say, the smooth complex valued functions h that are functions of the variables $(x_1, \ldots, x_n)$ alone. Moreover, every classical observable, that is, every function f in the variables $(\mathbf{x}, \mathbf{p})$ that is at most linear in $\mathbf{p}$ satisfies the hypothesis of (2.6.15) and hence is compatible with the vertical polarization. The above formulas (2.6.8) and (2.6.9) then yield the usual representation

$$\hat{x}_j(h) = x_j h, \tag{2.6.16}$$

$$\hat{p}_k(h) = i\,\frac{\partial h}{\partial x_j}, \tag{2.6.17}$$

$$\widehat{(\mathbf{x} \wedge \mathbf{p})}_{j,k} = -i\,\{x_j p_k - x_k p_j, \cdot\}, \tag{2.6.18}$$

on the smooth complex valued functions h in the variables $(x_1, \ldots, x_n)$ alone. Likewise, using the notation $z_j = x_j + i\,p_j$ etc., consider the $(\mathbf{C}, A \otimes \mathbf{C})$-sub Lie algebra $P \subseteq D_{\{\cdot,\cdot\}} \otimes \mathbf{C}$ generated by the differentials dz_j; we refer to it as the *antiholomorphic* polarization, for the following reason: The corresponding quantum module M^P, that is, the polarized elements or P-invariants in our prequantum module M, are the smooth complex valued functions h in the variables $(x_1, \ldots, x_n, p_1, \ldots, p_n)$ satisfying

$$0 = \chi(0, i\,dz_j)(h) = \{i\,z_j, h\} = \{-p_j + i\,x_j, h\} = \frac{\partial h}{\partial x_j} + i\,\frac{\partial h}{\partial p_j} = 2\frac{\partial h}{\partial \overline{z}_j},$$

where $1 \leq j \leq n$; in other words, the polarized elements M^P are the holomorphic functions h in the complex variables $(z_1, \ldots, z_n)$. It is common, see e. g. p. 102 of WOODHOUSE [57], to refer to the polarization spanned by the vector fields $\frac{\partial h}{\partial \overline{z}_j}$ as the antiholomorphic polarization. Thus we see that, under the obvious map induced by the Poisson structure, the polarization in our sense generated by the holomorphic (!) differentials passes to the antiholomorphic (!) polarization in the usual sense, whence our choice of terminology. We also note that it is customary in the literature to take the holomorphic polarization (say) P and refer to the holomorphic elements as those annihilated by its complex conjugate $\overline{P}$, see e. g. p. 134 of WOODHOUSE [57]. Moreover, we mention that the holomorphic Poisson potential (2.6.10) is *adapted to* P in the sense that it vanishes on P; indeed, by construction, we have

$$\vartheta^{\mathrm{hol}}(dz_j) = \vartheta^{\mathrm{hol}}(dx_j + i\,dp_j) = 0.$$

Having the antiholomorphic polarization at our disposal, we now look for those observables that are compatible with this polarization. Thus given a smooth function f in the variables $\mathbf{x}, \mathbf{p}$, in order for it to satisfy the hypothesis of (2.6.15), we must have that, for $j = 1, \ldots, n$, the function

$\{z_j, f\}$ is holomorphic in the variables $z_1, \ldots, z_n$. This amounts to the requirements that, for $1 \leq j, k \leq n$,

$$\{iz_k, \{iz_j, f\}\} = \frac{\partial}{\partial x_k}\left(\frac{\partial f}{\partial x_j} + i\frac{\partial f}{\partial p_j}\right) + i\frac{\partial}{\partial p_k}\left(\frac{\partial f}{\partial x_j} + i\frac{\partial f}{\partial p_j}\right) = 0$$

or, in other words, that

$$\frac{\partial^2 f}{\partial x_k \partial x_j} - \frac{\partial^2 f}{\partial p_k \partial p_j} = 0, \quad \frac{\partial^2 f}{\partial x_k \partial p_j} + \frac{\partial^2 f}{\partial p_k \partial x_j} = 0.$$

For example, the components $(\mathbf{x} \wedge \mathbf{p})_{j,k} = x_j p_k - x_k p_j$ of angular momentum satisfy this requirement and hence are compatible with the antiholomorphic polarization. Indeed, we have

$$\{x_\ell + i\, p_\ell, x_j p_k - x_k p_j\}$$
$$= x_j\{x_\ell, p_k\} - x_k\{x_\ell, p_j\} - i\, p_k\{x_j, p_\ell\} + i\, p_j\{x_k, p_\ell\}$$
$$= (x_j + i\, p_j)\, \delta_{k,\ell} - (x_k + i\, p_k)\, \delta_{j,\ell}.$$

Consequently the formula (2.6.18) yields a representation of the Lie algebra $so(n)$ on the space of holomorphic functions in the variables $(z_1, \ldots, z_n)$; this is indeed the usual representation, when $so(n)$ is viewed as a sub Lie algebra of $su(n)$ in the obvious way. Furthermore, the corresponding global representation of the group $SO(n)$ on the space of holomorphic functions in the complex variables $(z_1, \ldots, z_n)$ is the usual one, where $SO(n)$ is viewed as a sub Lie group of $SU(n)$ in the obvious way.

Likewise the "energy function" f given by

$$f(\mathbf{x}, \mathbf{p}) = \tfrac{1}{2}(\mathbf{x} \cdot \mathbf{x} + \mathbf{p} \cdot \mathbf{p}) = \tfrac{1}{2}\sum z_j \bar{z}_j$$

satisfies the above requirement. Hence it is represented by the operator $\hat{f}$ given by

$$\hat{f}(h) = -i\left\{\tfrac{1}{2}\sum z_j \bar{z}_j, h\right\} + (f - \vartheta^{\mathrm{hol}}(df))h$$

on the space of holomorphic functions h in the complex variables $(z_1, \ldots, z_n)$. Now

$$f - \vartheta^{\mathrm{hol}}(df) = \tfrac{1}{2}\left(\mathbf{x} \cdot \mathbf{x} + \mathbf{p} \cdot \mathbf{p} + 2i\,\mathbf{x} \cdot \mathbf{p} - 2\mathbf{p} \cdot \mathbf{p}\right)$$
$$= \tfrac{1}{2}\mathbf{z}\mathbf{z}.$$

Further, since h is holomorphic, we have $\{z_j, h\} = 0$ for $j = 1, \ldots, n$ whence

$$-i\left\{\tfrac{1}{2}\sum z_j \bar{z}_j, h\right\} = \tfrac{1}{2}\sum z_j\{-i\bar{z}_j, h\}$$
$$= \tfrac{1}{2}\sum z_j\{p_j - ix_j, h\}$$
$$= \tfrac{1}{2}\sum z_j\{-p_j - ix_j, h\}$$
$$= \tfrac{1}{2}\sum z_j\left(\frac{\partial h}{\partial x_j} - i\frac{\partial h}{\partial p_j}\right)$$
$$= \sum z_j \frac{\partial h}{\partial z_j};$$

220   J. Huebschmann

notice that the operator $\sum z_j \frac{\partial}{\partial z_j}$ is the usual EULER operator. Consequently we obtain the representation

$$\hat{f}(h) = \sum z_j \frac{\partial h}{\partial z_j} + \tfrac{1}{2}\mathbf{z}\mathbf{z}\, h.$$

If we take instead the Poisson potential $\widetilde{\vartheta}$ given by

$$\widetilde{\vartheta}(dz_j) = \tfrac{1}{2}z_j, \quad \widetilde{\vartheta}(d\bar{z}_j) = \tfrac{1}{2}\bar{z}_j,$$

or, what amounts to the same,

$$\widetilde{\vartheta}(dx_j) = \tfrac{1}{2}x_j, \quad \widetilde{\vartheta}(dp_j) = \tfrac{1}{2}p_j,$$

the energy operator $\hat{f}$ on the space of holomorphic functions h in the complex variables $(z_1, \ldots, z_n)$ looks like

$$\hat{f}(h) = \sum z_j \frac{\partial h}{\partial z_j}.$$

Whatever choice of polarization has been made, thereafter the usual inner product problem arises. This problem is independent of our approach to quantization and, as over a symplectic manifold, has to be examined in each case separately. For example, on the space of holomorphic functions in n variables we may take the inner product $< h, k >$ of two such functions h and k to be the integral of the product $h\bar{k}$ with respect to a suitable measure on $\mathbf{R}^{2n}$.

3. The graded case

The approach sketched so far is formal and can be generalized to graded Poisson algebras. We explain briefly a special case; more details may be found in [24]: Let $(A, \{\cdot, \cdot\})$ be a Poisson algebra, over a ground ring R that is not necessarily a field, let g be a Lie algebra, finitely generated and projective as an R-module, and let $\delta: g \to A$ be a *comomentum mapping*, cf. p. 373 of ARNOLD [3], that is to say, δ is a morphism of Lie algebras (over the ground ring R). For example, A could be the algebra $A = C^\infty(N)$ of smooth functions on a smooth finite dimensional Poisson manifold N, and δ could be the adjoint of a momentum mapping $J: N \to g^*$. These data give rise to a graded Poisson algebra $(\mathcal{A}, \{\cdot, \cdot\})$. To describe it, we denote by "s" the *suspension operator* so that, with g being viewed as concentrated in degree zero, sg is just g except that its elements have been regraded up by 1; the inverse operator "s^{-1}" is the *desuspension operator*. Now, as a graded algebra,

$$\mathcal{A} = \Lambda[s^{-1}g^*] \otimes \Lambda[sg] \otimes A, \tag{3.1.1}$$

and the Poisson structure is extended from A to $\mathcal{A}$ by means of

$$\{a, x\} = 0, \quad \{x, z\} = 0, \quad \{x, y\} = y(x), \quad \{y, w\} = 0, \qquad (3.1.2)$$

where $a \in A$, $x, z \in sg$, $y, w \in s^{-1}g^* = (sg)^*$, together with the extension to all of $\mathcal{A}$ by means of the Leibniz formula

$$\{ab, c\} = a\{b, c\} + (-1)^{|b||c|}\{a, c\}b, \quad a, b, c \in \mathcal{A}, \qquad (3.1.3)$$

in the graded sense. We note that, as an algebra, $\mathcal{A}$ is actually bigraded, whereas the Poisson structure is simply graded. Since g is assumed finitely generated projective over the ground ring, the graded algebra $\mathcal{A}$ can then be identified with the algebra

$$\mathrm{Alt}(g, \Lambda[sg] \otimes A), \qquad (3.2)$$

and what is called the corresponding *generator* $\Omega \in \mathcal{A} = \mathrm{Alt}(g, \Lambda[sg] \otimes A)$ of the classical BRST-transformation in the physics literature may be described as follows: Let

$$\Omega_0 : sg \xrightarrow{\ s^{-1}\ } g \xrightarrow{\ -\delta\ } A \qquad (3.3)$$

be the indicated composite, considered as an element $\Omega_0 \in \mathrm{Alt}(g, A \otimes \Lambda[sg])$ of degree -1, and let the morphism

$$\Omega_1 = \Omega_{[\cdot,\cdot]_g} : sg \otimes sg \longrightarrow sg \qquad (3.4)$$

of degree -1 be defined by the requirement that the diagram

$$
\begin{array}{ccc}
g \otimes g & \xrightarrow{\ [\cdot,\cdot]_g\ } & g \\[4pt]
{\scriptstyle s \otimes s}\Big\downarrow & & \Big\downarrow{\scriptstyle s} \\[4pt]
sg \otimes sg & \xrightarrow{\ \Omega_1\ } & sg,
\end{array}
\qquad (3.5)
$$

is commutative, Ω_1 being considered as an element $\Omega_1 \in \mathrm{Alt}(g, A \otimes \Lambda[sg])$. Then

$$\Omega = \Omega_0 + \Omega_1 \qquad (3.6)$$

is the generator of the classical BRST-transformation, and the following is folk-lore in the physics literature, cf. STASHEFF [49] or our paper [27].

Theorem 3.7. *The operator*

$$\{\Omega, -\} : \mathrm{Alt}(g, A \otimes \Lambda[sg]) \longrightarrow \mathrm{Alt}(g, A \otimes \Lambda[sg]) \qquad (3.7.1)$$

satisfies

$$\{\Omega, \{\Omega, -\}\} = 0. \qquad (3.7.2)$$

Furthermore, its cohomology is the differential graded Lie algebra cohomology of g with values in the Koszul complex $(A \otimes \Lambda[sg], \partial)$, where ∂ refers to the Koszul differential.

We note in passing that the operator (3.7.1) is what is called the *classical* BRST-*operator* in the physics literature.

We shall now sketch an approach to the quantization of graded Poisson algebras of the kind $\mathcal{A}$ above which includes in particular a quantization of Ω. More details may be found in our paper [24]. By quantization we mean here the obvious graded extension of the kind of quantization spelled out in the Introduction. For completeness we make it explicit as follows:

Definition 3.8. Let A be a graded real Poisson algebra. Then a *graded quantization* of A is a representation $a \mapsto \hat{a}$ of a certain graded sub Lie algebra $B \subseteq A$, A and B being viewed merely as graded Lie algebras, by linear operators $\hat{a}$ on a graded complex linear space $\mathcal{H}$ such that

- the graded Dirac condition $i\,[\hat{a}, \hat{b}] = \widehat{\{a, b\}}$ holds so that, in particular, for $a \in B$ of odd degree, $2\,i\,(\hat{a})^2 = \widehat{\{a, a\}}$;
- for a constant c, the operator $\hat{c}$ is given by $\hat{c} = c\,\mathrm{Id}$;
- a certain irreducibility condition is satisfied.

Once such a quantization has been carried out, with Ω included, that is, once such a representation of some graded sub Lie algebra B containing Ω has been obtained, a quantization $\widehat{\Omega}$ of Ω is an operator

$$\widehat{\Omega}: \mathcal{H} \longrightarrow \mathcal{H} \tag{3.9}$$

satisfying

$$\widehat{\Omega}\,\widehat{\Omega} = 0, \tag{3.10}$$

and hence its cohomology

$$\mathrm{H}^*(\mathcal{H}, \widehat{\Omega}) \tag{3.11}$$

is defined. The philosophy is now that the cohomology group

$$\mathrm{H}^0(\mathcal{H}, \widehat{\Omega}), \tag{3.12}$$

equipped with a suitable Hilbert space structure, yields a quantum state space for the system with symmetries described classically by N and J.

In our paper [24] the details have been worked out for a graded Poisson algebra $(\mathcal{A}, \{\cdot, \cdot\})$ of the kind (3.1). The problem of quantizing such a graded Poisson algebra can be split into two parts, that of quantizing A and $\Lambda[s^{-1}g^*] \otimes \Lambda[sg]$ separately and therafter combining the two constructions. We first make some remarks about the quantization of $\Lambda[s^{-1}g^*] \otimes \Lambda[sg]$: Let W be the direct sum $W = s^{-1}g^* \oplus sg$ in the graded sense. Then the obvious evaluation pairing on W yields a symmetric bilinear form

$$(\cdot, \cdot): W \otimes W \to R, \tag{3.13}$$

and the graded Poisson structure on Λ is determined by this form. Further, the concept of formal differentials carries over to the graded case in the obvious way; in the case at hand the corresponding module $D_{\Lambda[W]}$ of graded formal differentials is an induced module

$$D_{\Lambda[W]} = \Lambda[W] \otimes W; \tag{3.14}$$

and the form $(\cdot, \cdot)$ induces a bracket operation on $D_{\Lambda[W]}$ so that the pair $(\Lambda[W], D_{\Lambda[W]})$ is a graded Lie-Rinehart algebra. We write $D_{(\cdot,\cdot)}$ for $D_{\Lambda[W]}$ together with the bracket operation. Moreover, the structure determines an extension

$$0 \to \Lambda[W] \to \overline{L}_{(\cdot,\cdot)} \to D_{(\cdot,\cdot)} \to 0 \tag{3.15}$$

of graded Lie-Rinehart algebras. This is exactly the graded analogue of the extension (1.9) in the special case at hand. Furthermore, a prequantum module for the exterior algebra $\Lambda = \Lambda[W]$ looks like $M = \Lambda \otimes \mathbf{C}$, with a suitable structure morphism

$$\chi \colon \overline{L}_{(\cdot,\cdot)} \longrightarrow \operatorname{End}(M).$$

The notion of a polarization explained at the end of Section 2 carries readily over to the graded case, and after having chosen a suitable graded polarization P, taking invariants M^P with respect to the induced action of P on M, one obtains the following, cf. (5.15) in [24]:

Theorem 3.16. *The resulting quantum BRST-operator $\widehat{\Omega}$ on $M^P = \Lambda[s^{-1}g^*] \otimes \mathbf{C}$ is given by*

$$\widehat{\Omega} = \tfrac{1}{i}\, d,$$

where d is the usual differential computing Lie algebra cohomology of g with values in $\mathbf{R}$.

In the general case, with

$$\mathcal{A} = \Lambda[W] = \Lambda[s^{-1}g^*] \otimes \Lambda[sg] \otimes A$$

and a comomentum mapping $\delta \colon g \to A$, after having chosen a suitable prequantum module (M_A, χ_A) for A, the graded tensor product $M_{\mathcal{A}} = M_A \otimes \Lambda[W]$, with the appropriate structure map $\chi^{\otimes}$, is a graded prequantum module for $\mathcal{A}$. After having chosen a polarization $P_A \subseteq D_A$ that is compatible with the g-operation on M_A induced by the given comomentum mapping, one concocts a graded polarization P for $M_{\mathcal{A}} = M_A \otimes \Lambda[W]$ from P_A and the polarization that came into play in (3.16), and one obtains the following, cf. (8.30) in [24]:

Theorem 3.17. *The resulting quantum BRST-operator $\widehat{\Omega} = \widehat{\Omega}_0 + \widehat{\Omega}_1$ on the corresponding graded quantum module, that is, on the invariants $M^P = M_A^{P_A} \otimes \Lambda[s^{-1}g^*]$, is given by*

$$\widehat{\Omega} = \tfrac{1}{i}\, d,$$

where d is the usual differential computing Lie algebra cohomology of g with values in the quantum module $M_A^{P_A}$ for A, with g-module structure induced by the given comomentum mapping.

An example may be found in Section 8 of our paper [24]. It may be worthwhile pointing out that when the Lie algebra g is not unimodular, our theory differs from that in KOSTANT AND STERNBERG [35].

4. An example: angular momentum zero

As usual we take the cotangent bundle $T^*\mathbf{R}^n = \mathbf{R}^{2n}$, $n \geq 2$, with coordinates $(\mathbf{x}, \mathbf{p})$, as the phase space of a nonrelativistic particle moving in $\mathbf{R}^n$. The cotangent action of the rotation group $SO(n)$ on this phase space is

$$(A, \mathbf{x}, \mathbf{p}) \longmapsto (A\mathbf{x}, A\mathbf{p}), \quad A \in SO(n);$$

it is hamiltonian, and upon identifying $so(n)^*$ with $\mathbf{R}^{n(n-1)/2}$, the Ad^*-equivariant angular momentum map

$$J: \mathbf{R}^{2n} \longrightarrow so(n)^*$$

for this action assumes the form

$$J(\mathbf{x}, \mathbf{p}) = \mathbf{x} \wedge \mathbf{p};$$

here $so(n)^*$ refers to the dual of the Lie algebra $so(n)$, as usual. Details may be found in GOTAY-BOS [12], or in (5.11) and (7.10) of ARMS-GOTAY-JENNINGS [2]. For fixed angular momentum, a suitable reduction procedure yields a corresponding reduced classical system. For non-zero angular momentum this example can be treated by standard techniques, while for zero angular momentum, the origin is a singularity of the constraint set $J^{-1}(0)$, and the standard geometric quantization scheme is not applicable. We now indicate how our approach to the quantization of Poisson algebras works in this case. We proceed as follows.

For zero angular momentum, it is shown in Section 5 of [12] that, as a symplectic V-manifold, the geometrically reduced phase space $J^{-1}(0)/SO(n)$ (in the sense of Section 3 of [2]) looks like $\mathbf{R}^2/(\mathbf{Z}/2)$, with the obvious symplectic structure induced from that on $\mathbf{R}^2$ via the obvious projection $\mathbf{R}^2 \to \mathbf{R}^2/(\mathbf{Z}/2)$. Moreover, cf. p. 93 of [2], the corresponding geometrically reduced Poisson algebra actually coincides with that obtained by algebraic reduction à la SNIATYCKI-WEINSTEIN [47]; we write A_{red} for this reduced Poisson algebra. In view of what is said on p. 77 and p. 92 of [2], this algebra may be described as follows: Since $G = SO(n)$ is compact, in view of (5.12) of [2], as an algebra, the SNIATYCKI-WEINSTEIN reduced algebra is isomorphic to the quotient algebra

$$C^\infty(\mathbf{R}^{2n})^G/[I(J)]^G,$$

where $I(J)$ refers to the ideal of smooth functions generated by the components of J, and where $-^G$ denotes G-invariants as usual. With respect to the induced action, for $n \geq 3$, the subalgebra of G-invariants in the polynomial algebra $\mathbf{R}[x_1, p_1, \ldots, x_n, p_n]$ is the polynomial algebra generated by the three invariants $\tau_1 = \mathbf{x} \cdot \mathbf{x}$, $\tau_2 = \mathbf{x} \cdot \mathbf{p}$, $\tau_3 = \mathbf{p} \cdot \mathbf{p}$, while for $n = 2$,

$$\tau_4 = [\mathbf{x}, \mathbf{p}] = x_1 p_2 - p_1 x_2$$

is an additional invariant, subject to the relation

$$\tau_4^2 = \tau_1 \tau_3 - \tau_2^2;$$

see p. 77 of WEYL [56] for details, where $\mathrm{SO}(n)$ is written $\mathrm{O}^+(n)$. The three invariants τ_1, τ_2, τ_3 yield the smooth map

$$\tau = (\tau_1, \tau_2, \tau_3)\colon \mathbf{R}^{2n} \to \mathbf{R}^3,$$

and a result of G.W. SCHWARZ [45] implies that, for $n \geq 3$, $C^\infty(\mathbf{R}^{2n})^G = \tau^* C^\infty(\mathbf{R}^3)$. For $n = 2$ one must also take care of τ_4. In other words, for $n \geq 3$, the invariants in $C^\infty(\mathbf{R}^{2n})^G$ are functions f in the variables $\mathbf{x}$ and $\mathbf{p}$ such that

$$f(\mathbf{x}, \mathbf{p}) = F(\mathbf{x} \cdot \mathbf{x}, \mathbf{x} \cdot \mathbf{p}, \mathbf{p} \cdot \mathbf{p}) \qquad (4.1.1)$$

for a unique smooth function F in three variables. Actually, GOTAY-BOS [12] use the invariants

$$\sigma_1 = \mathbf{x} \cdot \mathbf{x} - \mathbf{p} \cdot \mathbf{p}, \quad \sigma_2 = 2\mathbf{x} \cdot \mathbf{p}, \quad \sigma_3 = \mathbf{x} \cdot \mathbf{x} + \mathbf{p} \cdot \mathbf{p},$$

and in [2] the notation $\mathbf{y}$ is used instead of $\mathbf{p}$ but this is of no account. The geometrically reduced Poisson algebra is now obtained by restricting these functions f to $J^{-1}(0)$, that is, by taking these functions modulo the ideal I of G-invariant C^∞-functions that are zero on $J^{-1}(0)$. In the case at hand, $J(\mathbf{x}, \mathbf{p}) = 0$ if and only if $\mathbf{p}$ is proportional to $\mathbf{x}$, see p. 92 of [2], and the ideal I coincides with the ideal $[I(J)]^G$ and is in fact the ideal generated by $\tau_2^2 = \tau_1 \tau_3$, cf. p. 93 of [2], (but in general we can only say that $[I(J)]^G \subseteq I$). Thus the reduced algebra A_{red} consists of classes of G-invariant functions f modulo $\tau_2^2 = \tau_1 \tau_3$. We note that this description is valid even for $n = 2$ since τ_4 then actually coincides with J and hence vanishes on the constraint set $J^{-1}(0)$.

To obtain a somewhat more explicit description of A_{red}, consider the algebra $C^\infty(\mathbf{R}^2)$ of smooth functions in the variables u, v with the usual Poisson structure given by $\{u, v\} = 1$ etc. What is said on p. 197 of [12] and p. 77 of [2] may be interpreted by saying that the association

$$f \mapsto h_f \in C^\infty(\mathbf{R}^2)$$

where h_f is the smooth function in two variables given by

$$h_f(u, v) = F(u^2, v^2, uv),$$

f and F being related by (4.1.1), induces an isomorphism

$$A_{\mathrm{red}} \longrightarrow C^\infty(\mathbf{R}^2)$$

of A_{red} onto the Poisson subalgebra of $C^\infty(\mathbf{R}^2)$ consisting of even functions, that is, of smooth functions h in the variables (u, v) satisfying

$$h(u, v) = h(-u, -v).$$

Here $C^\infty(\mathbf{R}^2)$ is endowed with the standard Poisson structure so that $\{u, v\} = 1$ etc., and it is straightforward to verify that the Poisson bracket of two even functions

$$\{h_1, h_2\} = \frac{\partial h_1}{\partial u}\frac{\partial h_2}{\partial v} - \frac{\partial h_1}{\partial v}\frac{\partial h_2}{\partial u}$$

is again an even function. Thus the reduced Poisson algebra for a non-relativistic particle moving in $\mathbf{R}^n$ with zero angular momentum may be described as that of $\mathbf{Z}/2$-invariants of the Poisson subalgebra of $C^\infty(\mathbf{R}^2)$ with respect to the $\mathbf{Z}/2$-action given by $(u, v) \mapsto (-u, -v)$. We denote the reduced Poisson structure by $\{\cdot, \cdot\}_{\mathrm{red}}$.

Our next aim is to construct a *prequantum module* for the reduced Poisson algebra $(A_{\mathrm{red}}, \{\cdot, \cdot\}_{\mathrm{red}})$. Actually, this is still a kind of toy example but yet it may well serve to illustrate the relevant ideas: We consider the corresponding extension (1.9) of Lie-Rinehart algebras which now looks like

$$0 \rightarrow A_{\mathrm{red}} \rightarrow \overline{L}_{\{\cdot,\cdot\}_{\mathrm{red}}} \rightarrow D_{\{\cdot,\cdot\}_{\mathrm{red}}} \rightarrow 0.$$

Write $A = (C^\infty(\mathbf{R}^2), \{\cdot, \cdot\})$, and let $\vartheta: D_A \longrightarrow A$ be the 1-form given by

$$\vartheta(du) = 0, \quad \vartheta(dv) = v.$$

This 1-form is a *Poisson potential*, cf. Section 2 above. Indeed,

$$\begin{aligned}
d\vartheta(du, dv) &= (du)^\sharp(\vartheta(dv)) - (dv)^\sharp(\vartheta(du)) - \vartheta[du, dv]\\
&= \{u, \vartheta(dv)\} - \{v, \vartheta(du)\} - \vartheta(d\{u, v\})\\
&= \{u, v\}
\end{aligned}$$

since $\vartheta(du) = 0$ and $\{u, v\} = 1$. However, for an even function $h \in A$,

$$\frac{\partial h}{\partial v}(-u, -v) = -\frac{\partial h}{\partial v}(u, v);$$

hence

$$\begin{aligned}
(\vartheta(dh))(u, v) &= \frac{\partial h}{\partial v}(u, v)v\\
&= -\frac{\partial h}{\partial v}(-u, -v)v\\
&= (\vartheta(dh))(-u, -v)
\end{aligned}$$

that is to say, the Poisson potential ϑ passes to a Poisson potential

$$\vartheta_{\mathrm{red}}: D_{\{\cdot,\cdot\}_{\mathrm{red}}} \longrightarrow A_{\mathrm{red}}$$

for the reduced Poisson algebra. Let $M_{\mathrm{red}} = (A_{\mathrm{red}} \otimes \mathbf{C}) < 1 >$, that is, M_{red} is the free $(A_{\mathrm{red}} \otimes \mathbf{C})$-module with the single basis element 1, and define a structure

$$\chi_{\mathrm{red}}: \overline{L}_{\{\cdot,\cdot\}_{\mathrm{red}}} \longrightarrow \mathrm{End}(M_{\mathrm{red}})$$

of an $(A_{\mathrm{red}}, \overline{L}_{\{\cdot,\cdot\}_{\mathrm{red}}})$-module on M_{red} by

$$(\chi_{\mathrm{red}}(0, dh))(1) = -i\,\vartheta(dh)$$
$$(\chi_{\mathrm{red}}(a, 0))(1) = i\,a.$$

It is readily seen that χ_{red} is a structure of a *prequantum module for* the reduced Poisson algebra $(A_{\mathrm{red}}, \{\cdot, \cdot\}_{\mathrm{red}})$. It may be shown that this prequantum module structure can be obtained by the reduction procedure in our paper [21], applied to the prequantum module for the unreduced data given in Section 2 above.

Alternatively, we may complexify and consider the 1-form

$$\vartheta^{\mathrm{hol}}: D_A \otimes \mathbf{C} \longrightarrow A \otimes \mathbf{C}$$

given by

$$\vartheta^{\mathrm{hol}}(du) = -i\,v, \quad \vartheta^{\mathrm{hol}}(dv) = v.$$

This 1-form is a again *Poisson potential*; as in Section 2 we refer to it as the *holomorphic* Poisson potential. The same argument as before shows that it passes to a Poisson potential

$$\vartheta^{\mathrm{hol}}_{\mathrm{red}}: D_{\{\cdot,\cdot\}_{\mathrm{red}}} \otimes \mathbf{C} \longrightarrow A_{\mathrm{red}} \otimes \mathbf{C}$$

for the reduced Poisson algebra. As earlier, let $M_{\mathrm{red}} = (A_{\mathrm{red}} \otimes \mathbf{C}) < 1 >$, that is, M_{red} is the free $(A_{\mathrm{red}} \otimes \mathbf{C})$-module with the single basis element 1, and define a structure

$$\chi^{\mathrm{hol}}_{\mathrm{red}}: \overline{L}_{\{\cdot,\cdot\}_{\mathrm{red}}} \otimes \mathbf{C} \longrightarrow \mathrm{End}(M_{\mathrm{red}})$$

of an $(A_{\mathrm{red}}, \overline{L}_{\{\cdot,\cdot\}_{\mathrm{red}}})$-module on M_{red} by

$$(\chi^{\mathrm{hol}}_{\mathrm{red}}(0, dh))(1) = -i\,\vartheta^{\mathrm{hol}}_{\mathrm{red}}(dh)$$
$$(\chi^{\mathrm{hol}}_{\mathrm{red}}(a, 0))(1) = i\,a.$$

This prequantum module can again be obtained by the reduction procedure in our paper [21], applied to the corresponding prequantum module for the unreduced data given in Section 2 above. Moreover, with the notation

$\zeta = u + i\,v$ etc., the *antiholomorphic* polarization is the $(\mathbf{C}, A_{\mathrm{red}} \otimes \mathbf{C})$-sub Lie algebra $P_{\mathrm{red}} \subseteq D_{\{\cdot,\cdot\}_{\mathrm{red}}} \otimes \mathbf{C}$ generated by the differential $d\zeta^2$. We shall show in [21] that this polarization can be obtained by the reduction procedure in our paper [20], applied to the corresponding polarization for the unreduced data given in Section 2 above. The corresponding quantum module $M_{\mathrm{red}}^{P_{\mathrm{red}}}$, that is, the polarized elements or P_{red}-invariants in our prequantum module M_{red}, are the smooth even complex valued functions h in the variables (u, v) satisfying

$$0 = \chi_{\mathrm{red}}^{\mathrm{hol}}(0, i\,d\zeta^2)(h) = 2\zeta\{i\,d\zeta, h\} = 2\zeta\{-v + i\,u, h\}$$
$$= 2\left(\frac{\partial h}{\partial u} + i\frac{\partial h}{\partial v}\right) = 4\zeta\frac{\partial h}{\partial\overline{\zeta}};$$

in other words, the resulting quantum module $M_{\mathrm{red}}^{P_{\mathrm{red}}}$ consists of the even holomorphic functions h in a single complex variable ζ. We now look for those observables that are compatible with the antiholomorphic polarization. Thus given a smooth even function f in the variables u, v, in order for it to satisfy the hypothesis of (2.6.15), we must have that the function $\{\zeta, f\}$ is holomorphic in the variable ζ. This amounts to the requirement that

$$\{i\zeta, \{i\zeta, f\}\} = \frac{\partial}{\partial u}\left(\frac{\partial f}{\partial u} + i\frac{\partial f}{\partial v}\right) + i\frac{\partial}{\partial v}\left(\frac{\partial f}{\partial v} + i\frac{\partial f}{\partial v}\right) = 0$$

or, in other words, that

$$\frac{\partial^2 f}{\partial u^2} - \frac{\partial^2 f}{\partial v^2} = 0, \quad \frac{\partial^2 f}{\partial u\partial v} = 0.$$

It is clear that, for example, the corresponding "energy function" f given by

$$f(u, v) = \tfrac{1}{2}\left(u^2 + v^2\right) = \tfrac{1}{2}\zeta\overline{\zeta}$$

satisfies this requirement. We obtain the representation $f \mapsto \hat{f}$ on the space $M_{\mathrm{red}}^{P_{\mathrm{red}}}$ of even holomorphic functions h in a single variable ζ given by

$$\hat{f}(h) = \zeta\frac{\partial h}{\partial\zeta} + \tfrac{1}{2}\zeta^2 h.$$

Notice that when h is an even holomorphic function so is $\hat{f}(h)$. If we take instead the reduced Poisson potential $\widetilde{\vartheta}_{\mathrm{red}}$ given by

$$\widetilde{\vartheta}_{\mathrm{red}}(d\zeta) = \tfrac{1}{2}\zeta, \quad \widetilde{\vartheta}_{\mathrm{red}}(d\overline{\zeta}) = \tfrac{1}{2}\overline{\zeta},$$

or, what amounts to the same,

$$\widetilde{\vartheta}(du) = \tfrac{1}{2}u, \quad \widetilde{\vartheta}_{\mathrm{red}}(dv) = \tfrac{1}{2}v,$$

the energy operator $\hat{f}$ is just the Euler operator, i. e. we have

$$\hat{f}(h) = \zeta \frac{\partial h}{\partial \zeta}, \quad h \in M_{\text{red}}^{P_{\text{red}}}.$$

Finally we may take the inner product $< h, k >$ of two such even functions h and k to be the integral of the product $h\overline{k}$ with respect to a suitable measure on $\mathbf{R}^2$.

It is interesting to compare this with the quantization of angular momentum obtained at the end of Section 2. Indeed, there the usual representation of $G = \mathrm{SO}(n)$ on the algebra of holomorphic functions in the complex variables $z_1 = x_1 + i\,p_1, \ldots, z_n = x_n + i\,p_n$, viewed as a complex vector space only, was obtained. We now assert that the subspace of G-invariants consists of functions f having the property that there is a unique holomorphic function H in a single complex variable so that

$$f(z_1, \ldots, z_n) = H(z_1^2 + \cdots + z_n^2).$$

Indeed, we already know that, for $n \geq 3$, the subspace of G-invariant *smooth* functions is the subspace (or subalgebra) of functions f that may be written

$$f(\mathbf{x}, \mathbf{p}) = F(\mathbf{x} \cdot \mathbf{x}, \mathbf{x} \cdot \mathbf{p}, \mathbf{p} \cdot \mathbf{p})$$

for a unique smooth function F in three variables. Now the additional requirement that f be a holomorphic function in the variables $z_1, \ldots, z_n$ implies that f should be a function of the holomorphic invariant

$$\sigma_1 + i\sigma_2, \quad \text{where} \quad \sigma_1 = \mathbf{x} \cdot \mathbf{x} - \mathbf{p} \cdot \mathbf{p}, \quad \sigma_2 = 2\mathbf{x} \cdot \mathbf{p}.$$

In other words, there is a unique holomorphic function H in a single complex variable so that

$$f(z_1, \ldots, z_n) = H(z_1^2 + \cdots + z_n^2)$$

as asserted. Suitably rephrased, this argument is also valid for $n = 2$. However, it is common to interpret the passage to the subspace of G-invariants as *reduction after quantization*. Thus we see that, in the case of the constrained system with singularities at hand, we obtain an affirmative answer to the question whether *quantization after reduction yields the same result as reduction after quantization*. It is interesting to examine this question in general. Results may be found in the literature for momentum mappings J having the property that 0 is a *regular value*. See GUILLEMIN-STERNBERG [15] and GOTAY [11].

5. Concluding remarks

(1) BRST-quantization extends what was referred to above as "reduction after quantization". The other approach, "quantization after reduction", requires a reduction procedure at the classical level, e. g. MARSDEN-WEINSTEIN reduction [41] for a symplectic manifold together with a momentum mapping or that of ŚNIATYCKI-WEINSTEIN [47] for a Poisson algebra together with a comomentum mapping. In our paper [20] we extended the ŚNIATYCKI-WEINSTEIN reduction procedure to prequantum modules. Our construction encapsules in particular a reduction procedure for prequantum bundles due to GUILLEMIN-STERNBERG [15]. Using our methods, we hope to be able to eventually solve the following conjecture which in GUILLEMIN-STERNBERG [15] is attributed to KIRILLOV: Let (N, σ) be a compact Kähler manifold, with its usual symplectic structure, assume the latter quantizable, let G be a compact Lie group with a Hamiltonian action on N and corresponding momentum mapping $J: N \to g^*$, and let $\mathcal{O} \subseteq g^*$ be an integral coadjoint orbit. Moreover, suppose that G preserves the holomorphic polarization, and let E_N be the complex representation of G obtained by holomorphic quantization applied to the given data. Then the conjecture says that the irreducible representation associated with $\mathcal{O}$ via the Borel-Weil Theorem occurs in E_N only if $\mathcal{O}$ is in the image of the momentum mapping J. In GUILLEMIN-STERNBERG [15] it is shown that the conjecture is true under the additional assumption that 0 is a regular value of the momentum mapping

$$J_{\mathcal{O}}: N \times \mathcal{O} \longrightarrow g^*, \quad J_{\mathcal{O}}(x, y) = J(x) - y.$$

(2) In a sense, the approach to the quantization of Poisson algebras in [18] and sketched in Section 2 above has been globalized in WEINSTEIN-XU [55].

(3) For smooth Poisson manifolds, our approach to the quantization of Poisson algebras has been reworked in a direct geometric way in VAISMAN [51].

REFERENCES

1. R. Abraham and J. E. Marsden, *Foundations of Mechanics*, Benjamin/ Cummings Publishing Company, 1978.
2. J. M. Arms, M. J. Gotay, and G. Jennings, *Geometric and algebraic reduction for singular momentum mappings*, Advances in Mathematics **79** (1990), 43–103.
3. V. I. Arnold, *Mathematical Methods of Classical Mechanics*, Graduate Texts in Mathematics, No. 60 (second edition), Springer, Berlin-Heidelberg - New York, 1989.

4. M. F. Atiyah, *Complex analytic connections in fibre bundles*, Trans. Amer. Math. Soc. **85** (1957), 181–207.

5. C. Becchi, A. Rouet, and R. Stora, *Renormalization of gauge theories*, Ann. Phys. **98** (1976), 287.

6. J. L. Brylinski, *A differential complex for Poisson manifolds*, J. of Diff. Geom. **28** (1988), 93–114.

7. H. Cartan and S. Eilenberg, *Homological Algebra*, Princeton University Press, Princeton, 1956.

8. C. Chevalley and S. Eilenberg, *Cohomology theory of Lie groups and Lie algebras*, Trans. Amer. Math. Soc. **63** (1948), 85–124.

9. P. A. M. Dirac, *Lectures on Quantum Mechanics*, Belfer Graduate School of Science, Yeshiva University, New York, 1964.

10. P. A. M. Dirac, *Generalized Hamiltonian systems*, Can. J. of Math. **12** (1950), 129–148.

11. M. J. Gotay, *Constraints, reduction, and quantization*, J. of Math. Phys. **27** (1986), 2051–2066.

12. M. J. Gotay and L. Bos, *Singular angular momentum mappings*, J. Diff. Geom. **24** (1986), 181–203.

13. V. Guillemin and S. Sternberg, *Symplectic techniques in Physics*, Cambridge University Press, London/New York, 1984.

14. V. W. Guillemin and S. Sternberg, *Homogeneous quantization and multiplicities of group representations*, J. Funct. Anal. **47** (1980), 344–380.

15. V. W. Guillemin and S. Sternberg, *Geometric quantization and multiplicities of group representations*, Invent. Math. **67** (1982), 515–538.

16. M. Henneaux, *Hamiltonian form of the path integral for theories with a gauge freedom*, Phys. Rep. **126** (1985), 1–66.

17. G. Hochschild, B. Kostant, and A. Rosenberg, *Differential forms on regular affine algebras*, Trans. Amer. Math. Soc. **102** (1962), 383–408.

18. J. Huebschmann, *Poisson cohomology and quantization*, J. für die Reine und Angew. Math. **408** (1990), 57–113.

19. J. Huebschmann, *Some remarks about Poisson homology*, Preprint Universität Heidelberg 1990.

20. J. Huebschmann, *Poisson reduction and quantization of constrained systems with singularities. I. Reduction of prequantum modules*, Preprint Universität Heidelberg 1990.

21. J. Huebschmann, *Poisson reduction and quantization of constrained systems with singularities. II. Reduction of polarizations and quantum modules*, In preparation.

22. J. Huebschmann, *Poisson reduction and quantization of constrained systems. III. Yang-Mills theory*, In preparation.

23. J. Huebschmann, *Poisson reduction and quantization of constrained systems with singularities. IV. Chern-Simons gauge theory*, In preparation.

24. J. Huebschmann, *Graded Lie Rinehart algebras, graded Poisson algebras, and BRST-quantization I. The finitely generated case*, Preprint Universität Heidelberg 1990.

25. J. Huebschmann, *Graded Lie Rinehart algebras and graded Poisson algebras*, In preparation.

26. J. Huebschmann, *Graded Lie Rinehart algebras, graded Poisson algebras, and BRST-quantization II. The general case*, In preparation.

27. J. Huebschmann, *Constraints, symmetries, homological perturbations, and the BRST-transformation*, Preprint 1989.

28. J. Huebschmann, *Extensions of Lie-Rinehart algebras. I. The Chern - Weil construction*, Preprint Universität Heidelberg 1989.

29. J. Huebschmann, *Extensions of Lie-Rinehart algebras. II. The spectral sequence and the Weil algebra*, Preprint Universität Heidelberg 1989.

30. J. Huebschmann, *Extensions of Lie-Rinehart algebras. III. Principal extensions*, In preparation.

31. D. Husemoller, J. C. Moore, and J. D. Stasheff, *Differential homological algebra and homogeneous spaces*, J. of Pure and Applied Algebra **5** (1974), 113–185.

32. A. A. Kirillov, *Unitary representations of nilpotent Lie groups*, Uspehi Mat. Nauk. **17** (1962), 57–101; Russ. Math. Surveys **17** (1962), 57–101.

33. B. Kostant, *Quantization and unitary representations*, In: Lectures in Modern Analysis and Applications, III, ed. C. T. Taam, Lecture Notes in Math. **170** (1970), 87–207, Springer, Berlin - Heidelberg - New York.

34. B. Kostant, *Graded manifolds, graded Lie theory, and prequantization*, in Diff. Geom. Meth. in Math. Physics, Bonn, 1975, Lecture Notes in Math., **570** (1977), 177–306, Springer, Berlin-Heidelberg-New York.

35. B. Kostant and S. Sternberg, *Symplectic reduction, BRS-cohomology, and infinite dimensional Clifford algebras*, Ann. of Physics (N. Y.) **176** (1987), 49–113.

36. J. L. Koszul, *Crochet de Schouten-Nijenhuis et cohomologie*, in E. Cartan et les Mathématiciens d'aujourd hui, Lyon, 25–29 Juin, 1984, Asterisque, **hors-serie**, (1985), 251–271.

37. A. Lichnerowicz, *Les variétés de Poisson et leurs algèbres de Lie associées*, J. Diff. Geo. **12** (1977), 253–300.

38. S. Lie, *Theorie der Transformationsgruppen*, Teubner, Leipzig, 1890.

39. K. Mackenzie, *Lie groupoids and Lie algebroids in differential geometry*, London Math. Soc. Lecture Note Series, vol. 124, Cambridge University Press, Cambridge, England, 1987.

40. M. P. Malliavin, *Algèbre homologique et opérateurs différentiels*, Ring theory, Proc. of a conference held at Granada, 1986, eds. J. L. Buesco P. Jara and B. Torrecillas, Lecture Notes in Math. vol. 1328 (1988), 172–186, Springer, Berlin-Heidelberg-New York-Tokyo.

41. J. Marsden and A. Weinstein, *Reduction of symplectic manifolds with symmetries*, Rep. on Math. Phys. **5** (1974), 121–130.

42. W. S. Massey and F. P. Petersen, *The cohomology structure of certain fibre spaces.I*, Topology **4** (1965), 47–65.

43. R. Palais, *The cohomology of Lie rings*, Amer. Math. Soc., Providence, R. I., Proc. Symp. Pure Math. **III** (1961), 130–137.

44. G. Rinehart, *Differential forms for general commutative algebras*, Trans. Amer. Math. Soc. **108** (1963), 195–222.

45. G.W. Schwarz, *Smooth functions invariant under the action of a compact Lie group*, topology **14** (1975), 63–68.

46. I. E. Segal, *Quantization of non-linear systems*, J. of Math. Phys. 1 (1960), 468–488.

47. J. Sniatycki and A. Weinstein, *Reduction and quantization for singular moment mappings*, Lett. Math. Phys. **7** (1983), 155–161.

48. J. M. Souriau, *Quantification géométrique*, Comm. Math. Physics **1** (1966), 374–398.

49. J. D. Stasheff, *Constrained Poisson algebras*, Bull. Amer. Math. Soc. **19** (1988), 287–290.

50. I. V. Tyutin, *Gauge invariance in field theory and statistical mechanics*, Lebedev preprint FIAN No. 39.

51. I. Vaisman, *On the geometric quantization of Poisson manifolds*, Preprint 1990.

52. V. S. Vladimirov and I. V. Volovich, *p-adic quantum mechanics*, Comm. Math. Phys. **123** (1989), 659–676.

53. A. Weinstein, *Poisson structures*, in E. Cartan et les Mathématiciens d'aujourd hui, Lyon, 25–29 Juin, 1984, Astérisque, **hors-serie**, (1985), 421–434.

54. A. Weinstein, *Symplectic groupoids and Poisson manifolds*, Bull. Amer. Math. Soc. **16** (1987), 101–104.

55. A. Weinstein and P. Xu, *Extensions of symplectic groupoids and quantization*, J. für die Reine und Angew. Math. (to appear).

56. H. Weyl, *The classical groups*, Princeton University Press, Princeton NJ, 1946.

57. N. Woodhouse, *Geometric quantization*, Clarendon Press, Oxford, 1980.

58. B. Kostant and S. Sternberg, *Anti-Poisson algebras and current algebras*, preprint.

UFR de Mathématiques
Université des Sciences et Techniques de Lille-Flandres-Artois
F-59655 Villeneuve d'Ascq CEDEX
France

Simple Quantization Formula

M. V. KARASEV

To Jean-Marie Souriau

The problem of quantization is closely connected with quasi-classical asymptotics. In particular, the Maslov asymptotic constructions for Lagrangian submanifolds contain many analogies with the Souriau–Kostant–Kirillov scheme of geomtric quantization [21, 28]. And both of these concepts are associated with the Schrödinger quantization rule. But there is also the well-known Wick rule which was developed by Bargmann and Berezin, using Gaussian dissipative packets instead of oscillating packets. Such packets were used in the theory of quasi-classical asymptotics [1–5] and in many other works. It was found that there is the conjunction of the Bargmann representation with geometric quantization (see [29,30,6,7]).

We may ask a natural question: is there an analogue of the Wick quantization for general Lagrangian submanifolds? Certain preliminary ideas were proposed in the works by V. M. Babich and his pupils (see the review in [8]), in the physical works by J. K. Klauder [9], S. W. McDonald [10] and in [11] (Chapter IV, p. 109). In these texts the Cauchy problem for quantum Hamiltonians was considered, and it was understood that using complex phase function in quasi-classical approximation allows us to exclude focal points, caustics, etc. Recently the papers by A. Voros [12] and J. Kurchan, P. Leboeuf, M. Saraceno [13] appeared with some interesting constructions in the stationary one-dimensional case, and by A. Laptev, Yu. G. Safarov [14] in the case of the Cauchy problem on Riemannian manifolds.

The general Wick-like quantization for an arbitrary Lagrangian submanifold was presented in the author's report at the Faddeev Conference at the Leningrad Mathematical Institute (November 1988) and published in the Proceedings of this Institute [15].

In the work given we discuss further generalizations of our basic formula [15]; see also [16–18].

1. The basic formula

Consider the phase space $\mathbb{R}^{2n} = \mathbb{R}^n_x \oplus \mathbb{R}^n_p$ with symplectic structure $dp \wedge dx$. Let Λ be a Lagrangian submanifold, dm a measure on Λ, and $m(\Lambda) = 1$. Denote $\Theta = pdx|_\Lambda$ and $J(\alpha) = \frac{\mathcal{D}(x(\alpha)+ip(\alpha))}{\mathcal{D}m(\alpha)}$ (where $\alpha \in \Lambda$); $J^2 : \Lambda \to \mathbb{C}$ is a smooth function. Then $j = (J/|J|)^2$ is the mapping

$j : \Lambda \to \mathbb{S}^1 \subset \mathbb{C}$ into the unit circle of the complex plane. Let $idz/2\pi z$ be a fundamental form on $\mathbb{S}^1$ and let

$$\mu = j^* \left(\frac{idz}{2\pi z} \right)$$

be the Arnold form [19] on Λ. The cohomology class $[\mu]$ is exactly the Maslov class [20]. We put

$$J(\alpha)^{1/2} = |J(\alpha)|^{1/2} \exp\left\{ -i\frac{\pi}{2} \int_{\alpha^0}^{\alpha} \mu \right\}, \quad \alpha^0 \text{ is fixed.}$$

Below we shall write

$$J^{1/2} \equiv \left(\frac{\mathcal{D}(x + ip)}{\mathcal{D}m} \right)^{1/2},$$

if we know from which initial point α^0 the argument of the Jacobian is counted.

Consider the function on $\mathbb{R}^n_x$:

$$\psi(x) = (2\pi\hbar)^{-n/2} \int_\Lambda J(\alpha)^{1/2} \exp\left\{ \frac{i}{\hbar} \int_{\alpha^0}^{\alpha} \Theta \right\}$$
$$\times \exp\left\{ -\frac{(x - x(\alpha))^2}{2\hbar} + \frac{i}{\hbar} p(\alpha)(x - x(\alpha)) \right\} dm(\alpha). \qquad (1)$$

For each real smooth function H on $\mathbb{R}^{2n}$ with no more than polynomial growth of all derivatives at infinity, we denote by $\widehat{H} = H(x, -i\hbar\partial/\partial x)$ the Weyl pseudodifferential operator. We also denote by $ad(H)$ the Hamiltonian field related to the function H.

Theorem 1 [15]. *Let* $H|_\Lambda = 0$ *and the measure* dm *be* $ad(H)$-*invariant. Let the quantization condition*

$$\frac{1}{2\pi\hbar}[\Theta] - \frac{1}{4}[\mu] \in H^1(\Lambda, \mathbb{Z}) \qquad (2)$$

hold on Λ. *Then*

$$\widehat{H}\psi = \mathcal{O}(\hbar^2), \qquad \|\psi\| = 1 + \mathcal{O}(\hbar)$$

in the sense of L^2-*norm.*

Example 1. The Cauchy problem is considered

$$i\hbar\partial G/\partial t = \widehat{H}G, \qquad G|_{t=0} = \delta(x - y^0).$$

236 M.V. Karasev

Denote by $\gamma_H^t : (y,\xi) \to (X(y,\xi,t), P(y,\xi,t))$ the trajectories of the field $ad(H)$ and put

$$\varphi_t(y,\xi) = \exp\left\{\frac{i}{\hbar}\int_0^t (P\dot{X} - H)dt - i\frac{\pi}{2}\int_0^t \mu\right\},$$

where the Arnold form μ is determined by the Jacobian

$$J(y,\xi) = \frac{\mathcal{D}(X(y,\xi,t) + iP(y,\xi,t))}{\mathcal{D}y}.$$

Formula (1) on the Lagrangian submanifolds $\Lambda_\xi = \{(X,P)|y \in \mathbb{R}^n\}$ yields

$$G(x,y^0,t) \simeq \frac{1}{(2\pi\hbar)^{3n/2}} \int\int J(\alpha,\xi)^{1/2} \varphi_t(y^0,\xi)\exp\left\{\frac{i}{\hbar}\int_{y^0}^y PdX\right\}$$

$$\times \exp\left\{-\frac{(x-X)^2}{2\hbar} + \frac{i}{\hbar}P(x-X)\right\} dyd\xi.$$

A similar formula (but different in certain essential details) was obtained by John Klauder in his heuristic investigation of path integrals [9].

Example 2. Quantization of Poisson transformations. Let D be a simply-connected domain in $\mathbb{R}^{2n}$ and $\gamma : D \to \mathbb{R}^{2n}$ be a smooth Poisson mapping:

$$\gamma : (y,\xi) \to (X,P), \qquad dP \wedge dX = d\xi \wedge dy.$$

Then $\Lambda = $graph (γ) is Lagrangian submanifold in the product of phase spaces $\mathbb{R}^{2n} \times \mathbb{R}^{2n}$ with symplectic structure $dp \wedge dx - d\xi \wedge dy$. Formula (1) gives a function on $\mathbb{R}_x^n \times \mathbb{R}_y^n$ which may be considered as the integral kernel of the operator $T(\gamma)$ in $L^2(\mathbb{R}^n)$. More precisely, let $\varphi \in C_0^\infty(D)$; denote

$$T(\gamma,\varphi)(x,\bar{y}) = \frac{2^{n/2}}{(2\pi\hbar)^{3n/2}} \int\int \left(\frac{\mathcal{D}(X+iP)}{\mathcal{D}(y+i\xi)}\right)^{1/2} \varphi(y,\xi)$$

$$\times \exp\left\{\frac{i}{\hbar}\int_{y_0,\xi_0}^{y,\xi} (PdX - \xi dy) + \frac{i}{\hbar}P(x-X) - \frac{i}{\hbar}\xi(\bar{y}-y)\right.$$

$$\left. - \frac{(x-X)^2}{2\hbar} - \frac{(\bar{y}-y)^2}{2\hbar}\right\} dyd\xi. \tag{3}$$

Then the operators $T(\gamma,\varphi)$ give the projective representation of the group of Poisson transformations:

$$T(\gamma_2,\varphi_2) \cdot T(\gamma_1,\varphi_1) = e^{iC} T(\gamma_2\gamma_1, \varphi_1\gamma_1^*\varphi_2) + \mathcal{O}(\hbar),$$

where $C = C(\gamma_2,\gamma_1)$ is a cocycle.

Example 3. Wick quantization. If we take $\gamma = id$, $\Lambda = \mathrm{diag}(\mathbb{R}^{2n} \times \mathbb{R}^{2n})$ in the previous example, then we get the Wick operator

$$T(id, \varphi) = \Phi(\overset{2}{a}, \overset{1}{a^*}).$$

Here $a_k = (x_k + \hbar \partial / \partial x_k)/\sqrt{2}$ are the annihilation operators and

$$\varphi(y, \xi) \equiv \Phi\left(\frac{y + i\xi}{\sqrt{2}}, \frac{y - i\xi}{\sqrt{2}} \right).$$

Note that the Maslov construction on the Lagrangian submanifold $\Lambda = \mathrm{diag}(\mathbb{R}^{2n} \times \mathbb{R}^{2n})$ gives the integral kernel of the Schrödinger operator $\varphi(\overset{2}{x}, -i\hbar \overset{1}{\partial}/\partial x)$.

2. Coulomb singularity

Consider the Hamiltonian

$$H = p^2 + \frac{1}{|x|}(V_0(x) + |x|V_1(x)), \quad x \in \mathbb{R}^3.$$

Suppose that there is a Lagrangian manifold $\Lambda \subset \mathbb{R}^6$ lying on the level $\{H(x, p = \lambda\}$ and that Λ possesses the $ad(H)$-invariant measure dm. Let the projection $\pi_x(\Lambda) \subset \mathbb{R}^3$ contain a singular point $x = 0$ in its closure. Then the ordinary semiclassical asymptotics cannot be used. But we can do the Souriau–Moser–Kustaanheimo spinor realization [21-23]:

$$\mathbb{R}^4 \to \mathbb{R}^3, \quad q \to {}^\sigma q \equiv \; <\begin{pmatrix} q_1 - iq_2 \\ q_3 - iq_4 \end{pmatrix}, \sigma \begin{pmatrix} q_1 + iq_2 \\ q_3 + iq_4 \end{pmatrix} >,$$

where $\sigma = (\sigma_1, \sigma_2, \sigma_3)$ are the Pauli matrices. Note that

$$\frac{\partial \psi}{\partial x}({}^\sigma q) = D(q) \frac{\partial}{\partial q} \psi({}^\sigma q),$$

where

$$D = \frac{1}{2|q|^2} \begin{bmatrix} q_3 & q_4 & q_1 & q_2 \\ -q_4 & q_3 & q_2 & -q_1 \\ q_1 & q_2 & -q_3 & -q_4 \end{bmatrix}.$$

Thus we have the mapping

$$a : \begin{pmatrix} q \\ \xi \end{pmatrix} \to \begin{pmatrix} {}^\sigma q \\ D(q)\xi \end{pmatrix}, \qquad a : \mathbb{R}^8_{q,\xi} \to \mathbb{R}^6_{x,p}.$$

Consider in $\mathbb{R}^8$ the function

$$D_0 = q_1\xi_2 - q_2\xi_1 + q_3\xi_4 - q_4\xi_3.$$

238 M.V. Karasev

The trajectories of the Hamiltonian flow $\gamma_{D_0}^t$ fibrate the level $D_0 = 0$ over some base N^6, and the mapping α is correctly reduced to the base

$$
\begin{array}{ccc}
& \mathbb{R}^8 & \\
& \cup & \searrow^{a} \\
\Lambda^{\#} \subset \{D_0 = 0\} & \mathbb{R}^6 & \supset \Lambda . \\
& {}^{b}\!\downarrow \quad \nearrow_{a} & \\
& N^6 &
\end{array}
$$

Denote $\Lambda^{\#} = b^{-1}a^{-1}(\Lambda)$. Then $\Lambda^{\#}$ is Lagrangian in $\mathbb{R}^8$. It lies on the level $\{H^{\#}(q,\xi) = 0\}$, where $H^{\#}(q,\xi) = \xi^2 - \lambda q^2 + V_0(^{\sigma}q) + q^2 V_1(^{\sigma}q)$ is a regularized Hamiltonian. There is the unique measure $dm^{\#}$ on Λ which is $ad(\mathbf{H}^{\#})$-and $ad(D_0)$-invariant and such that $a_* dm^{\#} = dm$. Formula (1) gives the function ψ on $\mathbb{R}_q^4$ for this case. But since it is $ad(D_0)$-invariant, it is a function on $\mathbb{R}_x^3$. We have

$$
\psi(x) = \frac{e^{-|x|/2\hbar}}{(2\pi\hbar)^2} \int_{\Lambda^{\#}} \left(\frac{\mathcal{D}(q+i\xi)}{\mathcal{D}m^{\#}} \right)^{1/2} \exp \left\{ -\frac{q^2}{2\hbar} - \frac{i}{h} q\xi + \frac{i}{\hbar} \int_{(q_0,\xi_0)}^{(q,\xi)} \xi dq \right\}
$$

$$
\times I_0 \left(\frac{1}{\hbar} \sqrt{ \frac{|x|}{2}(q+i\xi)^2 + \frac{x}{2} \cdot {}^{\sigma}(q+i\xi) } \right) dm^{\#}(q,\xi). \tag{4}
$$

Here we denote by I_0 the Bessel function

$$
I_0(s) \sim \frac{1}{\sqrt{2\pi s}} e^{-s}, \qquad s \to \infty.
$$

Theorem 2. *If $m^{\#}(\Lambda^{\#}) < \infty$ and the quantization condition (2) holds on $\Lambda^{\#}$, then the function (4) satisfies the equation $\widehat{H}\psi = \lambda\psi + \mathcal{O}(\hbar^2)$ in L^2-norm.*

3. Asymptotic quantization
on general symplectic manifolds

Let Ω be a closed symplectic manifold, ω its symplectic form, and let $\{\cdot,\cdot\}$ denote the Poisson brackets.

Symbols on Ω are the formal series $f = \sum_{k \geq 0} \hbar^k f_k$, where $f_k \in C^\infty(\Omega)$. The deformation quantization in the sense [24] is an associative operation $*$ on a symbol space, such that

$$
f * g = fg - \frac{i\hbar}{2}\{f,g\} + \mathcal{O}(\hbar^2), \qquad 1 * f = f * 1 = f.
$$

Theorem 3. (a result of De Wilde and Lecomte [31]). *The deformation quantization exists on each closed symplectic manifold.*

The second step in quantization is constructing the wave-packets space $\Gamma(\Omega)$ and irreducible representing the symbol algebra on this space: $f \to \widehat{f}$.

It can be done as follows. Let $\{\mathcal{U}_\varepsilon\}$ be an atlas on Ω with maps $\gamma_\varepsilon : \mathcal{U}_\varepsilon \to \mathbb{R}^{2n}_{x,p}$ for which $\gamma_\varepsilon^*(dp \wedge dx) = \omega\big|_{\mathcal{U}_\varepsilon}$. Then there are "glueing mappings" $\gamma_{\delta\varepsilon} = \gamma_\delta \cdot (\gamma_\varepsilon)^{-1}$. We can use the projective representation $\gamma \to T(\gamma)$ in order to quantize these glueings, and T can be constructed by means of the Maslov canonical operator or, as we know now, by the formula (3). Then we get the glueing operators

$$T_{\delta\varepsilon} = T(\gamma_{\delta\varepsilon}) \cdot (I + \hbar d_{\delta\varepsilon} + \hbar^2 \ldots),$$

where the corrections to the terms smaller with respect to the parameter $\hbar$ are taken so that the cocycle conditions

$$T_{\varepsilon\beta}T_{\beta\delta}T_{\delta\varepsilon} = e^{iC_{\beta\delta\varepsilon}} \cdot I + \mathcal{O}(\hbar^\infty), \quad T_{\varepsilon\delta}T_{\delta\varepsilon} = I + \mathcal{O}(\hbar^\infty)$$

hold. Here $\{C_{\beta\delta\varepsilon}\}$ is the 2-cocycle on Ω the cohomology class of which is $[\omega]/\hbar - \pi c_1$ (c_1 is the first Chern class). If this class is integer 2π, then we can take the operators $T_{\delta\varepsilon}$ for glueing sheaves of wave-packets under Ω.

The next theorem was proved in [25].

Theorem 4. *Let the quantization condition*

$$\frac{1}{2\pi\hbar}[\omega] - \frac{1}{2}c_1 \in H^2(\Omega, \mathbb{Z}) \tag{5}$$

holds on Ω. Then there exists a sheaf $\Gamma(\Omega)$ and representation $f \to \widehat{f}$ of the symbol algebra in $\mathrm{Hom}\Gamma(\Omega)$.

If Ω is provided with a polarization, then the representation $f \to \widehat{f}$ of symbols f, linear along the polarization, coincides with the Souriau–Kostant–Kirillov geometric quantization.

4. Index of two-dimensional surfaces
and invariant versions of the basic formula

As it is known from Maslov theory, the quantization condition has the form (2) on each Lagrangian submanifold $\Lambda \subset \mathbb{R}^{2n}$. But if we have a submanifold $\Lambda \subset \Omega$ in a general symplectic manifold Ω, then the question arises: how to write the quantization condition on Λ?

Let Ω be simply-connected. Then the Stokes theorem allows to write the first summand in (2) as $\oint \Theta \to \int_\Sigma \omega$, where Σ are two-dimensional surfaces drawn on 1-cycles on Λ, i.e., $\partial\Sigma \in H_1(\Lambda)$. Thus the problem is to find a two-dimensional analogue of the Maslov class $[\mu]$ in (2). This problem appears in the work [25].

As it turns out (see [26,27]), the quantization condition in the general case is

$$\frac{1}{2\pi\hbar}\int_\Sigma \omega - \frac{1}{4}\mathrm{Ind}_\Lambda(\Sigma) \in \mathbb{Z}, \quad \forall\Sigma \quad \partial\Sigma \subset \Lambda, \tag{6}$$

where Ind_Λ is an integer homologic invariant. If the surface Σ is closed, $\partial\Sigma = \emptyset$, then the number $\mathrm{Ind}_\Lambda(\Sigma)$ coincides with the double value of the Chern class on Σ and condition (6) transforms into (5).

If the manifold Ω possesses a real polarization, then the Maslov class is defined on Λ and $\mathrm{Ind}_\Lambda(\Sigma)$ coincides with the value of this class on the boundary $\partial\Sigma \subset \Lambda$. Thus, in this case the condition (6) tranforms into (2).

Now we can ask another question: what is the Arnold form μ on general Lagrangians $\Lambda \subset \Omega$?

Let $\prod$ be a real polarization and L a strictly positive complex polarization on Ω (see [28–30]). Consider the projection $l(\alpha) : \prod(\alpha) \to (T_\alpha\Lambda)^{\mathbb{C}}$ along $L(\alpha)$ at each point $\alpha \in \Lambda$. Then $\eta = dl \cdot l^{-1}$ is the connection form in the bundle $(T\Lambda)^{\mathbb{C}}$. A generalized Arnold form on Λ is defined as follows

$$\mu = \frac{2}{\pi}\mathrm{trace}(Im\,\eta).$$

Certainly, the class $[\mu]$ is a Maslov class on Λ; as we see it is independent of the polarizations $\prod$ and L if Ω is simply connected.

Now we shall construct an analogue of the wave function (1) under Ω. We suppose that Ω is simply connected and has a global positive complex polarization L. Let $\Lambda \subset \Omega$ be Lagrangian and dm be a measure on Λ. We consider a domain $\mathcal{U}_\varepsilon \subset \Omega$ and denote $\Theta_\varepsilon = \xi_\varepsilon^*(pdx)$. Introduce a local real polarization $\prod_\varepsilon = \xi_\varepsilon^{-1}(\mathbb{R}_p^n)$ on $\mathcal{U}_\varepsilon$. By means of those objects we determine a mapping $l = l_\varepsilon$ and the Arnold form $\mu = \mu_\varepsilon$ as above. We also have $dm(\alpha) = \kappa_\varepsilon(\alpha)d\alpha_1 \cdots d\alpha_n$ in local coordinates on Λ. Fix a certain point $\alpha_\varepsilon \in \Lambda \cap \mathcal{U}_\varepsilon$, real number C_ε, and determine the function

$$(\chi_L(\alpha))_\varepsilon(x) = \frac{e^{iC_\varepsilon}}{\sqrt{\kappa_\varepsilon(\alpha) \cdot |\det l_\varepsilon(\alpha)|}} \cdot \exp\left\{i\int_{\Gamma(\alpha,x)}\left(\frac{\Theta_\varepsilon}{\hbar} - \frac{\pi}{2}\mu_\varepsilon\right)\right\}, \tag{7}$$

where the path $\Gamma(\alpha, x)$ goes from α_ε to α on Λ, and then it goes in the plane $L(\alpha)$ up to the point which is projected into $x \in \mathbb{R}^n$ along $\prod_\varepsilon$ (see the figure).

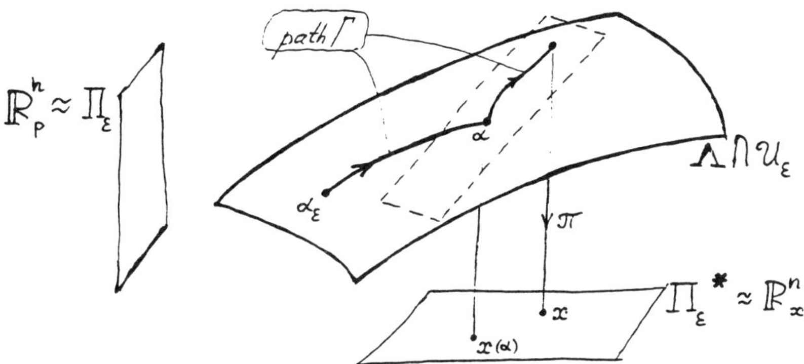

For each $f \in C^\infty(\Omega)$ denote by $\mathcal{V}_f$ the complex vector field on Λ defined by the formula

$$\mathcal{V}_f(\varphi) = <df|_\Lambda, \pi^{-1}l^*d\varphi>, \quad \varphi \in C_0^\infty(\Lambda)$$

Denote by $^m\mathcal{V}_f$ the operator transposed with respect to the measure dm, and put $b_f = \frac{1}{2}(\mathcal{V}_f - ^m\mathcal{V}_f)$.

Theorem 5. *If there exists a strictly positive complex polarization L under Λ, and the quantization conditions (6) hold on Λ. Then there is a cochain $\{c_\varepsilon\}$ such that functions (7) generate up to $\mathcal{O}(\hbar)$ a global section $\chi_L(\alpha)$ of the sheaf $\Gamma(\Omega)$ for each $\alpha \in \Lambda$. In this case the operator $K : C_0^\infty(\Lambda) \to \Gamma(\Omega)$ defined by*

$$K(\varphi) \overset{def}{=} (2\pi i\hbar)^{-n/2} \int_\Lambda \varphi(\alpha)\chi_L(\alpha)dm(\alpha) \tag{8}$$

satisfies the commutation formula

$$\widehat{f}K(\varphi) = K(f|_\Lambda \cdot \varphi) - i\hbar K(b_f\varphi) + \mathcal{O}(\hbar^2). \tag{9}$$

In particular, if $f|_\Lambda = \lambda = $const, then $\mathcal{V}_f = ad(f)|_\Lambda$. If, moreover, the measure dm is $ad(f)$-invariant and Λ is compact, then the section $\psi = K(1)$ is the asymptotic eigenvector of the operator $\widehat{f}$ in the sheaf $\Gamma(\Omega) : \widehat{f}\psi = \lambda\psi + \mathcal{O}(\hbar^2)$.

Remark. Formula (8) is an invariant form of writing (1) and the family of wave-packets $\chi_L(\alpha)$ is an analogue of those coherent states which are under the integral in (1). If Ω is a coadjoint orbit of a certain Lie group, then χ_L gives mod $\mathcal{O}(\hbar)$ the "Lagrangian version" of the coherent states of this group corresponding to Ω, and is closely connected with the Blattner kernels in geometric quantization [29].

The geometric objects which appear in our quantization scheme are not only Lagrangians Λ with measure dm, but also a polarization L over Λ and a connection on L.

If we restrict the accuracy to $\mathcal{O}(\hbar^2)$, then the choice of polarization and connection is not essential. But this plays an important role in the next term $\sim \hbar^2$ in (9).

Theorem 6. *Let Λ be compact, Lagrangian, $f|_\Lambda = \lambda =const$. Let the measure dm be $ad(f)-invariant$, the strictly positive polarization $L = \{L(\alpha)|\alpha \in \Lambda\}$ be supplied with the Hermitian connection and the Hamiltonian flow γ_f^t be geodesic on L. Then formula (9) is changed as follows*

$$\widehat{f}K(\varphi) = \lambda K(\varphi) - i\hbar K((ad(f) - i\hbar c_f)\varphi) + \mathcal{O}(\hbar^3),$$

where

$$c_f = \mathcal{W}_f + <\rho, ad(f)>,$$

$\mathcal{W}_f$ is a vector field on Λ, ρ is a certain closed 1-form on Λ. If, instead of (2) the quantization condition

$$\frac{1}{2\pi\hbar}[\Theta] - \frac{1}{4}[\mu] + \frac{\hbar}{2\pi}[\rho] \in H^1(\Lambda, \mathbb{Z})$$

holds on Λ, then the function $\psi = K(exp\{i\hbar \int_{\alpha_0}^{\alpha} \rho\})$ satisfies the relation $\widehat{f}\psi = \lambda\psi + \mathcal{O}(\hbar^3)$.

Note that the class $[\rho]$ is defined by Λ, L, and by the connection on L. It is an analogue of the invariant constructed in [16] for the symplectic germ under the isotropic submanifold.

REFERENCES

1. V. M. Babich, and V. S. Buldirev, *Asymptotic methods in problems of short wave diffraction.* Moscow, 1972 (in Russian).
2. V. P. Maslov, *Operational methods,* 1973, Moscow, (in Russian).
3. A. Melin, and J. Sjöstrand, *Fourier integral operators with complex-valued phase functions.* Lect. Notes Math., **459** (1975), 121–223.
4. V. P. Maslov, *Complex WKB-method in nonlinear equations,* 1976, Moscow, (in Russian).
5. A. S. Mishenko, B. Yu. Sternin, and V. E. Shatalov, *Lagrangian manifolds and methods of a canonical operator.* Moscow, 1978 (in Russian).
6. S. T. Ali, and G. G. Emch, *Geometric quantization: modular reduction theory and coherent states.* J. Math. Phys., **27**(12), (1986) 2936–2943.
7. G. M. Tuynman, *Generalized Bergman kernels and geometric quantization.* J. Math. Phys., **28**(3), (1987), 573–583.

8. V. M. Babich, *Multidimensional WKB-method and ray method: their analogues and generalizations.* Modern Probl. Math., **34** (1988), Moscow, (in Russian).

9. J. R. Klauder, *Global uniform, semiclassical approximation to wave equations.* Phys. Rev. Lett., **56**(9), (1986), 897–899.

10. S. M. McDonald, *Phase-space eikonal method for treating wave equations.* Phys. Rev. Lett., **54**(9), (1985), 1211–1214.

11. M. V. Karasev, *Problems in operator methods,* 1979, Moscow, MIEM, (in Russian).

12. A. Voros, and Wentzel–Kramers–Brillouin, *Method on the Bargmann representation.* Phys. Rev., A., **40**(12), (1989), 6814–6825.

13. J. Kurchan, P. Leboeuf, and M. Saraceno, *Semiclassical approximation in the coherent state representation.* Phys. Rev. A., **40**(12), (1989), 6800–6814.

14. A. Laptev and Yu. G. Safarov, *Global parametrization of Lagrangian manifolds and Maslov index.* Preprint of Linköping Univ., LITH - Mat - R -**89** -31, 13 pp.

15. M. V. Karasev, *Connections on Lagrangian submanifolds and some problems in quasiclassical approximation.* I. Zapiski Semin. LOMI, Leningrad, **172**, (1989), 41–54, (in Russian).

16. M. V. Karasev, *To the Maslov theory of quasi-classical asymptotics. Examples of new global quantization formula applications.* Preprint Inst. Theor, Phys., Kiev, ITP-89-78E, 32 pp.

17. M. V. Karasev, *New global asymptotics and anomalies in problem of quantization of adiabatic invariant.* Funct. Anal. and Priloz., **24**(2), (1990), (in Russian).

18. M. V. Karasev, and Yu. M. Vorobjev. *Integral representations of wave functions, generated by connection in the germ of Poisson structure.* Proc. XVIII Intern, Colloq. Theor. group methods in physics, Moscow, June 4–9 1990, Lect Notes Phys., 1991.

19. V. I. Arnold. *About characteristic classes in quantization conditions.* Funct. Anal. and Priloz., **1**(1), (1967), 1–14, (in Russian).

20. V. P. Maslov. *Perturbation theory and asymptotic methods.* Moscow Univ., 1965 (in Russian).

21. J. M. Souriau, *Structure des systèmes dynamiques.* Dunod, Paris, 1970.

22. J. K. Moser, Comm. Pure Appl. Math., **23** (1970), 609–636.

23. P. Kustaanheimo, *Spinor regularization of Kepler motion.* Ann. Univ. Turkuensis A., **73**(1), (1976), 3–7.

24. F. Bayen, M. Flato, C. Fronsdal, A. Lichnerowicz, and D. Sternheimer, *Deformation theory and quantization.* Ann. Phys., **111**(1), (1978), 61–151.

25. M. V. Karasev, and V. P. Maslov, *Pseudodifferential operators and canonical operator in general symplectic manifolds.* Investiya Akad. Nauks SSR, **47**(5), (1983), 999–1029, (in Russian).

26. M. V. Karasev, *Asymptotics of spectrum of mixed states for equations of selfconsistent field.* Theor. Mat. Phys., **61**(1), (1984), 118–127, (in Russian).

27. M. V. Karasev, *Poisson algebras of symmetries and asymptotics of spectral series.* Funct. Anal. and Priloz., **20**(1), (1986), 21–32, (in Russian).

28. B. Kostant, *Quantization and unitary representations.* Lect. Notes Math., **170** (1970), 87–208.

29. R. J. Blattner, *The metalinear geometry of non-real polarizations.* Lect. Notes Math., **570** (1977), 11–45.

30. A. A. Kirillov, *Geometric quantization.* Modern Probl. Math., Fundam. Directions, VINITI, (4), (1985), 141–178, Moscow, (in Russian).

31. M. De Wilde and P.B. Lecomte, *Existence of Star Products and of formal deformation of the Poisson Lie Algebra of arbitrary symplectic manifolds,* Lett. Math. Phys ,**7** (1983), 487–496.

MIEM
Moscow, USSR

Les groupes kählériens

ANDRÉ LICHNEROWICZ

A un grand scientifique et un ami

Abstract

A connected Kählerian Lie group is a product of a flat Kählerian Lie group and of a Kählerian Lie group which is, as Kähler manifold a complex bounded homogeneous domain.

Introduction

En Physique mathématique et Géométrie, les espaces homogènes symplectiques ou kählériens présentent un intérêt particulier sur lequel il est inutile de revenir. Il est donc naturel de s'intéresser aux groupes de Lie connexes admettant une structure symplectique ou kählérienne invariante à gauche. En collaboration avec A. Medina, j'ai rappelé dans [1] que ce problème avait fait depuis longtemps l'objet de recherches et avait conduit à deux intéressants articles [2] et [3] et nous avons commencé une étude géométrique des groupes kählériens.

Récemment, les espaces homogènes kählériens généraux ont fait l'objet d'articles importants de J. Dorfmeister et K. Nakajima [7]. Ces articles résolvent une conjecture faite dès 1967 par E.B. Vinberg et S.G. Gindikin [8] en liaison avec les travaux bien connus de I.I. Piatetski-Shapiro sur les domaines bornés homogènes complexes.

Le but de cet article est de caractériser les variétés des groupes kählériens par des moyens aussi géométriques que possible. J'établis ici que tout groupe kählérien connexe est le produit d'un groupe kählérien plat et d'un groupe kählérien qui, en tant que variété kählérienne, est un domaine borné homogène complexe. On obtient ainsi sous une forme plus précise et plus économique, un résultat qui pourrait s'établir à partir du résultat fondamental de [7].

Dans la partie I, je rappelle brièvement les énoncés nécessaires concernant les groupes de Lie à structure symplectique invariante à gauche (dits groupes symplectiques) et la décomposition de G. de Rham [4] des groupes munis d'une métrique riemannienne invariante à gauche (dits abusivement groupes riemanniens). La caractérisation due à Milnor [5] des groupes riemanniens plats est particulièrement utile.

La partie II porte sur la géométrie des groupes kählériens et conduit d'abord au théorème 1 de décomposition. Les groupes kählériens plats sont caractérisés. Dans la partie III, la proposition 6 concernant les conséquences géométriques, pour une algèbre de Lie kählérienne, de l'existence de deux sous-algèbres kählériennes orthocomplémentaires se révèle un instrument essentiel. Elle permet d'établir que le facteur non plat d'un groupe kählérien simplement connexe est un groupe kählérien qui est un espace d'Einstein-Kähler à courbure scalaire strictement négative. Cette approche permet aussi de simplifier de manière notable certains des raisonnements de [7]. On établit à partir d'un énoncé de [6] que tout groupe de Lie kählérien irréductible est *résoluble*.

On parvient ainsi au théorème 4 de décomposition et à son corollaire, qui constituent le résultat majeur de cet article.

I. PRELIMINAIRES

1. Groupes et algèbres de Lie symplectiques

Soit G un groupe de Lie connexe de dimension m, $\mathcal{G}$ son algèbre de Lie, C le tenseur de structure. Si $X \in \mathcal{G}$, nous notons X_+ (resp. X_-) le champ de vecteur de G invariant à gauche (resp. à droite) correspondant à X. Si $\{\ell_A\}$ $(A, B, \ldots = 1, \ldots, m)$ est une base de $\mathcal{G}$, le tenseur de structure C admet les composantes $\{C_{AB}^C\}$.

a)   Un groupe de Lie est *unimodulaire* s'il admet un élément de volume η biinvariant. Ou bien, pour un groupe de Lie, toute m-forme $\neq 0$ invariante à gauche est invariantement exacte et $\dim H^m(\mathcal{G}) = 0$ ou bien une telle forme n'est pas invariantement exacte et est biinvariante ; dans ce cas $\dim H^m(\mathcal{G}) = 1$ et G ou $\mathcal{G}$ *est unimodulaire*.

On sait que *pour que $\mathcal{G}$ soit unimodulaire, il faut et il suffit que* $\mathrm{Tr}(adX) = 0$ pour tout $X \in \mathcal{G}$. Si $\tau(\mathcal{G})$ est la 1-forme obtenue par contraction du tenseur de structure $(\tau_A = C_{AB}^B)$ il est équivalent de dire que $\tau(\mathcal{G}) = 0$.

b)   Un groupe de Lie connexe G de dimension paire $m = 2n$ admet *une structure symplectique invariante* σ s'il existe sur G une 2-forme fermée invariante à gauche σ de rang $2n$. En particulier G admet *une structure symplectique exacte invariante* s'il existe sur G une 1-forme invariante à gauche ρ telle que $\sigma = d\rho$ soit de rang $2n$. Ces deux propriétés ne dépendent que de l'algèbre de Lie $\mathcal{G}$. Par abus de langage, nous disons que G ou $\mathcal{G}$ est *symplectique* dans le premier cas, *symplectique exact* dans le second.

Un raisonnement cohomologique simple, portant sur la partie semi-simple d'une décomposition de Levi-Malcev de $\mathcal{G}$ permet d'établir ([1] p. 227 d'après [2])

Proposition 1. *Une algèbre de Lie symplectique unimodulaire est nécessairement résoluble. En particulier :*
1) une algèbre de Lie semi-simple ne peut être symplectique,
2) une algèbre de Lie réductive non abélienne ne peut être symplectique.

2. Eléments de géométrie riemannienne sur un groupe de Lie

Soit G un groupe de Lie connexe de dimension m et g une métrique riemannienne *invariante à gauche* sur G. Introduisons une base g-orthogonale $\{e_A\}$ $(A, B, \ldots = 1, \ldots, m)$ de l'algèbre de Lie $\mathcal{G}$ de G et soit $\{\theta^A\}$ la base correspondante de l'espace des 1-formes de G invariantes à gauche. On a :

$$g = g_{AB}\,\theta^A \otimes \theta^B \qquad (g_{AB} = \delta_{AB})$$

et les θ^A vérifient les équations de Maurer-Cartan

$$d\theta^A + \frac{1}{2}C^A_{BC}\,\theta^B \wedge \theta^C = 0 \qquad (2.1)$$

Nous notons $(\omega^A_B) = (\gamma^A_{BC}\theta^C)$ la 1-forme rapportée au corepère $\{\theta^C\}$ de la connexion riemannienne de $(G, g), (R^A{}_{B,CD})$ le tenseur de courbure, R_{AB} le tenseur de Ricci. Si V est un vecteur, nous posons $Ri(V, V) = R_{AB}V^A V^B$. Il résulte des équations de Maurer-Cartan (2.1) et de la nullité de la torsion que les coefficients constants de la connexion riemannienne vérifient :

$$\gamma^A_{BC} - \gamma^A_{CB} = -C^A_{BC} \qquad (\gamma^A_{BC} + \gamma^B_{AC} = 0) \qquad (2.2)$$

b) Soit $\Delta = d\delta + \delta d$ le laplacien de G. de Rham sur les formes, où δ est l'opérateur de codifférentiation. Si ξ est une 1-forme de G, on a :

$$(\Delta\xi)_B = -\nabla^A\nabla_A\xi_B + R_{AB}\xi^A \qquad (2.3)$$

où ∇ est l'opérateur de dérivation covariante dans la connexion riemannienne. Si (,) est le produit intérieur pour les formes :

$$(\Delta\xi, \xi) = -\nabla^A(\xi^B\nabla_A\xi_B) + \nabla^A\xi^B\nabla_A\xi_A\xi_B + Ri(\xi, \xi)$$

Prenons pour ξ une 1-forme harmonique ($\Delta\xi = 0$) invariante à gauche. On a $\xi^B\nabla_A\xi_B = 0$ et on obtient :

$$(\nabla\xi, \nabla\xi) + Ri(\xi, \xi) = 0$$

On voit que :

248 A. Lichnerowicz

Proposition 2. *Si (G,g) a une courbure de Ricci ≥ 0, toute 1-forme harmonique invariante à gauche de (G,g) est parallèle.*

En particulier $\tau(\mathcal{G})$ définit sur G une 1-forme fermée invariante à gauche notée τ. On a $\delta\tau = $ const. et τ est une forme harmonique. De plus on déduit de (2.2) que $\tau_A = C^B_{AB} = -\gamma^B_{AB}$. Il en résulte :

$$\delta\tau = (\tau, \tau)$$

Si $\delta\tau = 0$ on a $\tau = 0$ et G est unimodulaire.

Corollaire. *Si (G,g) a une courbure de Ricci ≥ 0, G est unimodulaire.*

c) Soit α (resp β) une p-forme (resp. $(p+1)$-forme) invariante à gauche de G. On a

$$(d\alpha, \beta) - (\alpha, \delta\beta) = \delta\lambda$$

où λ est la 1-forme donnée par $\lambda_B = -(1/p!)\alpha^{A_1 \cdots A_p}\beta_{BA_1 \cdots A_p}$.

Supposons G unimodulaire. Toute $(m-1)$-forme invariante à gauche est fermée et, par adjonction, toute 1-forme invariante à gauche est cofermée. On en déduit

$$(d\alpha, \beta) - (\alpha, \delta\beta) = 0 \tag{2.4}$$

Soit E l'espace des formes invariantes à gauche de G. D'après (2.4), dE et δE sont orthogonaux et une p-forme invariante à gauche vérifie $d\alpha = 0$, $\delta\alpha = 0$ et est orthogonale à dE et δE. Il en résulte que, *pour un groupe unimodulaire, on a une théorie harmonique pour les formes invariantes à gauche* [3]. En particulier une forme harmonique invariante à gauche de (G,g) qui est invariantement exacte est identiquement nulle.

3. Décomposition riemannienne d'un groupe de Lie et applications

a) Soit (G,g) un groupe de Lie *simplement connexe* muni d'une métrique g invariante à gauche. Introduisons la décomposition de G. de Rham de (G,g) en tant que variété riemannienne ; (G,g) étant complète, on a

$$(G,g) = (G_0, g_0) \times (G_1, g_1) \times \ldots \times (G_r, g_r) \tag{3.1}$$

où les (G_k, g_k) $(k = 0, 1, \ldots, r)$ sont des variétés riemanniennes telles que (G_0, g_0) admet un groupe d'holonomie réduit à $\{e\}$ et les (G_k, g_k) $(k = 1, 2, \ldots, r)$ ont des groupes d'holonomie irréductibles. Il résulte d'un théorème de Nomizu [4] que $(G_0, g_0), \ldots, (G_r, g_r)$ sont des groupes de Lie munis de métriques invariantes à gauche ; g_0 étant plate, G_0 est *unimodulaire* d'après le corollaire du paragraphe 2b.

b) *Supposons que (G, g) soit résoluble et admette une courbure de Ricci*
≥ 0. Dans la décomposition précédente, chaque facteur est résoluble et à
courbure de Ricci ≥ 0, donc unimodulaire.

Etudions par exemple (G_1, g_1) résoluble et unimodulaire ; G_1 étant
résoluble, on a dim $H^1(G_1) = \dim \mathcal{G}_1 - \dim \mathcal{G}_1' \neq 0$; G_1 étant unimodulaire, $H^1(G_1)$ est donné par les 1-formes fermées invariantes à gauche. Soit
$\varphi \neq 0$ une telle forme ; φ est harmonique, donc parallèle, ce qui contredit
l'irréductibilité. Les facteurs irréductibles ne peuvent exister.

Proposition 3. *Soit (G, g) un groupe résoluble muni d'une métrique
g invariante à gauche à courbure de Ricci ≥ 0. La métrique g est alors
nécessairement plate.*

c) Soit (G, g) un groupe de Lie admettant une métrique plate g invariante
à gauche. Pour abréger, nous dirons que le groupe G est *plat*. Ces groupes
ont été caractérisés par Milnor [5]. Si $X \in \mathcal{G}$ nous dirons que $ad(X)$ est
antiadjointe si cette transformation linéaire vérifie

$$(ad(X)Y, Z) + (Y, ad(X)Z) = 0 \tag{3.2}$$

Milnor a établi que G *est résoluble* et qu'on a par suite

Proposition 4. *Un groupe de Lie G admettant une métrique plate g
invariante à gauche a une algèbre de Lie qui est le produit semi-direct d'un
idéal abélien qui est son idéal dérivé $\mathcal{G}'$ et d'une algèbre de Lie abélienne
$\mathcal{H}$. L'algèbre de Lie $\mathcal{H}$ est l'orthocomplément de $\mathcal{G}'$ et pour tout $X \in \mathcal{H}$, la
transformation linéaire $ad(X)$ est antiadjointe.*

*Inversement, supposons $\mathcal{G}$ telle que $\mathcal{G} = \mathcal{G}' + \mathcal{H}$ où $\mathcal{G}'$ et $\mathcal{H}$ sont abéliens.
Donnons-nous une métrique plate sur $\mathcal{H}$ et supposons qu'il existe sur $\mathcal{G}'$
une métrique plate telle que, pour tout $X \in \mathcal{H}, ad(X)$ soit antiadjointe.
L'ensemble de ces deux métriques définit sur G une métrique plat invariante
à gauche.*

Supposons G munie d'une métrique plate g. On a $\mathcal{G} = \mathcal{G}' + \mathcal{H}$. Si $\{e_a\}$
$(a, b, \ldots = 1, \ldots, p)$ est une base orthonormée de $\mathcal{H}$, $\{e_i\}$ $(i, j, \ldots = 1, \ldots, q)$
une base orthonormée de $\mathcal{G}'$, $\{e_A\} = \{e_a, e_i\}$ est une base orthonormée de
$\mathcal{G}$. On a

$$[e_a, e_b] = 0 \qquad [e_i, e_j] = 0 \qquad [e_a, e_i] = C_{ai}^j e_j$$

et la relation (3.2) s'exprime par

$$C_{ai}^j + C_{aj}^i = 0 \tag{3.3}$$

Soit $\{\theta^A\} = \{\theta^a, \theta^i\}$ la base correspondante à $\{e_A\}$ de l'espace des
1-formes de G invariantes à gauche de telle sorte que :

$$g = g_{ab}\,\theta^a \otimes \theta^b + g_{ij}\,\theta^i \otimes \theta^j \qquad (g_{AB} = \delta_{AB})$$

Il résulte de (2.2), avec des notations évidentes :

$$2\gamma_{A,BC} = -C_{A,BC} - C_{B,CA} + C_{C,AB} \qquad (3.4)$$

On en déduit :

$$\gamma_{bc}^{a} = \gamma_{jk}^{i} = 0 \qquad\qquad \gamma_{bi}^{a} = \gamma_{jb}^{a} = \gamma_{ab}^{i} = 0 \qquad (3.5)$$

D'après (3.3), on a $\nabla X_{+} = 0$ pour $X \in \mathcal{H}$ et :

$$\gamma_{aj}^{i} = \gamma_{ij}^{a} = 0 \qquad\qquad \gamma_{ja}^{i} = C_{aj}^{i} \qquad (3.6)$$

II. ETUDE GEOMETRIQUE DES GROUPES KÄHLERIENS

4. Notion de groupe de Lie kählérien et d'algèbre de Lie kählérienne

a) Soit (G, σ) un groupe symplectique de dimension $m = 2n$ et 2-forme σ. Nous notons encore σ la 2-forme de rang $2n$ induite sur $\mathcal{G}$ par la forme symplectique ; son caractère fermé se traduit par :

$$\sigma([Y,Z],X) + \sigma([Z,X],Y) + \sigma([X,Y],Z) = 0 \qquad (X,Y,Z \in \mathcal{G}) \qquad (4.1)$$

Considérons sur $\mathcal{G}$ un opérateur j de structure complexe ($j^{2} = -\mathrm{Id}$) qui définisse sur $\mathcal{G}$ une métrique g donnée par

$$g(X,X) = \sigma(jX,X) \qquad\qquad g(X,X) > 0 \text{ pour } X \neq 0 \qquad (4.2)$$

Nous obtenons sur G une métrique g et une structure presque complexe $\mathcal{J}$ invariantes à gauche telles que $(g, \sigma, \mathcal{J})$ définisse sur G une structure *presque kählérienne* ; (G, σ, J) est dit *un groupe presque kählérien*, $(\mathcal{G}, \sigma, j)$ une *algèbre de Lie presque kählérienne*.

Pour que j définisse sur G une structure complexe intégrable $\mathcal{J}$, on sait qu'il faut et il suffit que l'on ait

$$[jX, jY] - [X, Y] = j[jX, Y] + j[X, jY] \qquad (X, Y \in \mathcal{G}) \qquad (4.3)$$

S'il en est ainsi $(G, g, \sigma, \mathcal{J})$ est une variété hermitienne à 2-forme fondamentale fermée et est donc kählérienne ; $(G, \sigma, \mathcal{J})$ admet une structure kählérienne invariante à gauche. Pour abréger $(G, \sigma, \mathcal{J})$ est dit *un groupe de Lie kählérien*. De même $(\mathcal{G}, \sigma, j)$ presque kählérienne est dite, si (4.3) est satisfaite, *une algèbre de Lie kählérienne*.

Soit $\mathcal{G}^c$ le compléxifié de $\mathcal{G}$. Si L^c et $\bar{L}^c$ sont les espaces propres de j, nous introduisons une base $\{e_\rho\}$ de L^c et la base complexe conjuguée $\{e_{\bar\sigma}\}$ de $\bar{L}^c$ $(\rho, \sigma, \ldots = 1, \ldots, n \; ; \bar\rho = \rho + n)$ et nous obtenons une base $\{e_A\} = \{e_\rho, e_{\bar\sigma}\}$ de $\mathcal{G}^c$ $(A, B, \ldots = 1, \ldots, 2n)$ pour laquelle j a pour composantes $j_\sigma^\rho = i\delta_\sigma^\rho$, $j_{\bar\sigma}^\rho = 0$; nous rapportons le tenseur de structure C de G à cette base. Nous notons $\{\theta^A\} = \{\theta^\rho, \theta^{\bar\sigma}\}$ les 1-formes à valeurs complexes invariantes à gauche de G correspondantes ; pour le corepère correspondant $\mathcal{J}$ a les mêmes composantes que j, g et σ pouvant s'écrire

$$g = 2g_{\rho\bar\sigma}\, \theta^\rho \otimes \theta^{\bar\sigma} \qquad\qquad \sigma = ig_{\rho\bar\sigma}\, \theta^\rho \wedge \theta^{\bar\sigma} \qquad\qquad (4.4)$$

où l'on peut supposer $g_{\rho\bar\sigma} = 0$ pour $\sigma \neq \rho$ et $g_{\rho\bar\rho} = 1$. Il résulte des équations de Maurer-Cartan que, pour que le système $\{\theta^\rho = 0\}$ soit intégrable, il faut et il suffit que

$$C_{\bar\sigma\bar\tau}^\rho = 0 \qquad\qquad (4.5)$$

ce qui est équivalent à (4.3).

b) Soit $(G, g, \mathcal{J})$ un groupe kählérien. Considérons la 1-forme $\{\omega_B^A\}$, rapportée au corepère $\{\theta^A\} = \{\theta^\rho, \theta^{\bar\sigma}\}$ de la connexion kählérienne ; (G, g) étant kählérien $\{\omega_B^A\} = \{\omega_\sigma^\rho, \omega_{\bar\sigma}^{\bar\rho}\}$. Posons $\omega_B^A = \gamma_{BC}^A \theta^C$; d'après (2.2), (3.4) les coefficients constants de la connexion vérifient

$$\gamma_{\sigma\tau}^\rho - \gamma_{\tau\sigma}^\rho = -C_{\sigma\tau}^\rho \qquad \gamma_{\sigma\bar\tau}^\rho = -C_{\sigma\bar\tau}^\rho \qquad \gamma_{\bar\sigma\tau}^\rho = \gamma_{\bar\sigma\bar\tau}^\rho = 0 \qquad (4.6)$$

Soit α la 2-forme de Ricci de la variété kählérienne $(G, g, \mathcal{J})$. Comme $R=$const., cette 2-forme est fermée et cofermée. Evaluons α d'après (4.6) : si $\{\Omega_\sigma^\rho\}$ est la 2-forme de courbure :

$$\alpha = i\Omega_\rho^\rho = d(i\omega_\rho^\rho)$$

Il résulte de (4.6) que :

$$\omega_\rho^\rho = C_{\bar\rho\sigma}^{\bar\rho}\, \theta^\sigma - C_{\rho\bar\sigma}^\rho\, \theta^{\bar\sigma}$$

Si ψ est la 1-forme réelle invariante à gauche vérifiant $\psi_\sigma = iC_{\bar\rho\sigma}^{\bar\rho}$, on a $\alpha = d\psi$. Il vient

Proposition 5. *La 2-forme de Ricci α d'un groupe kählérien G est fermée et cofermée et elle est invariantement exacte. Si G est unimodulaire, sa courbure de Ricci est nulle.*

Inversement *si un groupe kählérien $(G, g, \mathcal{J})$ a une courbure de Ricci ≥ 0, G est unimodulaire* d'après le corollaire du paragraphe 2b, *résoluble* d'après la proposition 1. Ainsi d'après la proposition 3, *sa métrique est nécessairement plate*. Nous dirons que *le groupe kählérien est plat*.

c) Soit $(G, g, \mathcal{J})$ un groupe kählérien *simplement connexe*. On déduit de résultats classiques concernant les variétés kählériennes que sa décomposition de G. de Rham (3.1) est une décomposition en un produit de groupes kählériens. Considérons un des facteurs irréductibles du produit $(G_1, g_1, \mathcal{J}_1)$: s'il est unimodulaire, sa courbure de Ricci est nulle d'après la proposition 5. Or nous avons vu que si une telle courbure est nulle, la métrique est plate, ce qui contredit l'irréductibilité. On a

Théorème 1. *Soit $(G, g, \mathcal{J})$ un groupe kählérien simplement connexe ; $(G, g, \mathcal{J})$ est en tant que variété riemannienne, un produit (3.1) de groupes kählériens tels que $(G_0, g_0, \mathcal{J}_0)$ soit plat et unimodulaire et que $(G_k, g_k, \mathcal{J}_k)$ soit riemanniennement irréductible, non unimodulaire et à courbure de Ricci non nulle.*

Nous allons étudier ces deux types de facteurs.

5. Etude des groupes kählériens plats

a) Soit $(G, g, \mathcal{J})$ un groupe kählérien plat et reprenons les notations du paragraphe 3. On a $\mathcal{G} = \mathcal{G}' + \mathcal{H}$, où $\mathcal{G}'$ est l'idéal dérivé de $\mathcal{G}, \mathcal{H}$ son orthocomplément et où $\mathcal{G}'$ et $\mathcal{H}$ sont des algèbres de Lie abéliennes. Pour $X \in \mathcal{H}$, on a d'après (3.5), (3.6) $\nabla X_+ = 0$. Par suite pour $X \in \mathcal{H}, Y \in \mathcal{G}$

$$[X_+, Y_+] = \nabla_{X_+} Y_+$$

Il en résulte :

$$\mathcal{J}[X_+, Y_+] = \nabla_{X_+}(\mathcal{J}Y_+) = [X_+, \mathcal{J}Y_+]$$

Ainsi pour $X, Y \in \mathcal{G}$, on a :

$$j[X, Y] = [X, jY] \tag{5.1}$$

Cette relation implique (4.3). D'après (5.1), on voit que j opère sur $\mathcal{G}'$ et $\mathcal{J}$ induit sur le sous-groupe G' correspondant une structure complexe. De même g étant hermitienne, j opère sur $\mathcal{H}$ et $\mathcal{J}$ induit sur le sous-groupe H correspondant une structure complexe. Ainsi les groupes H et G' sont des groupes commutatifs kählériens, donc de dimensions paires respectivement égales à $2p'$ et $2q'$.

b) Inversement, soit (G, g) un groupe "riemannien" plat avec

$$\dim \mathcal{H} = 2p' \qquad \dim \mathcal{G}' = 2q'$$

Choisissons sur $\mathcal{H}$ une structure complexe arbitraire $j_{\mathcal{H}}$ et sur $\mathcal{G}'$ une structure complexe $j_{\mathcal{G}'}$ vérifiant

$$j_j^k C_{ak}^i - j_k^i C_{aj}^k = 0 \qquad\qquad (5.2)$$

relation qui traduit (5.1) pour la structure $j = (j_{\mathcal{H}}, j_{\mathcal{G}'})$. Un calcul direct montre que l'opérateur $\mathcal{J}$ de structure complexe intégrable sur G correspondant à j vérifie $\nabla \mathcal{J} = 0$.

$\mathcal{J}$ et la métrique h donnée par :

$$h(X, Y) = (1/2)[g(X, Y) + g(jX, jY)] \qquad\qquad (5.3)$$

définissent sur G une métrique kählérienne plate invariante à gauche.

Théorème 2. *Un groupe de Lie $(G, g, \mathcal{J})$ kählérien plat a une algèbre de Lie $\mathcal{G}$ qui est le produit semi-direct d'un idéal abélien kählérien, l'idéal dérivé G', et d'une algèbre de Lie abélienne kählérienne $\mathcal{H}$. L'algèbre de Lie $\mathcal{H}$ est l'orthocomplément de $\mathcal{G}'$ et l'on a pour tout $X \in \mathcal{H}$:*

$$j[X, Y] = [X, jY] \qquad\qquad (X \in \mathcal{H}, Y \in \mathcal{G}')$$

Inversement soit (G, g) un groupe riemannien plat d'algèbre de Lie $\mathcal{G} = \mathcal{G}' + \mathcal{H}$, où $\mathcal{G}'$ et $\mathcal{H}$ sont de dimensions paires. Donnons-nous sur $\mathcal{H}$ un opérateur de structure complexe $j_{\mathcal{H}}$ et supposons qu'il existe sur $\mathcal{G}'$ un opérateur de structure complexe $j_{\mathcal{G}'}$ tel que :

$$j_{\mathcal{G}'}[X, Y] = [X, j_{\mathcal{G}'}Y] \qquad\qquad (X \in \mathcal{H}, Y \in \mathcal{G}')$$

L'ensemble $(j_{\mathcal{H}}, j_{\mathcal{G}'})$ définit sur G une structure complexe $\mathcal{J}$ invariante à gauche, à dérivée covariante nulle ; h donnée par (5.3) et $\mathcal{J}$ définissent G comme groupe kählérien plat.

c) L'exemple le plus simple d'algèbre de Lie d'un groupe kählérien plat non abélien est donné par l'algèbre de Lie $\mathcal{G}_4$ de dimension 4, engendrée par X, Y, Z, T vérifiant

$$[X, Y] = 0 \qquad [Z, X] = Y \qquad [T, X] = -Y$$
$$[Z, T] = 0 \qquad [Z, Y] = -X \qquad [T, Y] = X$$

$\mathcal{G}_4'$ abélien est donné par X, Y, $\mathcal{H}$ abélienne par Z, T. On voit facilement que $\mathcal{G}_4'$ est l'algèbre de Lie d'un groupe riemannien plat admettant la métrique :

$$g = (\omega^1)^2 + (\omega^2)^2 + (dx^3)^2 + (dx^4)^2$$

où $(\omega^1, \omega^2, dx^3, dx^4)$ est la base de l'espace des 1-formes réelles invariantes correspondant à la base (X, Y, Z, T) de $\mathcal{G}_4$. Introduisons j donné par :

$$jX = Y \qquad\qquad jZ = T$$
$$jY = -X \qquad\qquad jT = -Z$$

On vérifie immédiatement que (5.1) est satisfaite ; g et j définissent $\mathcal{G}_4$ comme algèbre de Lie kählérienne plate.

III. GROUPES KÄHLERIENS IRREDUCTIBLES ET APPLICATIONS

6. Formules liées à l'existence d'une sous-algèbre de Lie kählérienne. Conséquences

Soit $(G, g, \mathcal{J})$ un groupe kählérien simplement connexe, $(\mathcal{G}, g, j)$ son algèbre de Lie munie de la métrique g et de j, algèbre qui est dite kählérienne. Soit $\mathcal{H}$ une sous-algèbre de Lie de $\mathcal{G}$ non triviale telle que $j\mathcal{H} \subset \mathcal{H}$. Une telle sous-algèbre est dite une *sous-algèbre kählérienne*. La notion *d'idéal kählérien* est définie de manière analogue.

a) Soit $\mathcal{H}$ une sous-algèbre de Lie kählérienne de $(\mathcal{G}, g, j)$ de dimension $2q < 2n$, $\mathcal{K}$ l'orthocomplément de $\mathcal{H}$ de dimension $2p = 2n - 2q$, par rapport à g. A partir d'une base $\{e_a\}$ $(a, b, \ldots = 1, \ldots, 2p)$ de $\mathcal{K}$ et d'une base $\{e_i\}$ $(i, j, \ldots = 2p+1, \ldots, 2n)$ de $\mathcal{H}$, on obtient une base $\{e_a, e_i\}$ de $\mathcal{G}$ telle que :

$$[e_a, e_b] = C_{ab}^c e_c + C_{ab}^i e_i, \ [e_a, e_i] = C_{ai}^c e_c + C_{ai}^j e_j, \ [e_i, e_j] = C_{ij}^k e_k \qquad (6.1)$$

On note $\{\omega^a, \omega^i\}$ la base correspondante de l'espace des 1-formes réelles invariantes à gauche de G. De (6.1) et des équations de Maurer-Cartan, il résulte que le système $\{\omega^a = 0\}$ est intégrable et définit sur G un feuilletage P invariant à gauche. La feuille de P passant par e coïncide avec le sous-groupe de Lie connexe kählérien H d'algèbre de Lie $\mathcal{H}$. L'opérateur $\mathcal{J}$ de structure complexe opère sur P et induit sur chaque feuille une structure complexe. Nous notons $\{\theta^\rho\} = \{\theta^\alpha, \theta^\lambda\}$ $(\alpha, \beta, \ldots = 1, \ldots, p ;$ $\lambda, \mu, \ldots = p + 1, \ldots n)$ un système de 1-formes complexes invariantes à gauche correspondant à la décomposition $\mathcal{G} = \mathcal{H} \oplus \mathcal{K}$, le corepère $\{\theta^\rho, \theta^{\bar\rho}\}$ étant unitaire. On a :

$$g = 2g_{\alpha\bar\beta}\, \theta^\alpha \otimes \theta^{\bar\beta} + 2g_{\lambda\bar\mu}\, \theta^\lambda \otimes \theta^{\bar\mu} \qquad (6.2)$$

où $g_{\rho\bar\sigma} = 0$ pour $\sigma \neq \rho, g_{\rho\bar\rho} = 1$. Le feuilletage P étant holomorphe, le système $\{\theta^\alpha = 0\}$ est intégrable et l'on a $C_{\lambda\bar\beta}^\alpha = 0$. On en déduit $\gamma_{\lambda\bar\beta}^\alpha = 0$, donc $\gamma_{\alpha\beta}^\lambda = 0$ et il vient

$$C_{\alpha\beta}^\lambda = 0 \qquad (6.3)$$

Les équations de Maurer-Cartan pour les θ^λ s'écrivent d'après (6.3)

$$d\theta^\lambda + C_{\alpha\bar\mu}^\lambda\, \theta^\alpha \wedge \theta^{\bar\mu} + C_{\alpha\bar\beta}^\lambda\, \theta^\alpha \wedge \theta^{\bar\beta} = 0 \qquad (\text{mod } \theta^\mu = 0) \qquad (6.4)$$

b) Supposons que $\mathcal{K}$ soit aussi une *algèbre de Lie*, donc une sous-algèbre kählérienne, de $\mathcal{G}$ orthocomplément de $\mathcal{H}$. Le même raisonnement montre que le système $\{\theta^\lambda = 0\}$ est intégrable et il existe des cartes complexes $\{z^\rho\} = \{z^\alpha, z^\lambda\}$ de domaine U telles que :

$$g|_U = 2g_{\alpha\bar\beta}\, dz^\alpha \otimes dz^{\bar\beta} + 2g_{\lambda\bar\mu}\, dz^\lambda \otimes dz^{\bar\mu} \qquad (6.5)$$

(6.5) étant kählérienne, les $g_{\alpha\bar{\beta}}$ ne dépendent que des $\{z^c\} = \{z^\gamma, z^{\bar{\delta}}\}$ et les $g_{\lambda\bar{\mu}}$ des $\{z^k\} = \{z^\nu, z^{\bar{\rho}}\}$. La métrique g est donc *réductible*.

c) Supposons que $\mathcal{H}$ soit un *idéal kählérien* de $\mathcal{G}$. Dans (6.1), on a $C^c_{ai} = 0$. On en déduit, avec les notations du a, que $C^\alpha_{\beta\bar{\lambda}} = 0$ et $C^\alpha_{\beta\lambda} = 0$. De $C^\alpha_{\beta\bar{\lambda}} = 0$ il vient $\gamma^\alpha_{\beta\bar{\lambda}} = 0$ et par suite $\gamma^\alpha_{\beta\lambda} = 0$. Comme $C^\alpha_{\beta\lambda} = 0$, il vient $\gamma^\alpha_{\lambda\beta} = 0$, soit $\gamma^\lambda_{\alpha\bar{\beta}} = 0$, c'est-à-dire $C^\lambda_{\alpha\bar{\beta}} = 0$. Ainsi $C^i_{ab} = 0$ et $\mathcal{K}$ est sous-algèbre de Lie de $\mathcal{G}$. Nous sommes dans la situation du b et la métrique g est réductible. On a

Proposition 6. *Soit $(\mathcal{G}, g, j)$ une algèbre de Lie kählérienne. Si $\mathcal{H}$ est une sous-algèbre de Lie kählérienne non triviale de $\mathcal{G}$ et si son ortho-complément $\mathcal{K}$ dans $\mathcal{G}$ est aussi une sous-algèbre de Lie kählérienne, le groupe $(G, g, \mathcal{J})$ simplement connexe d'algèbre de Lie $\mathcal{G}$ est réductible en tant que variété kählérienne. Il en est en particulier ainsi si $\mathcal{H}$ est un idéal kählérien non trivial de $\mathcal{G}$.*

7. Un lemme et ses conséquences

a) Soit $(G, g, \mathcal{J})$ un groupe kählérien simplement connexe *irréductible* de 2-forme fondamentale σ et dimension $2n$. Sa 2-forme de Ricci α est $\neq 0$; si α n'est pas proportionnelle à σ, il existe une 2-forme fermée $\beta = \alpha - k\sigma \neq 0$ ($k \in \mathbf{R}$) de type $(1,1)$, invariante à gauche de rang $2p < 2n$. Soit $\mathcal{H}$ le sous-espace de $\mathcal{G}$ défini par les éléments $X \in \mathcal{G}$ tels que $i(X)\beta_e = 0$ et $\mathcal{K}$ l'orthocomplément de $\mathcal{H}$ dans $\mathcal{G}$. La $2p$-forme β^p étant fermée, les équations de Maurer-Cartan impliquent, avec des notations identiques à celles du paragraphe 6, qu'on a $C^\alpha_{ij} = 0$ et l'on voit que $\mathcal{H}$ est une sous-algèbre de Lie de $\mathcal{G}$. La 2-forme β étant de type $(1,1)$, $j\mathcal{H} \subset \mathcal{H}$ et $\mathcal{H}$ est une *sous-algèbre kählérienne non triviale de $\mathcal{G}$*.

b)

Lemme. *Sous les hypothèses du paragraphe a, on a $\gamma^\alpha_{\lambda\mu} = 0$.*

En effet posons $\chi = *(\beta^p)$; χ est une $2q$-forme invariante à gauche. D'après une formule classique, on a :

$$(\Delta\chi, \chi) = (\nabla\chi, \nabla\chi) + R(\chi)$$

avec

$$R(\chi) = \frac{1}{(2q-1)!} R_{k\ell}\chi^{ki_2\cdots i_{2q}}\chi^\ell{}_{i_2\cdots i_{2q}} - \frac{1}{2(2q-2)!} R_{k\ell,mn}\chi^{k\ell i_3\cdots i_{2q}}\chi^{mn}{}_{i_3}\cdots i_{2q}$$

$$(7.1)$$

Un calcul direct montre que :

$$(\delta d\chi, \chi) = (\nabla\chi, \nabla\chi)$$

et donc que

$$R(\chi) = 0$$

On voit immédiatement que si $\hat{R}_{kl}$ est le tenseur de Ricci de H, on a $R_{kl} = \hat{R}_{kl}$ et en explicitant $R_{kl,mn}$ dans le corepère choisi. on voit qu'il existe une constante $\mathcal{K}^2 > 0$ telle que

$$R(\chi) = \mathcal{K}^2 \sum_{\alpha, \gamma\mu} |\gamma^{\alpha}_{\lambda\mu}|^2 (\chi, \chi)$$

Ainsi $\gamma^{\alpha}_{\lambda\mu} = 0$.

c) $\gamma^{\alpha}_{\lambda\mu} = 0$ est équivalent à $\gamma^{\lambda}_{\alpha\bar\mu} = 0$, c'est-à-dire à $C^{\lambda}_{\alpha\bar\mu} = 0$ et (6.4) se réduit à :

$$d\theta^{\lambda} + C^{\lambda}_{\alpha\bar\beta} \theta^{\alpha} \wedge \theta^{\bar\beta} = 0 \qquad (\mathrm{mod}\ \theta^{\mu} = 0)$$

On en déduit que le système $\{\theta^{\lambda} = 0, \theta^{\bar\beta} = 0\}$ est intégrable. Le système $\{\theta^{\bar\beta} = 0\}$ étant lui-même intégrable ($\theta^{\bar\beta} = B^{\bar\beta}_{\bar\alpha} dz^{\bar\alpha}$), il existe q fonctions ζ^{μ} sur G, à valeurs complexes, telles que les $d\zeta^{\mu}$ et les $dz^{\bar\alpha}$ soient indépendantes et que

$$\theta^{\lambda} = B^{\lambda}_{\mu} d\zeta^{\mu} + B^{\lambda}_{\bar\alpha} dz^{\bar\alpha}$$

On a $\mathcal{J}\theta^{\lambda} = i\theta^{\lambda}$. Il en résulte :

$$B^{\lambda}_{\mu}(\mathcal{J}d\zeta^{\mu}) - iB^{\lambda}_{\bar\alpha} dz^{\bar\alpha} = iB^{\lambda}_{\mu} d\zeta^{\mu} + iB^{\lambda}_{\bar\alpha} dz^{\bar\alpha}$$

Cette relation étant valable pour toutes valeurs de $d\zeta^{\mu}, dz^{\bar\alpha}$, faisons $d\zeta^{\mu} = 0$. Il en résulte $B^{\lambda}_{\bar\alpha} = 0$ et il vient :

$$\theta^{\lambda} = B^{\lambda}_{\mu} d\zeta^{\mu}$$

Le système $\{\theta^{\lambda} = 0\}$ est donc intégrable et $C^{\lambda}_{\alpha\bar\beta} = 0$; $\mathcal{K}$ est donc toujours une sous-algèbre de Lie kählérienne de $\mathcal{G}$.

Proposition 7. *Si α n'est pas proportionnel à σ, $\mathcal{K}$ est une sous-algèbre kählérienne non triviale.*

D'après les propositions 6 et 7, la métrique g serait réductible, ce qui contredit l'hypothèse faite.

On voit qu'il existe $k \neq 0$ tel que :

$$\alpha = k\sigma$$

et $(G, g, \mathcal{J})$ est espace d'Einstein-Kähler ; α étant invariantement exacte, il en est de même de σ. La courbure scalaire $R \neq 0$ ne peut être positive, sinon G serait compact, ce qui est impossible pour un groupe symplectique simplement connexe. On a donc

$$R_{\rho\bar{\sigma}} = (R/2n)\, g_{\rho\bar{\sigma}} \qquad (R < 0)$$

Par homothétie, la métrique g peut être astreinte à

$$R_{\rho\bar{\sigma}} = -g_{\rho\bar{\sigma}}$$

en vertu d'un raisonnement classique. On a

Proposition 8. *Tout groupe kählérien irréductible est un espace d'Einstein-Kähler à courbure scalaire strictement négative et sa 2-forme σ est invariantement exacte.*

d) L'exemple le plus simple d'algèbre de Lie d'un groupe kählérien irréductible est donné par l'algèbre de Lie $\mathcal{G}_2$ de dimension 2 engendrée par X, Y vérifiant

$$[X, Y] = Y$$

Introduisons sur $\mathcal{G}_2$ l'opérateur de structure complexe j donné par :

$$jX = -Y \qquad jY = X$$

qui vérifie (4.3). Posons :

$$e_1 = X + iY \qquad e_{\bar{1}} = X - iY$$

et soit $(\theta^1, \theta^{\bar{1}})$ la base correspondante à $(e_1, e_{\bar{1}})$ des 1-formes complexes invariantes à gauche du groupe G_2 correspondant à $\mathcal{G}_2$. La métrique kählérienne est donnée dans cette base par $g_{1\bar{1}} = 2$. La 2-forme de Ricci α est donnée par :

$$\alpha = d\psi \qquad \psi = -i(\theta^1 - \theta^{\bar{1}})$$

et $R_{1\bar{1}} = -2$. Ainsi $Ri = -g$ et G_2 est kählérien irréductible à structure symplectique invariantement exacte.

e) Par produit de groupes kählériens irréductibles, nous obtenons à partir du Théorème 1

Théorème 3. *Tout groupe kählérien simplement connexe est, en tant que variété kählérienne, le produit d'un groupe kählérien plat et d'un groupe kählérien, espace d'Ein-stein-Kähler à courbure scalaire strictement* *négative dont la 2-forme est invariantement exacte. On peut astreindre cet espace d'Einstein à vérifier :*

$$R_{\rho\bar{\sigma}} = -g_{\rho\bar{\sigma}} \tag{7.3}$$

Soit $(G, g, \mathcal{J})$ un groupe kählérien simplement connexe qui soit espace d'Einstein-Kähler à $R < 0$. Un tel groupe n'admet pas de facteur plat et est produit de groupes kählériens irréductibles. Si l'algèbre de Lie $(\mathcal{G}, g, j)$ admet une sous-algèbre de Lie kählérienne $\mathcal{H}$ non triviale dont l'orthocomplément $\mathcal{K}$ dans $\mathcal{G}$ est aussi une sous algèbre de Lie, il résulte de la proposition 6 et du théorème précédent que G, en tant que variété kählérienne est produit des groupes kählériens correspondant à $\mathcal{H}$ et $\mathcal{K}$.

Corollaire. *Soit $(\mathcal{G}, g, j)$ l'algèbre de Lie kählérienne d'un groupe espace d'Einstein-Kähler à $R < 0$. Si $\mathcal{H}$ est une sous-algèbre kählérienne non triviale de $\mathcal{G}$ d'orthocomplément $\mathcal{K}$, qui est aussi une sous-algèbre de Lie de $\mathcal{G}$, la somme directe $\mathcal{H} + \mathcal{K}$ est l'algèbre de Lie d'un groupe kählérien admettant la même variété kählérienne que celle correspondant à $(\mathcal{G}, g, j)$.*

9. Caractérisation des groupes kählériens

Une algèbre de Lie kählérienne $(\mathcal{G}, g, j)$ est dite de *type plat* si elle est algèbre de Lie d'un groupe kählérien plat. Elle est dite de *type domaine* si elle est l'algèbre de Lie d'un groupe kählérien G simplement connexe qui, en tant que variété kählérienne, est un domaine borné homogène de $\mathbf{C}^n$ muni de sa métrique de Bergmann, sur lequel G opère de manière simplement transitive.

a)   Soit $(\mathcal{G}, g, j)$ un groupe kählérien simplement connexe *irréductible*. Soit $\mathrm{rad}(\mathcal{G})$ le radical de $\mathcal{G}$. Par "modification" au sens de [7] de $\mathrm{rad}(\mathcal{G})$, on peut substituer à $\mathcal{G}$ une algèbre kählérienne $\hat{\mathcal{G}}$ quasi-normale [7] telle que le groupe correspondant, connexe et simplement connexe, admette la même variété que $(\mathcal{G}, g, j)$ ([7] p.33–34).

Supposons donc $\hat{\mathcal{G}}$ quasi-normale. Il résulte de ([7]) p. 36, p. 46 et p. 54) que $\hat{\mathcal{G}}$ est résoluble, sinon il y aurait réductibilité. Comme la modification ne porte que sur le radical, l'algèbre $\mathcal{G}$ est résoluble.

Proposition 9. *Tout groupe kählérien irréductible est résoluble. A tout groupe kählérien simplement connexe correspond un groupe kählérien résoluble qui a la même variété kählérienne.*

D'après d'Atri, tout groupe kählérien résoluble qui est espace d'Einstein-Kähler à $R < 0$ est, en tant que variété kählérienne, un domaine borné homogène complexe de métrique homothétique à la métrique de Bergman du domaine. On en déduit

Théorème 4. *Tout groupe kählérien simplement connexe est, en tant que variété kählérienne, le produit d'un groupe kählérien plat par un groupe kählérien qui, en tant que variété kählérienne, est un domaine borné homogène de* $\mathbf{C}^p$, *pour une métrique d'Einstein-Kähler convenable.*

On démontre aisement

Corollaire. *Tout groupe kählérien est, en tant que variété kählérienne, le produit d'un groupe kählérien plat, simplement connexe ou non, par un groupe kählérien simplement connexe qui est, en tant que variété kählérienne, un domaine borné homogène complexe.*

REFERENCES

[1] A. Lichnerowicz et A. Medina, *On Lie groups with left-invariant symplectic or kählerian structures*, Lett. in Math. Phys. **16** (1988), 225–235.

[2] Bon-Yao Chu, *Symplectic homogeneous spaces*, Trans. Amer. Math Soc. **197** (1974), 145–159.

[3] J.I. Hano, *On Kählerian homogeneous spaces of unimodular Lie groups*, Amer. J. Math. **79** (1957), 885–900.

[4] K. Nomizu, *Studies on Riemannian homogeneous spaces*, Nagoya Math. J. **9** (1955), 43–56.

[5] J. Milnor, *Curvatures of left-invariant metrics on Lie groups*, Adv. Math. **21** (1976), 293–329.

[6] J. Dorfmeister, *Homogeneous Kähler manifolds admitting a transitive solvable group of automorphisms*, Ann Sci. Ecole Norm. Sup. **18** (1985), 143–188. *The radical conjecture for homogeneous Kähler manifolds*, CMS Conference Proc. **5** (1986), 189–208.

[7] J. Dorfmeister and K. Nakajima, *The fundamental conjecture for homogeneous Kähler manifolds*, Acta Mathem. **161** (1988), 23–70.

[8] E. B. Vinberg and S. G. Gindikin, *Kählerian manifolds admitting a transitive solvable automorphism group*, Math. Sb. **74 (116)** (1967), 333–351.

Collège de France
3, rue d'Ulm
75005 Paris
France

Géométrie des systèmes mécaniques
à liaisons actives

CHARLES-MICHEL MARLE

À Jean-Marie Souriau, dont le livre "Géométrie et relativité"
[15] a joué un rôle déterminant dans l'orientation professionnelle
que j'ai choisie, et dont l'autre livre "Structure des systèmes
dynamiques" [16] est pour moi une source constante d'inspiration,
en témoignage d'admiration et de respectueuse amitié.

1. Introduction

Nous dirons qu'un système mécanique est *à liaisons actives* s'il comporte certaines liaisons sur lesquelles un opérateur (qui peut soit faire partie du système, soit lui être extérieur) agit, influant ainsi sur le mouvement du système. Donnons quelques exemples:

- un enfant sur une escarpolette peut, en faisant jouer les articulations de son corps, mettre l'escarpolette en mouvement;
- un chat qui tombe en chute libre peut, en déformant son corps, modifier son orientation dans l'espace afin d'arriver au sol sur ses pattes;
- un satellite artificiel (tel que le télescope spatial Hubble) comporte des articulations auxquelles sont liés divers dispositifs (panneaux solaires, antennes, volants d'inertie, appareils d'optique, ...); on peut agir sur ces articulations afin d'orienter certains éléments du satellite dans une direction déterminée.

Nous nous proposons de mettre en place dans ce travail un cadre géométrique général pour l'étude des systèmes mécaniques à liaisons actives, et d'étudier la structure des équations du mouvement de ces systèmes. Cette étude nous a été suggérée par la lecture des travaux de Marsden, Montgomery et Ratiu [12] [13] [14], où le lecteur pourra trouver d'autres points de vue.

Nous nous limitons dans ce travail à l'étude des systèmes mécaniques dont les liaisons (actives ou autres) sont de type géométrique; nous voulons dire par là que ces liaisons ont pour effet de restreindre l'ensemble des positions des parties du système dans l'espace, ou des positions relatives de ces parties les unes par rapport aux autres; nous laissons de côté pour le moment les systèmes comportant des liaisons cinématiques (portant à la fois sur les positions et les vitesses des parties du système). Le cas de systèmes à liaisons cinématiques holonomes pourrait se ramener, sans trop de difficulté, à celui des systèmes à liaisons géométriques que nous étudions. Mais le cas des systèmes comportant des liaisons cinématiques non holonomes (comme, par exemple, un patineur sur glace) est beaucoup plus délicat: ainsi que l'ont signalé Woodhouse [19], paragraphe 2.2, et, plus récemment, Arnold, Kozlov et Neishtadt [3], chapitre 1, paragraphe 4, on peut hésiter entre plusieurs formulations non équivalentes des équations. Ainsi dans [3] les auteurs explicitent, parallèlement à la formulation la plus connue des équations du mouvement basée sur le principe de d'Alembert-Lagrange, une autre formulation qu'ils appellent "vakonomic equations". Les équations convenant pour un problème réel pourraient, selon ces auteurs, dépendre de propriétés fines du dispositif physique servant à réaliser les liaisons ([3], chapitre 1, paragraphe 6).

Certains résultats présentés en détail ici ont été annoncés dans la note [11].

2. La formulation mathématique

L'ensemble des configurations possibles du système considéré est une variété différentiable M, appelée *variété de configuration*. Les liaisons actives sur lesquelles l'opérateur peut agir sont représentées par une submersion surjective $\pi : M \to S$ de M sur une autre variété S, appelée *variété des états de liaison*. Chaque point $s \in S$ représente un état des liaisons actives, et l'ensemble des configurations possibles lorsque les liaisons sont dans l'état s est la sous-variété $M_s = \pi^{-1}(s)$ de M, appelée *variété de configuration à liaisons figées dans l'état s*.

Les propriétés dynamiques du système sont décrites par une fonction différentiable $L : TM \to \mathbf{R}$, définie sur le fibré tangent TM à la variété de configuration M, à valeurs réelles, appelée *lagrangien*.

Lorsque l'opérateur n'agit pas sur les liaisons actives, laissant celles-ci totalement libres de jouer, la donnée de M et de L suffit à déterminer les équations du mouvement: ce sont les équations de Lagrange associées à L. Celles-ci, appelées *équations du mouvement libre*, jointes à une donnée de Cauchy (point de TM représentant la position et la vitesse du système à l'instant initial) déterminent le mouvement du système, au moins lorsque le lagrangien L est suffisamment régulier.

Mais en fait l'opérateur agit sur les liaisons actives, de sorte que le mouvement réel diffère du mouvement libre. On supposera ici que cette action a pour effet de prescrire, à chaque instant t, l'état $s = \sigma(t)$ des liaisons actives. En d'autres termes, une courbe paramétrée $\sigma : t \mapsto \sigma(t)$ dans la variété S des états de liaison est donnée; elle décrit l'évolution de l'état des liaisons actives au cours du temps, imposée par l'opérateur; le mouvement du système au cours du temps est une courbe paramétrée

$t \mapsto x = c(t)$ dans la variété de configuration M vérifiant, à tout instant t,

$$\pi\big(c(t)\big) = \sigma(t).$$

Dans la suite de ce travail, nous allons formuler les équations du mouvement en supposant la courbe paramétrée $t \mapsto s = \sigma(t)$ donnée, et étudier leur structure. Remarquons cependant que la résolution de ce problème n'est qu'un premier pas dans l'étude des systèmes à liaisons actives. L'action de l'opérateur sur l'état des liaisons actives peut en effet être soumise à certaines limitations (par exemple, les efforts que l'opérateur peut exercer sur ces liaisons sont en pratique bornés, de sorte qu'il n'est pas toujours possible d'imposer à l'état des liaisons actives d'évoluer au cours du temps suivant une loi donnée d'avance (D. Holm [7]). Et surtout, si l'opérateur agit sur les liaisons actives, c'est afin d'obtenir un certain résultat: l'enfant sur une escarpolette cherche à augmenter l'amplitude des oscillations; le chat en chute libre cherche à amener ses pattes vers le bas; le programme d'ordinateur qui contrôle un télescope spatial doit orienter le miroir dans une direction donnée puis maintenir très précisément cette orientation. La loi $t \mapsto \sigma(t)$ d'évolution de l'état des liaisons n'est donc en général pas spécifiée d'avance: l'opérateur doit la choisir afin d'atteindre le but qu'il s'est fixé, résolvant ainsi un problème de contrôle (R. Montgomery [14]). Le cadre géométrique général décrit ci-dessous nous semble bien adapté à l'étude d'un tel problème. Ainsi qu'on le verra paragraphe 8, il englobe les trois exemples de systèmes mécaniques à liaisons actives donnés dans l'introduction (enfant sur une escarpolette, chat en chute libre, satellite artificiel).

Les variétés M, S, la submersion π, le lagrangien L, la courbe paramétrée σ, ainsi que toutes les applications considérées seront, sauf spécification contraire, supposées différentiables de classe C^∞.

3. Les équations du mouvement dans le formalisme lagrangien

À toute courbe paramétrée différentiable $c : [t_1, t_2] \to M$ dans la variété de configuration on associe l'intégrale d'action

$$I(c) = \int_{t_1}^{t_2} L\left(c(t), \frac{dc(t)}{dt}\right) dt. \tag{1}$$

D'autre part, une courbe paramétrée $\sigma : [t_1, t_2] \to S$ dans l'espace des états des liaisons actives est donnée; elle représente l'évolution au cours du temps de l'état de ces liaisons prescrite par l'opérateur.

Rappelons brièvement les quelques méthodes équivalentes qui permettent l'obtention des équations du mouvement.

3.1. Méthode basée sur le principe de moindre action.

Compte tenu des liaisons imposées variables avec le temps, l'espace-temps de configuration du système est le sous-ensemble de $\mathbf{R} \times M$:

$$N = \left\{ (t,x) \in \mathbf{R} \times M \mid t \in [t_1, t_2], \ \pi(x) = \sigma(t) \right\}. \tag{2}$$

Puisque π est une submersion surjective, N est une sous-variété à bord de $\mathbf{R} \times M$. La restriction à N de la projection de $\mathbf{R} \times M$ sur son premier facteur est une submersion surjective $\varpi : N \to [t_1, t_2]$.

Toute section $\widetilde{c} : [t_1, t_2] \to N$ de la submersion ϖ est de la forme $t \mapsto \big(t, c(t)\big)$, où $c : [t_1, t_2] \to M$ vérifie

$$\pi\big(c(t)\big) = \sigma(t) \quad \text{pour tout} \quad t \in [t_1, t_2]. \tag{3}$$

Réciproquement, à toute courbe paramétrée $c : [t_1, t_2] \to M$ vérifiant (3) correspond une section $\widetilde{c} = (\mathrm{id}_{[t_1,t_2]}, c)$ de la submersion π.

Soit $\widetilde{c} = (\mathrm{id}_I, c)$ une section différentiable locale de ϖ. Pour tout élément t de l'intervalle de définition I de $\widetilde{c}$ ($I \subset [t_1, t_2]$), le vecteur $\dfrac{dc(t)}{dt} \in T_{c(t)}M$ ne dépend que du jet $j^1\widetilde{c}(t)$ d'ordre 1 de $\widetilde{c}$ en t; par suite, $j^1\widetilde{c}(t) \mapsto \left(t, \dfrac{dc(t)}{dt}\right)$ est une application χ de l'espace $J^1\Gamma(\varpi)$ des jets d'ordre 1 de sections différentiables locales de ϖ dans $\mathbf{R} \times TM$. On vérifie que χ est injective et a pour image

$$\left\{ (t,v) \in \mathbf{R} \times TM \mid t \in [t_1, t_2], \ (1,v) \in T_{\varpi^{-1}(t)}N \right\}.$$

On pose $\widetilde{L} = L \circ \chi$ (on considère ici L comme une fonction définie sur $\mathbf{R} \times TM$ en l'identifiant implicitement à sa composée avec la projection de $\mathbf{R} \times TM$ sur TM). Pour toute section différentiable $\widetilde{c}$ de ϖ, on définit

$$\widetilde{I}(\widetilde{c}) = \int_{t_1}^{t_2} \widetilde{L}\big(j^1\widetilde{c}(t)\big)\, dt\,, \tag{4}$$

qui n'est qu'une autre manière d'écrire l'intégrale d'action (1).

Rappelons qu'une *variation* de la section différentiable $\widetilde{c}$ de ϖ est une application différentiable $\widetilde{k} :\,]-\epsilon, \epsilon[\times [t_1, t_2] \to N$, $\epsilon > 0$, telle que pour tout $s \in\,]-\epsilon, \epsilon[$, $\widetilde{k}_s : t \mapsto \widetilde{k}(s,t)$ soit une section de ϖ et que $\widetilde{k}_0 = \widetilde{c}$. Cette variation est dite *à extrémités fixées* si

$$\widetilde{k}(s, t_1) = \widetilde{c}(t_1) \quad \text{et} \quad \widetilde{k}(s, t_2) = \widetilde{c}(t_2) \quad \text{pour tout} \quad s \in\,]-\epsilon, \epsilon[. \tag{5}$$

En appliquant au système, considéré comme un système lagrangien dépendant du temps, le principe de moindre action de Hamilton, on voit que la courbe paramétrée différentiable $c : [t_1, t_2] \to M$ est un mouvement du système si et seulement si $\widetilde{c} = (\mathrm{id}_{[t_1,t_2]}, c)$ est une section de ϖ qui rend l'intégrale d'action (4) stationnaire pour toute variation $\widetilde{k}$ de $\widetilde{c}$ à extrémités fixées. Cela signifie que pour toute variation $\widetilde{k}$ de $\widetilde{c}$ à extrémités fixées, on a

$$\frac{d}{ds}\widetilde{I}(\widetilde{k}_s)\Big|_{s=0} = 0\,. \tag{6}$$

3.2. Méthode basée sur le principe de Hölder.

On revient à l'expression (1) de l'intégrale d'action. Une *variation* de la courbe paramétrée différentiable $c : [t_1, t_2] \to M$ est une application différentiable $k :\,]-\epsilon, \epsilon[\times [t_1, t_2] \to M$, $\epsilon > 0$, telle que $k(0, t) = c(t)$ pour tout $t \in [t_1, t_2]$. Cette variation est dite *à extrémités fixées* si

$$k(s, t_1) = c(t_1) \quad \text{et} \quad k(s, t_2) = c(t_2) \quad \text{pour tout} \quad s \in\,]-\epsilon, \epsilon[\,. \tag{7}$$

Une *variation infinitésimale* de c est une application différentiable $w : [t_1, t_2] \to TM$ telle que $w(t) \in T_{c(t)}M$ pour tout $t \in [t_1, t_2]$. Cette variation infinitésimale est dite *à extrémités fixées* si

$$w(t_1) = 0 \quad \text{et} \quad w(t_2) = 0\,. \tag{8}$$

À toute variation k de c est associée la variation infinitésimale $\delta k : t \mapsto \delta k(t) = \dfrac{\partial k(s, t)}{\partial s}\Big|_{s=0}$. Si k est à extrémités fixées, δk est à extrémités fixées.

Soit k une variation de c. Pour tout $s \in\,]-\epsilon, \epsilon[$, $k_s : t \mapsto k(s, t)$ est une courbe paramétrée différentiable dans M; on peut donc considérer l'intégrale d'action $I(k_s)$. On montre en calcul des variations que $s \mapsto I(k_s)$ est dérivable en $s = 0$ et que la valeur de cette dérivée ne dépend que de la variation infinitésimale δk associée à k. Cette dépendance est d'ailleurs linéaire. On notera donc $\langle \delta I(c), \delta k \rangle$ la dérivée de $I(k_s)$ par rapport à s pour $s = 0$. On montre qu'elle a pour expression

$$
\begin{aligned}
\frac{d}{ds}I(k_s)\Big|_{s=0} &= \langle \delta I(c), \delta k \rangle \\
&= \left\langle \mathcal{L}\left(\frac{dc(t)}{dt}\right)\Big|_{t=t_2}, \delta k(t_2) \right\rangle - \left\langle \mathcal{L}\left(\frac{dc(t)}{dt}\right)\Big|_{t=t_1}, \delta k(t_1) \right\rangle \\
&\quad - \int_{t_1}^{t_2} \langle [L]\big(j^2 c(t)\big), \delta k(t) \rangle\, dt\,. \tag{9}
\end{aligned}
$$

On a noté $\mathcal{L} : TM \to T^*M$ la transformation de Legendre associée à L. Rappelons que si on note (x^i) les coordonnées locales dans une carte de M, (x^i, v^i) et (x^i, p_i) les coordonnées locales dans les cartes associées, respectivement de TM et de T^*M, la transformation de Legendre $\mathcal{L}$ a pour expression

$$\mathcal{L} : (x^i, v^i) \mapsto \left(x^i, p_i = \frac{\partial L(x, v)}{\partial v^i}\right)\,. \tag{10}$$

On a noté $[L]$ la différentielle de Lagrange de L (voir [17]). C'est une application définie sur l'espace des jets d'ordre 2 d'applications différentiables de $\mathbf{R}$ dans M, à valeurs dans T^*M, fibrée au dessus de M, dont les composantes $[L]_i$ (correspondant aux coordonnées locales (p_i) sur T^*M dans les cartes considérées ci-dessus) ont pour expression

$$[L]_i\big(j^2 c(t)\big) = \frac{d}{dt}\frac{\partial L(x,v)}{\partial v^i} - \frac{\partial L(x,v)}{\partial x^i}, \tag{11}$$

avec $v(t) = \dfrac{dc(t)}{dt}$.

La courbe paramétrée c est dite *compatible avec les liaisons* si $\widetilde{c} = (\mathrm{id}_{[t_1,t_2]}, c)$ est une section de ϖ, c'est-à-dire si

$$\pi\big(c(t)\big) = \sigma(t) \quad \text{pour tout} \quad t \in [T_1, t_2]. \tag{12}$$

Lorsque c'est le cas, une variation k de c est dite *compatible avec les liaisons* si pour tout $s \in\,]-\epsilon, \epsilon[$, la courbe paramétrée k_s est compatible avec les liaisons, ce qui s'exprime par

$$\pi\big(k(s,t)\big) = \sigma(t) \quad \text{pour tous} \quad s \in\,]-\epsilon, \epsilon[,\ t \in [t_1, t_2]. \tag{13}$$

De même, supposant c compatible avec les liaisons, une variation infinitésimale w de c est dite *compatible avec les liaisons* si

$$T\pi\big(w(t)\big) = 0 \quad \text{pour tout} \quad t \in [t_1, t_2]. \tag{14}$$

Le *principe de Hölder* [3] consiste à affirmer qu'une courbe paramétrée différentiable $c : [t_1, t_2] \to M$ est un mouvement du système si et seulement si elle satisfait les deux conditions,

- la courbe c est compatible avec les liaisons,
- elle rend stationnaire l'intégrale d'action $I(c)$ pour les variations infinitésimales de c à extrémités fixées compatibles avec les liaisons, ce qui signifie qu'elle vérifie

$$\langle \delta I(c), w \rangle = 0 \tag{15}$$

pour tout variation infinitésimale w de c à extrémités fixées compatible avec les liaisons.

Remarque. Si la variation k de c est compatible avec les liaisons, la variation infinitésimale correspondante δk est compatible avec les liaisons. Réciproquement, pour des liaisons géométriques telles que celles considérées ici, si w est une variation infinitésimale de c compatible avec les liaisons, il existe une variation k de c compatible avec les liaisons telle que $\delta k = w$. On doit prendre garde au fait que cette dernière propriété n'est pas toujours vraie pour des liaisons cinématiques non holonomes.

La remarque ci-dessus permet, dans le cas considéré ici de liaisons géométriques, de reformuler le principe de Hölder en disant que la courbe paramétrée $c : [t_1, t_2] \to M$ est un mouvement du système si et seulement si elle satisfait les deux conditions:

- la courbe c est compatible avec les liaisons,
- pour toute variation k de c à extrémités fixées compatible avec les liaisons, elle vérifie

$$\frac{d}{ds} I(k_s)\big|_{s=0} = 0 \,. \tag{16}$$

En remarquant que les expressions (1) et (4) de l'intégrale d'action sont équivalentes, on voit alors que le principe de Hölder est équivalent au principe de moindre action appliqué, comme indiqué au paragraphe 3.1, au système considéré comme dépendant du temps.

3.3. Méthode basée sur le principe de d'Alembert-Lagrange.

Soit $c : [t_1, t_2] \to M$ une courbe paramétrée différentiable compatible avec les liaisons. On appelle *déplacement virtuel infinitésimal* à l'instant $t \in]t_1, t_2[$ tout vecteur $w_t \in T_{c(t)}M$. On dit que le déplacement virtuel infinitésimal w_t à l'instant t est *compatible avec les liaisons* s'il vérifie

$$T\pi(w_t) = 0 \,. \tag{18}$$

On appelle *travail virtuel des forces de liaison* pour le déplacement virtuel infinitésimal w_t à l'instant t, l'expression

$$\big\langle [L]\big(j^2 c(t)\big), w_t \big\rangle \,. \tag{18}$$

Le *principe de d'Alembert-Lagrange* [3] consiste à dire que la courbe paramétrée différentiable c, supposée compatible avec les liaisons, est un mouvement du système si et seulement si pour tout instant $t \in]t_1, t_2[$ et tout déplacement virtuel infinitésimal w_t à l'instant t compatible avec les liaisons, le travail virtuel des forces de liaison pour le déplacement virtuel infinitésimal w_t est nul.

Si $w : [t_1, t_2] \to TM$ est une variation infinitésimale de c à extrémités fixées, alors pour tout $t \in [t_1, t_2]$, $w(t)$ est un déplacement virtuel infinitésimal; w est compatible avec les liaisons si et seulement si $w(t)$ est compatible avec les liaisons pour tout $t \in]t_1, t_2[$. L'expression (9) de $\langle \delta I(c), w \rangle$ (dans laquelle on remplace δk par w), qui se simplifie car $w(t_1)$ et $w(t_2)$ sont nuls, w étant à extrémités fixées, montre que si la courbe paramétrée différentiable c satisfait les conditions imposées par le principe de d'Alembert-Lagrange, elle satisfait aussi les conditions imposées par le principe de Hölder. Réciproquement, le lemme fondamental du calcul des variations montre que si c satisfait les conditions imposées par le principe de Hölder, elle satisfait ausi celles imposées par le principe de d'Alembert-Lagrange. Ces deux principes sont donc bien équivalents.

On pourrait encore obtenir les équations du mouvement par application du principe de moindre courbure de Gauss ([**3**], chapitre 1, paragraphe 2, ou [**18**], chapitre 9, paragraphe 105), équivalent aux principes de Hölder et de d'Alembert-Lagrange. On ne développera pas cette méthode ici, la comparaison de différents principes pour l'obtention des équations du mouvement n'étant pas l'objet principal de ce travail. Le lecteur intéressé par ce sujet pourra se reporter à [**4**].

3.4. Expression des équations en coordonnées locales.

On considère une carte de M dans laquelle les coordonnées locales sont notées (x^i), $1 \leq i \leq m = \dim M$, et une carte de S dont les coordonnées locales sont notées (s^α); on numérotera ces dernières de manière un peu particulière, en imposant à l'indice α de vérifier $p + 1 \leq \alpha \leq m$, avec $p = m - \dim S$; cette convention se révèlera commode dans la suite. On suppose la projection $\pi(U)$ du domaine U de la carte considérée de M contenue dans le domaine V de la carte considérée de S. Dans ces cartes, l'application π a pour expression

$$s^\alpha = \pi^\alpha(x^1, \ldots, x^m), \quad p + 1 \leq \alpha \leq m.$$

On note (x^i, v^i) les coordonnées locales dans la carte de TM associée à la carte considérée de M.

On supposera que la courbe paramétrée donnée $\sigma : [t_1, t_2] \to S$, représentant l'évolution de l'état des liaisons actives imposée par l'opérateur, a son image contenue dans V. Son expression dans la carte de S considérée est

$$t \mapsto \left(s^\alpha = \sigma^\alpha(t)\right), \qquad p + 1 \leq \alpha \leq m.$$

Soit $c : [t_1, t_2] \to M$ une courbe paramétrée différentiable dont l'image est contenue dans U. Dans les cartes considérées, cette courbe et son relèvement canonique dans TM ont pour expressions respectives

$$t \mapsto \left(x^i = c^i(t)\right), \quad \text{et} \quad t \mapsto \left(x^i = c^i(t), \ v^i = \frac{dc^i(t)}{dt}\right), \quad 1 \leq i \leq m.$$

Soit $w : [t_1, t_2] \to TM$ une variation infinitésimale de c à extrémités fixées ($w(t_1) = w(t_2) = 0$). Dans les cartes considérées, elle a pour expression $t \mapsto \left(x^i = c^i(t), \ v^i = w^i(t)\right)$.

D'après (9), $\langle \delta I(c), w \rangle$ a pour expression

$$\langle \delta I(c), w \rangle = -\int_{t_1}^{t_2} \sum_{i=1}^{m} \left(\frac{d}{dt} \frac{\partial L}{\partial v^i} - \frac{\partial L}{\partial x^i}\right) w^i(t)\, dt. \tag{19}$$

Appliquons le principe de Hölder. On doit tout d'abord exprimer que c est compatible avec les liaisons, ce qui en coordonnées locales s'écrit

$$\pi^\alpha\big(c(t)\big) = \sigma^\alpha(t)\,, \quad p+1 \leq \alpha \leq m\,, \quad t \in [t_1, t_2]\,. \tag{20}$$

On doit encore exprimer que $\langle \delta I(c), w \rangle$ est nul pour toute variation infinitésimale w de c, à extrémités fixées, compatible avec les liaisons, ce qui en coordonnées locales s'exprime par

$$\sum_{i=1}^{m} \frac{\partial \pi^\alpha\big(c(t)\big)}{\partial x^i}\, w^i(t) = 0 \quad \text{pour tous} \quad t \in [t_1, t_2]\,, \; p+1 \leq \alpha \leq m\,. \tag{21}$$

Le lemme fondamental du calcul des variations et la théorie des multiplicateurs de Lagrange montrent que c satisfait le principe de Hölder (donc est un mouvement du système) si et seulement s'il existe des fonctions λ_α du temps t, $p+1 \leq \alpha \leq m$, telles que pour tout i, $1 \leq i \leq m$,

$$\frac{d}{dt}\frac{\partial L}{\partial v^i} - \frac{\partial L}{\partial x^i} - \sum_{\alpha=p+1}^{m} \lambda_\alpha \frac{\partial \pi^\alpha}{\partial x^i} = 0\,. \tag{22}$$

Ce sont les *équations de Lagrange avec multiplicateurs*. On doit bien entendu considérer que dans ces équations, $x = (x^i)$, $v = (v^i)$, sont remplacés, respectivement, par $\big(c^i(t)\big)$ et $\left(\dfrac{dc^i(t)}{dt}\right)$, $t \in [t_1, t_2]$.

3.5. Cas de coordonnées locales adaptées.

L'application $\pi : M \to S$ étant une submersion, on peut au voisinage de chaque point de M et au voisinage de la projection de ce point sur S, trouver des coordonnées locales (x^i), $1 \leq i \leq m$, et (s^α), $p+1 \leq \alpha \leq m$, de manière telle que π ait pour expression

$$(x^i) \mapsto (s^\alpha = x^\alpha)\,.$$

En d'autres termes, on peut faire en sorte que

$$\pi^\alpha(x^1, \ldots, x^m) = x^\alpha\,, \qquad p+1 \leq \alpha \leq m\,, \tag{23}$$

et par suite

$$\frac{\partial \pi^\alpha}{\partial x^i} = \delta_i^\alpha\,, \qquad 1 \leq i \leq m\,, \; p+1 \leq \alpha \leq m\,. \tag{24}$$

Les équations (20) exprimant que c est compatible avec les liaisons s'écrivent maintenant

$$c^\alpha(t) = \sigma^\alpha(t)\,, \qquad t \in [t_1, t_2]\,, \; p+1 \leq \alpha \leq m\,, \tag{25}$$

tandis que les équations de Lagrange avec multiplicateurs (22) deviennent

$$\frac{d}{dt}\frac{\partial L}{\partial v^i} - \frac{\partial L}{\partial x^i} = 0 \qquad \text{pour} \quad 1 \le i \le p\,, \tag{26}$$

$$\frac{d}{dt}\frac{\partial L}{\partial v^\alpha} - \frac{\partial L}{\partial x^\alpha} = \lambda_\alpha \qquad \text{pour} \quad p+1 \le \alpha \le m\,. \tag{27}$$

Les équations (26), jointes aux équations (25) exprimant les liaisons, déterminent le mouvement (au moins lorsque le langrangien est suffisamment régulier). Une fois celui-ci déterminé, les équations (27) donnent les valeurs des multiplicateurs de Lagrange λ_α. La connaissance de ceux-ci est utile pour le calcul de la puissance des forces de liaison, qui a pour expression

$$\sum_{\alpha=p+1}^{m} \lambda_\alpha \frac{d\sigma^\alpha}{dt}\,. \tag{28}$$

4. Les équations du mouvement dans le formalisme hamiltonien

On suppose pour simplifier le lagrangien L hyper-régulier [1], [6], c'est-à-dire tel que la transformation de Legendre $\mathcal{L} : TM \to T^*M$ qu'il définit soit un difféomorphisme. Cette hypothèse pourrait d'ailleurs être affaiblie (voir exemple 8.5 ci-après). Le hamiltonien $H : T^*M \to \mathbf{R}$ est défini par

$$H = \bigl(i(Z)dL - L\bigr) \circ \mathcal{L}^{-1}\,, \tag{29}$$

où Z désigne le champ de vecteurs de Liouville sur TM.

4.1. Expression des équations en coordonnées locales.

On considère une carte de M et une carte de S, ainsi que les cartes associées de TM et de T^*M. Les notations utilisées sont celles du paragraphe 3.4. On a déjà indiqué en (10) l'expression de la transformation de Legendre $\mathcal{L}$ au moyen des coordonnées locales (x^i, v^i) sur TM, (x^i, p_i) sur T^*M. Celle du hamiltonien est

$$H(x,p) = \sum_{i=1}^{m} p_i v^i - L(x,v)\,, \tag{30}$$

en considérant qu'au second membre, $v = (v^i)$ est exprimé au moyen de $x = (x^i)$ et de $p = (p_i)$ grâce à $\mathcal{L}^{-1}$. On remarque que, pour tout i $(1 \le i \le m)$,

$$\frac{\partial H(x,p)}{\partial x^i} = -\frac{\partial L(x,v)}{\partial x^i}\,, \tag{31}$$

$$\frac{\partial H(x,p)}{\partial p_i} = v^i\,. \tag{32}$$

270 Charles-Michel Marle

Comme $v^i = \dfrac{dx^i}{dt}$, cette équation donne la première équation de Hamilton

$$\frac{dx^i}{dt} = \frac{\partial H(x,p)}{\partial p_i}\,. \tag{33}$$

D'autre part, les équations de Lagrange avec multiplicateurs (22) deviennent

$$\frac{dp_i}{dt} = -\frac{\partial H}{\partial x^i} + \sum_{\alpha=1}^{m} \lambda_\alpha \frac{\partial \pi^\alpha}{\partial x^i}\,. \tag{34}$$

C'est la seconde équation de Hamilton avec multiplicateurs. Les équations de Hamilton (33) et (34), jointes aux équations (20) exprimant les liaisons, déterminent le mouvement du système.

4.2. Cas de coordonnées locales adaptées.

Comme au paragraphe 3.5, on suppose ici les coordonnées locales adaptées à la submersion $\pi : M \to S$. On utilise les mêmes notations que dans ce paragraphe. En dérivant par rapport à t l'équation (25) exprimant les liaisons on obtient

$$\frac{dx^\alpha}{dt} = \frac{d\sigma^\alpha}{dt}\,, \qquad p+1 \leq \alpha \leq m\,. \tag{35}$$

En tenant compte de ces relations dans les $m-p$ dernières équations de Hamilton (34), le système de Hamilton (33), (34) peut s'écrire

$$\begin{aligned}
\frac{dx^i}{dt} &= \frac{\partial H}{\partial p_i} \\
\frac{dp_i}{dt} &= -\frac{\partial H}{\partial x^i}
\end{aligned} \qquad \text{pour}\ \ 1 \leq i \leq p, \tag{36}$$

$$\frac{d\sigma^\alpha}{dt} - \frac{\partial H}{\partial p_\alpha} = 0 \qquad \text{pour}\ \ p+1 \leq \alpha \leq m\,, \tag{37}$$

$$\frac{dp_\alpha}{dt} + \frac{\partial H}{\partial x^\alpha} = \lambda_\alpha \qquad \text{pour}\ \ p+1 \leq \alpha \leq m\,. \tag{38}$$

Dans (37), le terme $\dfrac{d\sigma^\alpha}{dt}$ est connu, puisque la loi d'évolution σ de l'état des liaisons actives en fonction du temps est imposée par l'opérateur. Cette équation doit être utilisée afin de déterminer les coordonnées locales p_α ($p+1 \leq \alpha \leq m$) en fonction des x^i, x^α, p_i et des quantités connues $\dfrac{d\sigma^\alpha}{dt}$, ce qui est souvent possible, au moins localement (voir hypothèse paragraphe 5). Portant ces expressions de p_α dans les équations (36), on obtient un système d'équations qui suffit à déterminer le mouvement (les équations (38) ne sont pas nécessaires pour cela, elles servent seulement à déterminer les multiplicateurs de Lagrange λ_α).

5. Une expression géométrique intrinsèque des équations du mouvement

Soit $VM = \ker(T\pi)$ le sous-fibré vertical (relativement à la submersion $\pi : M \to S$) du fibré tangent TM. Soit V^*M le fibré dual de VM. On sait qu'il s'identifie au quotient de T^*M (fibré cotangent à M) par son sous-fibré $(VM)^0$, annulateur de VM (fibré des 1-formes sur M nulles sur VM). On a donc une projection canonique $\zeta : T^*M \to V^*M$. On sait que V^*M possède une structure naturelle de variété de Poisson [8] pour laquelle ζ est une application de Poisson: en effet V^*M s'identifie au qotient de T^*M par le feuilletage engendré par les champs de vecteurs hamiltoniens dont les hamiltoniens sont images réciproques (par la projection composée $T^*M \to M \to S$) de fonctions différentiables sur S. On note $q : V^*M \to M$ la projection canonique (déduite par quotient de la projection canonique $T^*M \to M$) et on pose $\widetilde{\pi} = \pi \circ q : V^*M \to S$. On supposera pour simplifier que pour tout état $s \in S$ des liaisons actives, l'espace de configuration $M_s = \pi^{-1}(s)$ pour des liaisons fixées dans l'état s est connexe. On voit alors que S s'identifie à l'ensemble des feuilles symplectiques de la variété de Poisson V^*M, car pour tout point $s \in S$, $\widetilde{\pi}^{-1}(s)$ est une feuille symplectique de V^*M canoniquement symplectomorphe à l'espace des phases T^*M_s du système lorsque les liaisons sont figées dans l'état s. Soit $T\pi : TM \to TS$ le prolongement aux vecteurs de la submersion $\pi : M \to S$, $\pi^*(TS)$ le fibré de base M image réciproque par π du fibré tangent TS de base S, et $\widetilde{T\pi} : TM \to \pi^*(TS)$ le morphisme de fibrés correspondant à $T\pi$. En le composant avec la transformation de Legendre inverse $\mathcal{L}^{-1} : T^*M \to TM$, on obtient une application $\widetilde{T\pi} \circ \mathcal{L}^{-1} : T^*M \to \pi^*(TS)$, fibrée au dessus de l'identité de M (mais pas nécessairement linéaire sur chaque fibre). De même, l'application $(\zeta, \widetilde{T\pi} \circ \mathcal{L}^{-1}) : T^*M \to V^*M \oplus \pi^*(TS)$ est fibrée au dessus de l'identité de M. On fera dans la suite l'hypothèse suivante.

Hypothèse. L'application $(\zeta, \widetilde{T\pi} \circ \mathcal{L}^{-1}) : T^*M \to V^*M \oplus \pi^*(TS)$ est un difféomorphisme.

Remarques.

1. Dans les coordonnées locales adaptées des paragraphes 3.5 et 4.2, (x^i, x^α, p_i), $1 \leq i \leq p$, $p + 1 \leq \alpha \leq m$, sont des coordonnées locales sur V^*M, et la projection $\zeta : T^*M \to V^*M$ a pour expression $(x^i, x^\alpha, p_i, p_\alpha) \mapsto (x^i, x^\alpha, p_i)$. De même, (x^i, x^α, v^i) sont des coordonnées locales sur $\pi^*(TS)$ et l'application $\widetilde{T\pi} : TM \to \pi^*(TS)$ a pour expression $(x^i, x^\alpha, v^i, v^\alpha) \mapsto (x^i, x^\alpha, v^\alpha)$. Enfin $(x^i, x^\alpha, p_i, v^\alpha)$ sont des coordonnées locales sur $V^*M \oplus$

$\pi^*(TS)$, et l'application $(\zeta, \widetilde{T\pi} \circ \mathcal{L}^{-1}) : T^*M \to V^*M \oplus \pi^*(TS)$ a pour expression $(x^i, x^\alpha, p_i, p_\alpha) \mapsto (x^i, x^\alpha, p_i, v^\alpha)$, avec $v^\alpha = \dfrac{\partial H(x,p)}{\partial p_\alpha}$.

2. L'hypothèse ci-dessus n'est pas toujours vérifiée, même lorsque le lagrangien L est hyper-régulier. Considérons par exemple le cas où $M = \mathbf{R}^2$ (coordonnées x^1, x^2), $S = \mathbf{R}$, la projection π étant $(x^1, x^2) \mapsto x^2$. Prenons pour lagrangien

$$L(x,v) = v^1 v^2 \,. \tag{39}$$

La transformation de Legendre a pour expression $(x^1, x^2, v^1, v^2) \mapsto (x^1, x^2, p_1 = v^2, p_2 = v^1)$; c'est un difféomorphisme. Le hamiltonien H a pour expression

$$H(x,p) = p_1 p_2 \,. \tag{40}$$

L'application $(\zeta, \widetilde{T\pi} \circ \mathcal{L}^{-1})$ a pour expression $(x^1, x^2, p_1, p_2) \mapsto (x^1, x^2, p_1, v^2 = p_1)$; ce n'est pas un difféomorphisme.

3. Cependant, cette hypothèse est automatiquement vérifiée si L est un lagrangien classique, de la forme

$$L(x,v) = \frac{1}{2} g_x(v,v) - P(x) \,, \quad x \in M \,, \; v \in T_x M \,, \tag{41}$$

où g est une métrique riemannienne et P une fonction différentiable sur M. Dans les coordonnées locales adaptées des paragraphes 3.6 et 4.2, la métrique g a pour expression

$$g_x(v,v) = \sum_{(i,j)} g_{ij} v^i v^j + 2 \sum_{(i,\alpha)} g_{i\alpha} v^i v^\alpha + \sum_{(\alpha,\beta)} g_{\alpha\beta} v^\alpha v^\beta \,, \tag{42}$$

où i et j prennent les valeurs $1, \ldots, p$, α et β les valeurs $p+1, \ldots, m$. Les g_{ij}, $g_{i\alpha}$ et $g_{\alpha\beta}$ sont fonctions des coordonnées locales (x^i, x^α). La métrique étant définie positive, la matrice $(m-p) \times (m-p)$ formée par ses composantes $g_{\alpha\beta}$ est inversible. Il est facile d'en déduire que $(\zeta, \widetilde{T\pi} \circ \mathcal{L}^{-1})$ est un isomorphisme de fibrés vectoriels de base M.

Comme au paragraphe 3, on suppose que l'action de l'opérateur sur l'état des liaisons actives est décrit par une courbe paramétrée différentiable $\sigma : [t_1, t_2] \to S$. Le mouvement du système est décrit par une courbe paramétrée différentiable $c : [t_1, t_2] \to M$ telle que $\pi(c(t)) = \sigma(t)$ pour tout $t \in [t_1, t_2]$. On note $\dfrac{dc}{dt} : [t_1, t_2] \to TM$ le relèvement naturel de c dans le fibré tangent à M. On compose $\dfrac{dc}{dt}$ avec la transformation de Legendre $\mathcal{L} : TM \to T^*M$ et avec la projection $\zeta : T^*M \to V^*M$. On obtient ainsi une courbe paramétrée $\widehat{c} = \zeta \circ \mathcal{L} \circ \dfrac{dc}{dt} : [t_1, t_2] \to V^*M$, qui représente l'évolution, au cours du temps, de l'état dynamique du système, compte tenu de l'évolution imposée aux liaisons actives.

Pour tout $t \in [t_1, t_2]$, on note $s_t = \sigma(t) \in S$ l'état des liaisons actives et $\dot{s}_t = \dfrac{d\sigma(t)}{dt} \in T_{s_t}S$ l'évolution infinitésimale (dérivée par rapport au temps) de cet état à l'instant t, imposés par l'opérateur. En composant le difféomorphisme $(\zeta, \widetilde{T\pi} \circ \mathcal{L}^{-1})^{-1}$ avec l'application injective $z \mapsto \big(z, \pi^*(\dot{s}_t)\big)$ de la feuille symplectique $\widetilde{\pi}^{-1}(s_t) \subset V^*M$ dans $V^*M \oplus \pi^*(TS)$, on obtient une immersion injective $\chi_{\dot{s}_t} : \widetilde{\pi}^{-1}(s_t) \to T^*M$, qui dépend évidemment du choix de $\dot{s}_t$. Dans les coordonnées locales adaptées des paragraphes 3.6 et 4.2, cette immersion a pour expression $(x^i, x^\alpha, p_i) \mapsto (x^i, x^\alpha, p_i, p_\alpha)$, où les p_α sont obtenus en résolvant l'équation

$$\dot{s}^\alpha_t = \frac{\partial H(x^i, x^\alpha, p_i, p_\alpha)}{\partial p_\alpha} \,. \tag{43}$$

Soit X_H le champ de vecteurs hamiltonien sur T^*M, relativement à sa structure symplectique canonique, associé au hamiltonien H. C'est le champ de vecteurs définissant, dans la formulation hamiltonienne, le mouvement libre du système (mouvement lorsque les liaisons actives sont laissées complètement libres). On restreint ce champ de vecteurs à l'image de l'immersion $\chi_{\dot{s}_t}$, et on projette cette restriction sur V^*M, par le prolongement aux vecteurs de la projection $\zeta : T^*M \to V^*M$. On obtient ainsi un champ de vecteurs $D_{\dot{s}_t}$ *le long* de la sous-variété $\widetilde{\pi}^{-1}(s_t)$ de V^*M (qui, en général, *n'est pas* tangent à cette sous-variété). On remarque que la projection de $D_{\dot{s}_t}$ sur S est automatiquement égale au vecteur $\dot{s}_t \in T_{s_t}S$.

Proposition. *Pour tout* $t \in [t_1, t_2]$, *on a*

$$\frac{d\widehat{c}(t)}{dt} = D_{\dot{s}_t}\big(\widehat{c}(t)\big) \,. \tag{44}$$

Cette proposition exprime le fait que le champ de vecteurs $D_{\dot{s}_t}$ le long de la sous-variété $\widetilde{\pi}^{-1}(s_t)$ représente l'évolution infinitésimale de l'état dynamique du système à l'instant t.

Démonstration. Les équations de Hamilton (36), (37) et (38) ne font qu'exprimer, dans les coordonnées locales adaptées des paragraphes 3.5 et 4.2, la construction géométrique de $D_{\dot{s}_t}$ décrite ci-dessus. $\square$

Remarque. On a construit ci-dessus, pour tout $t \in [t_1, t_2]$, un champ de vecteurs $D_{\dot{s}_t}$ le long de la sous-variété $\widetilde{\pi}^{-1}(s_t)$ de V^*M. Cela ne nous donne pas un champ de vecteurs partout défini sur V^*M. Si l'on souhaite décrire l'évolution infinitésimale du système au moyen d'un champ de vecteurs partout défini sur un espace d'évolution, on peut considérer l'image réciproque $\sigma^*(V^*M)$ de V^*M (considéré comme fibré sur la base S, par la projection $\widetilde{\pi}$) par l'application $\sigma : [t_1, t_2] \to S$. On voit que

274   Charles-Michel Marle

$\sigma^*(V^*M)$ est fibré sur $[t_1, t_2]$ et muni d'une structure canonique, au sens de Lichnerowicz [9]. En prenant pour chaque instant $t \in [t_1, t_2]$ l'image réciproque par σ du champ de vecteurs $D_{\dot{s}_t}$ le long de $\widetilde{\pi}^{-1}(s_t)$, on obtient un champ de vecteurs, noté $\sigma^*(D)$, partout défini sur $\sigma^*(V^*M)$, se projetant sur $[t_1, t_2]$ en le champ constant $\dfrac{\partial}{\partial t}$. Par la même méthode que celle utilisée plus loin (théorème du paragraphe 6), on montre que $\sigma^*(D)$ est un champ de vecteurs canonique au sens de Lichnerowicz [9], qui définit l'évolution infinitésimale de l'état dynamique du système. Cette approche avait été déjà utilisée en [10] pour la description des systèmes mécaniques à liaisons dépendant du temps.

6. Propriétés de la dynamique

Dans cette partie, au lieu de raisonner sur $\sigma^*(V^*M)$ comme dans la remarque précédente, on suppose un champ de vecteurs Y donné sur S. L'opérateur agit sur les liaisons actives de manière telle que l'évolution $\sigma : t \mapsto \sigma(t)$ de leur état au cours du temps soit une courbe intégrale de Y. La construction décrite dans le paragraphe précédent permet alors de définir, pour tout point $s \in S$, un champ de vecteurs $D_{Y(s)}$ le long de la sous-variété $\widetilde{\pi}^{-1}(s)$ de V^*M. La réunion de ces champs de vecteurs le long des feuilles symplectiques de V^*M est un champ de vecteurs D_Y sur V^*M, appelé *champ dynamique*, car la courbe paramétrée $\widehat{c} = \zeta \circ \mathcal{L} \circ \dfrac{dc}{dt}$ dans V^*M qui représente l'évolution, au cours du temps, de l'état dynamique du système, est une courbe intégrale de D_Y. On remarque que D_Y est projetable sur S et a pour projection Y.

Proposition.   *Pour tout champ de vecteurs Y sur S, le champ de vecteurs dynamique correspondant D_Y sur V^*M est un automorphisme infinitésimal de Poisson. Lorsque Y est identiquement nul, c'est-à-dire lorsque les liaisons actives sont figées, le champ dynamique correspondant, noté D_0, est hamiltonien (relativement à la structure de Poisson de V^*M), donc tangent au feuilletage symplectique.*

Démonstration. Soit $\chi : V^*M \to T^*M$ l'application composée de $(\zeta, \widetilde{T\pi} \circ \mathcal{L}^{-1})^{-1} : V^*M \oplus \pi^*(TS) \to T^*M$ et de l'application injective $z \mapsto \left(z, \pi^*Y\big(\widetilde{\pi}(z)\big) \right)$ de V^*M dans $V^*M \oplus \pi^*(TS)$. On pose $\widetilde{H} = H \circ \chi$. Dans les coordonnées locales adaptées du paragraphe 4.2, $\widetilde{H}(x^i, x^\alpha, p_i)$ s'obtient en remplaçant dans $H(x^i, x^\alpha, p_i, p_\alpha)$ les p_α par leurs expressions en fonction des (x^i, x^α, p_i) tirées de

$$\frac{\partial H(x^i, x^\alpha, p_i, p_\alpha)}{\partial p_\beta} = Y^\beta(x^\alpha). \tag{45}$$

On a donc

$$\frac{\partial \widetilde{H}(x^i, x^\alpha, p_i)}{\partial x^i} = \frac{\partial H(x^i, x^\alpha, p_i, p_\alpha)}{\partial x^i} + \sum_{\beta=p+1}^{m} \frac{\partial H(x^i, x^\alpha, p_i, p_\alpha)}{\partial p_\beta} \frac{\partial p_\beta}{\partial x^i}$$

$$= \frac{\partial H(x^i, x^\alpha, p_i, p_\alpha)}{\partial x^i} + \sum_{\beta=p+1}^{m} Y^\beta(x^\alpha) \frac{\partial p_\beta}{\partial x^i} \,,$$

et, de même,

$$\frac{\partial \widetilde{H}(x^i, x^\alpha, p_i)}{\partial p_i} = \frac{\partial H(x^i, x^\alpha, p_i, p_\alpha)}{\partial p_i} + \sum_{\beta=p+1}^{m} Y^\beta(x^\alpha) \frac{\partial p_\beta}{\partial p_i} \,.$$

Comme les $Y^\beta(x^\alpha)$ ne dépendent que des coordonnées x^α, non des (x^i, p_i), on peut écrire

$$\frac{\partial H(x^i, x^\alpha, p_i, p_\alpha)}{\partial x^i} = \frac{\partial}{\partial x^i} \left(\widetilde{H}(x^i, x^\alpha, p_i) - \sum_{\beta=p+1}^{m} Y^\beta(x^\alpha) p_\beta \right) \,,$$

$$\frac{\partial H(x^i, x^\alpha, p_i, p_\alpha)}{\partial p_i} = \frac{\partial}{\partial p_i} \left(\widetilde{H}(x^i, x^\alpha, p_i) - \sum_{\beta=p+1}^{m} Y^\beta(x^\alpha) p_\beta \right) \,,$$

en considérant bien entendu que les p_α sont remplacés par leurs expressions en fonction des (x^i, x^α, p_i). Observons cependant que les dérivations partielles $\dfrac{\partial H}{\partial x^i}$ (resp., $\dfrac{\partial H}{\partial p_i}$) figurant dans les membres de gauche de ces équations doivent être calculées pour x^α, p_j, p_α et x^k ($k \neq i$) constants (resp.,x^j, x^α, p_α et p_k ($k \neq i$) constants), avant remplacement des p_α par leurs expressions en fonction des (x^i, x^α, p_i).

Les équations de Hamilton (36) peuvent donc s'écrire

$$\frac{dx^i}{dt} = \frac{\partial}{\partial p_i} \left(\widetilde{H} - \sum_{\beta=p+1}^{m} Y^\beta p_\beta \right) \,,$$

$$\frac{dp_i}{dt} = -\frac{\partial}{\partial x^i} \left(\widetilde{H} - \sum_{\beta=p+1}^{m} Y^\beta p_\beta \right) \,. \tag{46}$$

Comme on a, d'autre part,

$$\frac{dx^\alpha}{dt} = Y^\alpha \,, \tag{47}$$

le champ de vecteurs D_Y a pour expression

$$D_Y = \sum_{\alpha=p+1}^{m} Y^\alpha \frac{\partial}{\partial x^\alpha} + \sum_{i=1}^{m} \left[\frac{\partial}{\partial p_i} \left(\widetilde{H} - \sum_{\beta=p+1}^{m} Y^\beta p_\beta \right) \frac{\partial}{\partial x^i} \right.$$
$$\left. - \frac{\partial}{\partial x^i} \left(\widetilde{H} - \sum_{\beta=p+1}^{m} Y^\beta p_\beta \right) \frac{\partial}{\partial p_i} \right]. \tag{48}$$

Or le tenseur de Poisson de $V^* M$ a pour expression

$$\Lambda = \sum_{i=1}^{p} \frac{\partial}{\partial p_i} \wedge \frac{\partial}{\partial x^i}. \tag{49}$$

Dans le domaine de la carte considérée, le champ de vecteurs D_Y s'exprime donc comme somme de $\sum_{\alpha=p+1}^{m} Y^\alpha \frac{\partial}{\partial x^\alpha}$, qui est un automorphisme infinitésimal de Poisson de $V^* M$, transverse aux feuilles symplectiques, et du champ de vecteurs hamiltonien associé à $\widetilde{H} - \sum_{\beta=p+1}^{m} Y^\beta p_\beta$, qui est aussi un automorphisme infinitésimal de Poisson de $V^* M$, tangent aux feuilles symplectiques. Dans le cas particulier où Y est nul, le champ de vecteurs correspondant D_0 est hamiltonien (relativement à la structure de Poisson de $V^* M$) et a pour hamiltonien $\widetilde{H}$. $\quad\square$

Remarque. La décomposition du champ dynamique D_Y utilisée dans la démonstration ci-dessus, en somme d'un champ de vecteurs transverse aux feuilles symplectiques et d'un champ hamiltonien taugent aux feuilles symplectiques, est locale; elle dépend du choix des cartes de M et de S adaptées à la submersion π utilisées dans la démonstration. On verra ci-dessous que dans le cas d'un système mécanique classique, il existe une décomposition globale remarquable du champ dynamique D_Y.

7. Cas d'un système mécanique classique

On suppose maintenant que le système considéré est un système mécanique classique, dont le lagrangien L est de la forme (41). Le terme $\frac{1}{2} g_x(v, v)$ est l'énergie cinétique du système et la fonction $P(x)$ son énergie potentielle. La métrique g permet de définir, de manière naturelle, une forme bilinéaire symétrique Θ sur le fibré vectoriel $\pi^*(TS)$. En effet, pour tout point $x \in M$, l'orthogonal de $V_x M = \ker(T_x \pi)$ relativement à g_x est un supplémentaire de $V_x M$ dans $T_x M$. Par suite, la restriction de $T_x \pi$ à cet orthogonal est un isomorphisme de cet orthogonal sur $T_{\pi(x)}S$. On peut donc définir Θ en posant

$$\Theta_x(\pi^*u_1, \pi^*u_2) = g_x(v_1, v_2)\,, \qquad x \in M\,,\ u_1 \text{ et } u_2 \in T_{\pi(x)}S\,, \qquad (50)$$

où π^*u_1 et π^*u_2 sont deux éléments de la fibre en x de $\pi^*(TS)$ correspondant aux vecteurs u_1 et u_2 tangents en $\pi(x)$ à S, et où v_1 et v_2 sont les vecteurs tangents en x à M orthogonaux au sous-espace vertical $V_xM = \ker(T_x\pi)$ (relativement à g_x), et tels que $T_x\pi(v_1) = u_1$, $T_x\pi(v_2) = u_2$.

Comme dans le paragraphe précédent on suppose un champ de vecteurs Y donné sur M. On peut alors définir une fonction $K_Y : M \to \mathbf{R}$ en posant, pour tout $x \in M$,

$$K_Y(x) = \frac{1}{2}\Theta_x\big(\pi^*(Y_{\pi(x)}), \pi^*(Y_{\pi(x)})\big)\,. \qquad (51)$$

La fonction K_Y s'interprète, dans une certaine mesure, comme *l'énergie cinétique des liaisons actives*. Par un léger abus de notations, on notera encore $K_Y : V^*M \to \mathbf{R}$ l'application composée de K_Y et de la projection canonique $q : V^*M \to M$. On note X_{K_Y} le champ de vecteurs hamiltonien sur V^*M (pour la structure de Poisson canonique de cette variété) qui lui est associé. On peut alors énoncer:

Théorème. *Pour tout champ de vecteurs Y sur S, le champ de vecteurs dynamique correspondant D_Y s'exprime comme somme de trois termes:*

$$D_Y = D_0 - X_{K_Y} + H(Y)\,. \qquad (52)$$

Le premier terme D_0 est le champ dynamique correspondant au cas où Y est identiquement nul. Le second terme $-X_{K_Y}$ est l'opposé du champ de vecteurs hamiltonien sur V^*M associé à la fonction K_Y (énergie cinétique des liaisons actives). Le troisième terme $H(Y)$ est le relèvement horizontal du champ de vecteurs Y sur S relativement à une connexion d'Ehresmann [5] sur le fibré $\widetilde{\pi} : V^*M \to S$, appelée connexion dynamique, entièrement déterminée par la submersion $\pi : M \to S$ et par la métrique riemannienne g sur M.

Démonstration. On utilise les coordonnées locales adaptées du paragraphe 4.2, les indices latins $i, j, \ldots$ prenant les valeurs $1, \ldots, p$, et les indices grecs $\alpha, \beta, \ldots$ les valeurs $p + 1, \ldots, m$. Pour alléger l'écriture, on utilise dans la suite la convention d'Einstein (sommation par rapport aux indices apparaissant une fois en position basse et une fois en position haute). On note $(g_{ij}, g_{i\beta}, g_{\alpha j}, g_{\alpha\beta})$ les composantes de la matrice représentant la métrique g, et $(g^{ij}, g^{i\beta}, g^{\alpha j}, g^{\alpha\beta})$ les composantes de la matrice inverse. Le lagrangien L et le hamiltonien H ont pour expressions

$$L = \frac{1}{2}\big(g_{ij}v^iv^j + g_{i\beta}v^iv^\beta + g_{\alpha j}v^\alpha v^j + g_{\alpha\beta}v^\alpha v^\beta\big) - P\,,$$

$$H = \frac{1}{2}\big(g^{ij}p_ip_j + g^{i\beta}p_ip_\beta + g^{\alpha j}p_\alpha p_j + g^{\alpha\beta}p_\alpha p_\beta\big) + P\,.$$

On note $\Theta_{\alpha\beta}$ les composantes de la matrice inverse de la matrice formée par les $g^{\alpha\beta}$. On vérifie aisément que les $\Theta_{\alpha\beta}$ sont bien les composantes de la forme bilinéaire Θ sur le fibré vectoriel $\pi^*(TS)$ dont on a donné ci-dessus une définition intrinsèque.

L'équation (45) s'écrit maintenant, compte tenu de la symétrie de g,

$$Y^\beta = \frac{\partial H}{\partial p_\beta} = g^{i\beta}p_i + g^{\alpha\beta}p_\alpha \,,$$

d'où

$$p_\alpha = \Theta_{\alpha\beta}(Y^\beta - g^{i\beta}p_i) \,.$$

La fonction $\widetilde{H}$ a donc pour expression

$$\widetilde{H} = \widetilde{H}_0 + \frac{1}{2}\Theta_{\alpha\beta}Y^\alpha Y^\beta \,,$$

où on a posé

$$\widetilde{H}_0 = \frac{1}{2}(g^{ij} - g^{i\alpha}\Theta_{\alpha\beta}g^{\beta j})p_i p_j + P \,.$$

On remarque que dans le cas particulier où Y est nul, $\widetilde{H}$ se réduit à $\widetilde{H}_0$. Par suite $\widetilde{H}_0$, tout comme $\widetilde{H}$, est une fonction définie de manière intrinsèque sur V^*M, qui ne dépend pas du choix des cartes dans lesquelles on l'a exprimée.

La fonction (localement définie, dans les cartes considérées) $\widetilde{H} - Y^\alpha p_\alpha$, qui intervient dans l'écriture (46) des équations de Hamilton, peut s'écrire sous la forme

$$\widetilde{H} - Y^\alpha p_\alpha = \widetilde{H}_0 + \Theta_{\alpha\beta}g^{i\beta}p_i Y^\alpha - \frac{1}{2}\Theta_{\alpha\beta}Y^\alpha Y^\beta \,.$$

En remarquant que les composantes de g et de Θ ne sont fonctions que des coordonnées locales (x^i, x^α), et que les Y^α ne sont fonctions que des (x^α), on voit que les équations de Hamilton (46) peuvent s'écrire

$$\frac{dx^i}{dt} = \frac{\partial \widetilde{H}_0}{\partial p_i} + \Theta_{\alpha\beta}g^{i\beta}Y^\alpha \,,$$

$$\frac{dp_i}{dt} = -\frac{\partial \widetilde{H}_0}{\partial x^i} - p_k Y^\alpha \frac{\partial(\Theta_{\alpha\beta}g^{k\beta})}{\partial x^i} + \frac{1}{2}Y^\alpha Y^\beta \frac{\partial \Theta_{\alpha\beta}}{\partial x^i} \,.$$

Le champ de vecteurs dynamique D_Y peut donc s'écrire

$$D_Y = \left(\frac{\partial \widetilde{H}_0}{\partial p_i}\frac{\partial}{\partial x^i} - \frac{\partial \widetilde{H}_0}{\partial x^i}\frac{\partial}{\partial p_i} \right)$$
$$+ Y^\alpha \left(\frac{\partial}{\partial x^\alpha} + \Theta_{\alpha\beta}g^{\beta k}\frac{\partial}{\partial x^k} - p_k\frac{\partial(\Theta_{\alpha\beta}g^{\beta k})}{\partial x^i}\frac{\partial}{\partial p_i} \right)$$
$$+ \frac{1}{2}Y^\alpha Y^\beta \frac{\partial \Theta_{\alpha\beta}}{\partial x^i}\frac{\partial}{\partial p_i} \,.$$

On voit que D_Y est somme de trois termes. Le premier,

$$\frac{\partial \widetilde{H}_0}{\partial p_i}\frac{\partial}{\partial x^i} - \frac{\partial \widetilde{H}_0}{\partial x^i}\frac{\partial}{\partial p_i},$$

n'est autre que D_0, champ dynamique dans le cas particulier où les liaisons sont bloquées. Le troisième terme, quadratique par rapport aux Y^α, s'écrit

$$\frac{1}{2}Y^\alpha Y^\beta \frac{\partial \Theta_{\alpha\beta}}{\partial x^i}\frac{\partial}{\partial p_i}.$$

Mais la fonction K_Y, définie par (51), a pour expression

$$K_Y = \frac{1}{2}\Theta_{\alpha\beta}Y^\alpha Y^\beta.$$

Elle n'est fonction que des coordonnées (x^i, x^α). Par suite le troisième terme, qui s'écrit

$$\frac{\partial K_Y}{\partial x^i}\frac{\partial}{\partial p^i},$$

n'est autre que $-X_{K_Y}$. Ce terme est donc défini de manière intrinsèque, indépendamment du choix des cartes dans lesquelles on l'a exprimé.

Puisque le premier et le troisième terme de D_Y, ainsi que D_Y lui-même, sont définis de manière intrinsèque, il en est de même de son second terme. Celui-ci a pour expression

$$Y^\alpha\left(\frac{\partial}{\partial x^\alpha} + \Theta_{\alpha\beta}g^{\beta k}\frac{\partial}{\partial x^k} - p_k \frac{\partial(\Theta_{\alpha\beta}g^{\beta k})}{\partial x^i}\frac{\partial}{\partial p_i}\right). \tag{53}$$

Il dépend linéairement des Y^α, et se projette sur S selon le champ de vecteurs Y. C'est donc le relèvement horizontal de Y relativement à une connexion d'Ehresmann sur V^*M. L'espression (53) montre que cette connexion est entièrement déterminée par la submersion π et la métrique riemannienne g sur M. $\quad\square$

Afin de donner une définition intrinsèque de la connexion dynamique sur le fibré $\widetilde{\pi} : V^*M \to S$, on remarque que la métrique riemannienne g permet de définir, de manière naturelle, une connexion d'Ehresmann sur le fibré $\pi : M \to S$, qu'on appellera *connexion cinétique* (car elle est définie au moyen de l'énergie cinétique). Par définition, pour tout point x de M et tout vecteur $u \in T_{\pi(x)}S$, le relèvement horizontal de u au point x relativement à la connexion cinétique est l'unique vecteur $v \in T_xM$ orthogonal (relativement à la métrique g) au sous-espace vertical $V_xM = \ker(T_x\pi)$, tel que $T_x\pi(v) = u$. On a alors:

Proposition. *La connexion dynamique est caractérisée par les deux propriétés suivantes.*

1. *Pour tout $z \in V^*M$ et tout $u \in T_{\widetilde{\pi}(z)}S$, le relèvement horizontal de u au point z relativement à la connexion dynamique a pour projection (par Tq) sur M le relèvement horizontal de u au point $x = q(z)$ relativement à la connexion cinétique.*

2. *Le relèvement horizontal sur V^*M, relativement à la connexion dynamique, de tout champ de vecteurs Y sur S, est un automorphisme infinitésimal de la structure de Poisson de V^*M, tangent à la section nulle de $q : V^*M \to M$.*

Démonstration. Soit $z \in V^*M$, $u \in T_{\widetilde{\pi}(z)}S$. On utilise les coordonnées locales adaptées du paragraphe 4.2. D'après (53), le relèvement horizontal de u au point z a pour expression

$$H(u) = u^\alpha \left(\frac{\partial}{\partial x^\alpha} + \Theta_{\alpha\beta} g^{\beta i} \frac{\partial}{\partial x^i} - p_j \frac{\partial(\Theta_{\alpha\beta} g^{\beta j})}{\partial x^i} \frac{\partial}{\partial p_i} \right) .$$

La projection v de $H(u)$ sur M a pour expression

$$v = u^\alpha \left(\frac{\partial}{\partial x^\alpha} + \Theta_{\alpha\beta} g^{\beta i} \frac{\partial}{\partial x^i} \right) .$$

Soit $w = w^i \dfrac{\partial}{\partial x^i}$ un vecteur vertical, élément de $\ker(T_x \pi)$. On a

$$g(v, w) = (g_{ij} \Theta_{\alpha\beta} g^{\beta i} + g_{\alpha j}) u^\alpha w^j .$$

Mais puisque $(g^{ij}, g^{i\beta}, g^{\alpha j}, g^{\alpha\beta})$ est la matrice inverse de $(g_{ij}, g_{i\beta}, g_{\alpha j}, g_{\alpha\beta})$,

$$g_{ij} g^{i\beta} + g_{\lambda j} g^{\lambda\beta} = \delta_j^\beta = 0 ,$$

d'où l'on déduit, puisque $(\Theta_{\alpha\beta})$ est la matrice inverse de $(g^{\alpha\beta})$,

$$g_{ij} g^{i\beta} \Theta_{\beta\alpha} + g_{\lambda j} g^{\lambda\beta} \Theta_{\beta\alpha} = g_{ij} g^{i\beta} \Theta_{\beta\alpha} + g_{\alpha j} = 0 .$$

Compte tenu de la symétrie de g et de Θ, ceci prouve que

$$g(v, w) = 0 .$$

Le vecteur v est donc bien le relèvement horizontal de u au point x relativement à la connexion cinétique.

Pour tout champ de vecteurs Y sur S, D_Y est un automorphisme infinitésimal de la structure de Poisson de V^*M. Comme D_0 et X_{K_Y} sont aussi des automorphismes infinitésimaux de cette structure, il en est de même de $H(Y) = D_Y - D_0 + X_{K_Y}$. De plus, l'expression (53) montre que $H(Y)$ est tangent à la section nulle de $q : V^*M \to M$.

On a donc prouvé que la connexion dynamique vérifie les propriétés 1 et 2. Soit H' l'opérateur de relèvement horizontal relativement à une autre connexion d'Ehresmann sur le fibré $\tilde{\pi} : V^*M \to S$, vérifiant aussi ces deux propriétés. Pour tout champ de vecteurs Y sur S, $H'(Y) - H(Y)$ est un automorphisme infinitésimal de Poisson tangent aux fibres de $q : V^*M \to M$. C'est donc un champ de vecteurs localement hamiltonien dont le hamiltonien local n'est fonction que des coordonnées locales (x^i, x^α) sur V^*M, non des coordonnées locales (p_i). Puisqu'il dépend linéairement de Y, son expression locale est de la forme

$$Y^\alpha \frac{\partial W_\alpha}{\partial x^i} \frac{\partial}{\partial p_i} \, ,$$

où les W_α sont fonctions des (x^i, x^α). Mais de plus, $H'(Y) - H(Y)$ doit s'annuler sur la section nulle de $q : V^*M \to M$. L'expression ci-dessus montre alors qu'il est identiquement nul. $\square$

8. Quelques exemples

On se propose de montrer dans ce paragraphe que divers systèmes mécaniques à liaisons actives rencontrés en pratique, notamment les exemples cités dans l'introduction, relèvent du formalisme développé dans ce travail.

8.1. Enfant sur une escarpolette.

L'espace de configuration du système est une variété différentiable sur laquelle le cercle T^1 agit librement. Cette action correspond à la rotation en bloc de l'escarpolette et de l'enfant autour de l'axe horizontal de l'escarpolette. La variété difffentiable S est l'ensemble des orbites de cette action, et la submersion π est l'application qui, à chaque point de M, associe l'orbite de ce point. La variété S représente l'ensemble des formes possibles du corps de l'enfant et des positions relatives de l'enfant par rapport à l'escarpolette.

On suppose que le centre de masse du système n'est jamais situé sur l'axe de l'escarpolette. Il existe alors une section globale de la submersion $\pi : M \to S$, qui, à chaque point $s \in S$, fait correspondre le point de $\pi^{-1}(s)$ pour lequel le centre de masse du système est situé le plus bas possible. Cette section permet d'identifier M au produit $T^1 \times S$, la submersion π s'identifiant à la projection sur le second facteur. On utilisera une carte locale quelconque de S et on notera (s^α) les coordonnées locales correspondantes $(2 \leq \alpha \leq m = \dim S + 1)$. Les coordonnées locales sur $M = T^1 \times S$ sont donc (θ, s^α). Elles sont adaptées à la submersion π au sens des paragraphes 3.5 et 4.2. On notera $(\theta, s^\alpha, \omega, v^\alpha)$ les coordonnées locales correspondantes sur TM. Le lagrangien L du système est de la forme (41), avec

$$P(\theta, s) = -D(s)\cos\theta. \tag{54}$$

On posera

$$g_{11}(s) = I(s), \quad g_{1\alpha}(s) = g_{\alpha 1}(s) = J_\alpha(s), \tag{55}$$

d'où l'expression de L

$$L(\theta, s^\alpha, \omega, v^\alpha) = \frac{1}{2}I(s)\omega^2 + J_\alpha(s)\omega v^\alpha + g_{\alpha\beta}(s)v^\alpha v^\beta + D(s)\cos\theta. \tag{56}$$

On remarque que les coefficients I, J_α, $g_{\alpha\beta}$ de la métrique riemannienne g ne dépendent que des variables $s = (s^\alpha)$, non de la variable θ, car le cercle T^1 agit sur M par isométries. Le terme D est lui aussi fonction seulement des variables $s = (s^\alpha)$; il représente le produit de l'accélération de la pesanteur et de la distance du centre de masse du système à l'axe de l'escarpolette.

8.2. Chat en chute libre.

L'espace de configuration est une variété différentiable $M = E \times N$, produit de l'espace euclidien E de dimension 3 et d'une autre variété N. Le facteur E correspond à l'ensemble des positions possibles du centre de masse G du chat dans l'espace, et le facteur N à l'ensemble des formes et des orientations possibles du corps du chat une fois la position du centre de masse fixée. On choisira arbitrairement une origine O dans E afin de pouvoir le considérer comme un espace vectoriel. On notera x_G le vecteur $\overrightarrow{OG}$, y un point courant de N; (x_G, y) désigne donc un point courant de M.

Le groupe G des déplacements de l'espace euclidien de dimension 3 agit librement sur M. Ce groupe est produit semi-direct de E (sous-groupe des translations) et de $\mathbf{SO}(3)$ (sous-groupe des rotations). On notera (l, r), $l \in E$, $r \in \mathbf{SO}(3)$, un élément de G. L'action de G sur M s'exprime par

$$(l, r).(x_G, y) = \bigl(l + r.x_G, \rho(r, y)\bigr), \tag{57}$$

où $\rho : \mathbf{SO}(3) \times N \to N$ est une action libre du groupe des rotations sur la variété N (correspondant à la rotation en bloc du corps du chat autour de son centre de masse).

La variété S est ici l'ensemble des orbites de l'action de $\mathbf{SO}(3)$ sur N: c'est l'ensemble des formes que peut prendre le corps du chat, indépendamment de sa position et de son orientation dans l'espace.

On sait que l'énergie cinétique du système est somme de l'énergie cinétique d'un point matériel de masse m égale à la masse totale, coïncidant avec le centre de masse, et de l'énergie cinétique du mouvement relatif du système par rapport au repère du centre de masse (repère en translation

par rapport auquel le centre de masse est fixe). D'autre part, l'énergie
potentielle du système dans le champ de pesanteur ne dépend que de la
position du centre de masse. Le lagrangien L du système est donc de la
forme

$$L = \frac{1}{2}m|\vec{v_G}|^2 - d(x_G) + \frac{1}{2}g_N(v_N, v_N)\,, \tag{58}$$

où d est le produit de l'altitude du centre de masse au dessus d'un plan
horizontal de référence, de la masse m du système, et du module de l'accélération de la pesanteur. Le terme $\frac{1}{2}g_N(v_N, v_N)$ est l'énergie cinétique du
mouvement relatif du système par rapport au repère du centre de masse;
g_N est une métrique riemannienne sur la variété N invariante par l'action ρ
de $\mathbf{SO}(3)$, et v_N désigne le vecteur tangent à N représentant la distribution
des vitesses relatives du système par rapport au repère du centre de masse.

L'expression du lagrangien L permet de retrouver un résultat bien
connu: le système des équations du mouvement se sépare en deux parties, correspondant l'une au mouvement du centre de masse, l'autre au
mouvement relatif par rapport au repère du centre de masse. La première
partie est

$$m\frac{d^2\vec{x_G}}{dt^2} + \overrightarrow{\mathrm{grad}}\,d(x_G) = 0\,, \tag{59}$$

tandis que la seconde partie est formée par les équations de Lagrange associées au lagrangien

$$L_N = \frac{1}{2}g_N(v_N, v_N)\,. \tag{60}$$

On est ainsi ramené à appliquer la méthode développée au paragraphe
4 au système réduit, dont l'espace de configuration est N et le lagrangien
L_N. La projection $\pi : N \to S$ est maintenant l'application qui associe à
chaque point de N son orbite sous l'action ρ du groupe $\mathbf{SO}(3)$.

8.3. Satellite artificiel.

Lorsqu'on suppose les dimensions du satellite petites auprès de son altitude au dessus de la terre, on peut admettre que son énergie potentielle
dans le champ attractif terrestre ne dépend que de la position de son centre
de masse. On voit alors que le lagrangien L du système est de même forme
que celui du chat en chute libre (la fonction $d(x_G)$ ayant cependant une
expression différente). Ici encore, le système des équations du mouvement
se sépare en une partie correspondant au mouvement du centre de masse
et une partie correspondant au mouvement relatif par rapport au repère
du centre de masse. Pour ce mouvement relatif, le problème du satellite
artificiel à liaisons actives est donc équivalent au problème du chat en chute
libre.

8.4. Anneau sur un cerceau.

Dans les trois exemples dont l'étude est esquissée ci-dessus, la fonction

K_Y, définie par l'équation (51), est constante sur chaque feuille symplectique de V^*M. Par suite, le terme $-X_{K_Y}$ qui apparaît dans la décomposition (52) du champ de vecteurs dynamique en somme de trois termes, est identiquement nul. On va donner un exemple, précédemment étudié par Aharonov et Anandan [2], dans lequel X_{K_Y} est non nul.

Le système considéré est constitué par un cerceau (tige mince rigide formant une courbe fermée plane différentiable, pas nécessairement circulaire) sur lequel est enfilé un anneau. Cet anneau, de dimensions négligeables, est assimilé à un point matériel P de masse m; il peut coulisser sans frottement sur le cerceau. L'opérateur contrôle la position du cerceau dans un plan (qu'on supposera horizontal, pour que la pesanteur n'intervienne pas dans le problème), mais non la position de l'anneau sur le cerceau. Pour simplifier, on supposera que l'ensemble des positions possibles du cerceau dépend d'un seul paramètre: l'opérateur peut faire tourner ce cerceau en bloc autour d'un point O de son plan. On supposera le cerceau de forme étoilée autour de O (cela signifie que toute demi-droite du plan issue de O coupe le cerceau en un point unique, l'intersection étant transverse). On choisit un point A lié au cerceau, et une orientation du plan, ce qui permet de repérer la configuration du système au moyen de deux paramàtres:

- l'angle $\theta = (OA, OP)$,
- l'angle β que fait OA avec une direction de référence.

La variété de configuration M est donc un tore de dimension 2, et la variété S un tore de dimension 1. La projection $\pi : M \to S$ s'exprime par $(\theta, \beta) \mapsto \beta$. On remarque que S est, comme dans les exemples précédents, l'ensemble des orbites d'une action d'un groupe sur M. Dans le cas présent, le groupe qui agit est le cercle; mais cette action ne correspond pas, comme dans les exemples précédents, à la rotation en bloc du système autour d'un axe ou d'un point; elle correspond plutôt au déplacement de l'anneau P le long du cerceau.

En pratique, pour repérer la position de P sur le cerceau, il est plus commode d'utiliser, plutôt que l'angle θ, l'abscisse curviligne ξ de P à partie de l'origine A. On doit considérer ξ comme définie modulo l (longueur du cerceau). L'angle θ est une fonction différentiable de ξ dont la dérivée ne s'annule en aucun point. La distance $r = OP$ est aussi une fonction différentiable de ξ, à valeurs strictement positives.

Le lagrangien L du système a pour expression

$$L = \frac{1}{2}m\big(r'^2\dot{\xi}^2 + r^2(\theta'\dot{\xi} + \dot{\beta})^2\big) + \frac{1}{2}I\dot{\beta}^2 \,. \tag{61}$$

On a noté (ξ, β) les coordonnées locales sur M, $(\xi, \beta, \dot{\xi}, \dot{\beta})$ les coordonnées locales correspondantes sur TM, r' et θ' les dérivées, respectivement de r et de θ, par rapport à ξ, m la masse de P et I le moment d'inertie du cerceau par rapport au point O. On vérifie d'ailleurs que I n'apparaît pas dans les équations finales, ce qui est en accord avec le fait que l'opérateur impose la position du cerceau; on ne l'a introduit que pour rendre le lagrangien hyper-régulier.

Le hamiltonien H du système a pour expression

$$H = \frac{1}{2(I + mr^2 r'^2)} \left(\frac{I + mr^2}{m} p_\xi{}^2 - 2r^2 \theta' p_\xi p_\beta + p_\beta{}^2 \right), \qquad (62)$$

où on a noté $(\xi, \beta, p_\xi, p_\beta)$ les coordonnées locales sut T^*M associées aux coordonnées locales (ξ, β) sur M.

Conformément aux hypothèses et notations du paragraphe 7, on suppose que l'opérateur contrôle le mouvement du cerceau de manière telle que l'évolution de l'angle β en fonction du temps soit une courbe intégrale de l'équation différentielle

$$\frac{d\beta}{dt} = Y\big(\beta(t)\big).$$

La fonction K_Y définie par (51) a pour expression

$$K_Y(\xi, \beta) = \frac{1}{2}(I + mr^2 r'^2)\big(Y(\beta)\big)^2. \qquad (63)$$

Les équations du mouvement s'écrivent

$$\begin{aligned}
\frac{d\xi(t)}{dt} &= \frac{1}{m} p_\xi - r^2 \theta' Y\big(\beta(t)\big), \\
\frac{dp_\xi(t)}{dt} &= (r^2 \theta')' p_\xi Y\big(\beta(t)\big) + \frac{m}{2}(r^2 r'^2)' \big(Y(\beta(t))\big)^2,
\end{aligned} \qquad (64)$$

où on a noté $'$ la dérivation par rapport à ξ. Les trois termes de la décomposition (52) du champ de vecteurs dynamique D_Y sur V^*M sont donc:

$$D_0 = \frac{1}{m} p_\xi \frac{\partial}{\partial \xi}, \qquad (65)$$

$$H(Y) = Y(\beta) \left(\frac{\partial}{\partial \beta} - r^2 \theta' \frac{\partial}{\partial \xi} + (r^2 \theta')' p_\xi \frac{\partial}{\partial p_\xi} \right), \qquad (66)$$

$$-X_{K_Y} = \frac{m}{2}(r^2 r'^2)'\big(Y(\beta)\big)^2 \frac{\partial}{\partial p_\xi}. \qquad (67)$$

On remarque que p_ξ est le produit par m de la mesure de la projection orthogonale de la vitesse absolue de l'anneau sur la tangente au cerceau, et que K_Y est la somme de l'énergie cinétique du cerceau et de la partie de l'énergie cinétique de l'anneau associée à la composante normale au cerceau de sa vitesse.

REFERENCES

[1]. R. Abraham and J. E. Marsden, "Foundations of mechanics", Benjamin, Reading, 1978.

[2]. Y. Aarhonov and J. Anandan, *Phase change during quantum evolution*, Phys. Rev. Letters **58** (1987), 1593–1596.

[3]. V. I. Arnold, V. V. Kozlov and A. I. Neishtadt, *Mathematical aspects of classical and celestial mechanics*, in "Dynamical systems III" (V. I. Arnold, ed.), Springer Verlag, Berlin, Heidelberg, 1988.

[4]. F. Cardin and G. Zanzotto, *On constrained mechanical systems: d'Alembert's and Gauss' principles*, J. Math. Phys. **30** (1989), 1473–1479.

[5]. C. Ehresmann, *Les connexions infinitésimales dans un espace fibré différentiable*, in "Colloque de Topologie, Bruxelles, 1950", p. 29–55, Masson, Paris, 1950.

[6]. C. Godbillon, "Géométrie différentielle et mécanique analytique", Hermann, Paris, 1969.

[7]. D. Holm, communication privée.

[8]. P. Libermann and C.-M. Marle, "Symplectic geometry and analytical mechanics", D. Reidel Publishing Company, Dordrecht, 1987.

[9]. A. Lichnerowicz, *La géométrie des transformations canoniques*, Bull. Soc. Math. de Belgique **31** (1979), 105–135.

[10]. C.-M. Marle, *Contact manifolds, canonical manifolds and the Hamilton-Jacobi method in analytical mechanics*, in "Proc. IUTAM-ISIMM Symposium on *Modern developments in analytical mechanics*" (S. Benenti, M. Francaviglia and A. Lichnerowicz, ed.), 255–277. Acta Academiae Scientiarum Taurinensis, Torino, Italy, 1983.

[11]. C.-M. Marle, *Sur la géométrie des systèmes mécaniques à liaisons actives*, C.R. Acad. Sc. Paris, **311**,I (1990), 839–845.

[12]. J. Marsden, R. Montgomery and T. Ratiu, *Reduction, symmetry and phases in mechanics*, Memoirs of the American Mathematical Society, **88** (1990), 436.

[13]. R. Montgomery, *The connection whose holonomy is the classical adiabatic angles of Hannay and Berry and its generalization to the nonintegrable case*, Commun. Math. Phys. **120** (1988), 269–294.

[14]. R. Montgomery, *Isoholonomic problems and some applications*, Commun. Math. Phys. **128** (1990), 565–590.

[15]. J.-M. Souriau, "Géométrie et relativité", Hermann, Paris, 1964.

[16]. J.-M. Souriau, "Structure des systèmes dynamiques", Dunod, Paris, 1969.

[17]. W. M. Tulczyjew, *Sur la différentielle de Lagrange*, C. R. Acad. Sc. Paris **280** A (1975), 1295–1298.

[18]. E. T. Whittaker, "A treatise on the analytical dynamics of particles and rigid bodies", Cambridge University Press, Cambridge, 1904 (fourth edition 1937, reissued with Foreword 1988).

[19]. N. Woodhouse, "Geometric quantization", Oxford University Press, Oxford, 1980.

Université Pierre et Marie Curie, Mathématiques
4, place Jussieu
F-75252 Paris Cedex 05
France

Structures de Poisson affines

ALBERTO MEDINA

A J.-M. Souriau

Abstract

Let G be a Lie group whose Lie algebra $\mathcal{G}$ is given by $\mathcal{G} = T_\varepsilon G$. Let $\rho(\sigma) = (Q(\sigma),\, Ad^*(\sigma))$ be an affine representation whose linear part is the coadjoint representation of G. Let us suppose that the differential $d_\varepsilon\, Q$ is antisymmetric. The tensor $d_\varepsilon\, Q$ defines in $\mathcal{G}^*$ a Poisson structure Λ whose the leaves are the orbits of ρ. The above Poisson structure is called pseudo-riemanian if ρ is $\mathcal{G}^* \times 0(p,q)$-valued, $p + q = \dim \mathcal{G}$; in that case every orbit F of ρ has a G-invariant foliation the leaves of which are affine subspaces of $\mathcal{G}^*$ whenever the dimension of F is maximal.

§0. Introduction. Principaux résultats

Soit V un espace vectoriel réel de dimension finie muni d'un tenseur de Poisson Λ. Nous dirons que Λ est affine si le crochet de Poisson de deux formes linéaires sur V est une forme affine [M1]. Se donner Λ revient à se donner une structure d'algèbre de Lie sur $\mathcal{G} = V^*$ et un 2-cocycle scalaire ω de $\mathcal{G}$. Si G est le groupe de Lie connexe et 1-connexe d'algèbre $\mathcal{G}$, les feuilles de Λ sont les orbites de la représentation $\rho_\omega : G \to GA(\mathcal{G}^*)$, par des transformations affines de G^*, dont $x \longmapsto (i_x\omega, ad^*(x))$ est la représentation infinitésimale associée.

Soit Λ affine. S'il existe sur V une forme quadratique non dégénérée q telle que les hamiltoniens ξ_f, $f \in V^* = \mathcal{G}$, soient des isométries infinitésimales de la variété pseudo-riemannienne (V, q) nous dirons que (V, Λ) est affine pseudo-riemannienne [M2]. Que Λ soit pseudo-riemannienne signifie que $\mathcal{G}$ est munie d'une forme bilinéaire symétrique non dégénérée $<, >$ pour laquelle les $ad(x)$ $x \in \mathcal{G}$ sont antisymétriques. On dit dans ce cas que $(\mathcal{G} <, >)$ est une algèbre de Lie orthogonale ([M-R1],[M-R2]). Si $\mathcal{G}$ est orthogonale les représentations $x \longmapsto (i_x\omega, ad^*(x))$ et $x \longmapsto D(x), ad(x))$ sont isomorphes où D et ω sont liés par la relation $\omega(a,b) =< Da, b >$.

Si Λ est affine et pseudo-riemannien, les feuilles pour Λ sont feuilletées par un feuilletage G-invariant (Théorème 2.1). Une feuille F de (V, Λ) est un espace homogène réductif si et seulement si le feuilletage de F est fait par des points ou ce qui est de même F est une sous-variété pseudo-riemannienne de (V, Λ) (Théorème 2.3). S'il existe une feuille de

(V, Λ) réductive et de dimension maximale, l'ensemble des points réductifs et réguliers au sens de Poisson de $V = \mathcal{G}$ coïncide avec l'ensemble des points réguliers de $\mathcal{G}$ (Théorème 2.6).

Que Λ est de rang nul en un point signifie que Λ est linéaire. Si Λ est linéaire et pseudo-riemannien les feuilles de $(\mathcal{G}, \Lambda)$ de dimension maximale sont feuilletées par des sous-espaces affines de $\mathcal{G}$ (Théorème 2.9). Ce résultat a été obtenu en collaboration avec P.Molino. Si une algèbre orthogonale $\mathcal{G}$ possède un point dont l'orbite pour la représentation adjointe de G est réductive et de dimension maximale alors toute sous-algèbre de Cartan de G est commutative maximale et de la forme $\ker(ad(x))$ pour x dans $\mathcal{G}$ (Théorème 2.6). Supposons Λ affine et ayant une feuille ouverte F. Alors G possède une forme symplectique invariante à gauche Ω qui est la pre-image par l'application orbitale de la forme symplectique de F. La variété symplectique (G, Ω) possède un feuilletage lagrangien invariant à gauche si G est complètement résoluble (Théorème 3.1). Ceci implique en particulier que toute nilvariété symplectique possède un feuilletage lagrangien.

Hormis les théorèmes 2.6 et 2.9 tous les résultats contenus dans cet article ont été présentés lors d'une conférence faite à Casablanca en mai 1987 à l'occasion du Séminaire International de Géométrie et Analyse.

Dans la suite $\widetilde{G}$ désigne le groupe de Lie connexe et 1-connexe d'algèbre de Lie $\mathcal{G}$.

§1. Structures de Poisson affines. Propriétés élémentaires

Dans la suite V est un espace vectoriel réel de dimension finie muni de sa structure de variété différentiable usuelle.

1.1. Définition : *Une structure de Poisson $\{\ \}$ sur V est dite affine si pour tout couple de formes linéaires f et h sur V et tout x dans V on a :*

$$\{f, h\}(x) = -\{f, h\}_L(x) + \omega(f, h) \tag{1}$$

où $\{f, h\}_L \in V^*$ et $\omega(f, h) \in \mathbf{R}$, $\mathbf{V}^*$ étant l'espace dual de V.

Si $\{\ \}$ est affine il est clair que l'espace des formes affines sur V est une sous-algèbre de dimension finie de l'algèbre de Lie $(C^\infty(V), \{\ \})$. De plus pour f, g et h dans cette sous-algèbre on a :

$$\{\{f, h\}, g\} = \{\{f, h\}_L, g\}$$

et dont l'identité de Jacobi implique que ω est un 2-cocycle réel de cette sous-algèbre.

Aussi si pour f, h dans V on pose $[f, h] = \{f, h\}_L$, il est immédiat de vérifier que $[,]$ est un crochet de Lie sur V^* et que ω est un 2-cocycle réel relatif à ce crochet. L'algèbre $A = V^*$ dont le produit est $[,]$ et le 2-cocycle

ω de A seront appelés respectivement l'algèbre de Lie et le cocycle de la structure de Poisson affine.

Réciproquement si $\mathcal{G}$ est une algèbre de Lie et ω est un 2-cocycle scalaire de $\mathcal{G}$, on définit une structure de Poisson afine sur la variété $V = \mathcal{G}^*$ en posant :

$$\{f, h\}(x) = -[f, h](x) + \omega(f, h) \tag{1'}$$

pour f, h dans $\mathcal{G}$ et x dans $\mathcal{G}^*$.

Il est évident que l'algèbre de Lie et le cocycle de cette structure sont respectivement $\mathcal{G}$ et ω. On a donc :

1.2 Lemme : *La donnée d'une structure de Poisson affine sur V équivaut à la donnée d'une structure d'algèbre de Lie sur $V^* = \mathcal{G}$ et d'un 2-cocyle réel de $\mathcal{G}$.*

Soit $GA(V) = V \times GL(V)$ le groupe des transformations affines de V, $aff(V) = V \times gl(V)$ son algèbre de Lie. Se donner un 2-cocycle ω de l'algèbre $\mathcal{G}$ équivaut à se donner un homomorphisme d'algèbres de Lie $\mathcal{G} \xrightarrow{\rho} aff(\mathcal{G}^*)$ $\rho(a) = (q(a), ad^*(a))$ dont q vérifie la condition

$$q(a)(b) = -q(b)(a) \qquad \text{pour } a, b \text{ dans } \mathcal{G}. \tag{2}$$

Si G est le groupe de Lie connexe 1-connexe d'algèbre $\mathcal{G}$, ρ détermine une représentation ρ_ω de G par des transformations affines de $\mathcal{G}^*$ par la formule

$$\rho_\omega(\sigma) = (Q(\sigma), Ad^*(\sigma)). \tag{3}$$

où $Q(\sigma) = (\sum_{k=1}^{\infty} \frac{1}{k!}(ad^*(a))^{k-1}).i_a\omega$ pour $\sigma = \exp(a)$, $\sigma \to Ad^*(\sigma)$ étant la représentation co-adjointe. Par conséquent les feuilles de la structure de Poisson affine sur $V = \mathcal{G}^*$, d'algèbre $\mathcal{G}$ et cocycle ω, sont les orbites de l'action de G sur $\mathcal{G}^*$ définie par ρ_ω.

Plus généralement soit G un groupe de Lie connexe d'algèbre $\mathcal{G}$ et $\rho(\sigma) = (Q(\sigma), Ad^*(\sigma))$, $\sigma \in G$, une représentation affine de G dans $\mathcal{G}^*$. Si $q = T_\varepsilon Q$ est alterné, c'est-à-dire vérifie (2), l'expression :

$$\{a, b\}(\nu) = -\nu([a, b]) + q(a)b \tag{4}$$

pour a, b dans $\mathcal{G}$ et ν dans $\mathcal{G}^*$ définit une structure de Poisson affine sur $V = \mathcal{G}^*$ d'algèbre $\mathcal{G}$ et cocycle $\omega(a, b) = q(a)(b)$. On a ainsi,

1.3 Lemme : 1. *Si V est muni d'une structure de Poisson affine d'algèbre $\mathcal{G}$ et cocycle ω, le groupe de Lie connexe et 1-connexe d'algèbre $\mathcal{G}$ agit sur V au moyen de ρ_ω de sorte que les orbites pour cette action sont les feuilles symplectiques de V.*

2. *Si G est un groupe de Lie connexe d'algèbre $\mathcal{G}$ et si $\rho(\sigma) = (Q(\sigma), Ad^*(\sigma))$ est une représentation de cocycle q alternée alors (4) définit une structure de Poisson affine sur $\mathcal{G}^*$.*

Evidemment les feuilles de V sont des espaces homogènes symplectiques pour l'action ρ_ω de G sur V. (voir [**S**], page 115).

1.4 Remarques. 1. Si $\omega' = \omega + \delta\lambda$ est un 2-cocycle de $\mathcal{G}$ cohomologue à ω, il est immédiat de constater que la translation $\beta \longmapsto \beta + \lambda$ est un isomorphisme entre les représentations ρ_ω et $\rho_{\omega'}$ de $\widetilde{G}$. Ainsi les structures de Poisson affines sur $V = \mathcal{G}^*$ dépendent des espaces de cohomologie $H^2(V, \mathbf{R})$ par rapport à chacune des structures d'algèbre de Lie sur l'espace V.

2. Une structure de Poisson affine sur V est de rang nul en un point si et seulement si elle est linéaire i,e ρ_ω est isomorphe à Ad^*.

En fait si $\rho = (Q, F)$ est une représentation affine continue de G dans V, les assertions suivantes sont équivalentes :

a) ρ est isomorphe à F.

b) ρ est isomorphe à une représentation linéaire de G.

c) Il existe u dans V fixe pour ρ.

d) Le cocycle Q est cohomologue à un 1-cobord, c'est-à-dire, il existe $v \in V$ tel que $Q(\sigma) = F_\sigma(v) - v$ quel que soit $\sigma \in G$.

1.5 Proposition : *Une structure de Poisson affine qui possède une feuille compacte est linéaire.*

En effet si l'orbite de $\nu \in \mathcal{G}^*$ pour G, $\quad M = G \cdot \nu$ est compacte et $c(M)$ est le centre de M (voir par exemple [**B**], page 85), l'ensemble $X = \{F \in GA(\mathcal{G}^*) ; \quad F(M) = M\}$ n'étant pas vide on aura $F(c(M)) = c(M)$ pour tout $F \in X$. En particulier $\rho(\sigma)(c(M)) = c(M)$ pour tout σ dans G et donc la proposition est une conséquence de la remarque 2 ci-dessus.

1.6 Définition : *Une structure de Poisson affine sur V est dite pseudo-riemannienne s'il existe une forme quadratique non dégénérée q sur V telle que pour tout $f \in V^*$, le hamiltonien ξ_f soit une isométrie infinitésimale de la variété pseudo-riemannienne (V, q).*

Supposons la structure de Poisson affine pseudo-riemannienne ; il est clair que pour $f \in V^*$, le linéarisé $\xi_f^L = \{f, \cdot\}_L$ du hamiltonien de f est une isométrie infinitésimale de (V, q). Désignons par $<, >$ la forme bilinéaire sur V associée à q et soit $\varphi : V \to V^*$ avec $\varphi(v) = <v, \cdot>$. Si q^* est la forme sur V^* transportée de q par φ alors $\varphi(\xi_f^L)$ est une isométrie infinitésimale de (V^*, q^*) autrement dit pour tout a b et c dans $\mathcal{G} = V^*$ on a :

$$< ad(a) \cdot b, c > + < b, ad(a) \cdot c >= 0, \tag{5}$$

où $<, >$ est la forme bilinéaire associée à q^*.

Une algèbre de Lie $\mathcal{G}$ munie d'une forme bilinéaire symétrique non dégénérée $<, >$ qui vérifie (5) est appelée une algèbre de Lie orthogonale (ou quadratique).

Dans [**M-R1**], [**M-R2**] nous avons fourni une classification, en dimension finie, de ces algèbres à l'aide de la notion de double extension. Dans une telle algèbre les 2-cocycles scalaires sont en bijection avec les dérivations antisymétriques de $\mathcal{G}$ par la formule

$$\omega(a,b) = <D(a), b>; \quad a,b \quad \text{dans} \quad \mathcal{G}. \tag{6}$$

En outre ω est un 2-cobord si et seulement si D est intérieure. De plus l'application $\varphi : \mathcal{G} \to \mathcal{G}^*$, $a \longmapsto <a, \cdot>$ est un isomorphisme entre les représentations adjointe et co-adjointe de $\mathcal{G}$. En fait si ω et D sont liées par (6) φ est un isomorphisme entre les représentations affines ρ_ω et ρ_D de $\widetilde{G}$, cette dernière étant définie pour $\sigma = \exp(a)$, $a \in \mathcal{G}$, par la formule

$$\rho_D(\sigma) = (Q_D(\sigma), Ad(\sigma) \tag{7}$$

avec $Q_D(\sigma) = (\sum_{k=1}^{\infty} \frac{1}{k!} (ad(a))^{k-1}) \cdot D(a)$.

Noter que la représentation infinitésimale associée à ρ_D n'est autre que l'application $a \longmapsto (D(a), ad(a))$. Ces remarques montrent en particulier que l'on a :

1.7 Proposition : 1. *Si Λ est un tenseur de Poisson affine d'algèbre de Lie $\mathcal{G}$ alors Λ est pseudo-riemannienne si et seulement si $\mathcal{G}$ est orthogonale.*

2. *Toute structure de Poisson affine d'algèbre de Lie orthogonale est pseudo-riemannienne.*

Soit ω un 2-cocycle scalaire de l'algèbre $\mathcal{G}$, considérons sur $\mathcal{G}^*$ la structure de Poisson affine :

$$\{a, b\}(\nu) = -\nu([a,b]) + \omega(a,b)$$

où $a, b \in \mathcal{G}$ et $\nu \in \mathcal{G}^*$. Si $f_\nu : \mathcal{G} \to \mathcal{G}^*$ est définie par $f_\nu(a) = -\nu([a, \cdot]) + \omega(a, \cdot)$ il est clair que l'algèbre d'isotropie $\mathcal{G}_\nu$ en ν est $\mathrm{Ker}(f_\nu)$ et que $\nu + \mathrm{Im} f_\nu$, est le tangent en ν à la feuille $\widetilde{G} \cdot \nu$. Si ν est P-régulier, c'est-à-dire si la feuille passant par ν est de dimension maximale, la proposition qui suit généralise le résultat de Duflo, Vergne ([**D-V**]) :

1.8 Proposition : *Si $\nu \in \mathcal{G}$ est P-régulier, relativement à une structure de Poisson affine sur $\mathcal{G}^*$ alors l'algèbre d'isotropie en ν est commutative.*

Voici une conséquence immédiate de ce résultat pour les algèbres de Lie orthogonales.

1.9 Corollaire : *Soit D une dérivation antisymétrique d'une algèbre de Lie orthogonale $\mathcal{G}$. Si l'algèbre $\mathcal{G}_a = \{b \in \mathcal{G} \;\; ; \;\; D(b) + [b,a] = 0\}$ est de dimension minimale parmi les algèbres $\mathcal{G}_x$, où x parcourt $\mathcal{G}$, alors $\mathcal{G}_a$ est commutative et presque toutes les algèbres $\mathcal{G}_x$ ont pour dimension celle de $\mathcal{G}_a$.*

Le corollaire résulte de la proposition et du fait que l'ensemble des points P-réguliers est un ouvert dense de $\mathcal{G}$. Pour prouver 2.8, suivons une démarche analogue à celle de [**D-V**] : choisissons un sous-espace B de $\mathcal{G}$ tel que $f_{\nu|B}$ soit injective et prenons ν' dans $\mathcal{G}^*$. L'ensemble $I = \{t \in \mathbf{R} \; ; \; f_{\nu+t\nu'|\mathbf{B}}$ est injective$\}$ est un ouvert de $\mathbf{R}$ qui contient l'origine et $\nu + t\nu'$ est P-régulier quel que soit t dans I. Soit $a \in \mathcal{G}_\nu$, si $t \in I$ il existe un unique $a(t) \in B$ tel que $a(t) + a \in \mathcal{G}_{\nu+t\nu'}$. On a donc une application $I \to B$, $t \to a(t)$, de différentiable nulle en zéro. Si l'on écrit $a(t) = tb + 0(t^2) \in B$ et l'on garde le coefficient en t de l'identité

$$\omega(a(t) + a, .) - \nu([a(t) + a, .]) - t\nu'([a(t) + a, .]) = 0$$

on obtient ,

$$\omega(b, .) - \nu([b, .]) - \nu'([a, .]) = 0.$$

Par conséquent si $c \in \mathcal{G}_\nu$ on aura $\nu'([a, c]) = 0$. Puisque ν' est quelconque $[a, c] = 0$ et la proposition est prouvée (voir [**D**]).

On dira que $\nu \in \mathcal{G}^*$ est réductif s'il existe un sous-espace m_ν de $\mathcal{G}$ tel que $\mathcal{G} = \mathcal{G}_\nu \oplus m_\nu$ avec $[\mathcal{G}_\nu, m_\nu] \subset m_\nu$. Si le groupe d'isotropie en ν est connexe, que ν soit réductif signifie que la feuille $G.\nu$ est un espace homogène réductif. Remarquer que si $\mathcal{G}_\nu$ est réductive dans $\mathcal{G}$ alors ν est réductif.

Si Λ est linéaire et si $\nu \in \mathcal{G}^*$ est réductif, la structure de Poisson transverse à $G.\nu$ est linéarisable ([**Mo**]). On a en fait :

1.10 Proposition : *Si $\nu \in \mathcal{G}^*$ est réductif, relativement à une structure de Poisson affine, alors la structure de Poisson transverse en ν à la feuille passant par ν est linéarisable.*

On vérifie sans peine que $T_\nu = \nu + m_\nu^0$, où m_ν^0 est l'annulateur de m_ν dans $\mathcal{G}^*$, est supplémentaire au tangent à $\widetilde{G}.\nu$ en ν. Soit $x \in \mathcal{G}_\nu$, regardons x comme l'application $x : T_\nu \to \mathbf{R}$, $\gamma \longmapsto (\gamma - \nu)(x)$. Désignons par $\widetilde{x}$ l'extension de x à $T_\nu + \mathcal{G}_\nu^0$ nulle sur $\mathcal{G}_\nu^0$. Montrons que le hamiltonien $\xi_{\widetilde{x}}$ est tangent à T_ν c'est à dire que pour $X \in \mathcal{G}_\nu^0$ et $\gamma \in T_\nu$ on a :

$$\{\widetilde{x}, X\}(\gamma) = \omega(\widetilde{x}, X) - \gamma([\widetilde{x}, X]) = 0.$$

Pour ceci posons $\gamma = \nu + \beta$ avec $\beta \in m_\nu^0$; puisque $[\mathcal{G}_\nu, m_\nu] \subset m_\nu$ on a alors :

$$\omega(\widetilde{x}, X) - \gamma([\widetilde{x}, X]) = \omega(\widetilde{x}, X) - \nu([\widetilde{x}, X]) - \beta([\widetilde{x}, X])$$
$$= \omega(\widetilde{x}, X) - \nu([\widetilde{x}, X])$$
$$= 0$$

car $x \in \mathcal{G}_\nu$.

Par conséquent si $x, y \in \mathcal{G}_\nu$ et $\gamma = \nu + \beta$ on aura :

$$\{\widetilde{x}, \widetilde{y}\}(\gamma) = \{\widetilde{x}, \widetilde{y}\}(\nu) + \{\widetilde{x}, \widetilde{y}\}(\beta)$$
$$= \omega(\widetilde{x}, \widetilde{y}) - \beta([\widetilde{x}, \widetilde{y}])$$

et donc la structure transverse à $\widetilde{G}.\nu$ en ν vient donnée par,

$$\{x, y\}(\gamma) = \{\widetilde{x}, \widetilde{y}\}(\gamma) = \omega(\widetilde{x}, \widetilde{y}) - \beta([\widetilde{x}, \widetilde{y}]).$$

En particulier $\{x, y\}(\nu) = \omega(\widetilde{x}, \widetilde{y})$ ce qui signifie que la structure induite sur la variété transverse (locale) T_ν est linéarisable. Ceci achève la preuve de 2.10.

§2. Structures de Poisson affines pseudo-riemanniennes

Considérons une structure de Poisson affine pseudo-riemannienne d'algèbre $\mathcal{G}$ et cocycle ω. Dans ce cas $\mathcal{G}$ est munie d'une structure orthogonale $<, >$ et $\varphi : \mathcal{G} \to \mathcal{G}^*$, $a \longmapsto < a, . >$ est un isomorphisme entre les représentations affines de $\mathcal{G}$, $a \longmapsto (\omega(a, .), ad^*(a))$ et $a \longmapsto (D(a), ad(a))$, où D vérifie (6). Ainsi la structure de Poisson donnée sur $\mathcal{G}^*$ s'identifie à la structure de Poisson affine sur $\mathcal{G}$ définie pour $f_1, f_2 \in \mathcal{G}^*$ et $c \in \mathcal{G}$ par

$$\{f_1, f_2\}(c) = < [a_1, a_2], c > + < Da_1, a_2 > \qquad (8)$$

où $f_i = \varphi(a_i)$, $\quad i = 1, 2$.

Les feuilles symplectiques sont les orbites de la représentation (7). Pour $a \in \mathcal{G}$ désignons par $T_a(G.a)$ et $\mathcal{G}_a$ le tangent à l'orbite $G.a$ en a et l'algèbre d'isotropie en a. On a :

2.1 Théorème : *Soit $\mathcal{G}$ une algèbre de Lie orthogonale. Tout tenseur de Poisson affine pseudo-riemannien sur $\mathcal{G}$ munit $\mathcal{G}$ d'un feuilletage de Sussmann-Stefan $\mathcal{F}$ dont $T_a(G.a) \cap \mathcal{G}_a = M_a$ est l'élément de contact en a. De plus le feuilletage $\mathcal{F}_a$ induit par $\mathcal{F}$ sur $G.a$ est régulier et invariant par G.*

Soit η_a l'endomorphisme de l'espace vectoriel $\mathcal{G}$ tel que $\eta_a(b) = D(b) + [b, a]$. On a : $\mathcal{G}_a = \ker \eta_a$ et $T_a(G.a) = a + \text{Im}(\eta_a)$. Le théorème résulte essentiellement de la proposition suivante :

2.2 Scolie : *Soit $\mathcal{G}$ une algèbre de Lie et η_a l'endomorphisme linéaire de $\mathcal{G}$ tel que $\eta_a(b) = D(b) + [b, a]$ où D est une dérivation de $\mathcal{G}$. On a :*

1. *Le noyau $\mathcal{G}_a = \ker \eta_a$ est une sous-algèbre de $\mathcal{G}$.*
2. *Si $u \in \mathcal{G}_a$ et $v = \eta_a(w)$, $[u, v] = \eta_a[u, w])$.*
3. *Le sous-espace $\mathcal{G}_a \cap \mathrm{Im}(\eta_a)$ est une sous-algèbre de $\mathcal{G}$.*

Démonstration : La première assertion est une conséquence du fait que $\mathcal{G}_a = \ker(D - ad(a))$. La troisième résulte de 1. et 2. Pour prouver 2. il faut montrer que l'on a :

$$[u, Dw + [w, a]] = D[u, w] + [[u, w], a]$$

sachant que $D(u) = [a, u]$. Or d'après l'identité de Jacobi ceci équivaut à

$$[u, Dw] = D[u, w] - [[a, u], w]$$

ou ce qui est de même

$$D[u, w] = [Du, w] + [u, Dw]$$

qui est vérifiée puisque D est une dérivation.

D'autre part sous les hypothèses du Scolie pour tout $\sigma \in G$ on a, $\mathcal{G}_{\sigma.a} = Ad(\sigma)(\mathcal{G}_a)$ où $\sigma.a = \rho_D(\sigma)(a)$. Par conséquent si $\sigma \in G_a$, groupe d'isotropie en a, alors $\mathcal{G}_a = Ad(\sigma)(\mathcal{G}_a)$ et $\mathrm{Im}(\eta_a) = Ad(\sigma)(\mathrm{Im}\ \eta_a)$ ce qui implique $M_a = Ad(\sigma)(M_a)$. Mais alors au moyen des éléments de G, considérés comme transformations de $G.a$ on définit sur $G.a$ un feuilletage régulier invariant par G qui n'est autre que celui induit par $\mathcal{F}$. Ceci prouve le théorème.

Si $\mathcal{G}$ est orthogonale, il est immédiat de vérifier que $\mathcal{G}_a$ et $\mathrm{Im}(\eta_a)$ sont orthogonaux entre eux. Ainsi l'orthogonal à $T_a(G.a)$ rencontre transversalement $G.a$ si et seulement si $\mathcal{G}_a \cap \mathrm{Im}(\eta_a) = \{0\}$ ou ce qui est de même si $\mathcal{G} = \mathcal{G}_a \oplus \mathrm{Im}(\eta_a)$.

Le feuilletage $\mathcal{F}_a$ de $G.a$ est dit trivial si ses feuilles sont des points. Nous avons :

2.3 Théorème : *Soit $\mathcal{G}$ une algèbre de Lie orthogonale et Λ un tenseur de Poisson affine pseudo-riemannien sur $\mathcal{G}$ On a :*

1. *Un point $a \in \mathcal{G}$ est réductif si et seulement si $M_a = \{0\}$.*
2. *Que $G.a$ soit un espace homogène réductif équivaut à l'une des conditions suivantes :*

    i) $G.a$ contient un point réductif,

    ii) La restriction de la métrique de $\mathcal{G}$ fait de $G.a$ une variété pseudo-riemannienne.

    iii) Le feuilletage $\mathcal{F}$ est trivial.

En outre si $G.a$ est réductif il est pseudo-riemannien.

295 296

Démonstration : Si $M_a = \{0\}$ c'est-à-dire si $\mathcal{G} = \mathcal{G}_a \oplus \operatorname{Im}(\eta_a)$ alors a est réductif d'après 2 de 2.2. Réciproquement soit $\mathcal{G} = \mathcal{G}_a \oplus m_a$ avec $[\mathcal{G}_a, m_a] \subset m_a$. il est immédiat de vérifier que la restriction de η_a à m_a est injective et puisque $\eta_a | \mathcal{G}_a$ est nilpotent on aura $A = \ker \eta_a \oplus \operatorname{Im}(\eta_a) = \ker(\eta_a)^n \oplus \operatorname{Im}(\eta_a)^n$ où $n = \dim \mathcal{G}$. Ainsi $\ker \eta_a = \ker(\eta_a)^n$ et $m_a = \operatorname{Im} \eta_a$. Ce qui montre 1. Supposons $G.a$ réductif. Si $x \in G.a$ alors $\mathcal{G} = \mathcal{G}_x \oplus m_x$ avec $[\mathcal{G}_x, m_x] \subset m_x$ et l'argument qui précède montre que $M_x = \{0\}$ ou ce qui est de même que la restriction de $<,>$ à $T_x(G.a)$ est non dégénérée ; aussi le feuilletage $\mathcal{F}_a$ sur $G.a$ est trivial car $\mathcal{F}_a$ est invariant par G. On a par conséquent i), ii) et iii).

Faisons l'hypothèse que $G.a$ ait un point x réductif. Alors la forme $<,>_x$ sur $\operatorname{Im} \eta_x$ induite par la forme orthogonale de $\mathcal{G}$ est non dégénérée. Aussi puisque $Ad(G_x)(\mathcal{G}_x) = \mathcal{G}_x$ on a $Ad(G_x)(\operatorname{Im} \eta_x) = \operatorname{Im} \eta_x$ car $Ad(\sigma)$, pour $\sigma \in G$, est une isométrie de $\mathcal{G}$. Or ceci implique que $G_x.(x + \mathcal{G}_x^\perp) = x + \mathcal{G}_x^\perp$ ce qui signifie que $G.a = G.x$ est un espace homogène réductif.

Finalement le groupe d'isotropie linéaire en x préserve la forme $<,>_x$, alors par transport par l'action de G, $<,>_x$ définit une métrique pseudo-riemannienne sur $G.a = G.x$. Ainsi si $G.a$ est réductif, $G.a$ est un espace homogène pseudo-riemannien. Noter que le théorème est une généralisation de 6.2 de [M2].

2.4 Remarques 1. Si $\mathcal{G}$ est réductive la représentation ρ_D admet un point fixe et par suite la structure de Poisson est linéaire. Que tout élément de $\mathcal{G}$ soit réductif équivaut à dire que l'idéal dérivé de $\mathcal{G}$ est compact ([M2]).

2. Supposons a réductif : $\mathcal{G} = \mathcal{G}_a \oplus \operatorname{Im} \eta_a$ et désignons par $\pi : \mathcal{G} \to \mathcal{G}_a$ la projection orthogonale. Si $f_1, f_2 \in \mathcal{G}_a$, les hamiltoniens $\xi_{\tilde{f_i}}$ où $\tilde{f_i} = f_i \circ \pi$ sont tangents à $\mathcal{G}_a$ en tout point de $\mathcal{G}_a$ proche de a. Considérons $\mathcal{G}_a$ muni de sa structure de Poisson linéaire qui est pseudo-riemannienne. Si l'on pose $f_i = < a_i, . >$ avec $a_i \in \mathcal{G}$ on a,

$$\{f_1, f_2\}_{\mathcal{G}_a} \circ \pi = \{\tilde{f_1}, \tilde{f_2}\}_{\mathcal{G}} - < Da_1, a_2 >$$

ce qui implique que π induit un isomorphisme entre la structure de Poisson transverse en a à $G.a$ et celle de $\mathcal{G}_a$. En fait si $G.a$ contient un point réductif, la structure de Poisson transverse à $G.a$ en x est isomorphe à celle linéaire de $\mathcal{G}_x$.

Plus généralement concernant la dimension de $\mathcal{F}_a$ on a :

2.5 Proposition : *Soit* Λ_D *un tenseur de Poisson affine pseudo-riemannien d'algèbre* $\mathcal{G}$ *et* $G.a$ *la feuille passant par* a. *Alors, la dimension de* $\mathcal{F}_a$ *est égale à la dimension de l'espace quotient* $\ker(\eta_a)^2/\mathcal{G}_a$.

Preuve : Soit $\pi : \mathcal{G} \to \mathcal{G}|_{\mathcal{G}_a}$, $\mathcal{G}_a = \ker \eta_a$, l'application quotient. Désignons par E_a l'image par π de $\operatorname{Im}(\eta_a)^2$ et définissons $\psi_a : E_a \to \mathcal{G}_a$ en posant $\psi_a(\bar{v}) = \eta_a(v)$. On a alors $\operatorname{Im} \psi_a = \ker \eta_a \cap \operatorname{Im} \eta_a$ et on vérifie sans peine que $\ker \psi_a = 0$. Ceci prouve la proposition.

Etudions la structure de Poisson linéaire sur une algèbre de Lie orthogonale $\mathcal{G}$. Les feuilles sont ici les orbites de la représentation adjointe d'un groupe de Lie connexe G d'algèbre de Lie $\mathcal{G}$. Un élément $a \in \mathcal{G}$ est dit régulier si $C = \ker(ad(a))^n$ où $n = \dim \mathcal{G}$, est une sous-algèbre de Cartan de $\mathcal{G}$. Nous avons

2.6 Théorème : *Soit $\mathcal{G}$ une algèbre de Lie orthogonale munie de sa structure de Poisson linéaire. S'il existe un point de $\mathcal{G}$ réductif et P-régulier alors l'ensemble des éléments réguliers de $\mathcal{G}$ coïncide avec l'ensemble des points P-réguliers et réductifs de $\mathcal{G}$.*

Signalons qu'il existe des algèbres orthogonales dont l'ensemble des éléments P-réguliers et réductifs est vide : celui-ci est le cas si $\mathcal{G}$ est nilpotente et non commutative (voir 6.3 de [**M2**]).

Démonstration : Soit $a \in \mathcal{G}$ réductif et P-régulier. Alors $\mathcal{G} = \ker(ad(a)) \oplus \mathrm{Im}\ ad(a)$ et $C = \mathcal{G}_a = \ker ad(a)$ est commutative compte tenu de 1.8. Désignons par N le normalisateur de C dans $\mathcal{G}$, montrons que $N \subset C$. Soient $u \in N$ et $v \in C$. Ecrivons $u = u_1 + u_2$ avec $u_1 \in C$ et $u_2 \in \mathrm{Im}\ ad(a)$. On a :

$$[u,v] = [u_1,v] + [u_2,v] = [u_2,v] \in C$$

mais aussi $[u_2,v] \in \mathrm{Im}\ ad(a)$ et par suite $[u,v] = 0$. En particulier pour tout u dans N on a $[u,a] = 0$ ce qui veut dire que $N \subset C$ et donc C est une sous-algèbre de Cartan de $\mathcal{G}$, ou bien a est régulier dans $\mathcal{G}$.

Réciproquement soit x régulier dans $\mathcal{G}$: $C' = \ker(ad(x))^n$ est une sous-algèbre de Cartan de $\mathcal{G}$. De plus on a, $\mathcal{G} = \ker(ad(x))^n \oplus \mathrm{Im}\ (ad(x))^n$. En passant au besoin à la complexifiée de $\mathcal{G}$ nous pouvons affirmer que C' est isomorphe à $C = \ker ad(a)$. Ainsi $C' = \ker ad(x)$ est commutative et $\mathcal{G} = \ker ad(x) \oplus \mathrm{Im}\ ad(x)$ ce qui veut dire que x est P-régulier et réductif.

Voici une conséquence importante du théorème.

2.7 Corollaire : *Soit $\mathcal{G}$ une algèbre de Lie orthogonale ayant un élément P-régulier et réductif. Si C est une sous-algèbre de Cartan de $\mathcal{G}$ alors $C = \ker ad(a)$ avec a régulier dans $\mathcal{G}$ et C est commutative et égale à son commutant dans $\mathcal{G}$. En particulier C est une sous-algèbre maximale abelienne de $\mathcal{G}$.*

D'après la preuve du théorème $C = \ker ad(x)$ où x est régulier dans $\mathcal{G}$ et C est abelienne. Par conséquent $C \subset \mathrm{Comm}_{\mathcal{G}}(C)$ où $\mathrm{Comm}_{\mathcal{G}}(C)$ est le commutant de C dans $\mathcal{G}$. Or comme $\mathrm{Comm}_{\mathcal{G}}(C) \subset N = C$, on a le résultat.

Observation : Remarquer que le théorème et le corollaire généralisent des résultats des algèbres de Lie semi-simples ([**M1**],[**M2**]).

Le lemme qui suit nous sera utile dans la suite.

2.8 Lemme : *Soit $Ad : G \to GL(\mathcal{G})$ la représentation adjointe de G et $\mathcal{G}_a = \ker ad(a)$ l'algèbre d'isotropie en $a \in \mathcal{G}$. Si $\mathcal{G}_a$ est commutative pour $u \in \mathcal{G}_a$ on a $\mathcal{G}_a \subset \mathcal{G}_u$. En particulier si $G.u$ est maximale on a $\mathcal{G}_a = \mathcal{G}_u$ et si $\mathcal{G}$ est orthogonale $M_a = M_u$.*

Démonstration : Soit $\bar{a}$ le champ fondamental sur $\mathcal{G}$ associé à a. On a $\bar{a}_b = [a, b]_b$ et en particulier $\bar{a}_b = 0$ si b appartient à $\mathcal{G}_a$. Considérons z dans $\mathcal{G}_a$, puisque $\bar{z}_b = [z, b]_b$, $\bar{z}_b = 0$ pour $b \in \mathcal{G}_a$ car $\mathcal{G}_a$ est commutative. En particulier $\bar{z}_u = [z, u]_u = 0$ c'est-à-dire que $z \in \mathcal{G}_u$.

2.9 Théorème : *Soit G un groupe de Lie orthogonal connexe, $Ad : G \to GL(\mathcal{G})$ la représentation adjointe. Si $G.x$ est maximale (P-régulière) alors*

1. *La trace de $G.x$ sur $\mathcal{G}_x$ est le sous-espace affine $x + Mx$ de $\mathcal{G}$.*
2. *Les feuilles du feuilletage $\mathcal{F}_x$ de $G.x$ sont les sous-espaces affines $u + M_u$ de $\mathcal{G}$, $u \in G.x$.*

Démonstration : Puisque x est P-régulier, $\mathcal{G}_x$ est commutative. D'après le lemme pour tout $u \in \mathcal{G}_x$, u P-régulier, on a, $M_x = M_u$ et donc $x + M_x$ et $u + M_u$ sont parallèles. Soit x' assez voisin de x, x' P-régulier et $x' \in \mathcal{G}_x$. D'après ce qui précède la trace du tangent à l'orbite $G.x'$ sur $\mathcal{G}_x$ est parallèle à $x + M_x$. Par conséquent si l'on suit en partant de x un chemin continu différentiable par morceaux dans $x + M_x$ il restera dans un voisinage de x tangent aux orbites et il sera donc tracé dans une seule orbite. Ceci prouve que $G.x \cap \mathcal{G}_x$ contient un voisinage de x dans $x + M_x$. Puisque cette intersection est analytique on aura $G.x \cap \mathcal{G}_x \supset x + M_x$. Mais puisque la dimension de l'intersection ne peut être supérieure à la dimension de l'intersection des espaces tangents on aura à fortiori $G.x \cap \mathcal{G}_x = x + M_x$.

2.10 Exemples : Soit $\mathcal{G} = s\ell(2, \mathbf{R})$. Si $\begin{pmatrix} \gamma & \alpha \\ \beta & \gamma \end{pmatrix} \in \mathcal{G}$ on a

$$det(\lambda I - ad(x)) = \lambda^3 - 4(\alpha\beta + \gamma^2)\lambda$$

Ainsi x est régulier si et seulement si $\alpha\beta + \gamma^2 \neq 0$ et la forme de killing k de $\mathcal{G}$ est donnée par $k(x, x) = 8(\gamma^2 + \alpha\beta)$. Tout point $x \neq 0$ appartenant au cône défini par $k(x, x) = 0$ est P-régulier et non réductif, son orbite qui est un demi-cône est une surface réglée plus précisément : $M_x = \ker ad(x) \cap$ Im $ad(x) = \ker ad(x)$. Si x est régulier son orbite est soit un hyperboloïde à deux nappes soit un hyperboloïde à une nappe. Un tel orbite rencontre transversalement l'algèbre d'isotropie en x qui est une sous-algèbre de Cartan de $\mathcal{G}$. Noter que $\mathcal{G}$ possède des sous-algèbres de Cartan non conjuguées par $\mathrm{Aut}(\mathcal{G})$: $\mathbf{Rx_0}$ et $\mathbf{Ry_0}$ avec

$$x_0 = \begin{pmatrix} 1 & 0 \\ 0 & -1 \end{pmatrix} \qquad y_0 = \begin{pmatrix} 0 & -1 \\ 1 & 0 \end{pmatrix}.$$

Exemple 2. Soit $G = \mathbf{R} \times \mathbf{C} \times \mathbf{R}$ munie du produit défini par

$$(u, z, t)(u', z', t') = (u + u' + e^{it}\tfrac{1}{2}\,\mathrm{Im}\,\bar{z}z', z + e^{it}z', t + t')$$

L'algèbre de Lie $\mathcal{G} = \mathrm{Vect}(e_0, e_1, e_2, d)$ de G a pour produit

$$[e_1, e_2] = e_0, \quad [d, e_1] = -e_2, \quad [d, e_2] = e_1$$

les autres produits étant nuls ou déduits de ceux-ci par antisymétrisation. En particulier $\mathbf{Re_0} = \mathbf{Z}(\mathcal{G})$ est le centre de $\mathcal{G}$.

Considérons sur $\mathrm{Vect}(e_1, e_2)$ le produit scalaire pour lequel (e_1, e_2) est une base orthonormée. Etendons cette forme de sorte que $\mathrm{Vect}(e_0, d)$ soit orthogonal à $\mathrm{Vect}(e_1, e_2)$ et $< e_0, d >= 1$, $< e_0, e_0 >=< d, d >= 0$. On vérifie que $(\mathcal{G}, <, >)$ est une algèbre de Lie orthogonale. De plus on constate ([M2]) pour $a \in \mathcal{G}$ que,

a est P-régulier si et seulement si $a \notin Z(\mathcal{G})$.

a est réductif si et seulement si $a \notin \mathcal{D}(\mathcal{G})$.

Par conséquent a est P-régulier et non régulier si et seulement si $a = \alpha e_0 + \beta e_1 + \gamma e_2$ avec β ou γ non nuls. Dans ce cas $M_a = \mathbf{Re_0}$ et $G.a$ est un cyclindre. Par contre l'orbite d'un point régulier est un paraboloïde de révolution.

Noter que $\ker(ad(d)) = \mathrm{Vect}(e_0, d)$ est une sous-algèbre de Cartan, elle est l'algèbre de Lie du sous-groupe $T = \{(u, 0, t) \;\; ; \;\; u, t \in \mathbf{R}\} = \mathbf{G_d}$. On a $G = \bigcup_{\alpha} \alpha T \alpha^{-1}$ où α parcourt G.

Exemple 3. Soit $\mathcal{G} = s\ell(3, \mathbf{R})$. Si $E_{ij} \in g\ell(3, \mathbf{R})$ désigne la matrice élémentaire alors $C = \mathrm{vect}(E_{11} - E_{22}, \quad E_{22} - E_{23})$ est une sous-algèbre de Cartan de $\mathcal{G}$. Par conséquent les orbites maximales sont de dimension 6. L'élément nilpotent $a = E_{12} + E_{23}$ est nilpotent principal c'est-à-dire $\ker(ad(a))$ est de dimension 2. On vérifie sans peine que $M_a = \ker(ad(a)) \cap \mathrm{Im}\, ad(a) = \mathrm{Vect}(a, E_{13})$. Ainsi l'orbite $SL(3, \mathbf{R}).a$ est feuilletée par des plans. D'autre part $b = E_{13}$ est un élément de Givental, c'est-à-dire la structure de Poisson transverse à $SL(2, \mathbf{R}).b$ n'est pas linéarisable. L'orbite $SL(2, \mathbf{R}).b$ est de dimension 4, elle est feuilletée par des feuilles de dimension 3 car $M_b = \mathrm{Vect}(E_{12}, E_{13}, E_{23})$.

3. Groupes de Lie à forme symplectique invariante

Soit Λ un tenseur de Poisson affine sur $V = \mathcal{G}^*$ d'algèbre $\mathcal{G}$ et cocycle ω et G le groupe connexe 1-connexe d'algèbre $\mathcal{G}$. Les feuilles de (V, Λ) sont les orbites de la représentation ρ_ω. Soit $\nu \in \mathcal{G}^*$, $\pi : G \to G.\nu$ l'application orbitale. Supposons $G.\nu$ ouverte alors π est un revêtement équivariant pour les actions de G sur G par les translations à gauche et de G sur $G.\nu$ pour ρ_ω. La forme symplectique de $G.\nu$ se relève par π en une forme symplectique sur G invariante c'est-à-dire stable par les translations à gauche.

Réciproquement si G admet une forme symplectique Ω invariante et $\omega = \Omega_\varepsilon$ où ε est l'élément neutre de G alors la représentation ρ_ω de G déduite de $x \longmapsto (i_x\omega, ad^*(x))$ possède une orbite ouverte à savoir celle de l'origine de $\mathcal{G}^*$.

Pour l'étude des groupes de Lie à forme symplectique invariante le lecteur est renvoyé à [L-M], [M-R3] et [D-M]. Ici nous nous contenterons de prouver le résultat suivant (voir 5.4 de [M1]).

3.1 Théorème : *Soit $\mathcal{G}$ un groupe de Lie d'algèbre de Lie $\mathcal{G}$ complètement résoluble. Si Ω est une forme symplectique invariante sur G alors (G, Ω) possède un feuilletage lagrangien invariant.*

Pour prouver le théorème rappelons le résultat suivant ([B-C]) dont la démonstration se fait par récurrence.

3.2 Lemme : *Soit E un espace vectoriel de dimension n sur un corps K commutatif de caractéristique nulle et soit $S = (E_i)$, $0 \le i \le n$, une suite de sous-espaces de E tels que*

$$\{0\} = E_0 \subset E_1 \subset \cdots \subset E_i \subset E_{i+1} \subset \cdots \subset E_{n-1} \subset E_n = E$$

et $\dim E_i = i$, $0 \le i \le n$.

Si ω est une forme bilinéaire alternée sur E, $\omega_i = \omega|E_i$ et $N(\omega_i)$ le noyau de ω_i alors

1. $P(\omega, S) = \displaystyle\sum_{i=1}^{n} N(\omega_i)$ est un espace totalement isotrope maximal pour ω.

2. $P(\omega, S) \cap E_j = P(\omega_j, S_j)$ ou $S_j = (E_i)$, $0 \le i \le j$.

Démonstration du théorème : Soit $\omega = \Omega_\varepsilon$ où ε est l'élément neutre de G. Puisque $\mathcal{G}$ est complètement résoluble, il existe une suite $S = (\mathcal{G}_i)$, $0 \le i \le n$ d'idéaux de $\mathcal{G}$ emboîtés comme dans le lemme avec $\dim \mathcal{G}_i = i$ et $[\mathcal{G}, \mathcal{G}_{i+1}] \subset \mathcal{G}_i$.

D'après le lemme $P(\omega, S) = \sum_{i=1}^{n} N(\omega_i)$ est un sous-espace totalement isotrope maximal pour ω. Montrons que $P(\omega, S)$ est une sous-algèbre de $\mathcal{G}$. Pour ceci observons d'abord que $N(\omega_i)$ est une sous-algèbre de $\mathcal{G}$ car $N(\omega_i)$ est une sous-algèbre de $\mathcal{G}_i$. D'autre part soient $x = x_1 + x_2 + \cdots + x_n$, $y = y_1 + y_2 + \cdots + y_n$ deux éléments de $P(\omega, S)$. Que $[x, y] \in P(\omega, S)$ résulte des remarques suivantes :

Les éléments $[x_i, y_i]$ appartiennent à $N(\omega_i)$ pour $1 \leq i \leq n$. Les crochets $[x_1, y_j]$, $[x_j, y_1]$ sont nuls car ils sont dans $\mathcal{G}_0$.

Pour $1 \leq j \leq n$, $[x_2, y_j] \in \mathcal{G}_1$ et la relation

$$\omega([x_2, y_j], \mathcal{G}_1) + \omega([y_j, \mathcal{G}_1], x_2) + \omega([\mathcal{G}_1, x_2], y_j) = 0$$

implique que $[x_2, y_j] \in N(\mathcal{G}_1)$ car il y a deux termes nuls dans cette formule.

Plus généralement si $i < j$ alors

$$\omega([x_i, y_j], \mathcal{G}_{i+1}) + \omega([y_j, \mathcal{G}_{i+1}], x_i) + \omega([\mathcal{G}_{i+1}, x_i], y_j) = 0$$

implique $\omega([x_i, y_j], \mathcal{G}_{i+1}) = 0$. En particulier $[x_i, y_j] \in N(\omega_{i-1})$ car $\mathcal{G}_{i-1} \subset \mathcal{G}_i \subset \mathcal{G}_{i+1}$.

Par conséquent $P(\omega, S)$ est une sous-algèbre lagrangienne de $(\mathcal{G}, \omega)$ c'est-à-dire elle est totalement isotrope maximal.

Regardons $\mathcal{G}$ comme l'algèbre de Lie des champs de vecteurs invariants à gauche sur G. La sous-algèbre $P(\omega, S)$ définit un feuilletage invariant à gauche sur G dont la feuille passant par ε est le sous-groupe de Lie connexe de G d'algèbre $P(\omega, S)$. Ce feuilletage est évidemment lagrangien c'est-à-dire ses feuilles sont lagrangiennes.

Voici une conséquence immédiate du théorème.

3.3 Corollaire : *Soit G un groupe de Lie nilpotent dont l'algèbre de Lie $\mathcal{G}$ est à constantes de structure rationnelle. Supposons G muni d'une forme symplectique Ω invariante à gauche. Si Γ est un sous-groupe discret co-compact de G, la variété quotient $M = G \setminus \Gamma$ est munie d'une forme symplectique déduite de Ω et d'un feuilletage lagrangien pour cette forme.*

Signalons que dans [**M-R3**] il est fournie une classification des groupes de Lie nilpotents à forme symplectique invariante.

Pour terminer remarquons que la preuve de 3.1 montre en fait le résultat plus général suivant :

3.4 Proposition : *Soit ω un 2 cocycle réel de l'algèbre de Lie $\mathcal{G}$. Si $\mathcal{G}$ est complètement résoluble, toute orbite de $\mathcal{G}^*$ pour la représentation ρ_ω est munie d'un feuilletage lagrangien invariant par G.*

REFERENCES

[B] M. Berger, "Géométrie vol.1", Cedic Nathan, Paris,1977.

[B-C] P. Bernat & N. Conze, "Représentations des groupes de Lie résolubles", Dunod, Paris, 1972.

[D-M] J.M. Dardie & A. Medina, *Groupes de Lie à structure symplectique invariante*, A paraître.

[D] P. Dazord, *Feuilletages et mécanique hamiltonnienne*, Publ. Dept. Math. Lyon (1983).

[D-V] M. Duflo & M. Vergne, *Une propriété de la représentation coadjointe d'une algèbre de Lie*, C.R. Acad. Sci. Paris (série A) **268** (1969), 583–585.

[L] A. Lichnerowicz, *Les variétés de Poisson et leurs algèbres de Lie associées*, J. Diff. Geom. **12** (1977), 253–300.

[L-M] A. Lichnerowicz & A. Medina, *On Lie groups with left-invariant symplectic or kählerian structures*, Lett. in Math. Phys. **16** (1988), 225–235.

[M1] A. Medina, *Structures de Poisson affines*, Colloque International, Casablanca (1987), Séminaire Montpellier (1987).

[M2] A. Medina, *Groupes de Lie munis de métriques bi-invariantes*, Tohoku Math. J. **37** (1985), 405–421.

[M-R1] A. Medina & Ph. Revoy, *Caractérisation des groupes de Lie ayant une pseudo-métrique bi-invariante*, in "Collection travaux en cours, Sud-Rhodanien III", 149–166, Hermann Paris, 1984.

[M-R2] A. Medina & Ph. Revoy, *Algèbres de Lie et produit scalaire invariant*, Ann. Ec. Norm. Sup, 4ème série (1985), 553–561.

[M-R3] A. Medina & Ph. Revoy, *Groupes de Lie à structure symplectique invariante*, in "Symplectic Geometry, Groupoids and Integrable Systems" (P. Dazord & A. Weinstein, eds.), Springer Verlag, New York, 1991 (MSRI Publications **20**).

[Mo] P. Molino, *Structure transverse aux orbites de la représentation coadjointe : orbites réductives*, Séminaire Montpellier (1984).

[S] J.-M. Souriau, "Structure des systèmes dynamiques", Dunod, Paris, 1970.

[W] A. Weinstein, *The local structure of Poisson manifolds*, J. Diff. Geom. **18** (1983), 523–557.

Département de Mathématiques (G.D.R.144, U.R.A. 1407 du C.N.R.S.)
Université d'Avignon
33, rue Louis Pasteur
84000 Avignon
France

Heisenberg and Isoholonomic Inequalities

R. MONTGOMERY

I would like to dedicate this paper to J.-M. Souriau, a pioneer in the development of geometric quantization. In this process one obtains Hilbert spaces from classical phase spaces in a geometrically natural way. The present paper is a kind of play on this theme of Souriau. We show how symplectic geometry can be born out of quantum mechanics.

0.1 Introduction

A basic object in quantum mechanics is a pure state. The space of all pure states forms the projective space, $\mathbf{P}\mathcal{H}$, of the Hilbert space $\mathcal{H}$ of the quantum system. The first half of this paper relates the geometry of $\mathbf{P}\mathcal{H}$ to the quantum physics.

Most of this first half is a review of material which can be found in a recent preprint of Aharonov and Anandan [1990] and the paper [1990] of the author. Perhaps the most striking result is a direct relation between the diameter of $\mathbf{P}\mathcal{H}$ and Heisenberg's time-energy uncertainty relation. Another result is the "isoholonomic inequality" which is a generalization of the isoperimetric inequality. It relates the length of a closed curve in projective space to its symplectic area.

In the second half of the paper we generalize to quantum statistical mechanics. We argue that the correct replacement for the manifold $\mathbf{P}\mathcal{H}$ of pure states is a manifold $\mathcal{O}$ of mixed states which consists of all density matrices conjugate to an initial density matrix. $\mathcal{O}$ is properly thought of as a co-adjoint orbit for the unitary group. We then proceed to relate the geometry of $\mathcal{O}$ to quantum statistical mechanics. Much of this material is new. Some of it appeared in [1990] and some errors there are corrected here. The main error was that I made the wrong choice for $\mathcal{S}$, where $\mathcal{S} \longrightarrow \mathcal{O}$ generalizes the Hopf fibration over $\mathbf{P}\mathcal{H}$. The Riemannian metric on $\mathcal{O}$ is induced from the metric on $\mathcal{S}$ by declaring that this generalized Hopf fibration be a Riemannian submersion. Thus this mistake in $\mathcal{S}$ led to an incorrect metric on $\mathcal{O}$ and consequently to an incorrect solution to the quantum statistical isoholonomic problem.

0.2 Credits

The arclength formula, equation (1) below, appeared in my paper [1990]. It was independently derived by Aharonov and Anandan [1990]. It is very likely, as Lichnerowicz remarked during my talk, that this formula has appeared elsewhere, probably several times, several decades ago, and in different guises.

The derivation of Heisenberg's uncertainty relation from this formula is due to Aharonov and Anandan [1990]. The equation for the element of phase interference is also due to Aharonov and Anandan [1987]. It is an extension of the fundamental work of Berry [1984] and Simon [1983] on the now-famous Berry's phase.

The isoholonomic inequality first appeared in my paper [1990]. That paper and indeed my interest in the subject was inspired by discussions with the physical chemist Alex Pines and his co-workers Joe Zwanziger, Marianne Koenig, and Karl Mueller. Pines asked me "What is the shortest loop with a given holonomy?". I recommend their recent review article [1990] for more inspiration and contact with experiment.

PART I

I.1 Overview

We will relate the differential geometry of the manifold of pure states $\mathbf{P}\mathcal{H}$ to the quantum physics of a system whose Hilbert space is $\mathcal{H}$. Recall that a point in $\mathbf{P}\mathcal{H}$ is a one-dimensional complex subspace of $\mathcal{H}$. Alternatively, it is the orthogonal projection operator onto such a subspace.

The first and perhaps most important relationship which we will derive is

$$ds = \frac{\Delta E dt}{\hbar} \tag{1}$$

Here ds is the arclength along any curve in $\mathbf{P}\mathcal{H}$ which is defined by solving Schrödinger's equation. ΔE is the instantaneous energy uncertainty. This is the root mean square variation of the possibly time-dependent operator which defines this Schrödinger equation. And t is the time parameter of the equation. A second relationship is

$$\text{infinitesimal element of phase interference} = -\Omega \tag{2}$$

where Ω is the Kähler form on $\mathbf{P}\mathcal{H}$.

These two equalities come with corresponding inequalities. The first is Heisenberg's time-energy uncertainty relation

$$(\Delta E)_{av}\Delta t \geq \frac{h}{4} \tag{3}$$

where Δt is the duration of time of an experiment and $(\Delta E)_{av}$ is the time average of the energy uncertainty. The second inequality is

$$L^2 \geq 2\pi\Phi - \Phi^2 \quad \text{with} \quad 0 \leq \Phi \leq 2\pi \tag{4}$$

In this inequality $L = \int_C ds$ is the length of any closed curve C in $\mathbf{P}\mathcal{H}$. $\Phi = -\int_D \Omega$ is the *symplectic area* enclosed by this curve. Thus D is any two-disc in $\mathbf{P}\mathcal{H}$ which is bounded by C and the integral is reduced modulo 2π to insure that $0 \leq \Phi \leq 2\pi$. If $\mathcal{H} = \mathbf{C}^2$, the Hilbert space of a two-level system, then $\mathbf{P}\mathcal{H} = \mathbf{P}^1 = S^2$ the two-sphere of radius $\frac{1}{2}$ and we have $\Phi = 2(\text{Area})$. Thus the isoholonomic inequality becomes the standard isoperimetric inequality on this two-sphere.

I.2 Hopf Fibration

Let $S(\mathcal{H}) \subset \mathrm{H}$ denote the unit sphere in Hilbert space. The Hopf fibration

$$\pi : S(\mathcal{H}) \to \mathbf{P}\mathcal{H}$$

is the map which assigns to each unit vector $\psi \in \mathcal{H}$ the complex line

$$\pi(\psi) = [\psi] := \mathbf{C}\psi$$

which it spans. Alternatively, if we think of $\mathbf{P}\mathcal{H}$ as the space of rank one projections then

$$\pi(\psi) = \psi \otimes \psi^* \tag{5}$$

I.3 Arclength

The quantum evolution is defined by Schrödinger's equation

$$\frac{d\psi}{dt} = -\frac{i}{\hbar}H(t)\psi(t)$$

Here $\psi(t) \in \mathcal{H}$, the Hilbert space of the system and $H(t)$ is the Hamiltonian of the system, a possibly time-dependent self adjoint operator on $\mathcal{H}$. We may take $\mathcal{H}$ to be finite-dimensional, thus avoiding questions regarding unboundedness and domains of $H(t)$. However, our results appear to be true when $\mathcal{H}$ is infinite-dimensional. The problems which originally motivated me came out of the field of nuclear magnetic resonance (NMR) in which $\mathcal{H}$ is usually finite-dimension. If n is this finite dimension then the system is said to be an n-level system.

The self-adjointness of the Hamiltonian H implies that Schrödinger's equation preserves the norm, $\frac{d}{dt}\langle\psi(t),\psi(t)\rangle = 0$, thus defining a non-autonomous dynamical system on the sphere $S(\mathcal{H})$. Since the Hamiltonian is complex-linear it also induces a nonautonomous dynamical system on projective space: if $C(0) = \pi(\psi(0))$ then $C(t) = \pi(\psi(t))$ is a solution curve of this system. Our first problem is to calculate the arclength along such a curve C of states in $\mathbf{P}\mathcal{H}$ in terms of the given Hamiltonian H.

In order to perform this calculation we must recall the canonical (Fubini-Study) metric on $\mathbf{P}\mathcal{H}$. It is defined by declaring the Hopf map $\pi : S(\mathcal{H}) \to \mathbf{P}\mathcal{H}$ to be a Riemannian submersion. We recall that a submersion $f : X \to B$ of one Riemannian manifold onto another is said to be "Riemannian" if its *restricted* differential

$$df_x : (ker(df_x))^{\perp} \subset T_x X \to T_{f(x)} B$$

is an isometry between the two inner product spaces. We define

$$Hor_x = (ker(df_x))^{\perp}$$

and call it the "horizontal space" at x. It is the orthogonal complement to the fiber $f^{-1}(f(x))$ through x.

In our situation we put the standard metric on $S(\mathcal{H})$, the one induced from $\mathcal{H}$,

$$ker(d_\psi \pi) = \text{real span of } i\Psi$$

and

$$T_\psi S(\mathcal{H})^{\perp} = \text{real span of } \Psi$$

from which it follows that

$$Hor_\psi = [\psi]^{\perp} \tag{6}$$

the orthogonal complement to the *complex* line through ψ. This implies that the arclength s along our curve C is given by

$$(\frac{ds}{dt})^2 = \langle \frac{d\psi}{dt}^{\perp}, \frac{d\psi}{dt}^{\perp} \rangle$$

where

$$\frac{d\psi}{dt}^{\perp} = \frac{d\psi}{dt} - \langle \frac{d\psi}{dt}, \psi \rangle \psi$$

is the orthogonal projection of $\frac{d\psi}{dt}$ onto Hor_ψ. (Our inner product convention is $\langle \lambda\psi, v \rangle = \lambda \langle \psi, v \rangle$.) Using Schrödinger's equation we obtain

$$\frac{d\psi}{dt}^{\perp} = -\frac{i}{\hbar}(H\psi - \langle H \rangle \psi)$$

where $\langle H \rangle = \langle \psi, H\psi \rangle$ is the expected value of the energy. Now plug this in to our previous formula for $(\frac{ds}{dt})^2$, take the square root and bring the dt over to the right hand side to obtain our arclength formula (1).

We take this opportunity to note that

$$\Delta E(t) = \sqrt{\sum_{i \neq 1} |H_{i1}(t)|^2}$$

where $H_{ij}(t) = \langle \psi_i, H\psi_j \rangle$ are the matrix elements of $H(t)$ in a (moving) orthonormal frame whose first element is $\psi_1 = \psi(t)$. Physically this says that ΔE measures the amount of energy required to knock $\psi(t)$ *out of its current state* $\pi(\psi(t))$.

When we integrate formula (1) for arclength we obtain the formula

$$L = \frac{1}{\hbar}(\Delta E)_{av}\Delta t \tag{7}$$

for the length $L = \int ds$ of our curve C. Here $(\Delta E)_{av} = \frac{1}{\Delta t}\int_{t_1}^{t_2} \Delta E(t)dt$ is the average energy uncertainty over the time interval $\Delta t = t_2 - t_1$ which parameterizes C.

The distance d between two states $p_1, p_2 \in \mathbf{P}\mathcal{H}$ is defined as usual in Riemannian geometry

$$d(p_1, p_2) = inf\{L(C) : C \text{ a smooth curve joining } p_1 \text{ to } p_2\}$$

and this distance is realized by a geodesic joining p_1 to p_2. Every geodesic on the base space B of a Riemannian submersion $X \to B$ is the projection of a *horizontal* geodesic on X. The general horizontal geodesic on $X = S(\mathcal{H})$ has the form

$$\gamma(s) = \cos(s)\psi_1 + \sin(s)\psi_2 \quad \text{where} \quad \langle \psi_1, \psi_2 \rangle = 0 .$$

It follows immediately that the distance d between two states $\pi(\psi)$ and $\pi(\phi)$ is given by

$$\cos(d) = |\langle \psi, \phi \rangle| \quad , \text{with} \quad 0 \leq d \leq \frac{\pi}{2} \tag{8}$$

In particular two states are orthogonal if and only if the distance between them is $\pi/2$.

I.4 Heisenberg's Inequality and a Review of Quantum Mechanics

We follow Aharonov and Anandan's derivation [1990] of Heisenberg's time-energy uncertainty principle from the formula for arclength. We begin by agreeing to call two states "distinguishable" if and only if they are orthogonal. We will argue this in a moment. The distance d between two distinguishable states is then $\pi/2$ as we have just shown. It follows from this and equation (7) for the length of a curve that if our quantum system is evolving according to Schrödinger and if $\pi(\psi(t_1))$, $\pi(\psi(t_2))$ are two distinguishable states along the evolution curve of states then $\Delta E_{av}\Delta t/\hbar \geq \pi/2$. Realizing that $\hbar = \frac{h}{2\pi}$ this becomes Heisenberg's inequality, equation (3).

We now argue the point regarding distinguishability of states. In order to do this we must recall the standard (Copenhagen) interpretation of quantum mechanics.

Measurements correspond to Hermitian operators, which we consequently call "observables". The observed values of a measurement of the observable A are its eigenvalues. The probability of observing a specific value a, given that we are in the state $\pi(\psi)$, is $p(a) = \langle \mathbf{P}_a\psi, \psi \rangle$ where $\mathbf{P}_a$ is the orthogonal projection onto a's eigenspace. If this value of the measurement is selected then immediately after the measurement the system is no longer in the state $[\psi]$ but instead it is in the new state $[\mathbf{P}_a(\psi)] = \pi(\frac{\mathbf{P}_a\psi}{\|\mathbf{P}_a\|})$. (Note that the probability that $\mathbf{P}_a = 0$ is zero.) This last fact is the mysterious "collapse of the wavepacket" phenomena. We will say that the new state is the "outcome" of the measurement of a.

We had defined two states to be "distinguishable" if and only if they are orthogonal. Now we see that this is equivalent to saying that two states are distinguishable if and only if they are possible outcomes of the measuement of some observable whose corresponding measured values are different.

For more on the topic of distinguishability of states see Datta et al [1988] and references therein. They argue that the distinguishability of non-orthogonal states leads to the instantaneous transmission of data — à la the EPR gedanken experiment — and hence to a violation of the special theory of relativity.

Aharonov and Anandan give another relation between the physics and geometry which we will take the opportunity to present now. Suppose that a is a simple eigenvalue of A and that ϕ is the corresponding normalized eigenvector. Then $\mathbf{P}_a\psi = \langle \psi, \phi \rangle \phi$ so that the probability $p(a)$ is $|\langle \psi, \phi \rangle|^2$. If a is measured (and selected, as in the Stern-Gerlach experiment, for instance) then we are in the state $\pi(\phi)$, so that we can call $p = p(a)$ the "probability of the transition" $\psi \to \phi$. It follows from equation (8) that the distance d and transition probability p between two states are related by

$$d = \arccos(\sqrt{p}) \ .$$

I.5 Phases and Holonomy

Formula (6) defines the horizontal distribution for the canonical connection on the Hopf fibration. The corresponding connection one-form is

$$\Gamma(\psi) = -\Im\langle \psi, d\psi \rangle$$

where "$\Im$" is plain TeX's funny way of saying "the imaginary part of". Its curvature form Ω is the Kähler form on $\mathbf{P}\mathcal{H}$:

$$\pi^*\Omega = d\Gamma$$

Suppose now that our curve $C(t) = \pi(\psi(t))$ is periodic with period Δt. Then

$$\psi(t_2) = \exp(i\Delta\theta)\psi(t_1)$$

for some phase factor $\Delta\theta$ defined modulo 2π. There is a formula for this phase factor:

$$\Delta\theta = -\int_D \Omega - \frac{E_{av}\Delta t}{\hbar} \ .$$

The surface integral is over any disc D which bounds our periodic curve C and $E_{av} = \frac{1}{\Delta t}\int_{t_1}^{t_2} E(t)dt$ is the average of the expected energy $E(t) = \langle\psi(t), H(t)\psi(t)\rangle$, around the loop C.

This formula is Aharonov and Anandan's generalization [1987] of Berry's [1983] phase formula. It is easy to derive, being just the integrated version of the decomposition of $\frac{d\psi}{dt}$ into horizontal and vertical parts. For details see Aharonov and Anandan [1987] or the author [1990], section 3. If $E_{av} = 0$, then we get

$$\Delta\theta = -\int_D \Omega$$

which is the integrated version of formula (2) for the element of "phase interference".

I.6 Isoholonomic Inequality

From the work thus far it follows that the two questions, "What is the shortest loop in $\mathbf{P}\mathcal{H}$ with a given holonomy?" and "How can we obtain a desired phase shift with the least average fluctuation in energy?" are the same, provided the average energy of the loop is zero. The physical chemist Alex Pines posed this question to me and in [1990] I solved it. I called the loops which solve this question "isoholonomic curves".

To describe these isoholonomic curves first consider the case $\mathcal{H} = \mathbf{C}^2$. Then $\mathbf{P}\mathcal{H} = S^2$, the two-sphere of radius $\frac{1}{2}$, and the Kähler form is the area form. Therefore our question is the classical isoperimetric problem, "What is the shortest loop enclosing a fixed area?", the answer to which is well-known. The solutions are geometric circles on the surface of the sphere.

Every such circle enjoys the following properties.

(1) It is the image under the Hopf projection of a geodesic in $S(\mathcal{H}\ )$.

(2) It is itself a geodesic (i.e. a great circle) if and only if the corresponding geodesic in $S(\mathcal{H})$ is *horizontal*.

(3) It can be generated by a *time-independent* Hamiltonian.

(4) Its length and (spherical) area are related by $L^2 = 4\pi A - 4A^2$. ($K = 4$ is the curvature of the sphere.)

Note in this last equality that it does not matter which of the two areas bounded by the circle are taken; the answer is the same.

I showed in [1990] that these properties (1)-(3) hold for the isoholonomic solution curves for a general Hilbert space. In fact, every such solution lies on some $S^2 = \mathbf{P}V \subset \mathbf{P}\mathcal{H}$ for some two-dimensional subspace V of $\mathcal{H}$ and is a isoholonomic minimum, and thus a geometric circle, for the restricted bundle-with-connection $S(V) \to \mathbf{P}V$. As for the final property (4), the symplectic area

$$\Phi = -\int_D \Omega$$

is independent modulo 2π of the choice of disc D bounded by C. Since the isoholonomic minima are geometric circles on some $\mathbf{P}V$ and since on this $\mathbf{P}V$ we have $\Omega = -2$ (area form) and the equality in property (4), we have demonstrated the validity of the Isoholonomic inequality (4) in the following sharp form. For any loop C on $\mathbf{P}\mathcal{H}$ with length L and enclosing a symplectic area Φ with $0 \le \Phi \le 2\pi$ the inequality

$$L^2 \ge 2\pi\Phi - \Phi^2$$

holds, with equality if and only if C is a geometric circle lying on some projective line $\mathbf{P}^1$ within $\mathbf{P}\mathcal{H}$.

I.7 Measuring Holonomy

In this section we will discuss methods for experimentally measuring holonomy. For further questions I recommend the review article by Zwanziger et al [to appear 1990].

Overall phase changes cannot be detected because experimental data, namely averages and probability distributions of measurements, are invariant under global phase changes $\psi \to e^{i\theta}\psi$. Only *relative phase changes* are measurable. This means that to observe holonomies the experimentalist must prepare a *superposition* of states.

Suppose that the normalized vector ψ_1 undergoes a projective cycle: $C_1(t) = \pi(\psi_1(t))$ under the influence of the Hamiltonian $H(t)$ and that another normalized vector ψ_2 undergoes a (different) projective cycle $C_2(t)$ of the same period T. Let $\Delta\theta_i = \gamma_i + \frac{E_i T}{\hbar}$, $i = 1, 2$ be the corresponding phases accumulated by the ψ_i during their cycles. The γ_i are the symplectic areas, i.e., the logarithms of the holonomies, of the loops. The E_i are their average expected energies. Now prepare the superposition

$$\psi = a\psi_1 + b\psi_2 \quad \text{with} \quad |a|^2 + |b|^2 = 1$$

and measure some observable M. After k periods the average value of the measurements will be

$$\langle \psi(kT), M\psi(kT) \rangle = A + B\cos\{k(\Delta\theta_1 - \Delta\theta_2) + \theta_0\}$$

where A, B, and θ_0 are constants independent of the number of periods k. (For example $A = |a|^2\langle\psi_1, M\psi_1\rangle + |b|^2\langle\psi_2, M\psi_2\rangle$.) In the particular case where $E_1 = E_2$ we have $\Delta\theta_1 - \Delta\theta_2 = \gamma_1 - \gamma_2$ and so the *difference* of the (logarithms) of the holonomies appears as the *frequency* of oscillation of observables.

In NMR experiments the x-component, $M = M_x$, of the bulk magnetization of a sample is measured as a function of a *frequency* ω. This frequency is related to the time t or kT by Fourier transform. The net result is that the difference of holonomies manifests itself as a shift (and perhaps a splitting) of a particular line (=peak) in the NMR spectrum (graph of M vs. ω).

Most (perhaps all to date) experiments measure the relative holonomies of pairs of curves with one of the two following types of geometries.

The first geometry consists of two curves for a two-level system. Thus $\mathbf{P}\mathcal{H} = S^2$ is the standard two-sphere in $\mathbf{R}^3$. The initial states ψ_1 and ψ_2 are prepared so as to be perpendicular to each other. This is equivalent to $\pi(\psi_2)$ being antipodal to $\pi(\psi_1)$ on the two-sphere. The two states remain antipodal, $C_1(t) = -C_2(t)$, since the evolution is unitary. The antipodal map is orientation reversing so that the two symplectic areas are negatives of each other. Consequently the observed frequency shift $\gamma_1 - \gamma_2$ will be the negative of the solid angle enclosed by C_1. This is the type of geometry incorporated in the NMR experiments of Tyko [1987], the optical experiments of Tomita and Chiao [1986], and the neutron interferometry experiments of Bitter and Dubbers [1987].

The second type of geometry concerns three-level systems so that $\mathbf{P}\mathcal{H} = \mathbf{P}^2$. C_1 lies entirely inside a two-level subsystem and C_2 is the constant state which represents the third level orthogonal to this subsytem. Thus $\gamma_2 = 0$ and the observed frequency shift will be $\gamma_1 - \gamma_2 = -\int_{D_1}\Omega$. This is the type of geometry tested in the experiment performed by Suter, Mueller, and Pines [1988].

There is a nice projective interpretation for this second type of experiment. $\mathcal{H} = \mathbf{C}^2 \oplus \mathbf{C}$ and the first curve travels in $\mathbf{P}^1 = \mathbf{P}(\mathbf{C}^2) \subset \mathbf{P}\mathcal{H} = \mathbf{P}^2$. The normal bundle N of $\mathbf{P}^1$ is naturally isomorphic to the tautological line bundle over $\mathbf{P}^1$. This is the line bundle whose unit vectors form the Hopf fibration $S^3 \to \mathbf{P}^1$. When we prepare a superposition of ψ_1 and ψ_2

we move off of the zero-section, $\mathbf{P}^1 \subset N$, and onto some non-zero vector $\pi(a\psi_1 + b\psi_2) \in N$. This vector evolves under Schrödinger as the *horizontal lift* of C_1 with respect to the natural connection on the tautological line bundle, and it is the logarithm of this holonomy which is measured.

PART II. GEOMETRY OF MIXED STATES

II.1 Overview

In this half of the paper we will see what happens to the relationship between physics and geometry when we replace quantum mechanics by quantum statistical mechanics. The basic objects are now density matrices. These can be thought of as elements in the dual, $u(\mathcal{H})^*$, of the Lie algebra of the unitary group, $U(\mathcal{H})$. Our replacement for $\mathbf{P}\mathcal{H}$ is the set $\mathcal{O} = \mathcal{O}(\rho)$ of all density matrices conjugate to a given density matrix ρ. In other words, $\mathcal{O}$ is the *co-adjoint orbit* through ρ.

Our arclength formula (1) no longer holds generally but it still holds for *horizontal* Schrödinger evolutions. The element of arclength on $\mathcal{O}$, and the connection (choice of horizontal) over $\mathcal{O}$ are still defined by declaring that a certain bundle projection $S \to \mathcal{O}$ is a Riemannian submersion. **Our main achievement here is the correct choice for the Riemannian manifold S**, the quantum statistical replacement for $S(\mathcal{H})$. We define S to be the fiber bundle over $\mathcal{O}$ whose fiber S_ρ over ρ consists of the set of orthogonal (*not orthonormal !*) frames, $\{\phi_i\}_{i=1,\dots,rank(\rho)}$, of eigenvectors for ρ. Thus $\rho\phi_i = c_i\phi_i$ (no sum) where the c_i are ρ's nonzero eigenvalues listed in (say) decreasing order. The normalization for the frames is according to these eigenvalues:

$$\langle \phi_i, \phi_j \rangle = c_i \delta_{ij} \qquad \text{(no sum)}$$

S sits inside the Hilbert space $k\,\mathcal{H}$, the direct sum of $k = rank(\rho)$ copies of $\mathcal{H}$. Its Riemannian metric is the one induced from this embedding.

$S \to \mathcal{O}$ forms a principal bundle which we will sometimes call the "generalized Hopf fibration". Its structure group $G = G(\rho)$ is the commutant of ρ, that is, the group of all unitaries which commute with ρ. The phase associated to a loop in $\mathcal{O}$ is thus the logarithm of an element of G. There is no "non-Abelian Stoke's formula"; consequently there is no simple formula such as Berry's for this phase in case G is non-Abelian. This much is well-known.

In the past authors (including myself) have chosen the frames comprising S to be orthonormal. This corresponds to inducing the metric on S from the bi-invariant metric on $U(\mathcal{H})$. There are at least two other well-known Riemannian metrics on a coadjoint orbit $\mathcal{O}$ of $U(\mathcal{H})$ for which

$U(\mathcal{H})$ acts by isometries. One of these is the induced metric, that is the one obtained by viewing the orbit as a submanifold of the Euclidean space formed by the (dual) Lie algebra. The other is the Kähler metric. This is the real part of the Kähler structure whose imaginary structure is the (Kirillov-Kostant-Souriau) symplectic structure. I claim that none of these choices are "correct" from the quantum statistical point of view. The correct metric is obtained by Riemannian submersion from the above described eigenbundle S. An argument for why this is correct appears in §II.6.

II.2 Review of Quantum Statistical Mechanics
The Generalized Hopf Fibration

I will now give a very brief review of the formalism of quantum statistical mechanics. See the books of Feynman (1972) and of Mackey (1963) for a more complete picture. In this review I completely ignore most of the central concepts such as the Boltzman distribution and temperature. The main goal is to convince you that the correct replacement for $\mathbf{P}\mathcal{H}$ is $\mathcal{O}$. We will also describe the generalized Hopf fibration $S \to \mathcal{O}$ in more detail.

Observables in quantum statistical mechanics are self-adjoint operators on $\mathcal{H}$, the same as in the standard non-relativistic mechanics. A *state* in quantum statistical mechanics is a non-negative Hermitian matrix ρ with trace one

$$\rho = \rho^* \qquad \rho \geq 0 \qquad tr(\rho) = 1 \ .$$

A state is also called a "density matrix" or a "mixed state". We can always express a state in the form

$$\rho = c_1 \psi_1 \otimes \psi_1^* + c_2 \psi_2 \otimes \psi_2^* + \ldots \qquad (II.1a)$$

with

$$c_i > 0 \ , \quad \sum c_i = 1 \ , \quad \text{and the } \psi_i \text{ an orthonormal frame.} \qquad (II.1b)$$

The c_i are of course the non-zero eigenvalues of ρ. The state is called "pure" if there is only one c_i, in which case it is $c_1 = 1$ and ρ has rank 1.

The set S of all states forms a convex bounded subset within the real vector space of all Hermitian matrices. Its extreme points are formed by the set S_{pure} of pure states. Moreover

$$S_{pure} = \mathbf{P}\mathcal{H}$$

with the identification map being defined by equation (5). In this manner, standard quantum mechanics is embedded in the quantum statistical mechanics.

If A is a bounded observable then $tr(\rho A)$ is its "expected value" relative to the state ρ. The assignment $A \mapsto tr(\rho A)$ is a positive $(A \geq 0 \Rightarrow tr(\rho A) \geq 0)$ normalized $(tr(\rho 1) = 1)$ linear functional on the space of bounded observables. Thus our states are states in the C^*-algebra sense, for the C^*-algebra $\mathcal{B}(\mathcal{H})$ of all bounded operators on $\mathcal{H}$.

We will henceforth identify the space of bounded observables, a real linear subspace of $\mathcal{B}(\mathcal{H})$, with the Lie algebra $u(\mathcal{H})$ of all skew-hermitian operators in the standard fashion: multiplication by i. Thus the set of states $\mathcal{S}$ forms a subset of $u(\mathcal{H})^*$, the dual of the Lie algebra of the unitary group. This subset is clearly invariant under the (co-) adjoint action $\rho \mapsto g\rho g^*$.

The quantum evolution of a state is defined by the equation

$$\frac{d\rho}{dt} = \frac{i}{\hbar}[\rho(t), H(t)]$$

which sometimes goes under the name of Liouville's equation. It looks just like Heisenberg's equation except for a sign change due to the fact that states transform contragrediently with respect to observables. $H(t)$ is the same as before: a time-dependent observable. These equations imply that

$$\rho(t) = g(t)\rho(0)g(t)^*$$

where $g(t)$ is a unitary matrix (called the "propagator" or fundamental solution). Consequently ρ remains on whatever co-adjoint orbit it began on.

This co-adjoint orbit is completely specified by ρ's non-zero eigenvalues. From now on we assume that

$$k = rank(\rho) < \infty$$

for simplicity. Let $\mathbf{c} = (c_1, c_2, c_3, \ldots, c_k)$ be the list of these eigenvalues, including multiplicity, and in increasing order: $c_1 \leq c_2 \leq c_3 \ldots$. Then the co-adjoint orbit is specified as

$$\mathcal{O}(\mathbf{c}) = \{\rho \in \mathcal{S} : \text{the nonzero spectrum of } \rho = \{\mathbf{c}\}\}$$

Of course, when we say "nonzero spectrum" we mean to include the information of multiplicities.

The spaces $\mathcal{O}(\mathbf{c})$ are the generalizations of $\mathbf{P}\mathcal{H} = \mathcal{O}(1)$. The generalization of $S(\mathcal{H})$ is

$$S(\mathbf{c}) = \{(\phi_i) : (\phi_i) \text{ is an orthogonal frame in } \mathcal{H}$$
$$\text{normalized according to } \langle \phi_i, \phi_j \rangle = c_i \delta_{ij}\}$$

as we said earlier. The number of elements in each such frame is $k =$ the rank of ρ. The generalized Hopf projection is

$$\pi(\phi_1, \phi_2, \ldots) = \sum \phi_i \otimes \phi_i^* \tag{II.2}$$

To relate this to equation (II.1a) for ρ, set $\phi_i = \sqrt{c_i}\psi_i$. In the final section I argue that this is the correct generalization of the Hopf fibration.

Let $c(1) = c_1 < c(2) = c_{m(1)+1} < \ldots$ be ρ's positive eigenvalues listed *without* multiplicities and let $m(i)$ be their corresponding multiplicities. Thus

$$\sum m(i)c(i) = 1$$

$$\sum m(i) = k$$

$\pi : S(\mathbf{c}) \to \mathcal{O}(\mathbf{c})$ is a principal bundle with structure group

$$G(\mathbf{c}) = U(m(1)) \times U(m(2)) \times \ldots$$

a product of unitary groups.

Example 1 $G(1) \to S(1) \to \mathcal{O}(1)$ is our friend the standard Hopf fibration.

Example 2 $G(\frac{1}{k}, \frac{1}{k}, \ldots, \frac{1}{k}) \to S(\frac{1}{k}, \frac{1}{k}, \ldots, \frac{1}{k}) \to \mathcal{O}(\frac{1}{k}, \frac{1}{k}, \ldots, \frac{1}{k})$ is the bundle of k-frames, also known as the Stieffel variety (except that our frames are normalized to be smaller). Its fiber is $U(k)$ and the base space is the Grassmannian of all k-planes in $\mathcal{H}$.

$S(\mathbf{c})$ inherits a Riemannian structure as a sub-manifold of $k\mathcal{H} = \mathcal{H} \oplus \ldots \oplus \mathcal{H}$ (k times). Now $U(\mathcal{H})$ and $U(k)$ act by isometries on $k\mathcal{H} = \mathcal{H} \otimes \mathbf{C}^k$. $U(\mathcal{H})$ and $G(\mathbf{c}) \subset U(k)$ leave $S = S(\mathbf{c})$ invariant and so act on it by isometries. The fibers of $\pi : S \to \mathcal{O}$ are the $G(\mathbf{c})$-orbits. This allows us to put a metric on $\mathcal{O}$ by declaring π to be a Riemannian submersion. We also get a connection on the bundle in this manner. (See the paragraph on Riemannian submersions near the beginning of the paper.)

With this structure of a Riemannian submersion in place, we can address all of the questions which we previously answered for $\mathbf{P}\mathcal{H}$. What is the (quantum statistical) meaning of the element of arclength on $\mathcal{O}$? Of the holonomy of a loop? How do we measure the holonomy? What is the shortest loop with a given holonomy? We will only address the first and last question.

II.3 Arclength

We begin by fixing $(c_1, c_2, \ldots) = \mathbf{c}$ and abbreviate $S(\mathbf{c}) = S$, and similarly for $\mathcal{O}$ and G. $U(\mathcal{H})$ acts transitively on S so that any tangent vector to S at $\phi = (\phi_1, \phi_2, \ldots)$ can be written

$$\left(\frac{d\phi}{dt}\right)_j = \frac{\sqrt{-1}}{\hbar} H\phi_j$$

for some self-adjoint operator H independent of the index j. This is to say, every tangent vector can be realized by some Schrödinger evolution. The squared length of this vector is

$$\|(\frac{d\phi}{dt})\|^2 = \sum \|\frac{d\phi_j}{dt}\|^2$$
$$= \frac{1}{\hbar^2} \sum \langle H\phi_j, H\phi_j\rangle$$
$$= \frac{1}{\hbar^2} \sum \langle \phi_j, H^2\phi_j\rangle$$
$$= \frac{1}{\hbar^2} tr(\rho H^2)$$

where $\rho = \pi(\phi)$ is the density matrix corresponding to the frame ϕ. The last line in this column of equalities follows from the general fact $tr(\rho A) = \sum \langle \phi_j, A\phi_j\rangle$. We can re-express this equation in the form

$$ds_S = \frac{dt}{\hbar}\sqrt{\langle H^2\rangle_\rho} \qquad (II.3)$$

Polarizing this formula we obtain the curious inner product formula

$$\langle iH_1(\phi), iH_2(\phi)\rangle_\phi = \frac{1}{\hbar^2} tr(\rho[H_1, H_2]_+)$$

where $iH(\phi)$ denotes the vector at ϕ generated by the Hamiltonian H and where $[H_1, H_2]_+ = H_1 H_2 + H_2 H_1$ denotes the anti-commutator.

We need to know the horizontal projection $T_\phi S \to (ker T_\phi \pi)^\perp$ in order to calculate the arclength on $\mathcal{O}$. The vertical projection $T_\phi S \to (ker T_\phi \pi)$ is

$$\delta\phi_j \mapsto \mathbf{P}_j(\rho)\delta\phi_j$$

where $\mathbf{P}_j(\rho) : \mathcal{H} \to \mathcal{H}$ is the orthogonal projection onto c_j's eigenspace. The horizontal projection is $1 -$ (this). In particular, if $\phi \in S$ evolves according to Schrödinger and is also a horizontal curve, then $\mathbf{P}_j(\rho)\frac{d\phi_j}{dt} = \mathbf{P}_j H\phi_j = 0$. Then $\langle \phi_j, H\phi_j\rangle = 0$ so that $tr(\rho H) = 0$. This yields our formula (1):

$$ds_\mathcal{O} = \frac{dt}{\hbar}(\Delta E)$$

provided that the evolution on S is by a horizontal Schrödinger evolution. For the case $\mathcal{O} = \mathbf{P}\mathcal{H}$ of part I we did not need to add the adjective "horizontal".

For a general Schrödinger evolution the element of arclength can be expressed as

$$ds_\mathcal{O}^2 = \frac{dt^2}{\hbar^2}(\langle H^2\rangle - \sum \langle \phi_j, H\mathbf{P}_j H\phi_j\rangle)$$

which is typically not equal to our previous expression for $ds_\mathcal{O}$ squared. In fact, no matter how the metric on $\mathcal{O}$ is chosen, $ds_\mathcal{O} \neq \frac{dt}{\hbar}(\Delta E)$ for general Schrödinger curves. This is because given any impure state ρ there exist H's which represent the zero-tangent vector to $\mathcal{O}$, i.e., for which $[\rho, H] = 0$, but for which $\Delta E_\rho \neq 0$.

A perhaps more illuminating formula for $ds_\mathcal{O}$ can be obtained by writing H in block form with respect to the eigenspaces $E(i)$ of the $c(i)$. (Recall that these are the eigenvalues listed without multiplicity.) Thus $H_{(ij)} = \mathbf{P}_j H \mathbf{P}_i : E_i \to E_j$ is an $m(i) \times m(j)$ matrix. We must also include blocks for the 0-eigenspace E_0 of ρ. We do this by adding an extra index 0 with $c(0) = 0$. Then one calculates that

$$(\frac{ds_\mathcal{O}}{dt})^2 = \frac{1}{\hbar^2}\sum_j c(j) \sum_{k \neq j} tr(H_{(jk)}H_{(kj)})$$

In words, the squared length is a weighted sum of the squared lengths of the *off-diagonal* blocks of the Hamiltonian, the weights being $\rho's$ eignevalues. As in part I, this arclength is a measure of the amount of energy required to change the current state to a new state.

It is illuminating to rewrite this formula:

$$(ds_\mathcal{O})^2 = \frac{1}{\hbar^2}\sum_{j>k}(c(j) + c(k))tr(H_{(jk)}H_{(kj)})(dt)^2$$

and then to compare it with the formula

$$(ds_{\mathcal{O},\text{Kähler}})^2 = \frac{1}{\hbar^2}\sum_{j>k}(c(j) - c(k))tr(H_{(jk)}H_{(kj)})(dt)^2.$$

This last formula is the formula for the Riemannian metric on $\mathcal{O}$ obtained by taking the real part of the corresponding Kähler metric, the one whose imaginary part is the natural (Souriau-Kostant-Kirillov) symplectic form on this co-adjoint orbit. See, for example, Besse.

As an example of this formula consider the case of the Grassmannian of k-planes, example 2 above. Then $\rho = \frac{1}{k}\mathbf{P}$ where $\mathbf{P}$ is a rank k projection operator. H is a 2×2 block matrix and

$$ds_\mathcal{O}^2 = \frac{1}{k\hbar^2}tr(H_{(12)}H_{(21)})$$

where $H_{(12)} = \mathbf{P}H(1 - \mathbf{P})$ and $H_{(21)} = H_{(12)}^*$ are the two off-diagonal blocks.

318   R. Montgomery

II.4 Phases and Curvature

There is no simple formula for the holonomy of a loop on $\mathcal{O}$. This is because there is no non-Abelian Stokes formula. However in the special case where G is Abelian, that is where all the c_i are distinct, there is a Stoke's formula.

The curvature of the connection can be calculated in terms of operator-valued forms. See Avron, Sadun, Segert and Simon [1988]. The result is

$$\Omega = \Sigma P_i(dP_i) \wedge (dP_i)P_i$$

where $P_i : \mathcal{O} \to u(\mathcal{H})$ is the projection-valued function which assigns to each density matrix $\rho \in \mathcal{O}$ the projection $P_i(\rho)$ onto the eigenspace for $c(i)$.

II.5 The Isoholomonic Problem and Some Corrections to the Literature

In [1990] I partially characterized the isoholomonic minimizers for $\pi : \mathcal{S} \to \mathcal{O}$. In this subsection we will review and **correct** this characterization.

In the general setting of Riemannian submersions it is simpler to try to characterize the **isoparallel extremals** instead of the isoholomonic minima. "Parallel" as opposed to "holonomic" because we will not be able to say that the curve on $\mathcal{O}$ is closed. (For a non-closed curve γ in $\mathcal{O}$ what we fix instead of the holonomy is the parallel translation map which is a G-automorphism from the fiber over the initial point $\gamma(t_1)$ to the fiber over the final point $\gamma(t_2)$. Alternatively, we fix the end points of the horizontal lift $\tilde{\gamma}$ of γ.) "Extrema" as opposed to "minimal" because we will only say that $\tilde{\gamma}$ extremizes length among all horizontal curves with these endpoints.

We will now state the analogues of properties (1) through (4) of the section "Isoholonomic Inequality" of part I. After the statements, we will discuss various errors which occured in my previous paper [1990].

(1') Any projected geodesic from $\mathcal{S}$ is an isoholonomic extremal on $\mathcal{O}$. For sufficiently small subarcs these extremals are isoholonomic minima.

(2') An isoparallel extremal is a geodesic on $\mathcal{O}$ if and only if it is the projection of a horizontal geodesic on $\mathcal{S}$.

(3') "Every isoparallel extremal is generated by a time-independent Schrödinger equation" is **simply false** for typical $\mathcal{S} \to \mathcal{O}$.

(1') and (2') are true for any Riemannian submersion **provided the fiber metric is covariantly constant**.

In [1990] and in my lecture I incorrectly stated that the converse to (1') holds in the general setting of Riemannian submersions. (I was quoting a theorem of another author.) But very recently I have found a counterexample to this alleged converse, that is, an isoparallel **minimum** which is not the projection of any geodesic from the total space. In this counterexample the Hörmander condition holds everywhere: the image of the curvature form together with all of its covariant derivatives spans the Lie algebra of G. In the example one more covariant derivative is needed for points over the counterexample curve in comparison to all other points. Such "pathologies" occur in our generalized Hopf fibrations but it is still possible that the the converse to (1') holds here. I do not know.

In [1990] I made the statement in quotes in (3'). My mistake was that I used the wrong metric on S. (I used a metric induced by Riemannian submersion from a bi-variant metric on U(H).)

To see that this statement in quotes in (3') is false, we note that the time-independent Schrödinger equation defined by H generates a geodesic in S **if and only if**

$$[\rho, H^2] = 0 \ .$$

But a dimension count shows that this last condition typically constrains the set of geodesic-generating H's to be smaller than the set of directions at φ. (If all the c_i's are distinct the dimension count yields dimension zero, so a finite set of directions.) Our claim follows: for typical S there are geodesics, and hence isoparallel extremals, which are not generated by any autonomous Schrödinger equation.

Regarding our statement concerning the condition $[\rho, H^2] = 0$. A curve $\varphi(t)$ in S is a geodesic if and only if $\frac{d^2\varphi}{dt^2} \perp T_\varphi S$. A direct calculation similar to our arclength calculation shows that this in turn is true if and only if $[\rho, H^2] = 0$, provided, of course, that the $\varphi(t)_j$'s evolve according to Schrödinger for this H.

II.6 How to use Dual Pairs to Ignore the Rest of the Universe

In this final section we will construct the bundle $S \to \mathcal{O}$ using symplectic reduction. This provides evidence that our bundle and in particular our normalization of the frame $\varphi \in S$ is the correct choice, both from the points of view of physics and of geometry.

A standard argument for the necessity of statistical mechanics proceeds are as follows. (See for example, Feynman [1972] Chapter2.) When we analyze any system we necessarily seperate it from the rest for the universe. But the true description involves the rest of the universe. Statistical mechanics provides a means for systematically ignoring the rest of the universe while still providing a reasonably accurate picture of the system of interest.

Restated in the formalism of quantum mechanics this argument begins as follows. The Hilbert space for the universe splits as a tensor product:

$$\mathcal{H}_u = \mathcal{H} \otimes \mathcal{H}_r$$

Here $\mathcal{H}$ is the Hilbert space of the system of interest and $\mathcal{H}_r$, "r" for "rest", is the Hilbert space for the rest of the universe. (Recall that in quantum mechanics when two systems are coupled we take their tensor product, not their direct sum.) Alternatively, simply set $\mathcal{H}_u = k\mathcal{H} = \mathcal{H} \otimes \mathbf{C}^k$ of the previous section and *interpret* $\mathbf{C}^k$ as the rest of the universe.

We require a method for ignoring all of the extraneous or perhaps inaccessible information in $\mathcal{H}_r$. Given the nature and participants of this conference, there is really only one possible method to use: symplectic reduction!

The group to reduce by is $U(\mathcal{H}_r) = U_r$. It acts transitively on the set of directions in $\mathcal{H}_r$, and so reduction by its action should get rid of all $\mathcal{H}_r$ information.

$U(\mathcal{H}) = U$ and U_r form a Howe dual pair within the large group $Sp(\mathcal{H}_u)$ of all real-linear symplectic transformations of $\mathcal{H}_u$. This means that they are reductive and are each other's commutants within the large group. Our pair of groups (U, U_r) also satisfies an additional technical property labeled (Q) below. Any dual pair enjoying property (Q) has the following remarkable property. *Every symplectic reduced space for one group is isomorphic to the closure of a co-adjoint orbit in the dual of the other group's Lie algebra.* In our case this means that our mixed-state manifolds $\mathcal{O}(\rho)$ are reduced spaces for this action of U_r on $\mathcal{H}_u$.

To be honest, the italicized statement above is only known to be true when one element of the dual pair is compact and when this is the group by which we are reducing. This was proved recently by Lerman, Montgomery, and Sjamaar [1991]. (I suspect it is true for general Howe pairs.) The reason the proof is so recent is that it requires the machinery of stratified symplectic reduction to make sense of the possibly singular reduced space. This machinery was developed very recently by Sjamaar and Sjamaar-Lerman.

In any case, for our dual pair, both elements are compact. For this reason there is also no need to take the closure in the italicized statement above: orbits are already closed.

We now outline the correspondence "orbits $\leftrightarrow$ reduced spaces" for a general dual pair (G, G_r) within $Sp(E)$, where E is a symplectic vector space. For more on dual pairs see Kazhdan, Kostant and Sternberg [1978] Our particular dual pair appears on the last page of their article.

Write down the momentum maps, J_r and J, for the two groups :

$$Lie(G_r)^* \longleftarrow E \longrightarrow Lie(G)^* \ .$$

Take $\rho \in Lie(G)^*$ with the property that there is a $\varphi \in E$ with $J(\varphi) = \rho$. Construct the reduced space:

$$M_{\mathbf{c}} := J_r^{-1}(\mathbf{c})/(G_r)_{\mathbf{c}}$$

for

$$\mathbf{c} = J_r(\varphi)$$

Here $(G_r)_{\mathbf{c}}$ denotes the isotropy group of $\mathbf{c}$. Since the G and G_r actions on E commute, J is a G_r-invariant map. It follows that the map J, upon restriction to $J_r^{-1}(\mathbf{c})$, induces a map

$$i_{\mathbf{c}} : M_{\mathbf{c}} \longrightarrow \mathcal{O}(\rho)$$

whose target is the co-adjoint orbit through ρ. It maps onto $\mathcal{O}$ because J_r is G - invariant so that $G(J_r^{-1}(\mathbf{c})) = J_r^{-1}(\mathbf{c}))$.

We need to know when $i_{\mathbf{c}}$ is one-to-one. This is where the technical property (Q) comes in . The epimorphism $i_{\mathbf{c}}$ is one-to-one iff any G_r - invariant functions on E can be expressed as a smooth function of the components of J. (Recall that the reduced space is isomorphic to $J_r^{-1}(G_r\mathbf{c})/G_r$.) The fact that G and G_r form a dual pair implies that every quadratic form on E which is invariant under G_r can be expressed in terms of the components of J which are themselves homogeneous invariant quadratic polynomials on E. (Identify the Lie algebra of $Sp(E)$ with the space of homogeneous quadratic functions on E.) Thus we need to know whether or not these are all of the invariants. This is the technical property which we need to continue :

the space of G_r-invariant polynomials on E are
generated by the homogeneous quadratic invariants
(PROPERTY Q)

A theorem of Weyl [1973] says that this condition does indeed hold for our dual pair.

In our case the dual pair diagram is:

$$u_r^* \longleftarrow H_u \longrightarrow u^*$$

The arrows are the momentum maps J_r and J for the respective groups. As usual, we identify the Lie algebras, u, u_r with the vector space of Hermitian operators on the respective spaces. Their dual spaces are identified with Hermitian operators of trace class by using the trace functional. We can write any $\varphi \in \mathcal{H}_u$ in the form:

$$\varphi = \sum c_{\alpha j} |\psi_\alpha > |j >$$

where we used Dirac's notation. Thus $\{|\psi_\alpha >\}$ form a basis for $\mathcal{H}$, $\{|j >\}_{j=1,2\ldots}$ for $\mathcal{H}_r$, and $\{|\psi_\alpha > |j >\}$ for $\mathcal{H}_u$. We calculate that the images of φ under the two momentum maps are

$$J(\varphi) = \sum_i c_{\alpha i}\bar{c}_{\beta i} |\psi_\alpha >< \psi_\beta|$$

and

$$J_r(\varphi) = \sum_\alpha c_{\alpha i}\bar{c}_{\alpha j} |i >< j| .$$

The formula for $J(\varphi)$ is exactly the formula (2.4) for a density matrix found in Feynman [1972].

From the formula for J we see that

$$J(\varphi) \geq 0$$

$$trace(J(\varphi)) = \|\varphi\|^2$$

$$J(\varphi) = 0 \Leftrightarrow \varphi = 0$$

$$rank(J(\varphi)) = rank(\varphi) \ .$$

In the last equality we view φ as a linear map $\mathcal{H}_r \to \mathcal{H}$. (Alternatively, $rank(\varphi)$ is the minimal number of indecomposable elements, $\psi \otimes i$, which make up φ.) It follows that the image of the sphere of normalized states in $\mathcal{H}_u$ consists of the set of density matrices with rank less than or equal to k, where $k = dim\mathcal{H}_r$. In particular, if $dim\mathcal{H}_r \geq (dim\mathcal{H})$ then *every* density matrix has the form $\rho = J(\varphi)$.

We will now take a density matrix with rank less than or equal to k and walk through the dual pair construction. First diagonalize ρ as in equation (II.1a)

$$p = \sum c_i |\psi_i\rangle \langle \psi_i|$$

where ψ_i forms an orthonormal basis for the image of ρ. Then set

$$\varphi = \sum \sqrt{c_i} \ |\psi_i\rangle |i\rangle$$

so that $J_r(\varphi) = diag(c_1, c_2, ...) = $ c. Here $\mathcal{H}_u = \mathcal{H} \otimes \mathcal{H}_r \simeq k\mathcal{H} = \mathcal{H} \oplus ... \oplus \mathcal{H}$ by the isomorphism $\sum |\varphi(i)\rangle|i\rangle \longmapsto (\varphi(1), \varphi(2), ...)$. (Note : this map depends on the choice of basis for $\mathcal{H}_r$ but not for $\mathcal{H}$.) Under this isomorphism $J_r^{-1}(c) = S(c_1, c_2...)$. Also $J : J_r^{-1}(c) \longrightarrow \mathcal{O} = \mathcal{O}(\rho)$ is our previous submersion $\pi : S(c_1, c_2...) \longrightarrow \mathcal{O}(c_1, c_2...)$. In summary, the generalized Hopf fibration is the canonical reduced space submersion

$$J_r^{-1}(c) \longrightarrow J_r^{-1}(\mathbf{c})/(U_r)_{\mathbf{c}} = M_{\mathbf{c}}$$

Some amusing consequences follow from this last fact. For instance, the problem of finding a "non-Abelian phase formula" is a special case of the general problem of reconstructing the dynamics on a momentum level set given the dynamics on its reduced space.

II.7 Conclusion : A Flaw and an Oversight

We end the paper by pointing out a flaw and an oversight in our geometrization of quantum statistical mechanics.

Flaw: The flaw is that in choosing $\mathcal{O}$ we have ignored the measurement process. There are two types of measurement processes; those in which no particular value of the measurement is selected, and those in which a particular value is selected from among the possible observed values. The latter are termed "filtering measurement" and the canonical example is the Stern-Gerlach experiment in which an apparatus acts as a spin polarizer. (See, for example, the discussion in Cohen-Tannoudji et al [1977].) Upon making the first type of measurement the state ρ transforms according to

$$\rho \longmapsto \sum \mathbb{P}_i \rho \mathbb{P}_i$$

where $\{\mathbb{P}_i\}$ is the spectral decomposition of the observable $A = \sum a_i \mathbb{P}_i$ being measured. In a filtering measurement

$$\rho \longmapsto \mathbb{P}_1 \rho \mathbb{P}_1 / trace(\mathbb{P}_1 \rho \mathbb{P}_1)$$

where a_1 is the value of the measurement which is selected. I am indebted to Jeeva Anandan for enlighting me as to these two types of measurements and their effects on density matrices.

Either type of measurement will, in general, knock our state out of its co-adjoint orbit $\mathcal{O}$. Thus measurements cannot be accounted for by working within a single co-adjoint orbit $\mathcal{O}$.

Any state can be achieved by applying some non-filtering measurement to an initial pure state. Thus, to account for these measurements we would have to work on all of $\mathcal{S}$. I do not know what the correct geometric structure on $\mathcal{S}$ is; correct in the sense of correctly reflecting the measurement process. It is almost certainly not Riemannian since it should not be even infinitesimally isotropic: directions along co-adjoint orbits (quantum evolutions) are much different from directions transverse to them (measurements, or wave packet collapses).

For filtering measurements we do not need all of $\mathcal{S}$, but rather those matrices ρ' which can be reached from an initial ρ by a sequence of filtering measurement. This is some part (perhaps all) of the subvariety of matrices with rank less than or equal to the rank of ρ.

A series of filtering measurements leads to a series $\mathcal{O} \to \mathcal{O}' \to \mathcal{O}'' \to ...$ of co-adjoint orbits. If any of the measurements is non-degenerate (that is, the corresponding eigenvalue is nondegenerate) then the series terminates at $\mathbb{P}H = \mathcal{S}_{pure}$. Different series of filtering measurements lead to different series of orbits. Thus the correct generalization of $\mathbb{P}H$ might not be a single orbit $\mathcal{O}$, but instead a "cascade" of orbits:

Oversight: This is very serious! In our "geometrization" we have completely ignored the most important concepts in statistical mechanics, for example, temperature, entropy, thermal equilibrium, and the Gibbs states. Perhaps some reader will pursue this in the future.

REFERENCES

Aharonov, Y. and Anandan, J. [1990], *Geometry of Quantum Evolution*, Preprint.

Aharonov, Y. and Anandan, J. [1987], *Phase Change During Cyclic Quantum Evolution*, Phys. Rev. Letters **58** (1987), 1593–1667.

Avron, J.E., Sadun, L., Segert, J., Simon, B. [1988], *Chern Numbers and Berry's Phases in Fermi Systems*, Comm. Math. Phys. **124** (1988), 595–627.

Berry, M. [1984], *Quantal Phase Factors Accompanying Adiabatic Changes*, Proc. Roy. Soc. London Ser. A **392** (1984), 45.

Besse, A.L. [1987], "Einstein Manifolds", Ergenbnisse Math. (3) vol. 10, Springer, Berlin, 1987.

Bitter, T. and Dubbers, D. [1987], Phys. Rev. Letters **59** (1987), 251–254.

Cohen-Tannoudji, C., Diu, B., Laloë, F. [1977], "Quantum Mechanics", Vol. 1, Ch. *III*, Wiley-Interscience, 1977.

Datta, A., Home, D., and Raychadhuri, A. [1988], *Is Quantum Mechanics with CP nonconservation incompatible with Einstein's locality condition at the statistical level?*, Physics Lett. A **130**, no. 4-5 (1988), 187–191.

Feynman, R.P. [1972], "Statistical Mechanics a Set of Lectures", Benjamin-Cummings, 1972.

Lerman, E., Montgomery, R. and Sjamaar, R. [1991], *Examples of Stratified Symplectic Reduction*, in the proceedings of summer conference in Warwick, England, 1991, on Singularity Theory, D. Salomon, ed.

Mackey, G.W. [1963], "Mathematical Foundations of Quantum Mechanics", W.A. Benjamin Inc., Reading, Mass, 1963.

Kazhdan, D., Kostant, B. and Sternberg, S. [1978], *Hamiltonian Actions and Dynamical Systems of Calogero Type*, Comm. of Pure and Applied Math. **31** (1978), 481–508.

Montgomery, R. [1990], *Isoholonomic Problems and Some Applications*, Comm. Math. Phys. **128** (1990), 565–592.

Simon, B. [1983], *Holonomy, the Quantum Adiabatic Theorem, and Berry's Phase*, Phys. Rev. Letters **51** (1983), 2167–2170.

Suter, D., Mueller, K. and Pines, A. [1988], *Study of Aharonov-Anandan Quantum Phase by NMR Interferometry*, Phys. Rev. Letters **60** (1988), 1218–1220.

Tomita, A. and Chiao, R. [1986], *Observations of Berry's Topological Phase by Use of an Optical Fiber*, Phys. Rev. Letters **57** (1986), 937–940.

Tycko, R. [1987], *Adiabatic Rotational Splittings and Berry's Phase in Nuclear Quadrupole Resonance*, Phys. Rev. Letters **57** (1987), 2281–2284.
Weyl, H. [1973] "The Classical Groups", Princeton Univ. Press, 2nd ed., 1973.
Zwanziger, J., Koenig, M. and Pines, A. [to appear 1990], *Berry's Phase*, Annual Rev. of Phys. and Chem.

Mathematics Department
UC Santa Cruz
Santa Cruz, CA 95064
USA

Internal Supersymmetry and Superconnections

YUVAL NE'EMAN and SHLOMO STERNBERG

A Jean-Marie Souriau,
en signe d'affection

1   Introduction

In [NS80] we introduced a family of representations for the superalgebras $sl(m/n)$ depending on a parameter b (which is a non-negative integer for $n \geq 2$ but could take on any complex value for $n = 1$). For the case $n = 1$, the representations with $b = (m - 1)/2$ have the special property that there is a bilinear form which is invariant under $sl(m/1)$. For each value of $m \geq 5$, we showed how to use these representations to describe a theory of colored quarks and leptons recurring in 2^{m-5} generations. We showed how to embed a weak su(2) and color su(3) in $su(m/1)$ and also an internal $su(m - 5/1)$ which is responsible for the generational symmetry. The weak hypercharge is required to commute with these three subalgebras and hence, *a priori* , depends on two free parameters. We showed how, by the appropriate choice of these parameters we could arrange that the correct values of the weak hypercharge be achieved for all 2^m particles. For example, if we take $m = 7$, we have 128 particles consisting of four generations. Each generation consists of sixteen particles and sixteen antiparticles, the sixteen consisting of the right and left handed electron like particles and their neutrinos, and two types of right and left handed quarks, each occurring in three colors. Notice that the right handed neutrinos are included, but our choice of weak hypercharge and isospin generators determines their weak isospin and hypercharge to be zero, and hence they do not participate (to lowest orders) in the electroweak interaction. We will review the definitions of these representations in Section **4** and the computations specific to the case $su(7/1)$ in Section **5**. The next issue is that of the dynamics. Our proposal is that the Yang-Mills theory, in which the interactions are coded into a connection on the underlying principal bundle together with a gauge invariant Lagrangian be generalized as follows: For a system with internal supersymmetry such as the one we propose, replace connections by superconnections as introduced by Quillen [Q85]. In our theory, this has the effect of having the Higgs field appear as the degree zero term of the superconnection; in other words as a natural geometric object, rather than an extraneous field introduced "by hand" in the standard model. Furthermore, in such a theory, the Higgs field appears quadratically in the expression for the supercurvature. Hence a Lagrangian which is quadratic in the supercurvature will be a fourth order polynomial in the Higgs field, another assumption of an *ad hoc* nature for spontaneous symmetry breaking in the standard model. Also, replacing connections by superconnections to yield a super Yang Mills Lagrangian gives the third order coupling between the Higgs and the matter fields necessary for fermionic mass acquisition in spontaneous symmetry breaking. These proposals were announced in [NS90]. In Section **6** we review the basics of Quillen's theory of superconnections and we describe how superconnections can be used in a theory with internal supersymmetry. In sections **2** and **3** we make some general remarks concerning the nature of theories of elementary particles, their interactions and the method of spontaneous symmetry breaking.

One clear prediction in our theory is that the number of generations be a power of two, and, more specifically, if we choose $m = 7$, that there be four generations. Currently, the number of generations is estimated by the width of the Z_0 decay. As of the present writing, it is generally accepted that the experiments indicate three generations, which, if correct, would vitiate our theory. For a recent report see [D90]. In this experiment, for example, the estimate on the number, N_ν, of neutrino families from the total Z-width (using the standard model) is 3.3 and from the peak hadronic cross-section is 3.01, indicating three generations. However the neutrinos are assumed to be light. There is the possibility of a heavy mass for the neutrino in the fourth generation, as suggested by recent experiments for the tau neutrino. Also, considering the recent indications of non-zero masses for all neutrinos (from solar neutrino experiments) it appears that a fourth generation is not excluded and might even be plausible. Finally, the total width computation depends on the quark masses, so heavy quarks in the third and fourth generation affect the conclusions based on total Z width. These matters will be discussed elsewhere.

2 Elements and Interactions

Ever since the fundamental paper of Wigner [W39] on the irreducible representations of the Poincaré group, it has been a (perhaps implicit) definition in physics that an elementary particle "is" an irreducible representation of the group, G, of "symmetries of nature". The search for the elements then has three components:

1. Determine the group G (a problem in physics).

2. Find the irreducible representations of G (a problem in mathematics).

3. Determine which irreducible representations of G occur as elementary particles. (a problem in physics).

Of course these three steps interact with one another in that putative solutions of steps 2 and 3 go into the determination of the group G, etc.

But it may be useful to begin by saying something about this remarkable definition.

Let us begin with a brief description of the historical background. The notion of an "element" has a venerable tradition in science, going back to the ancient Greeks who had a rather abstract notion of the fundamental constituents of matter. A key step in the development of eighteenth century chemistry was Thomas Boyle's operational definition of a chemical element as a substance that could not be decomposed into other substances by a series of well defined chemical operations. Notice that this definition had nothing to say about the "particle like" properties of the elements and was quite independent of the atomic hypothesis which did not win general acceptance until much later. Boyle's definition led to the search for the chemical elements which culminated with the discovery of the periodic table by Mendele'ev toward the end of the nineteenth century. But the periodicity of the table remained unexplained. Also shortly thereafter came the discovery of radioactivity which destroyed the notion of the immutability of the chemical elements, and Boyle's chemical definition could no longer be regarded as fundamental. Finally, the discovery of the electron opened the way to the notion of subatomic particle, finally culminating in the nuclear theory of the atom in the 1930's with the fundamental particles being the electron, proton and neutron. Each of these is localizable in space (in other words a "particle") and capable of an independent existence. With the proliferation of particles and resonances came a search for more fundamental building blocks, culminating in the "standard model" of the 1970's in which the elements are leptons and quarks, the leptons consisting of electron like particles and neutrinos, each quark coming in "three colors". These particles also exhibit a periodicity. That is they come in families or "generations", each generation consisting of an electron like particle, a neutrino and two types of colored quarks. In addition there are the "force carriers" consisting of the photon and various massive vector bosons. The quarks have not been observed as separate particles with an independent existence. Experiments in which energetic photons probe the inside of nucleons have detected point-like electrical charges with fractional values fitting the quarks. It has been conjectured that quarks are "confined" and the equations of Quantum Chromodynamics, the inter-quark force, have been shown to lead to confinement in the approximation where continuous space-time is approximated by a discrete lattice. Also the material particles acquire their masses by "spontaneous symmetry breaking" some mathematical aspects of which we will review in Section **3**. A detailed summary of the experimental evidence for the development of these theories, presented in a non-technical fashion, can be found in [NK]. The existence of the generational repetition of the quarks and leptons has led to the idea that there is yet a more fundamental level of matter and that the quarks and leptons themselves are in some sense composite. There have been a number of proposed theories in this direction of which ours is one.

Let us now discuss the definition given above from a conceptual viewpoint. One of the principal discoveries of quantum mechanics is that classical (Aristotelian) logic must be replaced by quantum logic in which propositions are closed subspaces of a separable complex Hilbert space, H, implication is subspace inclusion, the logical connective *or*, usually denoted by $\vee$, is sum of subspaces, and $\wedge$, the logical connective *and*, is intersection of subspaces,and the negation of a proposition is the orthocomplement of the subspace. See the basic paper [BN36] or the books [M] and [V]. For a general discussion of various competing formulations of the foundations of quantum mechanics, see the chapter on quantum metaphysics in [Su]. It is a fundamental theorem of Wigner (analogous to the fundamental theorem of projective geometry) that (if dim $H \geq 3$) any automorphism of a quantum logic can be implemented by an operator U on H where U is either unitary or antiunitary and is determined only up to a phase factor. Hence to say that a group G acts as automorphisms of the underlying logic of our quantum system means that we are given a projective representation of G on H by unitaries or antiunitaries. If the representation is highly reducible, as would be the case for a complex system with many interactions, the difference between a projective and an ordinary representation tends to disappear with the increase in the number of summands, as only one choice of overall arbitrary phase is allowed. Hence one tends to look for ordinary representations of G by unitaries or antiunitaries. Of course this means that the connected component of G has an ordinary, unitary representation. In any event, from a mathematical point of view, it now becomes reasonable to suggest that the fundamental building blocks of the theory are irreducible representations of G.

One of the requirements on G is that it be related to the geometry of space- time. For example, if we have a theory which does not include gravitational interaction, we would demand that G acts on space time so as to include all the isometries. In a flat space time this amounts to the requirement that we be given a homomorphism of G onto the Poincaré group. (One might want to replace the Poincaré group by one of the other isometry groups for a homogeneous space time, such as the deSitter or anti-deSitter groups.) For example, in Wigner's original paper, G is taken to be the Poincaré group itself, i.e. $G = Sl(2, \mathbf{C}) \ltimes \mathbf{R}^{1,3}$, the semidirect product of $G = Sl(2, \mathbf{C})$ with the group of translations, $\mathbf{R}^{1,3}$. His celebrated result was that the physically relevant irreducible representations of the Poincaré group are parametrized by a non-negative parameter, m , the mass, and a half integer parameter, s, the spin, where $s = 0, \frac{1}{2}, 1, \ldots$, if $m > 0$ while $s = 0, \pm\frac{1}{2}, \pm 1, \ldots$ if $m = 0$. Since the mass and the spin are not the only physical parameters describing particles, and since the observed values of the masses can not be explained by the Poincaré group alone, this suggests the need for a larger group.

An important issue is that of localizability, that is the existence of a "current": Let M denote space time, and let $Z = Z^3(M)$ denote the space of closed three forms on M (which vanish sufficiently rapidly at spacelike infinity if we take M to be ordinary Minkowski space). Suppose that we have a unitary representation of G on a Hilbert space H. Let V be some representation (usually finite dimensional) of G. A V - valued current on H is a sesquilinear form

$$J : H \times H \to V \otimes Z$$

which is equivariant for the action of G. The fact that $J(\phi, \psi)$ is closed implies that its integral over any spacelike surface, $\int_S J(\phi, \psi)$, is a V - valued sesquilinear form on H which is independent of the choice of S. If V is the trivial one dimensional representation of G then J is simply called a current. In case H is irreducible, then $\int_S J(\phi, \psi)$ is an invariant scalar valued sesquilinear form which then must be some multiple of the scalar product (ϕ, ψ). For example, if we take G to be the Poincaré group, and consider mass zero representations, then the representations with $s = 0$ and $s = \frac{1}{2}$ have currents, but the representations with $s \geq 1$ do not, cf.[WW80]. For $s = 1$ this expresses the well known fact that the photon can not be localized. On the other hand, if we take $V = \mathbf{R}^{1,3*} = M^*$ and let G act on V via the homomorphism from G to $Sl(2, \mathbf{C})$ then a V-valued current exists. Thus although the photon can not be localized, its energy or momentum can be localized, a fact that was known to Poincaré. For $s > 1$ such a V valued current does not exist either. The mass zero representations have the property that they all extend to irreducible unitary representations of $SU(2, 2)$ the (fourfold cover of the connected component of) the group of conformal transformations of conformally completed Minkowski space. The $s = 0$ representation does not have a $SU(2, 2)$ current, but the $s = \pm\frac{1}{2}$ representations do. This suggests that "matter" be built up out of spin $\frac{1}{2}$ particles. One of the reason why the currents are of importance is that to lowest order in perturbation theory, many of the standard interactions involve the currents associated to the in and out states of the scattered particles.

In order to describe interactions, it is assumed that the irreducible representations are realized on a space of sections of some vector bundle on space time. That is, it is assumed that H is the Hilbert space completion of a space $\mathbf{E}$ of smooth sections of a vector bundle $E \to M$, where $\mathbf{E}$ is characterized as being the space of solutions of a differential operator expressed in terms of a flat connection on E. For example, for particles whose behavior under the Poincaré group is that of a spin $\frac{1}{2}$ particle, the vector bundle $\mathbf{E}$ would be of the form $W \otimes S$ where S is the spin bundle and W is a vector bundle with a flat connection. Then gauge theory of forces says that interactions are manifested by the presence of a nonflat connection on W. More specifically, suppose that W is an associated bundle to a principal bundle P with structure group K. Let $\mathcal{C}$ denote the space of all connections on P and $\mathcal{A}^0(M, E)$ the space of sections of E. Then the interactions are expressed in terms of a Lagrangian which is a local map from $C \times \mathcal{A}^0(M, E)$ to functions on M. For example, the Yang-Mills Lagrangian is given by

$$L(A, s) = \frac{1}{2}(\|\mathbf{F}\|^2 + (\mathbf{D}_A s - ms, s)) \tag{1}$$

where $\mathbf{F} = \mathbf{F}(A)$ denotes the curvature of the connection A, and where $\mathbf{D}_A s$ denotes the Dirac operator associated to the connection A. The metric and scalar product on the right hand side of (1) are determined by the Lorentzian metric on M, by an invariant metric on W and by an invariant bilinear form on the Lie algebra of K, and m denotes the mass matrix. The relation with the representation point of view is then the following: We consider the Lagrangian L in the neighborhood of the the trivial connection on P and the zero section of E. Then the quadratic terms in (1) will give a quadratic Lagrangian,

$$L_0 = L_{01} + L_{02}. \tag{2}$$

The terms on the right hand side of (2) correspond to the two terms on the right hand side of (1). Here L_{02} does not involve the connection A and so is a quadratic Lagrangian on E. The Euler Lagrange equations of L_{02} then define the representation space $\mathbf{E}$. Upon restriction to the Poincaré group, $\mathbf{E}$ decomposes into a direct sum of spin $\frac{1}{2}$ representations with masses given by the eigenvalues of the mass matrix, m. Similarly, L_{01} does not involve s, it is purely a function of A. In fact, the space, C, of all connections is an affine space whose associated linear space is $\mathcal{A}^1(k)$ the space of one forms on M with values in k where k is the Lie algebra of K. Hence L_{01} is a quadratic Lagrangian on $\mathcal{A}^1(k)$. The corresponding Euler-Lagrange equations define a subspace, $\mathbf{C}$ of the space of connections which is a representation space for G. Upon restriction to the Poincaré group this decomposes into a direct sum of dim k copies of the $m = 0, s = \pm 1$ representations. These are regarded as the "force carriers" in that the forces giving rise to bound states and to the scattering matrix are determined from the higher order terms in (1) according to the Feynman rules. A detailed discussion of the Feynman rules from the point of view of quantum field theory can be be found in any text on the subject, for example [IZ]. For a clear and elementary presentation of these rules without invoking the entire machinery of quantum field theory (but essentially dealing with the lowest order in perturbation theory, the tree diagrams) cf. [G].

The above described theory, although beautiful from a mathematical point of view, must be modified in order to conform to physical reality. The trouble is that it predicts too many (dim k) massless vector bosons, i.e. particles with $m = 0, s = \pm 1$. For example, in the theory of strong interactions where $k = su(3)$ the theory would predict $8 = $ dim $su(3)$ such particles and in electroweak interactions where $k = su(2) \times u(1)$ it predicts 4. But only the photon is observed. This problem is dealt with differently in the two domains. In the theory of strong interactions, it is believed that these particles are not observed as free particles because of "confinement". In the electroweak theory, the three extra vector bosons are "given a mass" by the Higgs mechanism of spontaneous symmetry breaking. In addition, the Higgs mechanism gives masses to some of the massless fermions. That is, one starts with a theory with $m = 0$ in (1) but the Higgs mechanism ends up with some of the fermions having mass. The mathematics of confinement is not yet understood. We will review some of the mathematical aspects of the Higgs mechanism in the next section, in the geometrical language of reduction of principal bundles.

In this paper we will modify this Yang-Mills-Higgs picture by replacing vector bundles by supervector bundles and connections by superconnections. In other words, in the "free particle" situation, we have replaced the idea of a group acting as symmetries by a supergroup. But our "superization" is relatively mild, in that our supergroup is purely internal so we don't replace the underlying manifold by a supermanifold. In all likellihood, a fully supersymmetric theory would be a desirable generalization of our current program. Similarly, the question of what happens to the general program when "group" is replaced by "quantum group" remains to be explored.

3   The Higgs Mechanism

The standard treatment of spontaneous symmetry breaking and the Higgs mechanism from the point of view of quantum field theory is the classic paper [AL73] which is extremely clear and coherent. In this section we wish to emphasize one aspect of this subject, since we hope to be able to apply it to our model. The exposition in this section is taken from [RS85].

Let $P_K \to M$ be a principal bundle with structure group K over the base manifold M. There are various geometrical objects that we can associate with P_K. For example, if the vector space F is a K module, then we can construct the associated bundle to F and then consider the space of j forms on the base with values in this vector bundle. Let us denote the space of all such j forms by $\mathcal{A}^j(M, F)$ or $\mathcal{A}^j(F)$. Let us denote the direct sum (over j) of all the $\mathcal{A}^j(M, F)$ spaces by $\mathcal{A}^*(M, F)$. We may want to consider various different modules F. Summing over various different F's that come into consideration gives us a space which we shall denote by $\mathcal{A}^*(P_K)$. We can also consider the space of all connections on P_K which we have denoted by $\mathcal{C}$. By a *field* we shall mean an element of the space $\mathcal{A}^*(P_K) \times \mathcal{C}$. We shall denote the general field by ϕ.

Let $\mathrm{Aut}P_K$ denote the group of all diffeomorphisms of P_K which commute with the action of K. Then $\mathrm{Aut}P_K$ acts on the space of all fields, $\mathcal{A}^*(P_K) \times \mathcal{C}$. We can construct Lagrangian densities

$$L : \mathcal{A}^*(P_K) \times \mathcal{C} \to C^\infty(M)$$

out of geometric data which will be equivariant for the action of $\mathrm{Aut}P_K$, or at least for the action of a large subgroup of $\mathrm{Aut}P_K$. For example, the Lagrangian density given in (1) will be equivariant under the subgroup of $\mathrm{Aut}P_K$ which induces a transformation of M which is an isometry. More generally, suppose that M has a (pseudo)Riemann metric and that the Lie algebra of K is given a K invariant scalar product as is each of the representation spaces, F, that enter into our definition of the geometrical fields. Then we can consider various scalar valued functions on M built out of the geometrical fields such as

$$
\begin{aligned}
\|\omega\|^2 \qquad & \omega \in \mathcal{A}^k(F) \\
\|d_A\omega\|^2 \qquad & \omega \in \mathcal{A}^k(F),\ A \in \mathcal{C} \\
\|curv(A)\|^2 \qquad & A \in \mathcal{C} \\
V(f) \qquad & f \in \mathcal{A}^0(F),\ V : F \to \mathbf{R},
\end{aligned}
$$

where in the last item, V is a K invariant function on F. If we have a projection of P_K onto a subbundle of the bundle of orthonormal frames of M, so that $T(M)$ and $T^*(M)$ are associated bundles to P_K, then we might also have expressions such as

$$
W(f,\omega)
$$

where W is some K invariant polynomial.

In any event, we can build some Lagrangian density, L, out of these objects and then integrate $L(\phi)$ with respect to some volume density, $dvol$, to obtain

$$
\mathbf{L}(\phi) = \int L(\phi)dvol. \tag{3}
$$

In quantum field theory one starts from the "functional integral"

$$
\int e^{i\mathbf{L}(\phi)}\mathcal{D}(\phi) = \int e^{i\int L(\phi)dvol}\mathcal{D}(\phi). \tag{4}
$$

Of course these functional integrals are not very well defined in a mathematical sense, but via perturbation theory give rise to the Feynman rules of physics. However, even within the framework of perturbation theory there is a problem of interpretation due to the fact that the integrand in (4) is invariant under a very large group of symmetries. Indeed, let $\mathrm{Gau}(P_K)$ denote the subgroup of $\mathrm{Aut}(P_K)$ consisting of those automorphisms which induce the identity transformation on M. It is what the mathematicians call the gauge group, and what the physicst call the group of local gauge transformations. Then (3) is invariant under $\mathrm{Gau}(P_K)$. One of the ideas of spontaneous symmetry breaking is to remove some of the redundancy in (4) by providing a partial cross section for the action of $\mathrm{Gau}(P_K)$.

In order to describe how this works, we first recall an alternative description of the group $\mathrm{Gau}(P_K)$. Let K act on itself by conjugation. Then the associated bundle of P_K relative to this conjugation action is a fiber bundle over M each of whose fibers is a group. This makes the space of smooth sections of this associated bundle into a group. This group of sections can be identified with $\mathrm{Gau}(P_K)$ as follows: A section of the bundle associated to the conjugation action can be thought of as a function $\tau : P_K \to K$ which satisfies the identity

$$\tau(pb) = b^{-1}\tau(p)b. \tag{5}$$

To each such τ one associates the diffeomorphism ψ_τ of P_K defined by

$$\psi_\tau(p) = p\tau(p). \tag{6}$$

It is easy to check that the map $\tau \mapsto \psi_\tau$ is a homomorphism from the group of sections τ onto $\mathrm{Gau}(P_K)$.

Next, one singles out a particular space, U on which K acts, a section of the associated bundle to the space U being called a *Higgs field*. (In the Weinberg-Salam model the group K is taken to be $SU(2) \times U(1)$ and the space U is then taken to be $\mathbf{C}^2$. One of our proposals in this paper is to take U to be the odd piece, k_1, of an internal superalgebra, $k = k_0 + k_1$, or, more generally, to take U to be built out of k_1 in some geometric way, for example as a Grassmann or flag variety.)

We assume the following about the action of K on U: That outside of a small G invariant singular set $S \subset U$ there is a cross-section, Γ, for the K action. Thus we assume that

$$U - S = \Gamma \times (K/H) \tag{7}$$

as K spaces,where H is the common isotropy group of all points of Γ. For example, in the case of the Weinberg-Salam model with $K = SU(2) \times U(1)$ and $U = \mathbf{C}^2$ we could take $S = 0$ and Γ to consist of all vectors of the form $\begin{pmatrix} 0 \\ \rho \end{pmatrix}$ with ρ real and positive, and H is isomorphic to $U(1)$.

Let $\mathcal{B}(U)$ denote the space of all sections of the the bundle associated to U which do not intersect the sub-bundle determined by S. Thus $\mathcal{B}(U)$ consists of those smooth maps $f : P_K \to U$ which satisfy the equivariance condition

$$f(pa) = a^{-1}f(p), \ \forall a \in K, \ p \in P_K \tag{8}$$

and such that

$$f(P_K) \subset U - S. \tag{9}$$

In view of the decomposition (7), condition (9) implies that we can write f as

$$f = (g, r)$$

where $g : P_K \to \Gamma$ and $r : P_K \to K/H$, where each of the components, g and r satisfies the equivariance condition (8). Since the action of K on Γ is trivial, condition (8) simply says that g is a Γ-valued function on M.

For r, condition (8) says that r is a section of the bundle associated to the homogeneous space K/H. Now a section of this bundle is the same as a reduction of the principal bundle, P_K to an H subbundle. Indeed, define the sub-bundle P_H by

$$P_H^r = r^{-1}(H) = \{p \in P_K \mid r(p) \in H \in K/H\}. \tag{10}$$

Note that $P_K = P_H \times_H K$. This explains why we say that we have reduced the structure group of P_K to H. This observation is the starting point of the theory of Cartan connections, cf. [RS85] where this subject is reviewed in the context of the study of the interaction of spin and torsion in gravitational theories. The group $\mathrm{Gau}(P_K)$ acts on all geometric objects, in particular on the space of all smooth sections of the bundle associated to K/H, i.e. on the space of all reductions to an H-subbundle. We will assume that this space is non-empty, and that $\mathrm{Gau}(P_K)$ acts transitively on it. For example, in the case that $H = e$ is the trivial subgroup, this action is always transitive. Indeed, let $r_i, i = 1, 2$ be two maps $r_i : P_K \to K$ satisfying (8). If we then define $\tau : P_K \to K$ by

$$\tau(p) = r_1(p)r_2(p)^{-1}$$

then τ satisfies (5) and hence defines an element of $\mathrm{Gau}(P_K)$. Also

$$r_1(p\tau(p)) = r_2(p)r_1(p)^{-1}r_1(p) = r_2(p)$$

and so ψ_τ carries r_1 into r_2.

Under our assumptions of the existence of a cross-section, r, and the transitivity of the action of $\mathrm{Gau}(P_K)$, all of the bundles bundles P_H^r are carried into one anther by an element of $\mathrm{Gau}(P_K)$. In particular they define the same abstract principal H bundle which we will denote by P_H. From this point of view, the choice of an $f \in \mathcal{B}(U)$ and the decomposition $f = (g, r)$ then determmines a specific embedding

$$r : P_H \to P_K \tag{11}$$

with $r(P_H) = P_H^r$. If F is any representation space of K, it becomes a representation space of H by restriction. Hence r determines a pullback map

$$r^* : \mathcal{A}^* P_K \to \mathcal{A}^* P_H \tag{12}$$

which is consistent with the various geometrical Lagrangians, so that, for example

$$\|r^*\omega\|^2(q) = \|\omega\|^2(r(q)), \quad q \in P_H. \tag{13}$$

Noticc that r^* is one to one since $P_K \times_K F = P_H \times_H K \times_K F = P_H \times_H F$. Now let us examine how connections pull back under r. Let h denote the Lie algebra of H so that h is a subalgebra of k, the Lie algebra of K. Let us assume that h has an H invariant vector space complement, l in k so that

$$k = h + l. \tag{14}$$

It is always possible to choose such a complement if H is compact or semisimple. Now a connection, A, on P_K is a k valued one form on P_K satisfying an equivariance condition

$$R_b^* A = Ad_b^* A \tag{15}$$

together with the interior product condition

$$\iota(\xi_{P_K})A = \xi, \ \xi \in k \tag{16}$$

where ξ_{P_K} is the vector field on P_K coming from the right action of K. Now on $r(P_H) = P_H^r$ only the subgroup H acts. The decomposition (14) implies that A decomposes as $A = A_H + \theta_l$. I.E., if we restrict to $\xi \in h$ the decomposition (14) implies that the restriction of A decomposes as a sum of a connection on P_H and an element of $\mathcal{A}^1(P_H, l)$, the space of one forms on M with values in the vector bundle over M associated to the principal bundle P_H and the representation, l, of H. In other words, if $\mathcal{C}(P_K)$ denotes the space of connections on P_K and $\mathcal{C}(P_H)$ the space of connections on P_H the map r defines a pullback map

$$r^* : \mathcal{C}(P_K) \to \mathcal{C}(P_H) \times \mathcal{A}^1(P_H, l). \tag{17}$$

For example, if $K = SU(2) \times U(1)$ and $H = U(1)$, a K connection gives rise to a $U(1)$ connection A_H which represents an electromagntic field together with a one form with values in a three dimentional bundle giving the W_+, W_-, and Z_0 particles. The relation between the curvature of the connection A on P_K at the point $p = r(q)$ and the two components of its pullback under (17) is given by

$$F_A(p) = F_{A_H}(q) + d_{A_H}\theta_l(q) - \frac{1}{2}[\theta_l, \theta_l]. \tag{18}$$

Thus we can write down a geometric Lagrangian $L' : \mathcal{C}(P_H) \times \mathcal{A}^1(P_H, l) \to C^\infty(M)$ such that

$$L'(r^*(A)) = \|F_A\|^2.$$

Proposition 1 (Reduction of the principal bundle.) *Let r be a section of the bundle associated to K/H. Then r induces a pullback map of the fields:*

$$r^* : \mathcal{C}(P_K) \times \mathcal{A}^*(P_K) \to \mathcal{C}(P_H) \times \mathcal{A}^*(P_H). \tag{19}$$

For any geometrical Lagrangian, L, on P_K there is a Lagrangian, L', on P_H such that

$$L' \circ r^* = L.$$

In particular this holds for the r component of a Higgs field $f = (g, r)$.

In this way the $\mathrm{Gau}(P_K)$ redundancy inherent in L has been reduced to a $\mathrm{Gau}(P_H)$ redundancy inherent in L'.

The "mass aquisition" occurs in Proposition 1 as follows: Consider the function

$$\mathcal{C}(P_K) \times \mathcal{A}^0(U) \to C^\infty(M)$$

given by

$$(A, f) \mapsto \|d_A f\|^2.$$

Suppose that f avoids S, so $f \in \mathcal{B}(U)$ can be written as $f = (g, r)$. Then

$$r^* f = (g, H)$$

where the second component, H, is constant, and $g : M \to \Gamma$. Thus along the subbundle $r(P_H)$ we can identify df with $dg : TM \to T\Gamma$. Since h acts trivially on $T(K/H)_H$, we have

$$A \cdot f(p) = \theta_l \cdot f \in T(O_{g(m)})$$

where $m \in M$ is the image of p under the projection of P_K to M and $O_{g(m)}$ denotes the K orbit through the point $g(m) \in \Gamma$ in U. Suppose that U carries a K invariant Riemann metric and that the cross section, Γ, has been chosen so that the tangent spaces $T\Gamma_{g(m)}$ and $TO_{g(m)}$ are perpendicular. For example, in the Weinberg-Salam example, Γ is a radial ray passing through "north pole" of the three spheres $O_{g(m)}$ for all m. Then the orthogonal decomposition

$$TU_{g(m)} = T\Gamma_{g(m)} \oplus T(O_{g(m)})_{g(m)}$$

gives

$$\|d_A f\|^2(p) = \|dg(p)\|^2 + \|\theta_l \cdot g(m)\|^2.$$

Now the action of K gives us an identification of l with the tangent space $T(O_{g(m)})_{g(m)}$ and hence the metric on $T(O_{g(m)})_{g(m)}$ pulls back to a metric on L (depending on the point $g(m)$). Let us denote this metric by $\|\cdot\|^2_{g(m)}$. Then we can write the preceding equation as

$$\|d_A f\|^2(p) = \|dg(p)\|^2 + \|\theta_l\|^2_{g(m)}. \tag{20}$$

Let us assume that h and l are perpendicular for the scalar product on k. Then substituting into (18) gives

$$\|F_A\|^2 = \|dA_H\|^2 + \|d\theta_l\|^2 + \dots \tag{21}$$

along $r(P_H)$ where $\dots$ consist of terms cubic or higher in the fields. Let us combine (20) with (21). Suppose that our original Lagrangian on P_K contains (as a summand) a term of the form

$$L = \|F_A\|^2 + a\|d_A f\|^2 + V(f) \tag{22}$$

where V is some K invariant function defined on U. (Here a is some constant.) If we expand (22) about zero in f and A (about a flat connection and the zero section of U) then the quadratic terms in A decouple as dim k copies of the Lagrangian for Maxwell's equations. In other words, we get dim k particles of mass 0 and spin ± 1. But suppose that V takes on a minimum along some non-zero orbit O_{min} where $O_{min} \subset U - S$ and we choose f_0 to be a section of O_{min}. According to our assumptions such an f_0 exists, and the group $\text{Gau}(P_K)$ acts transitively on the space of such sections. Let

$$\|\theta_l\|_0^2 = \|\theta_l\|_{g_0(m)}^2$$

where $f_0 = (g_0, r_0)$ so that g_0 is the constant function which maps P_K to $O_{min} \cap \Gamma$. By the transitivity of the $\text{Gau}(P_K)$ action and the invariance of our Lagrangian, we may assume that we have chosen $f_{(0)}$ with $r_0 \equiv H$. If we then write $L = L' \circ r^*$ we find that, ignoring overall constants,

$$L' = \|d\theta_l\|^2 + a\|\theta_l\|_0^2 + \|F_{A_H}\|^2 + \dots. \tag{23}$$

The quadratic terms in (23) are those of the Maxwell equations in the A_H component but the Proca Lagrangian, (positive mass, spin 1) in the l component. To summarize, we have proved

Proposition 2 (Mass acquisition for vector bosons.) *Suppose that we expand (22) about a non-zero minimum section under our assumptions as to the existence of the crosssection and the action of $Gau(P_K)$. This then leads to a reduction of the principal bundle as in Proposition 1, and the replacement of (22) by (23) so that the l components of A acquire mass.*

In the standard model, the representation space U has to be supplied as part of the theory. It is one of our proposals here that in a theory with internal supersymmetry, U can be chosen as the the odd component of a superalgebra, and the Higgs field, f, as the exterior degree 0 piece of a superconnection, see Section **6**.

4   Special representations of $sl(m/n)$

We begin by recalling the definition of these superalgebras. (For general facts about Lie superalgebras we refer the reader to the book [Sch] or the article [CNS75].) Let

$$V = V_0 \oplus V_1$$

be a supervector space with dim $V_0 = m$ and dim $V_1 = n$. The Lie superalgebra $sl(V_0/V_1)$ is the set of all endomorphisms of W with supertrace zero. A typical such endomorphism can be written as

$$\begin{pmatrix} A & B \\ C & D \end{pmatrix}, tr A = tr D. \tag{24}$$

Here $A \in \mathrm{Hom}(V_0, V_0)$, $B \in \mathrm{Hom}(V_1, V_0)$, $C \in \mathrm{Hom}(V_0, V_1)$ and $D \in \mathrm{Hom}(V_1, V_1)$. Those endomorphisms which preserve the grading (i.e. those with $B = C = 0$) are called even, and those which reverse the grading (i.e. those with $A = D = 0$) are called odd. In case $V_0 = \mathbf{C}^m$ and $V_1 = \mathbf{C}^n$ we denote this algebra by $sl(m/n)$. In what follows we shall assume that V_0 and V_1 are finite dimensional so that we may identify $\mathrm{Hom}(V_1, V_0)$ with $V_0 \otimes V_1^*$. In particular if $v \in V_0$ and $\xi \in V_1^*$ then $v \otimes \xi$ is identified the rank one linear transformation given by

$$(v \otimes \xi) w = \langle \xi, w \rangle v$$

where $\langle \xi, w \rangle$ denotes the value of the linear function ξ on the vector w. These rank one linear transformations span $\mathrm{Hom}(V_1, V_0)$. Similar identifications will be made for each of the four spaces corresponding to the four "matrix" entries of (24). For example, we can compute the supercommutator

$$\left[\begin{pmatrix} 0 & v \otimes \xi \\ 0 & 0 \end{pmatrix}, \begin{pmatrix} 0 & 0 \\ x \otimes \mu & 0 \end{pmatrix} \right] = \begin{pmatrix} \langle \xi, x \rangle v \otimes \mu & 0 \\ 0 & \langle \mu, v \rangle x \otimes \xi \end{pmatrix}.$$

To save space we shall write the above bracket relations (and similar ones) as follows: We have

$$sl(V_0, V_1)_0 = V_0 \otimes V_0^* \bigoplus V_1 \otimes V_1^* \tag{25}$$

and

$$sl(V_0, V_1)_1 = V_0 \otimes V_1^* \bigoplus V_1 \otimes V_0^*. \tag{26}$$

Then we would write the preceding bracket relation as

$$[v \otimes \xi, x \otimes \mu] = \langle \xi, x \rangle (v \otimes \mu) \oplus \langle \mu, v \rangle (x \otimes \xi). \tag{27}$$

The Lie superalgebra $sl(V_0/V_1)$ has a natural representation on the space $\bigwedge(V)$. Perhaps the best way to see this representation is by imbedding $sl(V_0/V_1)$ in the orthosymplectic algebra as the centralizer of a one dimensional subalgebra. This "Howe pair" point of view is explained by Howe in his original paper [H77]. In [NS82] we used this description in conjunction with the method of dimensional reduction. But let us describe this representation directly here. Recall that, by definition,

$$\bigwedge(V) = \bigwedge(V_0) \bigotimes S(V_1)$$

where $S(V_1)$ denotes the symmetric algebra, so

$$S(V_1) = \bigoplus S^k(V_1)$$

and $S^k(V_1)$ consists of homogeneous polynomials of degree k on V_1^*. Each $x \in V_1$ defines a multiplication operator

$$m_x : S^k(V_1) \to S^{k+1}(V_1)$$

given by

$$(m_x f)(\eta) = \langle \eta, x \rangle f(\eta), \ \forall \eta \in V_1^*. \tag{28}$$

Each $\xi \in V_1^*$ defines a derivation D_ξ determined by

$$D_\xi(fg) = (D_\xi f)g + fD_\xi g, \ D_\xi 1 = 0 \tag{29}$$

and

$$D_\xi x = \langle \xi, x \rangle, \ \forall x \in V_1 = S^1(V_1). \tag{30}$$

The standard Fock commutation relations hold, i.e.

$$D_\xi m_x - m_x D_\xi = \langle \xi, x \rangle id. \tag{31}$$

Thus, equations (28) - (31) give the usual Fock representation of the Heisenberg algebra, $\mathrm{Hei}(V_1 \oplus V_1^*)$, by creation and annihilation operators. But these equations continue to make sense if we replace the spaces $S^b(V_1)$ by the space $F^b(V_1)$ of all smooth functions defined on some cone in V_1 and homogeneous of degree b. If $\dim V_1 > 1$ these spaces will be infinite dimensional. But if $\dim V_1 = 1$, and, to fix the ideas, we choose an identification of V_1 with $\mathbf{C}$, then we can think of the "function" x^b as a formal symbol. That is, we let $S^b(V_1)$ denote the one dimensional space with basis element p_b, and define

$$m_x : S^b(V_1) \to S^{b+1}(V_1), m_x p_b = x p_{b+1} \tag{32}$$

and

$$D_\xi : S^b(V_1) \to S^{b-1}(V_1), D_\xi p_b = b\xi p_{b-1}. \tag{33}$$

The commutation relations (31) continue to hold.

Let

$$\bigwedge V_0 = \bigoplus \wedge^k(V_0)$$

be the exterior algebra of V_0. Each $v \in V_0$ defines an operation of exterior multiplication

$$e_v : \wedge^k(V_0) \to \wedge^{k+1}(V_0), e_v w = v \wedge w \tag{34}$$

and each $\mu \in V_0^*$ defines a superderivation of $\bigwedge V_0$,

$$i_\mu : \wedge^k(V_0) \to \wedge^{k-1}(V_0)$$

determined by

$$i_\mu(w_1 \wedge w_2) = (i_\mu w_1) \wedge w_2 + (-1)^{deg w_1} w_1 \wedge i_\mu w_2,$$

$$i_\mu v = \langle \mu, v \rangle, v \in V_0 = \wedge^1(V_0),$$

and

$$i_\mu c = 0, c \in \mathbf{C} = \wedge^0(V_0).$$

It is then easy to check the (super)commutation relations

$$[e_{v_1}, e_{v_2}] = e_{v_1} e_{v_2} + e_{v_2} e_{v_1} = 0,$$

$$[i_{\mu_1}, i_{\mu_2}] = i_{\mu_1} i_{\mu_2} + i_{\mu_2} i_{\mu_1} = 0,$$

and

$$[e_v, i_\mu] = e_v i_\mu + i_\mu e_v = \langle \mu, v \rangle.$$

In short, m and D are Bose-Einstein creation and annihilation operators while e and i are Fermi-Dirac creation and annihilation operators. For any $A \in V_0 \otimes V_0^* = \mathrm{End}(V_0)$ we will let A continue to denote its induced derivation of $\wedge(V_0)$, and similarly for $D \in V_1 \otimes V_1^*$ we will let D continue to denote its induced derivation on $S(V_1)$. We then get a representation of $sl(V_0/V_1)$ on $\wedge(V_0) \otimes S(V_1)$ given by

$$\rho(A, D) = A \otimes I \oplus I \otimes D, \ A \in V_0 \otimes V_0^*, \ D \in V_1 \otimes V_1^* \qquad (35)$$

for the even elements and

$$\rho(v \otimes \xi, x \otimes \mu) = e_v \otimes D_\xi + i_\mu \otimes m_x. \qquad (36)$$

The operators in (35) preserve degrees in $\wedge(V_0)$ and in $S(V_1)$. The first operator on the right of (36) lowers the symmetric degree by one and raises the exterior degree by one, and the second operator on the right does the opposite. Thus all operators perserve the total degree. So the total degree parametrizes the decomposition of $\wedge(V_0) \otimes S(V_1)$ into invariant subspaces under $sl(V_0/V_1)$. Now let us specialize to the case $V_1 = \mathbf{C}$, and let us write S^b for $S^b(V_1)$. Now b can be any complex number and (35) and (36) define a representation, ρ_b, of $sl(V_0, V_1)$ on

$$(\wedge^0(V_0) \otimes S^b) \oplus \ldots \oplus (\wedge^m(V_0) \otimes S^{b-m}), \qquad (37)$$

where $m = \dim(V_0)$. Since all the spaces S^a are one dimensional, the dimension of the representation space in (37) is always 2^m, the same as the dimension of the exterior algebra $\wedge(V_0)$, but of course the actual representation depends on b. Notice that we have an identification of $S^r \otimes S^s \to S^{r+s}$ determined by $p_r \otimes p_s \mapsto p_{r+s}$. Tensored with exterior multiplication, this gives a map

$$(\wedge^k(V_0) \otimes S^r) \bigotimes (\wedge^l(V_0) \otimes S^s) \to \wedge^{k+l}(V_0) \otimes S^{r+s}.$$

In particular, taking $k + l = m$, we see there is a bilinear map

$$(\wedge^k \otimes S^{b-k}) \bigotimes (\wedge^{m-k} \otimes S^{b-m+k}) \to \wedge^m \otimes S^{2b-m}$$

between "opposite" summands in (37). Now the supertrace zero condition in the definition of $sl(V_0/V_1)$ implies that the action of the even part of $sl(V_0/V_1)$ on $\wedge^m \otimes S^{-1}$ is trivial. Hence if we choose

$$b = \frac{m-1}{2},\tag{38}$$

then the above pairing defines a bilinear form on the representation space which is (super)invariant under $sl(V_0, V_1)$. So (38) will be our choice of b.

5   The case for $su(7/1)$

Our plan is now as follows: we will regard the weak isospin $su(2)$ and the color $su(3)$ as (commuting) subalgebras of the even part of $sl(m/1)$. On the other hand, the generational behavior will be produced by an $sl(m-5/1)$ subLiesuperalgebra. In [NS80] we explained how this works for any $m \geq 5$. We shall work here in the case $m = 7$ which will produce four generations. We will write our matrices according to the block decomposition of $\mathbf{C}^7 = V_0$ given by

$$\mathbf{C}^7 = \mathbf{C}^2 \oplus \mathbf{C}^2 \oplus \mathbf{C}^3\tag{39}$$

where the first $\mathbf{C}^2$ component corresponds to weak isospin, the second $\mathbf{C}^2$ is generational and the last $\mathbf{C}^3$ corresponds to color. Thus, for example, I_3, the third component of weak isospin will have the matrix form

$$I_3 = \left(\begin{array}{ccccccc|c}
\frac{1}{2} & 0 & 0 & 0 & 0 & 0 & 0 & 0 \\
0 & -\frac{1}{2} & 0 & 0 & 0 & 0 & 0 & 0 \\
0 & 0 & 0 & 0 & 0 & 0 & 0 & 0 \\
0 & 0 & 0 & 0 & 0 & 0 & 0 & 0 \\
0 & 0 & 0 & 0 & 0 & 0 & 0 & 0 \\
0 & 0 & 0 & 0 & 0 & 0 & 0 & 0 \\
0 & 0 & 0 & 0 & 0 & 0 & 0 & 0 \\
\hline
0 & 0 & 0 & 0 & 0 & 0 & 0 & 0
\end{array}\right).\tag{40}$$

The weak hypercharge, Y, is taken to be an even element which commutes with all three subalgebras. Hence it must have the form

$$Y = \left(\begin{array}{ccccccc|c}
x & 0 & 0 & 0 & 0 & 0 & 0 & 0 \\
0 & x & 0 & 0 & 0 & 0 & 0 & 0 \\
0 & 0 & y & 0 & 0 & 0 & 0 & 0 \\
0 & 0 & 0 & y & 0 & 0 & 0 & 0 \\
0 & 0 & 0 & 0 & z & 0 & 0 & 0 \\
0 & 0 & 0 & 0 & 0 & z & 0 & 0 \\
0 & 0 & 0 & 0 & 0 & 0 & z & 0 \\
\hline
0 & 0 & 0 & 0 & 0 & 0 & 0 & d
\end{array}\right).$$

The fact that Y commutes with the full superalgebra $sl(2/1)$ implies that $d = y$. On the other hand, the supertrace condition implies that $d = 2x + 2y + 3z$. Thus

$$y = -2x - 3z$$

so the only free parameters in the choice of Y are x and z. We shall see that the choice

$$x = -\frac{1}{3}, z = 0$$

gives "correct" values for the hypercharge of 128 particles occuring as eigenvalues of Y in the representation (37) with $b = 3$. In other words, we are going to choose Y as

$$Y = \left(\begin{array}{ccccccc|c}
-\frac{1}{3} & 0 & 0 & 0 & 0 & 0 & 0 & 0 \\
0 & -\frac{1}{3} & 0 & 0 & 0 & 0 & 0 & 0 \\
0 & 0 & \frac{2}{3} & 0 & 0 & 0 & 0 & 0 \\
0 & 0 & 0 & \frac{2}{3} & 0 & 0 & 0 & 0 \\
0 & 0 & 0 & 0 & 0 & 0 & 0 & 0 \\
0 & 0 & 0 & 0 & 0 & 0 & 0 & 0 \\
0 & 0 & 0 & 0 & 0 & 0 & 0 & 0 \\
\hline
0 & 0 & 0 & 0 & 0 & 0 & 0 & \frac{2}{3}
\end{array}\right). \tag{41}$$

Let $e_1, \ldots, e_7$ be the standard basis of $\mathbf{C}^7$. If $I = \{i_1, \ldots, i_k\}$ is any ordered subset of $1, \ldots, 7$ then we will let let e_I denote the element

$$e_I = e_{i_1} \wedge \cdots \wedge e_{i_k}$$

of $\bigwedge^k(V_0)$ and let

$$f_I = e_I \otimes p_{3-k} \in \bigwedge^k \otimes S^{3-k}.$$

If X is a diagonal matrix,

$$X = diag(x_1, \ldots, x_7 | d), d = x_1 + \cdots + x_7,$$

then f_I is an eigenvector of X with eigenvalue

$$x_{i_1} + \cdots + x_{i_k} + (3 - k)d.$$

Thus, for example, $f_\emptyset \in \bigwedge^0 \otimes S^3$ is an eigenvector of Y with eigenvalue $2 = 3 \times \frac{2}{3}$ and an eigenvector of I_3 with eigenvalue 0. Similarly f_{134} is an eigenvector of Y with eigenvalue $1 = -\frac{1}{3} + \frac{2}{3} + \frac{2}{3} + 0$, and an eigenvector of I_3 with eigenvalue $\frac{1}{2}$. So we can now tabulate the eigenvalues of Y and I_3. On $\bigwedge^0 \otimes S^3$ we have

$$\begin{array}{c|c}
\bigwedge^0 \otimes S^3 & f_\emptyset \\
\hline
Y & 2 \\
I_3 & 0
\end{array} \tag{42}$$

On $\bigwedge^1 \otimes S^2$ we have

$\bigwedge^1 \otimes S^2$	f_1	f_2	f_3	f_4	f_5	f_6	f_7
Y	1	1	2	2	$\frac{4}{3}$	$\frac{4}{3}$	$\frac{4}{3}$
I_3	$\frac{1}{2}$	$-\frac{1}{2}$	0	0	0	0	0

$$(43)$$

On $\bigwedge^2 \otimes S^1$ we have

$\bigwedge^2 \otimes S^1$	f_{12}	f_{13}	f_{14}	f_{15}	f_{16}	f_{17}	f_{23}
Y	0	1	1	$\frac{1}{3}$	$\frac{1}{3}$	$\frac{1}{3}$	1
I_3	0	$\frac{1}{2}$	$\frac{1}{2}$	$\frac{1}{2}$	$\frac{1}{2}$	$\frac{1}{2}$	$-\frac{1}{2}$

$$(44)$$

$\bigwedge^2 \otimes S^1$	f_{24}	f_{25}	f_{26}	f_{27}	f_{34}	f_{35}	f_{36}
Y	1	$\frac{1}{3}$	$\frac{1}{3}$	$\frac{1}{3}$	2	$\frac{4}{3}$	$\frac{4}{3}$
I_3	$-\frac{1}{2}$	$-\frac{1}{2}$	$-\frac{1}{2}$	$-\frac{1}{2}$	0	0	0

$\bigwedge^2 \otimes S^1$	f_{37}	f_{45}	f_{46}	f_{47}	f_{56}	f_{57}	f_{67}
Y	$\frac{4}{3}$	$\frac{4}{3}$	$\frac{4}{3}$	$\frac{4}{3}$	$\frac{2}{3}$	$\frac{2}{3}$	$\frac{2}{3}$
I_3	0	0	0	0	0	0	0

On $\bigwedge^3 \otimes S^0$ we have

$\bigwedge^3 \otimes S^0$	f_{123}	f_{124}	f_{125}	f_{126}	f_{127}	f_{134}	f_{135}	f_{136}
Y	0	0	$-\frac{2}{3}$	$-\frac{2}{3}$	$-\frac{2}{3}$	1	$\frac{1}{3}$	$\frac{1}{3}$
I_3	0	0	0	0	0	$\frac{1}{2}$	$\frac{1}{2}$	$\frac{1}{2}$

$$(45)$$

$\bigwedge^3 \otimes S^0$	f_{137}	f_{145}	f_{146}	f_{147}	f_{156}	f_{157}	f_{167}	f_{234}	f_{235}
Y	$\frac{1}{3}$	$\frac{1}{3}$	$\frac{1}{3}$	$\frac{1}{3}$	$-\frac{1}{3}$	$-\frac{1}{3}$	$-\frac{1}{3}$	1	$\frac{1}{3}$
I_3	$\frac{1}{2}$	$\frac{1}{2}$	$\frac{1}{2}$	$\frac{1}{2}$	$\frac{1}{2}$	$\frac{1}{2}$	$\frac{1}{2}$	$-\frac{1}{2}$	$-\frac{1}{2}$

$\Lambda^3 \otimes S^0$	f_{236}	f_{237}	f_{245}	f_{246}	f_{247}	f_{256}	f_{257}	f_{267}	f_{345}
Y	$\frac{1}{3}$	$\frac{1}{3}$	$\frac{1}{3}$	$\frac{1}{3}$	$\frac{1}{3}$	$-\frac{1}{3}$	$-\frac{1}{3}$	$-\frac{1}{3}$	$\frac{4}{3}$
I_3	$-\frac{1}{2}$	$-\frac{1}{2}$	$-\frac{1}{2}$	$-\frac{1}{2}$	$-\frac{1}{2}$	$-\frac{1}{2}$	$-\frac{1}{2}$	$-\frac{1}{2}$	0

$\Lambda^3 \otimes S^0$	f_{346}	f_{347}	f_{356}	f_{357}	f_{367}	f_{456}	f_{457}	f_{467}	f_{567}
Y	$\frac{4}{3}$	$\frac{4}{3}$	$\frac{2}{3}$	$\frac{2}{3}$	$\frac{2}{3}$	$\frac{2}{3}$	$\frac{2}{3}$	$\frac{2}{3}$	0
I_3	0	0	0	0	0	0	0	0	0

In view of the invariant pairing, the eigenvalues on on the remaining four summands are the negatives of the eigenvalues listed above. That is, the eigenvalues on $\Lambda^4 \otimes S^{-1}$ are the negatives of those occuring in $\Lambda^3 \otimes S^0$, the eigenvalues on $\Lambda^5 \otimes S^{-2}$ are the negatives of those on $\Lambda^2 \otimes S^1$, the eigenvalues on $\Lambda^6 \otimes S^{-3}$ are the negatives of those of $\Lambda^1 \otimes S^2$ and the eigenvalues on $\Lambda^7 \otimes S^{-4}$ are the negatives of those on $\Lambda^0 \otimes S^3$.

We now recall the table of values for weak hypercharge and isospin in the standard model (in one generation with the quarks Cabbibo rotated):

$particle$	e_R	ν_{eL}	e_L	u_L	d_L	u_R	d_R	
Y	-2	-1	-1	$\frac{1}{3}$	$\frac{1}{3}$	$\frac{4}{3}$	$-\frac{2}{3}$	(46)
I_3	0	$\frac{1}{2}$	$-\frac{1}{2}$	$\frac{1}{2}$	$-\frac{1}{2}$	0	0	

The quarks occur with color multiplicity three, and there is no right handed neutrino in the standard model. But we may include a right handed neutrino if we assign to it weak hypercharge and isospin zero, so that it does not contribute to the lowest terms in the electroweak interaction. With these modifications, and if we include the antiparticles, we see that each set of eigenvalues of (46) is repeated four times in our 128 dimensional representation space. The four fold (generational) repetition is mediated by the $sl(2/1)$ subalgebra, and hence we would expect that the Cabibbo rotation and the Kobayashi-Maskawa matrix are somehow related to such a subsuperalgebra. Also the precise relation to chirality is not clear to us at the present, but we expect that it might be related to dimensional reduction via a Howe pair mechanism, cf. the discussion in [NS82].

In order to construct a Lagrangian, we will want a "compact" form of the superalgebra $sl(m/n)$. This will be the Lie superalgebra $su(m/n)$ whose definition we briefly recall, cf. [SW78] and [SaVS]. Let $V = V_0 \oplus V_1$ be an (m, n) dimensional supervector space. A *superhermitian structure* on V is by definition a map

$$H : V \times V \to \mathbf{C}$$

satisfying the following properties, where u, v, w represent homogeneous elements of V, i.e. elements of either V_0 or V_1 and $|u|$ denotes the degree of U etc.:

$$
\begin{aligned}
&(i) && \forall u, v \in V, \ |u| \neq |v| && H(u, v) = 0 \\
&(ii) && \forall u, v \in V && H(u, v) = (-1)^{|u||v|} \overline{H(v, u)} \\
&(iii) && \forall u, v, w \in V, \ \forall a, b \in \mathbf{C} && H(au + bv, w) = aH(u, v) + bH(v, w) \\
&(iv) && H(u, v) = 0 \ \forall v \in V && \Longrightarrow u = 0
\end{aligned}
$$

It is easy to see that the above conditions say that H has the form

$$H(u_0 + u_1, v_0 + v_1) = H_0(u_0, v_0) + iH_1(u_1, v_1) \tag{47}$$

where H_0 and H_1 are pseudohermitian forms on V_0 and V_1 respectively, and where $u = u_0 + u_1$ is the decomposition of u into its homogeneous components and similarly for v. We now define the subalgebra

$$su(V_0/V_1) \subset sl(V_0/V_1)$$

by defining it as a subsuperalgebra, where the condition on homogeneous components ξ is given by

$$H(\xi u, v) + (-1)^{|\xi||u|} H(u, \xi v) = 0, \tag{48}$$

where u and v are homogeneous elements of V. In terms of the matrix decomposition (24), this says that $A \in u(V_0)$ and $D \in u(V_1)$ relative to the pseudohermitian forms H_0 and H_1 (and of course the trace condition of (24) still holds). For the odd elements (48) is the same as

$$H_1(Cu_0, v_1) = iH_0(u_0, Bv_1), \forall u_0 \in V_0, v_1 \in V_1$$

or

$$H_0(Bu_1, v_0) = iH_1(u_1, Cv_0), \forall u_1 \in V_1, v_0 \in V_0,$$

either of these conditions implying the other. We have thus proved

Proposition 3 *We have a natural isomorphism of the even part, $su(V_0, V_1)_0$, of $su(V_0, V_1)$ with the subalgebra of*

$$u(V_0) \oplus u(V_1)$$

consisting of those pairs of operators satisfying the trace condition of (24). We also have a natural identification of the odd part,

$$su(V_0, V_1)_1 = \mathrm{Hom}(V_1, V_0)$$

as an $su(V_0, V_1)_0$ module.

In the case of interest to us, $V_0 = \mathbf{C}^m$ with its standard positive definite Hermitian form, and $V_1 = \mathbf{C}$ with its standard positive definite Hermitian form. Then our proposition asserts that $su(V_0, V_1)_1$ can be identified with $\mathrm{Hom}(\mathbf{C}, \mathbf{C}^m) \sim \mathbf{C}^m$. For example, in a theory with just leptons and no generations, so we choose $m = 2$, as in [F79] or [N79], the odd part of our superalgebra would be $\mathbf{C}^2$, which is precisely the space for the Higgs field in the Weinberg-Salam model. Our proposal, to be presented in the next section, is that in a theory with internal supersymmetry, Quillen's theory of superconnections suggests that the Higgs field always take its values in the odd part of the internal superalgebra.

Let us also point out that if we drop the supertrace condition of (24), that is if we consider the algebra $u(m/1)$ instead of $su(m/1)$, then we are in the situation of a Hermitian Lie algebra as described in [SW78]. That is, the assertion of Prop.3 holds, without the trace condition. We can think of the algebra $u(m/1)_0 = u(m) \oplus u(1)$ as the subalgebra of $u(m+1)$ which preseves a splitting $\mathbf{C}^{m+1} = \mathbf{C}^m \oplus \mathbf{C}$. Then the ordinary Lie algebra $u(m+1)$ decomposes as a direct sum

$$u(m+1) = [u(m) \oplus u(1)] \oplus \mathbf{C}^m = g_0 \oplus g_1$$

with the action of g_0 on g_1 the same as in the superalgebera case. The difference lies in the bracket back into g_0. But these two brackets are related as the real and imaginary parts of a Hermitian form with values in the complexification of g_0. For details see [SW78]. We can use this identification to restrict the choice of possible Lagrangians.

6   Superconnections

Recall [CNS75] or [Sch] that if $A = A_0 \oplus A_1$ and $B = B_0 \oplus B_1$ are superalgebras, then the superalgebra $A \otimes B$ is defined by

$$(A \otimes B)_0 = (A_0 \otimes B_0) \bigoplus (A_1 \otimes B_1), \quad (A \otimes B)_1 = (A_1 \otimes B_0) \bigoplus (A_0 \otimes B_1) \tag{49}$$

with multiplication (on homogeneous elements) given by

$$(a \otimes b)(a' \otimes b') = (-1)^{|b||a'|} aa' \otimes bb' \tag{50}$$

where $|a|$ denotes the degree of a. For example, suppose that A is the (supercommutative) superalgebra of all differential forms on a manifold, M, and that $B = \mathrm{End}\, E$ where $E = E_0 \oplus E_1$ is a supervector space. Thus B_0 consists of all "matrices" of the form

$$\begin{pmatrix} R & 0 \\ 0 & S \end{pmatrix}$$

with $R \in \mathrm{End}(E_0)$, $S \in \mathrm{End}(E_1)$ while B_1 consists of all "matrices" of the form

$$\begin{pmatrix} 0 & K \\ L & 0 \end{pmatrix}$$

with $K \in \mathrm{Hom}(E_1, E_0)$, $L \in \mathrm{Hom}(E_0, E_1)$ as in Section **4**. If we choose bases of E_0 and E_1 then we can think of R, S, K and L as actual matrices. We can then think of elements of $A \otimes B$ as matrices whose entries are differential forms forms, but we must remember the rules (49) and (50). For example, if ω_0 and ω_1 are matrices of differential forms of odd exterior degree then

$$\begin{pmatrix} \omega_0 & 0 \\ 0 & \omega_1 \end{pmatrix}$$

is an odd element of $A \otimes B$. Similarly, if L_{01} and L_{10} are matrices of forms of even exterior degree, then

$$\begin{pmatrix} 0 & L_{01} \\ L_{10} & 0 \end{pmatrix}$$

is an odd element of $A \otimes B$. Then rule (50) says that

$$\begin{pmatrix} \omega_0 & 0 \\ 0 & \omega_1 \end{pmatrix} \begin{pmatrix} 0 & L_{01} \\ L_{10} & 0 \end{pmatrix} = \begin{pmatrix} 0 & \omega_0 \wedge L_{01} \\ \omega_1 \wedge L_{10} & 0 \end{pmatrix} \tag{51}$$

while

$$\begin{pmatrix} 0 & L_{01} \\ L_{10} & 0 \end{pmatrix} \begin{pmatrix} \omega_0 & 0 \\ 0 & \omega_1 \end{pmatrix} = \begin{pmatrix} 0 & -L_{01} \wedge \omega_1 \\ -L_{10} \wedge \omega_0 & 0 \end{pmatrix}. \tag{52}$$

The minus sign in (52) arises from passing the odd elements of B through the differential forms of odd exterior degree as prescribed by (50). In this example we can consider the supervector space $A \otimes E$ with grading as in (49). Then $A \otimes E$ is a (left) module for $A \otimes B$ where we apply the sign rule analogous to (50). Thus

$$A \otimes B \subset \mathrm{End}(A \otimes E).$$

We can think of A as embedded in $A \otimes B$ as $A \otimes I$. This makes $A \otimes E$ into an A module where the action is the obvious one. Since $A \otimes E$ is a supervectorspace, $\mathrm{End}(A \otimes E)$ is a superalgebra. It is easy to see that an element of $\mathrm{End}(A \otimes E)$ belongs to $A \otimes B$ if and only if it supercommutes with all elements of A. In other words,

Proposition 4 *$A \otimes B$ is the supercentralizer of A in* $\mathrm{End}(A \otimes E)$.

We can define the (odd) operator $d \in \mathrm{End}(A \otimes E)$ by

$$d(\alpha \otimes e) = d\alpha \otimes e \tag{53}$$

where $d\alpha$ is the usual exterior derivative of the differential form α. We can write this definition symbolically as defining d as the "matrix"

$$\begin{pmatrix} d & 0 \\ 0 & d \end{pmatrix}.$$

If, by abuse of notation, we let α denote multiplication by α as an element of of $\text{End}(A \otimes E)$ then the supercommutator of d with α is given by

$$[d, \alpha] = d\alpha, \tag{54}$$

where the right hand side denotes multiplication by $d\alpha$. More generally, for any $\alpha \otimes b \in A \otimes B$ we have

$$[d, \alpha \otimes b] = d\alpha \otimes b. \tag{55}$$

So if, for $\sigma \in A \otimes B$ we define $d\sigma$ by

$$d(\alpha \otimes b) = d\alpha \otimes b, \tag{56}$$

then we can write (55) as

$$[d, \sigma] = d\sigma, \sigma \in A \otimes B. \tag{57}$$

In particular, if ω is an odd element of $A \otimes B$ then

$$(d + \omega)^2 = d^2 + [d, \omega] + \omega^2 = d\omega + \omega^2. \tag{58}$$

Notice that the right hand side of (58) is an element of $A \otimes B$. If we write ω out as a "matrix"

$$\omega = \begin{pmatrix} \omega_0 & L_{01} \\ L_{10} & \omega_1 \end{pmatrix}, \tag{59}$$

then we can write (58) as

$$(d+\omega)^2 = \begin{pmatrix} d\omega_0 + \omega_0 \wedge \omega_0 + L_{01} \wedge L_{10} & dL_{01} + \omega_0 \wedge L_{01} - L_{01} \wedge \omega_1 \\ dL_{10} + \omega_1 \wedge L_{10} - L_{10} \wedge \omega_0 & d\omega_1 + \omega_1 \wedge \omega_1 + L_{10} \wedge L_{01} \end{pmatrix}. \tag{60}$$

We should remember that in (60) the ω's and L's are matrices of even and odd forms respectively, but not necessarily homogeneous with respect to exterior degree. Thus, if the base space is four dimensional then

$$\omega_0 = A_0 + C_0 \tag{61}$$

where A_0 is a matrix of one forms and C_0 is a matrix of three forms, and similarly for ω_1. Also

$$L_{01} = h_{01} + B_{01} + D_{01} \tag{62}$$

350   Y. Ne'eman and S. Sternberg

where h_{01} is a matrix of functions, B_{01} is a matrix of two forms and D_{01} is a matrix of four forms, and similarly for L_{10}. The $\wedge$ occurring in (60) denotes matrix multiplication where the matrix entries are multiplied via exterior multiplication. Now let $E \to M$ be a supervector bundle over an ordinary manifold M. So $E = E_0 \oplus E_1$ where E_0 and E_1 are ordinary vector bundles. Let $\mathcal{A}(M)$ denote the ring of smooth differential forms on M and $\mathcal{A}(M, E)$ the space of smooth E valued forms. Then $\mathcal{A}(M, E)$ is a module for $\mathcal{A}(M)$ as before. We can consider $\text{End}(\mathcal{A}(M, E))$ and $\mathcal{A}(M, \text{End}(E))$ so that $\mathcal{A}(M, \text{End}(E)) \subset \text{End}(\mathcal{A}(M, E))$.

The analogue of Prop.4 is

Proposition 5 $\mathcal{A}(M, End(E))$ *is the centralizer of* $\mathcal{A}(M)$ *in* $End(\mathcal{A}(M, E))$.

In fact, $\mathcal{A}(M, \text{End}(E))$ is the centralizer of the ring of functions, $\mathcal{A}^0(M)$, in $\text{End}(\mathcal{A}(M, E))$.

An odd element $D \in \text{End}(\mathcal{A}(M, E))_1$ is called a *superconnection* if

$$[D, \alpha] = d\alpha, \ \forall \alpha \in \mathcal{A}(M). \tag{63}$$

In other words,

$$D : \mathcal{A}(M, E)_0 \to \mathcal{A}(M, E)_1, \ D : \mathcal{A}(M, E)_1 \to \mathcal{A}(M, E)_0,$$

and

$$D(\alpha \wedge \sigma) = d\alpha \wedge \sigma + (-1)^{|\alpha|} \alpha \wedge D\sigma \ \forall \alpha \in \mathcal{A}(M), \ \sigma \in \mathcal{A}(M, E).$$

The *curvature*, $F = F(D)$ of the superconnection is defined as

$$F = D^2. \tag{64}$$

Notice that for any function f we have

$$[D^2, f] = [D, f]D + D[D, f] = (df)D + D(df) = [D, df] = ddf = 0.$$

Thus

$$F \in \mathcal{A}(M, End(E))_0. \tag{65}$$

The difference, $D_1 - D_2$, between two superconnections is an element of $\mathcal{A}(M, \text{End}(E))$ by (63). Hence, in terms of a local trivialization of E, the most general superconnection can be written locally as

$$D = d + \omega, \ \omega \in \mathcal{A}(M, End(E))_1. \tag{66}$$

This means that the local expression for the curvature is given by (58), or, in "matrix" language, by (60). If p is any polynomial (or entire function) of one variable and Str denotes the supertrace then [Q85] or [BGV], $\text{Str}(p(F))$ is a closed form, i.e.

$$Str(p(F)) \in \mathcal{A}(M)$$

and
$$dStr(p(F)) = 0.$$

Furthermore, up to an exact form, $Str(p(F))$ is independent of the choice of the superconnection, i.e.

$$Str(p(F(D_1))) - Str(p(F(D_2))) = d\alpha(D_1, D_2)$$

where $\alpha(D_1, D_2)$ is a differential form which has a simple expression in terms of D_1 and D_2. Thus, for example, the Chern character corresponds to $p(z) = e^{-z}$. For all this see the above mentioned references, [Q85] and [BGV]. Now $Str(ab)$ is antisymmetric in a and b if a and b are both odd elements of $End(E)$. Hence if α and β are odd forms, the expression $Str(\alpha \otimes a)(*\beta \otimes b)$ is anti-symmetric as a function of $\alpha \otimes a$ and $\beta \otimes b$. Furthermore, on the even terms, an expression such as $StrF * F$ in the Lagrangian will lead to negative kinetic energy terms for the dynamics (unless E_0 or E_1 is trivial). Hence we proceed as follows: choose an invariant bilinear form b on the Lie algebra $End(E)_0$. Here invariant means invariant under the "even group" $Aut(E_0) \otimes Aut(E_1)$. As the adjoint representation of this group is not irreducible, there will be some choices here, beyond overall scale. Then the Lagrange density for the purely Yang Mills part of the theory is

$$\text{Y-M}(D) = b(F, *F). \tag{67}$$

The vector bundle we want to consider is associated to the 128 dimensional representation of $su(7/1)$ so it is useful to to have the principal bundle definition of connection form which we adopt with a slight modification from [BGV]: Let $k = k_0 \oplus k_1$ and let K be a Lie group whose Lie algebra is k_0. Let us suppose that there is a representation of K on k whose induced (derivative) action is the adjoint action of k_0. Suppose that $P_K \to M$ is a principal K bundle over M. We define a superconnection form to be an odd element, A, of $\mathcal{A}(P, k)$ satisfying (15) and (16). The meaning of (16) is now the following: A is no longer a form of exterior degree one, but rather

$$A = A_{[0]} + A_{[1]} + \cdots + A_{[dim(M)]}, \tag{68}$$

where $A_{[i]}$ is a form of exterior degree i with values in k_1 if i is even and with values in k_0 if i is odd. Then condition (16) says that $A_{[1]}$ is an ordinary connection form, and that the interior product of any vertical vector with any of the other components vanishes. Thus (15) implies that each of the components $A_{[i]}, i \neq 1$ is an element of $\mathcal{A}^i(M, k)$ of total odd degree. Now suppose that F is a superspace which is a (K, k) module. That is, suppose that F is a k module in the sense of superalgebras and is a representation space for the group K and that these actions are consistent in the obvious sense. Let ρ denote the representation of k on F. We can form the vector bundle associated to the representation of K on F and the principal bundle P_K. Then $\mathcal{A}^*(M, F)$ can be identified with F valued forms on P whose interior product with any vertical vector vanish, and which are K equivariant. Then a superconnection form A defines a superconnection via the formula

$$\mathbf{D}_A = d + \rho(A) : \mathcal{A}^{\pm}(M, F) \to \mathcal{A}^{\mp}(M, F). \tag{69}$$

It is easy to check that (15) and (16) imply that $d + \rho(A)$ carries forms whose interior product with vertical vectors vanish and which are equivariant, into forms whose interior product with vertical vectors vanish and which are equivariant, and so (69) is well defined. Furthermore,

$$(d + \rho(A))^2 = \rho(dA) + \rho(A)^2 = \rho(dA) + \frac{1}{2}[\rho(A), \rho(A)],$$

since if a is an odd element of a superalgebra then $a^2 = \frac{1}{2}[a, a]$. But the fact that ρ is a representation of the superalgebra k implies that $[\rho(A), \rho(A)] = \rho([A, A])$. Thus

$$F = \mathbf{D}_A^2 = \rho(dA + \frac{1}{2}[A, A]). \tag{70}$$

So if we define

$$\mathbf{F} = \mathbf{F}(A) = (dA + \frac{1}{2}[A, A]), \tag{71}$$

then we can write (70) as

$$F = \rho(\mathbf{F}).$$

So from now on we will drop the ρ and the distinction between F and $\mathbf{F}$. Notice that A enters quadratically in $\mathbf{F}$ and hence to order four in a Lagrangian which is quadratic in $\mathbf{F}$. In the case of interest to us, where the internal superalgebra is $su(7/1)$, the even exterior components of A take values in the odd piece of $su(7/1)$ which is $\mathbf{C}^7$. In other words, in the decomposition (68), $A_{[0]}$ is a section of the bundle associated to the representation of $K \sim U(7)$ on $\mathbf{C}^7$. It is a suitable candidate for the Higgs field, and super Yang Mills Lagrangian will involve it to the fourth order. We don't expect this to be the whole story, since applying the Higg's mechanism to $A_{[0]}$ will only cut down from $U(7)$ to $U(6)$. Of course it is tempting to also use $A_{[2]}$ and $A_{[4]}$. Now $A_{[2]}$, being a two form on four dimensions has six components, which is too many. However it is conceivable that some form of self-duality constraint arising from dynamical considerations might cut down on the number of components. For example, a dynamically produced self duality will give four (independent) components instead of the eight arising from degrees zero, two and four. Applying the Higgs mechanism would then cut down from $U(7)$ to $U(3)$ which could then be the electromagnetic force together with an unbroken color $SU(3)$. At the moment, we do not see how to carry this out, or to explain the relation to chirality as we mentioned in the preceding section. We should also point out, that if S denotes the spin bundle, then the connection A modifies the Dirac Lagrangian on the bundle (associated to) $F \otimes S$ in such a way as to introduce a cubic term which is quadratic in the matter field and linear in the Higgs field. Thus the acquisition of mass by the fermions is built into the superconnection formalism.

REFERENCES

[AL73] Abers, E.S. and Lee, B., *Gauge theories*, Physics Reports **9** (Section C of Physics Letters) (1973), 1–141.

[BN36] Birkhoff, G. and Von Neumann, J., *The logic of quantum mechanics*, Annals of Math.**37** (1936), 823–843.

[BGV] Berline, N., Getzler, E. and Vergne, M., "Heat Kernels and the Dirac operator", Springer Verlag, Heidelberg, New York, 1991.

[CNS75] Corwin, L., Ne'eman, Y. and Sternberg, S., *Graded Lie algebras in mathematics and physics*, Rev. Mod. Phys.**47** (1975), 573–604.

[D90] Decamp, D. et al, ALEPH collaboration, *A precise determination of the number of families with light neutrinos and of the Z boson partial widths*, Physics Letters **B235** (1990), 399–411.

[F79] Farlie, D.B., Phys. Lett. **B 82** (1979), 97–100.

[G] Griffith, D., "Introduction to elementary particles", John Wiley and Sons, New York, 1987.

[H77] Howe, R., *Remarks on classical invariant theory*, Preprint (1977).

[IZ] Itzykson, C. and Zuber, J.B., "Quantum Field Theory", McGraw Hill, New York, 1980.

[M] Mackey, G., "The mathematical foundations of quantum mechanics", Benjamin, New York, 1963.

[N79] Ne'eman, Y., Phy Lett. **B 81** (1979), 190–194.

[NK] Ne'eman, Y. and Kirsh, Y, "The particle hunters", Cambridge University Press, Cambridge, London, New York, 1986.

[NS80] Ne'eman, Y., and Sternberg, S., *Internal supersymmetry and unification*, Proc. Natl. Acad. Sci. USA **77** (1980), 3127–3131.

[NS82] Ne'eman, Y. and Sternberg, S., *Internal supersymmetry and dimensional reduction*, in "Gauge theories: fundamental interactions and rigorous results", eds. Dita. P., Georgescu, V. and Purice, R., 103–142, Birkhauser, Boston, 1982.

[NS90] Ne'eman, Y. and Sternberg, S., *Superconnections and internal supersymmetry dynamics*, Proc. Natl. Acad. Sci. USA **87** (1990), 7875–7877.

[Q85] Quillen, D., *Superconnections and the Chern character*, Topology **24** (1985), 89–95.

[RS85] Rapoport, D. and Sternberg, S., *On the interaction of spin and torsion*, Annals of Physics **158** (1985), 447–475.

[SaVS] Sanchez-Valenzuela, O. and Sternberg, S., *The automorphism group of a Hermitian superalgebra*, Lecture Notes in Mathematics **1251**, 1–48, Springer Verlag, 1985.

[Sch] Scheunert, M., "The theory of Lie superalgebras", Lecture Notes in Mathematics **716**, Springer Verlag, Heidelberg, New York, 1979.

[S77] Sternberg, S., *Some recent results on the metaplectic representation*, in "Group theoretical methods in physics", Lecture Notes in Physicis **79**, Springer Verlag, Heidelberg, New York, 1977.

[SW78] Sternberg, S. and Wolf, J., *Hermitian Lie algebras and metaplectic representations*, Trans. Amer. Math. Soc. **238** (1978), 1–43.

[Su] Sudbery, A., "Quantum mechanics and the particles of nature", Cambridge University Press, Cambridge, London, New York, 1986.

[V] Varadarajan, V.S., "Geometry of quantum mechanics", Springer Verlag, Heidelberg, New York, 1985.

[WW80] Weinberg, S. and Witten, E *Limits on classical particles*, Physics Letters **97B** (1980), 59.

[W39] Wigner, E. *On unitary representations of the inhomogeneous Lorentz group*, Annals of Math. **40** (1939), 149–204.

Sackler Institute of Advanced Studies
University of Tel Aviv
ISRAEL

Yuval Ne'eman   Shlomo Sternberg
Ministry of Science   Department of Mathematics
Jerusalem   Harvard University
Israel   Cambridge, MA 02138
                                        USA

Quelques remarques sur les variétés de Poisson-Nijenhuis

R. OUZILOU

A J.-M. Souriau

Introduction

La recherche de coordonnées adaptées à une structure géométrique donnée a conduit Nijenhuis [N] à introduire une notion nouvelle alors : celle de torsion d'un tenseur de type $(1,1)$. Cette idée a été reprise par F. Magri en vue d'étudier les systèmes complètement intégrables; un tenseur de Nijenhuis apparait alors comme un opérateur de récursion d'un système hamiltonien avec lequel il est compatible.

Cet exposé qui regroupe plusieurs propriétés, qu'on trouvera dans les articles récents cités en bibliographie, et notamment dans l'article [K-M] de Y. Kosmann-Schwarzbach et F. Magri, fait aussi la comparaison entre les opérateurs de Nijenhuis et les r-matrices de Reyman et Semenov-Tian-Shansky. La situation géométrique décrite ici est locale : en considérant un ouvert d'un espace localement convexe en dualité avec un autre e.l.c. on obtient une algèbre de Lie différentielle (appelée ici *algèbre composite*) sur laquelle la notion de compatibilité entre un tenseur de Nijenhuis et un tenseur de Poisson s'exprime et s'exploite aisément.

1. Crochet de Richardson-Nijenhuis

Soient K un anneau commutatif à unité et E un K-module. On définit un K-module $\mathbb{Z}$-gradué :

$$\mathcal{E} = \bigoplus_{n=-\infty}^{+\infty} E^{(n)}$$

en posant $E^{(n)} = \{0\}$ pour $n < -1$, $E^{(-1)} = E$ et, pour $n \geq 0$,

$$E^{(n)} = \Lambda^{n+1}(E, E) \, ,$$

ensemble des applications multilinéaires alternées de E^{n+1} dans E.

Pour tout $u \in E^{(p)}$ on définit un endomorphisme gradué de $\mathcal{E}$, i_u : $E^{(q)} \to E^{(q+p)}$ en posant pour tout $v \in E^{(q)}$: $i_u v = 0$, si $q \leq -1$ ou

$p \leq -2$, et :

$$i_u v(x_1, \ldots, x_{p+q+1}) =$$

$$\sum_{\substack{\sigma(1)<\cdots<\sigma(p+1), \\ \sigma(p+2)<\cdots<\sigma(p+q+1)}} \varepsilon_\sigma \cdot v(u(x_{\sigma(1)}, \ldots, x_{\sigma(p+1)}), x_{\sigma(p+2)}, \ldots, x_{\sigma(p+q+1)})$$

si $q \geq 0$ et $p \geq -1$ (ε_σ est la signature de la permutation σ).

L'image de l'application injective i de $\mathcal{E}$ dans l'ensemble des morphismes gradués de $\mathcal{E}$ est une algèbre de Lie graduée, i.e., pour tout couple (u, v) d'éléments de $\mathcal{E}$ il existe un élément $[u, v] \in \mathcal{E}$ tel que

$$i_v \circ i_u - (-1)^{\deg(u) \cdot \deg(v)} i_u \circ i_v \equiv [i_u, i_v] = i_{[u,v]}$$

et la somme des degrés de u et v est égal au degré de $[u, v]$.

De façon précise, le crochet de Richardson-Nijenhuis de u et v est, pour $u \in E^{(p)}$, $v \in E^{(q)}$:

$$[u, v] = i_v u - (-1)^{pq} i_u v \, .$$

On arrive ainsi à faire de $\mathcal{E}$ une algèbre de Lie $\mathbb{Z}$-graduée qui est une extension naturelle de l'algèbre de Lie

$$E^{(0)} = gl(E) \, .$$

Exemples : **1°.** Soit $\mu \in E^{(1)}$, i.e., on a une application bilinéaire alternée $\mu : E \times E \to E$. La condition $[\mu, \mu] = 0$ signifie que μ vérifie l'identité de Jacobi.

2°. Si $\mu \in E^{(1)}$ et $N \in E^{(0)}$ alors $[\mu, N]$ est l'application bilinéaire alternée $N \cdot \mu : E \times E \to E$ définie par :

$$(N \cdot \mu)(x, y) = \mu(N \cdot x, y) + \mu(x, N \cdot y) - N \cdot (\mu(x, y)) \, .$$

La condition $[\mu, N] = 0$ signifie que N est une dérivation de l'algèbre (E, μ).

Remarque : On démontre aisément ([K-M] et [O]) que, si N est le générateur infinitésimal d'un groupe à un paramètre Φ_t à valeur dans $E^{(0)}$, i.e. :

$$\frac{d}{dt} \Phi_t = (\Phi_t)_* N \, ,$$

alors $[\mu, N]$ est le générateur infinitesimal du groupe à un paramètre $\Phi_t^* \mu$ à valeur dans $E^{(1)}$, en ce sens que :

$$\frac{d}{dt}(\Phi_t^* \mu) = [\Phi_t^* \mu, N] \, .$$

2. Operateurs de Nijenhuis

2.1 Invariance par un flot.

Soient X (resp. N) un tenseur de type $(0,1)$ (resp. $(1,1)$) sur une variété $\mathcal{M}$ (i.e., X est un champ de vecteurs et N un champ d'opérateurs linéaires sur $\mathcal{M}$ qu'on supposera de classe C^∞). Notons Φ_t le flot de X. Rappelons que N est dit *invariant* par X si, pour tout $t \in \mathbb{R} : \Phi_t^* N = N$. Algébriquement ceci signifie que $\mathcal{L}_X N = 0$, i.e., pour tout champ de vecteur Y sur $\mathcal{M}$:

$$0 = \mathcal{L}_X(N{\cdot}Y) - N{\cdot}[X,Y] = [X, N{\cdot}Y] - N{\cdot}[X,Y] \ .$$

Dans ces conditions, les fonctions symétriques des valeurs propres de N, et par conséquent les fonctions

$$\operatorname{tr} N, \ldots, \operatorname{tr} N^k, \ldots \ ,$$

sont des intégrales premières de X et N est un opérateur de *récursion* pour X, d'où :

$$[X, N^k{\cdot}X] = 0 \qquad , \qquad k \in \mathbb{N} \ .$$

Il est naturel de s'interroger sur le comportement du champ de vecteurs $N{\cdot}X$ par rapport à N. Or, de $(\mathcal{L}_{N{\cdot}X} N){\cdot}Y = [N{\cdot}X, N{\cdot}Y] - N{\cdot}[N{\cdot}X, Y]$ et de $(N \circ \mathcal{L}_X N){\cdot}Y = N{\cdot}[X, N{\cdot}Y] - N^2{\cdot}[X, Y]$ on déduit que :

$$\mathcal{L}_{N{\cdot}X} N = N \circ \mathcal{L}_X N \tag{1}$$

si, et seulement si, la torsion de Nijenhuis de N, i.e., le tenseur anti-symétrique de type $(2,1)$ $\tau(N)$ donné par :

$$\tau(N)(X,Y) = [N{\cdot}X, N{\cdot}Y] - N{\cdot}([N{\cdot}X, Y] + [X, N{\cdot}Y] - N{\cdot}[X,Y]) \tag{2}$$

est identiquement nulle. Dans ces conditions, on a, pour tout couple d'entiers (k, ℓ) :

$$\mathcal{L}_X N = 0 \implies [N^k{\cdot}X, N^\ell{\cdot}X] = 0 \ .$$

Définition : Si les conditions (équivalentes!) (1) et (2) sont vérifiées, on dit que N est un *tenseur de Nijenhuis*.

De cette première ébauche il résulte immédiatement que si N est de Nijenhuis sur $\mathcal{M}$ alors, pour tout champ de vecteurs X préservant N, les champs de vecteurs $N{\cdot}X, \ldots, N^k{\cdot}X, \ldots$ préservent les champs d'opérateurs $N, \ldots, N^k, \ldots$; par conséquent, les traces des N^k sont des intégrales premières de tous ces champs de vecteurs.

2.2 Premiers exemples.

Etant donnés un K-module E et une application K-bilinéaire alternée $\mu : E \times E \to E$, un opérateur K-linéaire N sur E est dit de Nijenhuis pour μ si, pour tout couple d'éléments de E :

$$\mu(N{\cdot}X, N{\cdot}Y) + N^2{\cdot}\mu(X, Y) = N{\cdot}(\mu(N{\cdot}X, Y) + \mu(X, N{\cdot}Y)) \ .$$

Il est clair que ceci signifie que N est un morphisme de $(E, N{\cdot}\mu)$ dans (E, μ).

Exemple 1 : Tout opérateur A d'"entrelacement" sur (E, μ) [R-S] (i.e., si cette algèbre est de Lie, A est dans le commutant de la représentation adjointe de E : $A{\cdot}\mu(x,y) = \mu(A{\cdot}x,y))$ est un opérateur de Nijenhuis. De façon plus générale, si N est un opérateur de Nijenhuis pour (E, μ) alors pour tout opérateur d'entrelacement A, commutant avec N, l'opérateur $N \circ A$ est de Nijenhuis pour (E, μ).

Remarque 1 : Cette propriété est à rapprocher d'une propriété des matrices de Yang-Baxter [R-S] : une telle matrice R composée avec une matrice d'entrelacement A, redonne une r-matrice $R \circ A$.

Théorème. *(démontré dans* [O]*)*
i) *Soit N un opérateur de Nijenhuis sur une K-algèbre de Lie $\mathfrak{g}$ de dimension finie. Alors :*
  a) *Pour tout K-polynôme p à une indéterminée, $p(N)$ est un opérateur de Nijenhuis.*
  b) *La composante semi-simple et la composante nilpotente de N sont des opérateurs de Nijenhuis sur $\mathfrak{g}$.*
ii) *Un opérateur linéaire semi-simple sur $\mathfrak{g}$ est de Nijenhuis si, et seulement si, ses espaces spectraux, ainsi que leurs sommes deux à deux, sont des sous-algèbres de $\mathfrak{g}$.*

Exemple 2 : Un projecteur P sur $\mathfrak{g}$ tel que $\mathfrak{g} = \mathfrak{a} \oplus \mathfrak{b}$, avec $\mathfrak{a} = \operatorname{Ker} P$ et $\mathfrak{b} = \operatorname{Im} P$, est donc de Nijenhuis si, et seulement si, $\mathfrak{a}$ et $\mathfrak{b}$ sont des sous-algèbres de $\mathfrak{g}$. La formule :

$$2N = I + R \qquad (I : \text{identité de } \mathfrak{g}) \ ,$$

définit une correspondence biunivoque entre les r-matrices R involutives $(R^2 = I)$ et les opérateurs de Nijenhuis idempotents sur $\mathfrak{g}$.

2.3 Sur l'espace dual d'une algèbre de Lie.

Soit N un opérateur linéaire sur une algèbre de Lie réelle $\mathfrak{g}$. On peut alors définir sur l'algèbre de Lie-Poisson $C^\infty(\mathfrak{g}^*)$ un opérateur $\mathcal{N}$ en posant pour toute fonction différentiable $f : \mathfrak{g}^* \to \mathbb{R}$ et tout $\xi \in \mathfrak{g}^*$:

$$(\mathcal{N}f)(\xi) = f({}^t N{\cdot}\xi) \ .$$

Dans [K-M] on démontre que, si N est un opérateur de Nijenhuis idempotent, i.e., $\mathfrak{a} = \operatorname{Ker} N$ et $\mathfrak{b} = \operatorname{Im} N$ sont des sous-algèbres de Lie de $\mathfrak{g}$ supplémentaires, alors $\mathcal{N}$ est lui aussi de Nijenhuis. Ceci signifie que, dans la décomposition en produit tensoriel :

$$C^\infty(\mathfrak{g}^*) = C^\infty(\mathfrak{a}^*)\widehat{\otimes}C^\infty(\mathfrak{b}^*)\,,$$

la sous-algèbre associative $C^\infty(\mathfrak{a}^*)$, identifiée à l'ensemble des fonctions sur $\mathfrak{g}$ qui ne dépendent que de la variable $\xi \in \mathfrak{a}^*$, est aussi une sous-algèbre de Lie-Poisson de $C^\infty(\mathfrak{g}^*)$ ainsi que l'ensemble $C^\infty(\mathfrak{a}^*)'$ des fonctions $f(\xi,\eta)$, $\xi \in \mathfrak{a}^*$, $\eta \in \mathfrak{b}^*$ qui s'annulent avec η. On a une situation analogue avec $\mathfrak{b}$ au lieu de $\mathfrak{a}$, i.e., avec le projecteur de Nijenhuis $M = I - N$ et son associé $\mathcal{M}$ (qui n'est pas le supplémentaire de $\mathcal{N}$!). En résumé, on peut dire avec [G-S] que $\mathfrak{a}^*$ et $\mathfrak{b}^*$ sont deux sous-espaces co-isotropes de la variété de Poisson $\mathfrak{g}^*$ dès que $\mathfrak{g}$ est la somme directe de deux sous-algèbres $\mathfrak{a}$ et $\mathfrak{b}$.

Plus généralement, on peut montrer grâce au théorème (2.2), en procédant par récurrence sur le nombre de valeurs propres de N que :

Proposition. *Si N est un opérateur de Nijenhuis semi-simple sur $\mathfrak{g}$, alors $\mathcal{N}$ est lui-aussi de Nijenhuis sur $C^\infty(\mathfrak{g}^*)$.*

Application : Soit $\mathfrak{g} = \mathfrak{a} \oplus \mathfrak{b}$ une algèbre de Lie double au sens de [L-W]. Désignons par i (resp. p) l'injection canonique (resp. la projection canonique) de $\mathfrak{a}$ dans $\mathfrak{g}$ (resp. de $\mathfrak{g}$ sur $\mathfrak{a}$), alors, comme il vient d'être vu, l'opérateur $N = i \circ p$ est de Nijenhuis et il est clair que, pour tout couple $(f,g) \in C^\infty(\mathfrak{g}^*) \times C^\infty(\mathfrak{g}^*)$:

$$\{\,\mathcal{N}f, \mathcal{N}g\,\}_{\mathfrak{g}^*} = \{\,f|_{\mathfrak{a}^*}, g|_{\mathfrak{a}^*}\,\}_{\mathfrak{a}^*} \circ {}^t i\,,$$

du fait que ${}^t i : \mathfrak{g}^* \to \mathfrak{a}^*$ est une morphisme de Poisson.

Dire que $\mathcal{N}$ est de Nijenhuis signifie, de façon explicite que, pour tout couple (u,v) de fonctions différentiable sur $\mathfrak{a}^*$, on a, en désignant par u^c et v^c les prolongements canoniques de ces fonctions à $\mathfrak{g}^* = \mathfrak{a}^* \oplus \mathfrak{b}^*$ et en prenant des prolongements arbitraires u^a et v^a de u et v à $\mathfrak{g}^*$,

$$\{u,v\} = (\{u^c, v^a\} + \{u^a, v^c\} - \{u^a, v^a\})|_{\mathfrak{a}^*}\,,$$

ce qui permet de retrouver le lemme de Kostant-Symes : "S'il existe des prolongements de Casimir u^a et v^a alors u et v sont en involution sur $\mathfrak{a}^*$".

3. Algèbres composites

On entendra par là une K-algèbre de Poisson $(\mathcal{P}, \cdot, [\,.\,,.\,])$ munie d'une $\mathbb{Z}$-graduation $\mathcal{P}_n$ telle que :

i) $n \notin \{-1, 0\} \Longrightarrow \mathcal{P}_n = \{0\}$

et, pour tout couple d'entiers (p, q) :

ii) $[\mathcal{P}_p, \mathcal{P}_q] \subset \mathcal{P}_{p+q}$

iii) $\mathcal{P}_p \cdot \mathcal{P}_q \subset \mathcal{P}_{p+q+1}$

Exemple : Soit M une variété différentielle. Notons $\mathcal{A}$ (resp. $\mathcal{X}$) l'algèbre associative (resp. l'algèbre de Lie) des fonctions différentiables (resp. des champs de vecteurs) sur M. On a alors une algèbre composite

$$\mathcal{P}_{-1} = \mathcal{A} \quad , \quad \mathcal{P}_0 = \mathcal{X} \, ,$$

prolongeant naturellement le produit ordinaire des fonctions et le crochet de Lie de des champs de vecteurs, en posant, pour tout $X \in \mathcal{X}$ et tout $f \in \mathcal{A}$:

$$[X, f] = \mathcal{L}_X f \quad , \quad X \cdot f = f \cdot X \, .$$

De façon plus générale, on peut associer canoniquement à tout couple (E, F) d'e.l.c. en dualité une algèbre composite constituée par une algèbre de Lie $\mathcal{E}$, une algèbre associative de fonctions $\mathfrak{A}$ et un $\mathfrak{A}$-module $\mathcal{F}$ en dualité avec $\mathcal{E}$ et qui joue vis-à-vis de $\mathcal{E}$ (assimilé à un module de champ de vecteurs) le rôle des modules des 1-formes différentielles sur une variété.

Voici, précisément, comment une dualité de Mackey : $F \times E \to K$ ($K = \mathbb{R}$ ou $\mathbb{C}$) conduit à une algèbre composite sur un ouvert U de E.

Définitions : Une fonction $f : U \to K$ de classe C^∞ sera dite *admissible* pour cette dualité si, pour tout $x \in U$, sa différentielle $df(x)$ appartient à l'image canonique de F dans le dual topologique E' de E. Ces fonctions admissibles constituent évidemment un anneau qui sera noté $\mathfrak{A}$.

Une forme différentielle de degré r sur U sera dite *admissible* si c'est une somme de formes différentielles $f_0 \, df_1 \wedge \cdots \wedge df_r$ où $f_0, f_1, \ldots, f_r$ sont des fonctions admissibles. Ces formes différentielles constituent un $\mathfrak{A}$-module qui sera noté $\mathcal{F}^r$. On posera, pour simplifier les notations, $\mathcal{F} = \mathcal{F}^1$.

Un champ de vecteurs X de classe C^∞ sur U sera dit *admissible* pour la dualité considérée si $\mathcal{L}_X \mathfrak{A} \subset \mathfrak{A}$, ce qui revient à dire que :

$$\theta \in \mathcal{F} \quad \Longrightarrow \quad \langle \theta, X \rangle \in \mathfrak{A} \, .$$

Ces champs de vecteurs constituent une algèbre de Lie, notée $\mathcal{E}$, en $\mathfrak{A}$-dualité naturelle avec $\mathcal{F}$: $\langle ., . \rangle : \mathcal{F} \times \mathcal{E} \to \mathfrak{A}$. $\mathcal{E}$ est évidemment un $\mathfrak{A}$-module et on a une algèbre composite :

$$\mathcal{A} = \mathcal{A}_{-1} \oplus \mathcal{A}_0 \quad \text{avec} \quad \mathcal{A}_{-1} = \mathfrak{A} \, , \, \mathcal{A}_0 = \mathcal{E} \, .$$

La règle de Leibnitz permet de prolonger $\mathcal{L}_X$, $X \in \mathcal{E}$, aux formes différentielles admissibles :

$$\langle \mathcal{L}_X \theta, Y \rangle = \mathcal{L}_X \langle \theta, Y \rangle - \langle \theta, [X, Y] \rangle \, .$$

3.2 Déformation par un opérateur de Nijenhuis.

Etant donnée une K-algèbre composite $\mathcal{P}$, on dira qu'un opérateur K-linéaire sur $\mathcal{P}_0$ est *tensoriel* s'il est $\mathcal{P}_{-1}$-linéaire. Une partie $\mathcal{Q}$ de $\mathcal{P}_0$ engendre *tensoriellement* $\mathcal{P}_0$ si, par définition :

$$\mathcal{P}_0 = \mathcal{P}_{-1}{\cdot}\mathcal{Q} \ .$$

Voici un résultat fort utile, aisé à vérifier :

Lemme. *Soit $\mathcal{Q}$ une sous-algèbre de Lie commutative de $\mathcal{P}_0$ engendrant $\mathcal{P}_0$ tensoriellement. Alors, un opérateur linéaire tensoriel N sur $\mathcal{P}_0$ est de Nijenhuis si, et seulement si, pour tout couple (u, v) d'éléments de $\mathcal{Q}$:*

$$[N{\cdot}u, N{\cdot}v] = N{\cdot}([N{\cdot}u, v] + [u, N{\cdot}v]) \ .$$

Exemple : Dans le cas de l'algèbre composite d'une dualité de Mackey cette condition se traduit par :

$$N'_x(v; N{\cdot}u) - N'_x(u; N{\cdot}v) = N{\cdot}(N'_x(v; u) - N'_x(u; v)) \ ,$$

avec

$$N'_x(u; v) = \frac{d}{dt}(N_{x+tv}{\cdot}u)|_{t=0} \ .$$

Rappel : Soient $\mathcal{P} = (\mathcal{A}, \mathcal{E}, \mathcal{F})$ l'algèbre composite attachée à une dualité (E, F) et à un ouvert de E, et N un opérateur tensoriel de Nijenhuis sur $\mathcal{E}$. On peut alors déformer le crochet de Lie de $\mathcal{E}$ en posant :

$$[X, Y]_N = [N{\cdot}X, Y] + [X, N{\cdot}Y] - N{\cdot}[X, Y] \ ,$$

et il s'en suit (cf. [K-M] et [O]) une déformation par N de toute l'algèbre composite $\mathcal{P}$ en posant pour tout $X \in \mathcal{E}$:

$$\begin{aligned}
(\mathcal{L}_X f)_N &= \mathcal{L}_{N{\cdot}X} f &,&\quad f \in \mathcal{A} \\
(\mathcal{L}_X \alpha)_N &= \mathcal{L}_{N{\cdot}X} \alpha - \mathcal{L}_X({}^t N)\alpha &,&\quad \alpha \in \mathcal{F} \ ,
\end{aligned}$$

à condition que N admette un transposé sur $\mathcal{F}$.

4. Opérateurs de Poisson

Considérons comme précédemment une algèbre composite $(\mathcal{A}, \mathcal{E}, \mathcal{F})$ provenant d'une dualité de Mackey. A tout opérateur antisymétrique P de $\mathcal{F}$ dans $\mathcal{E}$ on associe un "crochet"

$$\{.,.\}^\mu_P \ ,$$

où μ désigne le crochet de Lie de $\mathcal{E}$ (si aucune confusion n'est à craindre on ne mentionnera plus cette lettre μ en indice). Ce crochet est déterminé par les propriétés suivantes :

i) Pour toute $f \in \mathcal{A}$ et tout couple (α, β) d'éléments de $\mathcal{F}$:

$$\{\alpha, f\beta\}_P = f\{\alpha, \beta\}_P + (\mathcal{L}_{P\alpha}f)\beta \ .$$

ii) Pour tout couple (f, g) d'éléments de $\mathcal{A}$:

$$\{df, dg\}_P = -d\{f, g\}_P$$

où le crochet de fonctions $\{f, g\}_P$ est défini par : $\{f, g\}_P = \langle df, P dg \rangle$. De façon plus explicite :

$$\{\alpha, \beta\}_P = \mathcal{L}_{P\alpha}\beta - \mathcal{L}_{P\beta}\alpha + d\langle \alpha, P\beta \rangle \ .$$

A partir de là on définit "le crochet de Schouten" $[P, P]$ par :

$$[P, P](\alpha, \beta) = P\{\alpha, \beta\}_P - [P\alpha, P\beta]$$

et on montre que ([K-M], [D-S] et [O])

$$\langle \alpha, [P, P](\beta, \gamma) \rangle = \langle \mathcal{L}_{P\alpha}\gamma, P\beta \rangle + \cdots + \dots$$
$$\mathcal{L}_{[P,P](\alpha,\beta)}\gamma = \{\{\alpha, \beta\}_P, \gamma\}_P + \cdots + \cdots + d\langle \alpha, [P, P](\beta, \gamma) \rangle$$

(les points de suspension désignent les termes obtenus par permutation circulaire sur les 1-formes α, β, γ). De ceci il résulte clairement le résultat suivant (devenu classique pour une variété différentiable de dimension finie) : *P est un morphisme de $(\mathcal{F}, \{., .\}_P)$ dans $(\mathcal{E}, [., .])$ si, et seulement si, $[P, P] = 0$, et alors $\{., .\}_P$ vérifie l'identité de Jacobi.* Notons que la condition $[P, P] = 0$ s'exprime localement par :

$$\langle \alpha, P'_x(\beta, P_x\gamma) \rangle + \cdots + \cdots = 0 \ .$$

Définition fondamentale : Un opérateur de Nijenhuis $N : \mathcal{E} \to \mathcal{E}$ est dit *compatible* avec un opérateur de Poisson $P : \mathcal{F} \to \mathcal{E}$ si les deux conditions suivantes sont satisfaites :

i) $\quad N{\cdot}P = P{\cdot}^t N$

ii) $\quad \{., .\}^\mu_{N{\cdot}P} = \{., .\}^{N{\cdot}\mu}_P$

Commentaire : La condition i) signifie que $N{\cdot}P$ est antisymétrique, ce qui permet de définir le crochet $\{., .\}^\mu_{N{\cdot}P}$ associé au crochet de Lie μ de

E par l'opérateur antisymétrique $N \cdot P$. Le second crochet de ii) est obtenu en déformant au préalable le crochet de Lie μ par l'opérateur N avant de faire opérer P. De façon précise, cette condition ii) s'explicite, si N est compatible avec P, en :

$$\text{ii')} \quad P \cdot (\mathcal{L}_X({}^t N \cdot \alpha) - \mathcal{L}_{N \cdot X}(\alpha)) = \mathcal{L}_{P \cdot \alpha}(N) X$$

pour toute 1-forme admissible $\alpha \in \mathcal{F}$ et tout champ admissible $X \in \mathcal{E}$.

Remarque : Cette condition est automatiquement assurée dès que N est un champ constant d'opérateurs sur $U \subset E$.

Exemples : 1°. Sur le dual $\mathfrak{g}^*$ d'une algèbre de Lie, la condition i) signifie que pour tout $\xi \in \mathfrak{g}^*$, N_ξ est dans le commutant de la représentation coadjointe de $\mathfrak{g}^*$.

2°. Soit $\Omega : \mathcal{E} \to \mathcal{F}$ un opérateur $\mathcal{A}$-linéaire antisymétrique, alors, en posant $N = P \circ \Omega$, i) est évidemment vérifié. Le cas le plus intéressant est celui d'un opérateur symplectique Ω, i.e.,

$$0 = d\Omega(X, Y) \equiv (\mathcal{L}_X \Omega) Y - (\mathcal{L}_Y \Omega) X + d \langle \Omega X, Y \rangle \ .$$

N est de Nijenhuis si, et seulement si, $P \cdot \Omega$ est aussi symplectique et, dans ces conditions, N est compatible avec P, i.e., ii) est vérifiée automatiquement (cf. [M-M]).

5. Harmonie et Hiérarchies

Les notations sont celles du paragraphe précédent.

Définition : Une 1-forme fermée $\alpha \in \mathcal{F}$ sera dite *en harmonie* avec un opérateur $\mathcal{A}$-linéaire $N : \mathcal{E} \to \mathcal{E}$ si la 1-forme admissible ${}^t N \alpha$ est elle aussi fermée, ce qui se traduit par :

$$\mathcal{L}_X({}^t N \alpha) = \mathcal{L}_{N \cdot X}(\alpha)$$

pour tout champ de vecteurs admissible $X \in \mathcal{E}$.

Dans ces conditions, on montre ([M-M]) que, si N est de Nijenhuis :

i) toutes les itérées $({}^t N)^k \alpha$ sont en harmonie avec N,

ii) Les 1-formes fermées en harmonie avec N forment une sous-algèbre de $(\mathcal{F}, \{ \, . \, , \, . \, \}_P)$ et, sur cette sous-algèbre, ${}^t N$ est un opérateur d'entrelacement : ${}^t N \{\alpha, \beta\}_P = \{{}^t N \alpha, \beta\}_P$. Dans [K-M] on montre que ${}^t N$ est de Nijenhuis pour le crochet de Poisson des 1-formes. (En ce qui nous concerne, nous nous sommes limités à deux formes admissibles pour une dualité de Mackey.)

Hiérarchies : Si l'on part d'une 1-forme admissible $\alpha \in \mathcal{F}$ on obtient, en posant pour tout entier k :

$$\alpha_k = ({}^tN)^k\alpha \quad , \qquad X_k = P\alpha_k ,$$

une famille de 1-formes et une famille de champs en orthogonalité :

$$\langle \alpha_k, X_\ell \rangle = 0 ,$$

et toutes deux involutives :

$$[X_k, X_\ell] = 0 \quad , \qquad \{\alpha_k, \alpha_\ell\}_P = 0 .$$

Exemple : En dimension finie, on obtient une hiérarchie de formes différentielles exactes $\alpha_k = d\,I_{k+1}$ à partir des traces des itérés d'un tenseur de Nijenhuis $I_k = \frac{1}{k}\operatorname{tr} N^k$; on a, en effet, pour tout champ de vecteurs X :

$$\langle dI_{k+1}, X \rangle = \mathcal{L}_X(\operatorname{tr} \frac{1}{k+1}N^{k+1}) = \operatorname{tr}(N^k \circ \mathcal{L}_X N)$$
$$= \operatorname{tr}(N^{k-1} \circ N \circ \mathcal{L}_X N)$$

qui, du fait que $N \circ \mathcal{L}_X N = \mathcal{L}_{NX} N$ devient

$$\langle d\,I_{k+1}, X \rangle = \operatorname{tr}(N^{k-1} \circ \mathcal{L}_{NX} N) = \mathcal{L}_{NX} \operatorname{tr} \frac{1}{k}N^k ,$$

soit $\langle d\,I_{k+1}, X \rangle = \langle d\,I_k, N.X \rangle$, d'où :

$$d\,I_{k+1} = {}^tN.d\,I_k .$$

Ce calcul reste valable pour un champ d'opérateurs de Nijenhuis traçable sur un espace de Hilbert et donne les relations fondamentales :

$$\{\operatorname{tr} N^k, \operatorname{tr} N^\ell\} = 0 .$$

Une question pour conclure : Lu et Weinstein [L-W] ont montré récemment que les orbites de la représentation coadjointe d'un groupe de Lie compact G semi-simple, sont munies d'une structure de Poisson canonique différente de la structure symplectique habituelle de Kirillov. De ce fait on disposerait, pour chacune de ces orbites, en cas de compatibilité, d'un tenseur de Nijenhuis. Il serait naturel alors de se demander si on ne peut pas "collectiviser" ces tenseurs de Nijenhuis sur tout le dual $\mathfrak{g}^*$ de l'algèbre de Lie $\mathfrak{g}$ de G; en d'autres termes, existe-t-il sur la variété de Poisson linéaire $\mathfrak{g}^*$ un tenseur de Nijenhuis compatible qui, par réduction aux orbites suivant le procédé de [M-M], redonnerait les structures de Poisson-Bruhat des orbites coadjointes?

REFERENCES

[[D-S]] P. Dazord & D. Sondaz, *Groupes de Poisson affines*, Preprint Dépt. Math. Lyon (1989).

[[G-S]] V. Guillemin & S. Sternberg, *Symplectic techniques in physics*, Cambridge UP, Cambridge, MA, 1984.

[[K-M]] Y. Kosmann-Schwarzbach & F. Magri, *Poisson-Nijenhuis structures*, Ann. Inst. H. Poincaré Phys. **53 n°1** (1990).

[[L-W]] J.H. Lu & A. Weinstein, *Poisson-Lie groups, dressing transformations and Bruhat decomposition*, J. Diff. Geom. **31** (1990), 501–506.

[[M-M]] F. Magri & C. Morosi, *A geometrical characterisation of integrable Hamiltonian systems through the theory of Poisson-Nijenhuis manifolds*, Quaderino S 19. University of Milan..

[[N]] A. Nijenhuis, X_{n-1}-*forming sets of eigen vectors*, Indagationes Math. **13** (1951), 200–212.

[[O]] R. Ouzilou, *Tenseurs de Nijenhuis I : propriétés de base*, Preprint Dépt. Math. Saint-Etienne (1990).

[[R-S]] A.G. Reyman & M.A. Semenov-Tian-Shansky, *Compatible Poisson structures for Lax equations : an r-matrix approach*, Phys. Lett A **130** (1988), 456–460.

Université Jean Monnet
Faculté des Sciences et Techniques
23, rue du Docteur Paul Michelon
F-42023 Saint-Etienne Cedex 2
France

Deformation Quantization of Kähler Manifolds

JOHN RAWNSLEY

To Jean-Marie Souriau

Kähler manifolds form a special class of symplectic manifolds and often appear as classical models for internal symmetries of particles. In this talk I want to show how the data normally used in the Geometric Quantization method [6], [9] can be used to construct a star-product [1] by exploiting the coherent states of Berezin [2] when applied to the quantization of a Kähler manifold. I begin with some notation.

Let M be a Kähler manifold with Kähler 2-form ω. The latter is a symplectic form and we suppose that (M, ω) is quantizable. This means that there is a hermitian line bundle with connection (L, ∇) over M such that the curvature of ∇ is $-2\pi i\omega$. L can be given the structure of a holomorphic line bundle in such a way that the (local) holomorphic sections are those which are covariant constant in the antiholomorphic directions (the polarized sections). If vol denotes the Liouville volume on M, which in this case coincides with the Riemannian volume, then we have a norm on sections given by

$$\|s\|^2 = \int_M |s|^2 \, vol.$$

We denote by $\mathcal{H}$ the Hilbert space of holomorphic sections of L for which this norm is finite. If M is compact, then this is the space of all holomorphic sections of L.

Prequantization associates to each smooth function ϕ on M a differential operator $Q(\phi)$ on the smooth sections of L by

$$Q(\phi) = \frac{i}{2\pi}\nabla_{X_\phi} + \phi$$

where X_ϕ denotes the Hamiltonian vector field associated to ϕ by

$$i(X_\phi)\omega = d\phi.$$

These operators satisfy

$$Q(\phi)Q(\psi) - Q(\psi)Q(\phi) = \frac{i}{2\pi}Q(\{\phi, \psi\}).$$

However $Q(\phi)$ does not preserve $\mathcal{H}$ unless the $(1,0)$ part X'_ϕ of X_ϕ is an infinitesimal holomorphic automorphism. We denote by $\mathcal{Q}$ the space of these functions and call it the space of quantizable functions. If M is compact then $\mathcal{Q}$ is finite dimensional, being a one-dimensional extension of the Lie algebra of Killing fields. Unless M admits lots of symmetries, the space of quantizable functions can reduce to the constants alone.

Berezin [2] has introduced a method of quantization in this context which produces a much larger space of quantizable functions. This is based on the notion of coherent states [8]. These arise as follows. It is easy to see that pointwise evaluation of holomorphic sections is continuous in the Hilbert space topology on $\mathcal{H}$, so is represented by inner product with vectors in the Hilbert space. We have to take into account the fact that the values are in the fibres of L rather than the complex numbers $\mathbb{C}$, so we must also choose basis elements for each fibre. Thus, for each q in $L_x\backslash\{0\}$ we have a vector e_q in $\mathcal{H}$ such that

$$s(x) = \langle s, e_q \rangle q, \qquad \forall s \in \mathcal{H}.$$

We call the vectors e_q the coherent states. It is clear that they depend very simply on the change in basis of the fibre, namely

$$e_{cq} = \bar{c}^{-1} e_q, \qquad c \in \mathbb{C}^*.$$

Moreover the fact that the sections of L are holomorphic, and e_q occurs in the antilinear argument of the inner product means that e_q depends antiholomorphically on q. It is clear from the definition that finite linear combinations of coherent states are dense in $\mathcal{H}$, and so the coherent states essentially determine the structure of $\mathcal{H}$. In particular the Hilbert space structure of $\mathcal{H}$ will be determined by the inner product of pairs of coherent states. This will be a function holomorphic in one variable and antiholomorphic in the other, so determined by analytic continuation by its restriction to the diagonal. Hence the structure is determined by the function $\|e_q\|^2$. Note that the product $|q|^2\|e_q\|^2$ depends only on the base-point x, so we can define a function

$$\epsilon(x) = |q|^2\|e_q\|^2, \qquad q \in L_x\backslash\{0\}.$$

It follows from what we have said that this function ϵ and the Hermitian structure in L together give the Hilbert space structure in $\mathcal{H}$ very explicitly. On the other hand it is easy to see that

$$\epsilon = \sum_i |s_i|^2$$

for any orthonormal basis s_i of $\mathcal{H}$. Thus automorphisms of the quantization, which act unitarily in $\mathcal{H}$, will preserve ϵ. In particular, if M is a coadjoint orbit then ϵ is invariant under a transitive group and so is constant. This allows the coherent states to be read directly from the Hermitian structure.

Example The round 2-sphere S^2 with its complex structure as the Riemann sphere is a Kähler manifold with Kähler 2-form ω the standard area form. In terms of a standard affine coordinate z on S^2 we have

$$\omega = \frac{i}{2} \frac{dz \wedge d\bar{z}}{(1 + z\bar{z})^2}.$$

It is easy to see that the holomorphic line bundle with Chern number n has a section s_0 with $|s_0|^2 = (1 + z\bar{z})^{-n}$ on the coordinate patch. Since this example is homogeneous under $SU(2)$, ϵ is constant so analytic continuation gives the inner product of coherent states as

$$\langle e_{s_0(w)}, e_{s_0(z)} \rangle = \epsilon(1 + z\bar{w})^n$$

and so

$$e_{s_0(w)}(z) = \epsilon(1 + z\bar{w})^n s_0(z).$$

In [8] it is shown that ϵ also determines whether the map $\sigma \colon M \to P(\mathcal{H})$ defined by $\sigma(x) = \mathbb{C}e_q$, $q \in L_x \backslash \{0\}$, is symplectic. If $\partial\bar{\partial}\log\epsilon = 0$ then σ is symplectic and the quantization data is all induced by σ.

Now take any linear operator A on $\mathcal{H}$. Then the ratio

$$\widehat{A}(x, \bar{y}) = \frac{\langle Ae_{q'}, e_q \rangle}{\langle e_{q'}, e_q \rangle}, \qquad q \in L_x \backslash \{0\}, q' \in L_y \backslash \{0\}$$

depends only on the base-points x, y and so defines a function $\widehat{A}(x, \bar{y})$ on an open set in $M \times M$ which is holomorphic with respect to x, antiholomorphic with respect to y. It follows that $\widehat{A}$ is determined by analytic continuation from its restriction to the diagonal. This restriction we also denote by $\widehat{A}$, and call the covariant symbol of A. It will be globally defined provided that for each point x the coherent states e_q are non-zero. This will be the case if ϵ vanishes nowhere. This holds for example, if $\mathcal{H}$ is non-trivial and ϵ is constant. We shall assume from now on that ϵ is nowhere vanishing. Then the covariant symbols are all globally defined smooth functions, and their analytic continuations are defined except, possibly, for a closed subvariety of $M \times M$ which depends only on the geometry and not the particular operator chosen.

If s_i denotes an orthonormal basis for $\mathcal{H}$, then for $A \in \text{End}(\mathcal{H})$ we have

$$
\begin{aligned}
Tr\, A &= \sum_i \langle As_i, s_i \rangle \\
&= \sum_i \int_M (As_i(x), s_i(x))\, vol_x \\
&= \sum_i \int_M (\langle As_i, e_q \rangle q, \langle s_i, e_q \rangle q)\, vol_x, \qquad q \in L_x \backslash \{0\} \\
&= \sum_i \int_M \langle s_i, A^* e_q \rangle \langle e_q, s_i \rangle |q|^2\, vol_x \\
&= \int_M \langle e_q, A^* e_q \rangle |q|^2\, vol_x \\
&= \int_M \widehat{A}(x) \epsilon(x)\, vol_x.
\end{aligned}
$$

When $\mathcal{H}$ is finite dimensional we can substitute the identity operator in the above formula and deduce

$$
\dim \mathcal{H} = \int_M \epsilon(x)\, vol_x.
$$

In the special case where ϵ is constant we conclude that

$$
\epsilon = \dim \mathcal{H} / vol(M).
$$

Denote by $\mathcal{E}(L)$ the set of symbols of linear operators on $\mathcal{H}$. It is a subspace of $C^\infty(M)$. Moreover, we have argued that the symbol determines the off-diagonal values and so the matrix elements $\langle Ae_{q'}, e_q \rangle$. Since the finite linear combinations of the coherent states are dense in $\mathcal{H}$, it is clear that these matrix elements determine the original operator A. Thus the map

$$
A \mapsto \widehat{A} : \text{End}(\mathcal{H}) \to \mathcal{E}(L)
$$

is a bijection. Its inverse gives Berezin's quantization rule which attaches to each function $\widehat{A}$ in $\mathcal{E}(L)$ the operator A of which it is the symbol.

Theorem. [3] *If $\phi \in \mathcal{Q}$ then*

$$
\widehat{Q(\phi)} = \frac{i}{2\pi} X'_\phi (\log \epsilon) + \phi
$$

so if ϵ is constant then $\mathcal{Q} \subset \mathcal{E}(L)$ and $\widehat{Q(\phi)} = \phi$.

Hence when ϵ is constant Berezin's quantization rule is an extension of Geometric Quantization, but when ϵ is not constant it will give different results. Notice also that $\dim \mathcal{E}(L) = (\dim \mathcal{H})^2$ depends very much on the bundle L unlike the space of quantizable functions for Geometric Quantization, and grows polynomially with the Chern class of L. In [3] we also prove that when all the ϵ functions involved are constant then under pointwise multiplication of functions we have $\mathcal{E}(L_1)\mathcal{E}(L_2) \subset \mathcal{E}(L_1 \otimes L_2)$ so that the set $\mathcal{C}_L$ of all symbols for all the powers of a given bundle L forms a subalgebra of $C^\infty(M)$. By the Kodaira embedding theorem this subalgebra separates points so is dense.

Following Berezin we define a new multiplication $*$ on $\mathcal{E}(L)$ by transfering the composition of operators from $\mathrm{End}(\mathcal{H})$ to $\mathcal{E}(L)$. Unlike pointwise multiplication, which by the above remarks has $\mathcal{E}(L)\mathcal{E}(L) \subset \mathcal{E}(L^2)$, we have $\mathcal{E}(L) * \mathcal{E}(L) \subset \mathcal{E}(L)$. Moreover, by using the reproducing property of the coherent states an easy calculation gives an integral formula for this multiplication:

$$(\phi * \psi)(x) = \int_M \phi(x, \bar{y})\psi(y, \bar{x})F(x, y)\epsilon(y)\, vol_y$$

where

$$F(x, y) = \frac{|\langle e_q, e_{q'}\rangle|^2}{\|e_q\|^2 \|e_{q'}\|^2}, \qquad \pi(q) = x, \pi(q') = y.$$

Note that it follows from the Cauchy-Schwartz inequality that $F(x, y) \leq 1$ with equality where $\sigma(x) = \sigma(y)$. If the line bundle is sufficiently positive that σ is a Kodaira embedding, then we shall have $F(x, y) < 1$ except on the diagonal.

The multiplication $*$ on $\mathcal{E}(L)$ gives an algebra structure to a space of smooth functions on M. Of course this algebra is just a copy of $\mathrm{End}\,\mathcal{H}$ from the way it was constructed. Since the symbol of the identity operator is the constant function 1, it follows that

$$1 * f = f * 1 = f, \qquad \forall f \in \mathcal{E}(L).$$

Moreover, since holomorphic automorphisms τ of L and (M, ω) act unitarily on $\mathcal{H}$ by an operator $Q(\tau)$, it follows that

$$\{Q(\tau)AQ(\tau)^{-1}\}\widehat{}} = \widehat{A} \circ \tau^{-1}$$

and hence that the operation $*$ is invariant under the group of automorphisms. Further, since elements of $\mathcal{Q}$ generate such automorphisms, with the operator corresponding with a function f given by the prequantization formula, it follows easily that, when ϵ is constant,

$$\phi * \psi = \phi\psi + \frac{i}{2\pi} X'_\phi(\psi)$$

where $\phi \in \mathcal{Q}$ and $\psi \in \mathcal{E}(L)$.

Now we fix a bundle L and consider its powers L^k as in [7]. The bundle L^k will be a quantization for $k\omega$. We denote the quantities ϵ, F, $*$, etc. associated with L^k by $\epsilon^{(k)}$, $F^{(k)}$, $*^{(k)}$, etc., and assume that all the $\epsilon^{(k)}$ are constant. Then $\mathcal{E}(L^{k_1})\mathcal{E}(L^{k_2}) \subset \mathcal{E}(L^{k_1+k_2})$ and $1 \in \mathcal{E}(L^k)$ gives us inclusions $\mathcal{E}(L^k) \subset \mathcal{E}(L^{k+1})$ with the union $\mathcal{C}_L = \cup_k \mathcal{E}(L^k)$ a dense subalgebra of $C^\infty(M)$ Note also that if $\phi, \psi \in \mathcal{E}(L^l)$ and $k \geq l$ then $\phi, \psi \in \mathcal{E}(L^k)$ and so $\phi *^{(k)} \psi$ is defined. Hence we obtain a family of products

$$*^{(k)} \colon \mathcal{E}(L^l) \times \mathcal{E}(L^l) \to \mathcal{C}_L.$$

Theorem. *[4] Let (M,ω) be a compact Kähler manifold with quantization L and suppose all $\epsilon^{(k)}$ are constant, then $\phi *^{(k)} \psi$ has an asymptotic expansion in powers of $1/k$ as $k \to \infty$*

$$\phi *^{(k)} \psi \sim \sum_{r \geq 0} k^{-r} C_r(\phi, \psi)$$

where the coefficients C_r are smooth bidifferential operators without constant term and where

$$C_0(\phi, \psi) = \phi\psi, \qquad C_1(\phi, \psi) - C_1(\psi, \phi) = \frac{i}{2\pi}\{\phi, \psi\}.$$

This theorem depends crucially on the fact that $F^{(k)} = F^k$, and the fact that $F < 1$ except on the diagonal. This makes the integral, asymptotically, concentrate near the diagonal and gives an immediate proof of the formula for C_0. A version of the Morse Lemma due to E. Combet allows us to prove the rest of the theorem.

We can use the coefficients of this expansion to define a formal product on the formal power series in $1/k$ with values in $C^\infty(M)$ by setting

$$\phi *_k \psi = \sum_{r \geq 0} C_r(\phi, \psi)$$

since differential operators can act on all smooth functions ϕ and ψ on M.

It is natural to ask if this is a star-product in the sense of [1]? That is, is this product associative? It is reasonable to expect it will be since the coefficients are obtained by expanding a family of associative products. In fact there are no results which allow us to conclude that the Taylor expansion of a family of associative products is always an associative product on formal power series, and it turns out to be very difficult to prove associativity in general. So far we have only been able to prove the associativity by using results from representation theory when M is a compact coadjoint orbit.

Theorem. [4] *If M is an integral coadjoint orbit of a compact semisimple Lie group then $*_k$ is a formal star-product.*

Our first attempt to prove this was by means of a stronger result which we have not been able to establish in this generality, namely that the formal series for $*_k$ actually converges on pairs of elements of $\mathcal{C}_L$ to $*^{(k)}$ provided that k is sufficiently large.

Theorem. [4] *If M is a compact Hermitian symmetric space with semisimple isometry group then for each l there is a $k(l) \geq l$ such that $\phi *_k \psi = \phi *^{(k)} \psi$ for ϕ, ψ in $\mathcal{E}(L^L)$ if $k \geq k(l)$.*

Observe that from the relationship $\epsilon^{(k)} = \dim \mathcal{H}^{(k)}/k^n vol(M)$ that the Riemann-Roch formula shows $\epsilon^{(k)}$ is a polynomial in $1/k$ and so is certainly a convergent series. So this term can be removed from the formulas in considering the convergence of $*_k$. Next observe that to prove this theorem it is enough to establish it for symbols of rank one operators determined by coherent states. Writing out the product of two such symbols it is immediate that it is enough to prove that

$$\int_M F(x,y)^p F(y,z)^q \, vol_y$$

is a rational function of p.

This integral is invariant under the isometry group so in the case of the 2-sphere we can chose one point to be the origin leaving an integral of the form

$$\int_{\mathbb{C}} \frac{1}{(1+y\bar{y})^p} \left(\frac{|1+y\bar{z}|^2}{1+y\bar{y}} \right)^q \frac{dyd\bar{y}}{(1+y\bar{y})^2}.$$

Expanding the numerator, and switching to polar coordinates we end up with integrals of the form

$$\int_0^\infty \frac{r^{2s+1}dr}{(1+r^2)^{p+q+2}}$$

where $s \leq q$. The integral can now be evaluated explicitly and seen to be rational in p.

To evaluate the integral for a general Hermitian symmetric space we use invariance to express the integrand in terms of radial coordinates and reduce to products of integrals of the same form as the 2-sphere case. This use of radial coordinates limits this approach to the hermitian symmetric case. I have, however, calculated the integral for the flag manifold of $SU(3)$ and found it also to be rational, so I conjecture that for all compact semisimple coadjoint orbits the products $*^{(k)}$ are rational functions of k and so the formal series converges.

REFERENCES

1. F. Bayen et al., *Quantization by deformations*, Ann. Phys. **111** (1978), 61.
2. F. Berezin, *Quantization*, Math. USSR Izvestija **8** (1974), 1109.
3. M. Cahen, S. Gutt, J. Rawnsley, *Quantization of Kähler manifolds, I*, J. Geom. Phys., (To appear).
4. M. Cahen, S. Gutt, J. Rawnsley, *Quantization of Kähler manifolds, II*, Trans. Amer. Math. Soc., (To appear).
5. E. Combet, *Intégrales exponentielles*, Lecture Notes in Mathematics, Springer-Verlag, Berlin, 1982.
6. B. Kostant, *On the definition of quantization*, Géométrie symplectique et physique mathématique, CNRS **237** (1975), 187-210.
7. C. Moreno, ** products on some Kähler manifolds*, Lett. Math. Phys. **11** (1986), 361-372.
8. J. Rawnsley, *Coherent states and Kähler manifolds*, Quart. J. Math. Oxford (2) **28** (1977), 403.
9. J.-M. Souriau, *Structure des systèmes dynamiques*, Dunod, Paris, 1970.

Mathematics Institute
University of Warwick
Coventry CV4 7AL
United Kingdom
e-mail jhr@maths.warwick.ac.uk

Algèbres de Lie graduées et quantification

CLAUDE ROGER

"Au grand quantificateur..."

Introduction

Ce travail est consacré principalement à l'étude de quantification par déformations, à l'aide de méthodes d'algèbres de Lie graduées.

Dans la partie I nous donnons les définitions essentielles concernant les algèbres de Lie graduées, leurs cohomologies et l'utilisation de celles-ci pour les déformations de structures. Nous suivons ici l'approche de P. Lecomte [24]. Une structure associée à une algèbre de Lie graduée L^* est un élément $c \in L^1$ qui vérifie $[c, c] = 0$. Le crochet gradué par c définit alors une différentielle $\partial_c : L^* \to L^*$ par $\partial_c(x) = [c, x]$ et on obtient la cohomologie associée à c. Cette cohomologie permet d'étudier les déformations formelles des structures: si c_t est une série formelle à valeurs dans L^1 telle que $c_t = c + tc_1 +.$, la relation $[c_t, c_t] = 0$ qui exprime la condition que c_t est une déformation de la structure c, se décompose suivant les puissances de t, et on obtient les relations: $[c, c_1] = 0$, $2[c, c_2] + [c_1, c_1] = 0$ etc... qui s'expriment dans la cohomologie de l'algèbre de Lie graduée L^*. Ces résultats sont utilisés et développés dans toute la suite de l'article.

La partie II est l'illustration de la précédente ; nous donnons des exemples explicites d'algèbres de Lie graduées et de leurs structures associées. Un certain nombre d'entre eux sont classiques, sinon bien connus (crochets gradués de Schouten, Fröhlicher–Nijenhuis, Richardson–Nijenhuis). La présentation unifiée que nous en donnons pourra contribuer à clarifier et populariser le concept d'algèbre de Lie graduée. Nous décrivons un certain nombre d'autres exemples qui seront utiles dans la suite (algèbres de Hopf, groupes quantiques, structures de Poisson quadratiques).

Pour chacun de ces exemples, nous décrivons les structures et les cohomologies associées: on obtient notamment les structures associatives et la cohomologie de Hochschild, les structures d'algèbres de Lie et la cohomologie de Koszul–Chevalley–Eilenberg, les structures de Poisson et la Λ-cohomologie. D'autres structures sont développées dans les parties suivantes.

La partie III est consacrée aux bigèbres de Lie et groupes quantiques: nous publions ici les détails des démonstrations des résultats annoncés dans la note [25]. Ces résultats ont été obtenus en collaboration entre P. Lecomte et l'auteur. En III.a nous construisons une algèbre de Lie graduée dont les structures associées sont précisément les structures de bigèbres de Lie. Le paragraphe III.b sur la cohomologie des bigèbres de Lie détaille la théorie cohomologique due à Drinfeld [10], vue comme théorie de classification et d'obstruction à l'existence des groupes quantiques associés à une bigèbre de Lie. Les paragraphes III.c et III.d contiennent les calculs explicites pour le cas semi simple de dimension finie et le cas du groupe de Heisenberg de dimension 3 respectivement.

Les équations de Yang-Baxter sont introduites dans la partie IV. Les travaux de Grabowski nous permettent d'interpréter l'équation de Yang-Baxter quantique dans le cadre des algèbres de Lie graduées et structures associées ; ce type de formalisme nous amène à proposer un cadre général pour décrire la quantification algébrique.

Enfin la partie V contient les développements récents sur le crochet de Fröhlicher-Nijenhuis ; outre des résultats de Vinogradov, nous proposons une nouvelle définition de ce crochet dans un cadre général et purement algébrique, qui permet de donner un formalisme pour une quantification de l'équation de Yang-Baxter modifiée (selon Semenov-Tian-Chansky).

L'auteur doit remercier ici P. Lecomte, dont les idées ont été essentielles pour le développement des résultats de cet article, et dont la collaboration a permis d'élaborer les résultats de la partie III ; Y. Kosmann–Schwarzbach pour des discussions toujours stimulantes et son séminaire qui a permis l'exposition de versions préliminaires de ce travail, et enfin tous les auditeurs du groupe d'études organisé sur ce thème durant l'année académique 1989-1990 au sein de l'U.R.A. 746 - G.D.R. 144.

I. Méthodes d'algèbres de Lie graduées et déformations de structures

Définition: Une algèbre de Lie graduée est un espace vectoriel $\mathbb{Z}$-gradué L^*, muni d'une application bilinéaire:

$$L^* \times L^* \xrightarrow{\quad [,] \quad} L^* \text{ telle que}$$

1) $[L^a, L^b] \subset L^{a+b}$ (le crochet est gradué)
2) Si $A \in L^a, B \in L^b$ alors $[A, B] = (-1)^{ab+1}[B, A]$ (antisymétrie graduée)
3) Si $A \in L^a, B \in L^b, C \in L^c$, on a l'identité de Jacobi graduée:

$$(-1)^{ac}[[A, B], C] + (-1)^{bc}[[C, A], B] + (-1)^{ab}[[B, C], A] = 0.$$

Définition: Une structure associée à l'algèbre de Lie graduée L^* est un élément $c \in L^1$ tel que $[c, c] = 0$.

A chaque structure on peut associer un opérateur $\partial_c : L^* \to L^*$ de degré $+1$ défini par $\partial_c(X) = [c, X]$.

Lemme: ∂_c *est un opérateur cohomologique, c'est-à-dire* $\partial_c \circ \partial_c = 0$.

Démonstration: Pour $X \in L^p$, on peut appliquer l'identité de Jacobi graduée au triplet (c, c, X) ce qui donne

$$-[[c, X], c] + (-1)^p[[c, c], X] + (-1)^p[[X, c], c] = 0$$

de $[c, c] = 0$ et $[X, c] = (-1)^{p+1}[c, X]$ on déduit :

$$2[[c, X], c] = 0, \ \text{soit} \ (\partial_c \circ \partial_c)(X) = 0.$$

On notera $H_c^*(L)$ la cohomologie obtenue.

Proposition: *Le crochet de* L^* *induit sur* $H_c^*(L)$ *une structure d'algèbre de Lie graduée.*

Démonstration: Soient $X \in L^X$ et $Y \in L^Y$ deux cocycles ; on peut appliquer l'identité de Jacobi graduée à (X, Y, c):

$$(-1)^X[[X, Y], c] + (-1)^Y[[c, X], Y] + (-1)^{XY}[[Y, c], X] = 0.$$

Donc $\partial_c([X, Y]) = 0$ et $[X, Y]$ est aussi un cocycle.

D'autre part si $X \in L^X$ et $Y = \partial_c(Z)$ avec $Z \in L^Z$, alors l'identité de Jacobi graduée nous donne

$$(-1)^X[[X, Z], c] + (-1)^Z[[c, X], Z] + (-1)^{XZ}[[Z, c], X] = 0.$$

Soit $\partial_c([X, Z]) = (-1)^{XZ+1}[Y, X]$. Par conséquent, le crochet d'un cocycle et d'un cobord est un cobord, et on obtient une structure bien définie sur la cohomologie.

Ce formalisme est bien adapté à l'étude des déformations des structures. Une déformation d'une structure c est une famille à un paramètre d'éléments de L^1, notée c_t, telle que $[c_t, c_t] = 0$ et $c_0 = c$. Si t est un paramètre scalaire, c_t est une déformation vraie. Si t est un paramètre formel (i.e. $c_t \in L^1[[t]]$) alors c_t est une déformation formelle.

Si on écrit le développement de $c_t : c_t = c + tc_1 + \cdots$, on déduit de $[c_t, c_t] = 0$ l'équation $[c, c_1] = 0$ et donc c_1 est un 1-cocycle de c. La partie d'ordre 1 d'une déformation (vraie ou formelle) est donc associée à un 1-cocycle de la cohomologie de c. La déformation sera dite triviale si il existe une famille a_t dans L^0 (formelle ou non suivant le cas) telle que $c_t = [c, a_t]$.

On a alors $c_1 = [c, a_1]$, soit $c_1 = \partial_c(a_1)$: le cocycle associé à la déformation infinitésimale est alors un cobord. On peut donc dire que le premier espace de cohomologie $H_c^1(L)$ classifie les déformations infinitésimales de c modulo les déformations triviales.

Définition: Une structure $c \in L^1$ sera dite rigide si toute déformation de c est triviale.

En particulier si $H_c^1(L) = 0$ alors c est rigide ; la réciproque étant fausse (voir par exemple [14] pour des algèbres de Lie de dimension finie, [23] en dimension infinie).

On peut vouloir construire une déformation vraie ou déformation formelle par étapes successives, en partant de la déformation infinitésimale c_1: si $c_t = c + t c_1 + t^2 c_2$ est le début d'une déformation formelle, alors le terme de degré 2 dans $[c_t, c_t]$ s'annule et on a:

$$2[c, c_2] + [c_1, c_1] = 0 \text{ soit } \partial_c(c_2) = -\frac{1}{2}[c_1, c_1]$$

On voit apparaitre ici une obstruction en général non triviale: le crochet $[c_1, c_1] \in L^2$ doit être un cobord ; remarquons que c_1 étant bien définie modulo un cobord, la classe de cohomologie de $[c_1, c_1]$ dans $H_c^2(L)$ est bien définie. On peut interpréter l'obstruction comme le crochet gradué de la classe de cohomologie de c_1 avec elle même dans $H_c^2(L)$. L'application $x \to [x, x]$ de $H_c^1(L)$ dans $H_c^2(L)$ est notée Sq par certains auteurs, et la structure d'algèbre de Lie graduée de $H_c^*(L)$ est appelée opération cohomologique dans le livre de D.B. Fuks sur la cohomologie des algèbres de Lie.

On rencontre d'autres obstructions du même type si on essaie d'itérer cette construction: $c_t = c + \sum_{i=1}^{k} t^i c_i$ est une déformation à l'ordre k de c si $[c_t, c_t] = 0$ modulo les termes de degré $> k$ en t. c_t sera triviale à l'ordre k s'il existe $a_t = \sum_{i=1}^{k} t^i a_i$ telle que $c_t = [c, a_t]$ modulo les termes de degré $> k$. On a alors comme précédemment:

$$\partial_c(c_n) = -\frac{1}{2} \sum_{\substack{i+j=n \\ i>0 \, j>0}} , [c_i, c_j] = P_n(c, c_1, \ldots, c_{n-1}) \text{ si } n \leq k.$$

On vérifie facilement que la classe de cohomologie de $P_n(c_1, \ldots, c_{n-1})$ est bien déterminée: si c_t est triviale à l'ordre k, chacun des P_n est un cobord. Nous avons donc trouvé, pour chaque ordre de déformation, une obstruction au prolongement à un ordre supérieur, bien définie dans $H_c^2(L)$.

- Si $H_c^2(L) = 0$ il n'existe aucune obstruction, toute déformation à l'ordre n se prolonge à l'ordre $n+1$, et toute déformation infinitésimale se prolonge en une déformation formelle.

- Si $H_c^2(L)$ est non nul, seule une étude détaillée de chaque cas permet de conclure ; il faut noter qu'à chaque étape, le terme d'ordre n peut être modifié en ajoutant un cocycle, qui pourra modifier les obstructions P_n à un ordre supérieur.

Cette approche de la théorie algébrique des déformations a ses origines dans les travaux de M. Gerstenhaber [13] et de A. Nijenhuis et R. Richardson [32] ; elle a été développée sous cette forme dans le travail de M. De Wilde et P. Lecomte [22] auquel nous renvoyons le lecteur pour plus de détails.

Remarque: On aurait pu traiter les objets de carré nul en degré p quelconque: on aurait obtenu une théorie analogue en multipliant tous les degrés par p.

Récemment, P. Lecomte a construit une théorie universelle des déformations, obtenue à l'aide de la cohomologie de l'algèbre de Lie graduée (cf [24]). On définit les cochaines sur L^* à valeurs dans la représentation adjointe: une p-cochaine est une application p linéaire de L^* dans L^* qui est alternée au sens gradué: la permutation de $X \in L^X$ avec $Y \in L^Y$ introduit un facteur $(-1)^{XY+1}$. Une p-cochaine c est de poids q si le degré de $c(X_1, \ldots, X_p)$ vaut $x_1 + \ldots + x_p + q$ (avec $\deg(X_i) = x_i$). On notera $c_q^p(L^*)$ l'espace des p-cochaines graduées de poids q sur L^*. Le bord ∂ : $c_q^p(L^*) \to c_q^{p+1}(L^*)$ est défini par la formule:

$$\partial c(X_0, \ldots, X_p) = \sum_{i=0}^{p} (-1)^{\alpha_i} [X_i; c(X_0, \hat{X}_i, \ldots, X_p)] +$$

$$\sum_{0 \leq i < j \leq p} (-1)^{\alpha_{i,j}} c([X_i, X_j], \ldots X_p)$$

avec $\alpha_i = i + x_i(x_0 + \ldots + x_{i-1}) \alpha_{ij} = x_i + x_j + x_i x_j$.

La cohomologie du sous complexe formé par les cochaines de poids q sera notée $H_{gr}^*(L^*)_q$.

Si on a maintenant $c \in L^1$ une structure et $[\gamma] \in H_{gr}^p(L^*)_q$ la classe de cohomologie de la p-cochaine graduée γ, alors $\gamma(c, \ldots, c)$ est dans L_{p+q}, et $\partial \gamma = 0$ implique que $[c, \gamma(c, \ldots, c)] = 0$ donc $\gamma(c, \ldots, c)$ est un cocycle pour la cohomologie de la structure c. Ceci définit un homomorphisme caractéristique:

$$H_{gr}^p(L^*)_q \xrightarrow{\quad I \quad} H_c^{p+q}(L)$$

$$I([\gamma]) = [\gamma(c, \ldots, c)]$$

On obtient finalement l'homomorphisme

$$\bigoplus_{p=0}^{\infty} H_{grad}^p(L^*)_{k-p} \xrightarrow{\quad I \quad} H_c^k(L).$$

Un calcul simple utilisant $\partial_\gamma = 0$ permet de montrer que si $X \in L^*$ vérifie $\partial_c(X) = 0$ alors $[I(\gamma), X] = 0$ dans $H_c^*(L)$. Les classes universelles $I(\gamma)$ sont donc dans le centre de l'algèbre de Lie graduée $H_c^*(L)$; par conséquent, les classes de $H_c^1(L)$ obtenues par l'image par I de $\bigoplus_{p=0}^{\infty} H_{grad}^p(L_{1-p}^*)$ engendrent des déformations formelles (voir [24] pour plus de détails). On peut donc parler de déformations universelles engendrées par la cohomologie graduée.

II. Exemples d'algèbres de Lie graduées
Structures associées et cohomologie

Dans cette partie, nous donnons une suite d'exemples d'algèbres de Lie graduées, en décrivant pour chaque cas les structures et la cohomologie obtenues. Nous espérons que cette approche contribuera à clarifier et populariser les principales structures existant déjà dans la littérature, pour certaines depuis longtemps (cf. par exemple A. Nijenhuis [31]).

II.a. L'algèbre $M(E)$

E étant un espace vectoriel, on notera $M^p(E)$ l'espace des applications $(p+1)$ linéaires de E à valeurs dans E. Pour $A \in M^a(E), B \in M^b(E)$, on a le produit intérieur $i_A(B) \in M^{a+b}(E)$ défini par la formule:

$$i_A(B)(x_0, \ldots, x_{a+b}) =$$
$$\sum_{k=0}^{b}(-1)^{ak} B(x_0, \ldots, x_{k-1}, A(x_k, \ldots, x_{k+a}), \ldots, x_{a+b})$$

Ce produit intérieur vérifie la formule suivante:

$$i(i(C)\cdot B)\cdot A - i(C)\cdot i(B)\cdot A = (-1)^{bc}(i(i(B)\cdot C)A - i(B)\cdot i(C)\cdot A) \quad (+)$$

On en déduit facilement que: $A\Delta B = i(B) \cdot A + (-1)^{ab+1}i(A) \cdot B$ définit un crochet de Lie gradué sur $M^*(E)$. On a en outre la formule qui nous sera utile par la suite:
$i(B\Delta C)A = i(C) \cdot i(B) \cdot A + (-1)^{bc+1}i(B) \cdot i(C) \cdot A$. Cette algèbre de Lie graduée a été construite par Lecomte et de Wilde [22] comme généralisation de la fameuse algèbre de Richardson et Nijenhuis (voir plus bas).

Si $C \in M^1(E)$, alors $C\Delta C = 2i(C) \cdot C$ et
$i(C) \cdot C(a, b, c) = C(C(a, b), c) - C(a, C(b, c))$. Donc $C\Delta C = 0$ si et seulement si C détermine une structure d'algèbre associative sur E. Si on note

$A = (E, C)$ l'algèbre associative obtenue, il est alors facile de voir que ∂_C n'est autre que l'application cobord des cochaines de Hochschild et la cohomologie s'identifie alors à la cohomologie de Hochschild de A, avec un décalage d'indices d'une unité.

$$H_c^k(M^*(E)) = H_h^{k+1}(A).$$

La cohomologie de Hochschild a été utilisée pour les déformations de structure associative pour la première fois par Gerstenhaber [13].

II.b. L'algèbre de Lie graduée associée à une algèbre de Hopf

Soit H une algèbre de Hopf cocommutative (voir plus loin pour les généralités sur les algèbres de Hopf, ou se reporter à [1]). On peut lui associer une algèbre de Lie graduée U^*.

$$U^{-1} = k \ U^p = \underbrace{H \otimes \ldots \otimes H}_{p+1 \text{ termes}} \ U^p = 0 \text{ si p } < -1$$

On notera fonctionnellement les éléments de U^*: un élément $A \in U^a$ est déterminé par $A(X_0, \cdots, X_a)$, les X_i pouvant s'interpréter comme des éléments de l'algèbre de Hopf duale H^0.

La comultiplication de $H, \Delta : H \to H \otimes H$ induit $\Delta_i : U^p \to U^{p+1}$, $i = 0, \ldots, p$ par $\Delta_i(\alpha_0 \otimes \ldots \otimes \alpha_p) = \alpha_0 \otimes \ldots \otimes \Delta_i(\alpha_i) \otimes \ldots \otimes \alpha_p$. Nous noterons ceci fonctionnellement. $\Delta_i(A)(X_0, \ldots, X_{p+1}) = A(X_0, \ldots, X_i + X_{i+1}, \ldots, X_{p+1})$. On peut alors définir un produit intérieur:

$$i(B).A(X_0, \ldots, X_b) = A(X_0 + \cdots + X_b)B(X_0, \ldots, X_b) \ si \ A \in U^0$$

et pour le cas général, on définit $i(B).A$ de façon à avoir une dérivation graduée de U^*, algèbre associative pour le produit tensoriel:

$$i(B).(A_1.A_2) = i(B)A_1.A_2 + (-1)^{a_1} A_1.i(B)A_2.$$

Ce qui donne la formule:

$$i(B).A(X_0, \ldots, X_{a+b}) =$$

$$\sum_{i=0}^{a}(-1)^{ib} A(X_0, \ldots, X_{i-1}, X_i + \ldots + X_{i+b}, \ldots, X_{a+b}).B(X_i, \ldots, X_{i+b})$$

On prend alors comme crochet $[A, B] = i(B)A + (-1)^{ab+1}i(A).B$.

Ce produit intérieur et ce crochet vérifient les mêmes propriétés formelles que ceux du paragraphe précédent, et en particulier la formule (+).

On peut maintenant expliciter les éléments de carré nul: si $F \in U^1 = H \otimes H$, alors $[F, F] = 2i(F).F$ et
$i(F).(F)(X, Y, Z) = F(X + Y, Z)F(X, Y) - F(X, Y + Z)F(Y, Z)$
En terme de la comultiplication Δ, la nullité de $[F, F]$ peut s'écrire

$$[(\Delta \otimes 1)F].(F \otimes 1) = [(1 \otimes \Delta)F](1 \otimes F).$$

L'interprétation de cette formule en terme d'équation de Yang-Baxter quantique sera donnée dans la suite. Remarquons simplement que si l'algèbre de Hopf cocommutative H est engendrée comme algèbre associative par ses éléments primitifs $\mathbb{G}$ alors H s'identifie à $U(\mathbb{G})$ l'algèbre enveloppante de l'algèbre de Lie $\mathbb{G}$ (cf. [15]). D'autre part, on voit immédiatement que $I = 1 \otimes 1 \in H \otimes H$ est un élément de carré nul; la cohomologie associée est donnée par l'application $\partial_I : U^a \to U^{a+1}$,

$$\partial_I(A) = A \otimes 1 + \sum_{i=1}^{a}(-1)^{a+i+1}\Delta_i(A) + (-1)^a 1 \otimes A$$

(cf. Gerstenhaber [12]). On peut alors montrer que $H_I^a(U^*) = \Lambda_{a+1}(\mathbb{G})$ l'algèbre extérieure de l'algèbre de Lie $\mathbb{G}$ (cf. [15]). Cette algèbre de Lie graduée a été introduite récemment par Janusz Grabowski ([15]). Pour ces méthodes de calcul dans les algèbres de Hopf, le lecteur peut se reporter à [9].

Remarque générale: De même que les dérivations d'une algèbre associative constituent une algèbre de Lie, les dérivations graduées d'une algèbre associative graduée constituent une algèbre de Lie graduée. Soit A^* une algèbre associative graduée. Une application $D : A^* \to A^*$ sera une dérivation graduée de degré k si $D(A^p) \subset A^{p+k}$ et $D(XY) = D(X)Y + (-1)^{kx}X.D(Y)$. L'espace $Der(A^*)$ des dérivations graduées est alors une algèbre de Lie graduée pour le commutateur gradué: Si D_i est de degré d_i, $i = 1, 2$; alors $[D_1, D_2] = D_1 \circ D_2 + (-1)^{d_1 d_2 + 1}D_2 \circ D_1$; on vérifie facilement que $[D_1, D_2]$ est une dérivation graduée de degré $d_1 + d_2$, ainsi que l'identité de Jacobi graduée. Dans le cas précédent la formule du produit intérieur définit une inclusion $U^* \hookrightarrow Der(U^*)$ qui permet d'interpréter le crochet gradué comme celui des dérivations.

Les dérivations d'algèbres associatives graduées fournissent beaucoup d'exemples d'algèbres de Lie graduées, et notamment ceux qui vont suivre, avatars divers de la célèbre construction de Richardson-Nijenhuis.

II.c. L'algèbre de Richardson-Nijenhuis $A(E)$ (*)

Pour un espace vectoriel E, nous noterons E' son dual. L'algèbre extérieure $\Lambda^*(E')$ est naturellement une algèbre associative graduée. On

---

(*) [32], [22] pour la présentation ci-dessous.

a $A(E) = \otimes_{a=-1}^{\infty} A^a(E)$ où $A^a(E)$ est l'espace des applications $(a+1)$ linéaires alternées de E à valeurs dans E. On peut donc écrire $A^a(E) = \Lambda^{a+1}(E') \otimes E$. Un élément de $A^a(E)$ peut s'écrire sous la forme $\alpha \otimes X$ avec $X \in E$ et $\alpha \in \Lambda^{a+1}(E')$. On a l'inclusion naturelle $A^a(E) \stackrel{i}{\hookrightarrow} Der^a(\Lambda^*(E'))$ par la formule $i(\alpha \otimes X).\beta = \alpha_\wedge i(X).\beta$.

Le crochet des dérivations graduées définit alors sur $A^*(E)$ le crochet de Richardson-Nijenhuis, que l'on peut exprimer par la formule:

$$[\alpha \otimes X, \beta \otimes Y] = \beta_\wedge i_Y \alpha \otimes X + (-1)^{ab+1} \alpha_\wedge i_X \beta \otimes Y \qquad (++)$$

si $\alpha \in \Lambda^{a+1}(E')$ et $\beta \in \Lambda^{b+1}(E')$.

Si $c \in A^1(E)$ alors $[c,c] \in A^2(E)$ est défini par la formule:

$$\frac{1}{2}[c,c](X_1, X_2, X_3) = \Sigma \varepsilon_{ijk} c(c(X_i, X_j), X_k).$$

Donc $[c,c] = 0$ si et seulement si c vérifie l'identité de Jacobi et est donc une structure d'algèbre de Lie sur E. On notera $\mathbb{G} = (E, c)$ l'algèbre de Lie obtenue ; un calcul simple montre ensuite que la cohomologie s'identifie à celle de $\mathbb{G}$ à coefficients dans la représentation adjointe. On a:

$$H_c^k(A^*(E)) = H_{Lie}^{k+1}(\mathbb{G}).$$

On retrouve ainsi l'approche bien connue des déformations d'algèbres de Lie. Les exemples qui vont suivre sont pour une bonne part des variantes de $A^*(E)$.

Remarque 1: Les exemples II.a et II.c nous ont fourni des structures d'algèbres de Lie graduées sur l'espace des applications multilinéaires (resp. multilinéaires alternées) de E à valeurs dans E. Si l'on considère les applications multilinéaires symétriques de E à valeurs dans E et on essaie une formule analogue à $(++)$, on obtient l'algèbre de Lie des champs de vecteurs formels sur E.

Remarque 2: Le crochet de $A(E)$ peut s'obtenir à partir de celui de $M(E)$ par l'antisymétrisation $\alpha : M(E) \to A(E)$

$$\alpha(c)(X_1, \ldots, X_a) = \sum_{\sigma \in \Sigma a} \frac{\varepsilon(\sigma)}{a!} c(X_{\sigma(1)}, \ldots, X_{\sigma(a)}).$$

On a alors pour $A \in A^a(E)$, $B \in A^b(E)$, $[A, B] = \frac{(a+b+1)!}{(a+b)!(b+1)!} \alpha(A \Delta B)$.

On peut obtenir d'autres crochets de Lie gradués sur diverses sous-algèbres de $M(E)$ en utilisant les constructions de D. Gurevich [16]. Soit $S : E \otimes E \to E \otimes E$ une solution unitaire de l'équation de Yang-Baxter quantique (voir plus loin): on a $S \circ S = Id(E \otimes E)$ et $(S \otimes Id) \circ (Id \otimes S) \circ (S \otimes Id) = (Id \otimes S) \circ (S \otimes Id) \circ (Id \otimes S)$ dans les endomorphismes de $E \otimes E \otimes E$.

On définit $S_i : \overset{p+1}{\otimes} \to \overset{p+1}{\otimes} E$, $\quad i = 0, \ldots, p-1$, $S_i = Id^i \otimes S \otimes Id^{p-i+1}$.

Une application $c \in M(E)$ sera dite S-alternée si:

$$(c \circ S_i)(X_0, \ldots, X_p) = -c(X_0, \ldots, X_p).$$

On a ainsi un sous-espace $A_S^p(E) \subset M^p(E)$. Une vérification directe quoique fastidieuse montre que $A_S^*(E)$ est une algèbre de Lie graduée. Si $S = P$ défini par $P(x \otimes y) = y \otimes x$ alors $A_S^*(E) = A^*(E)$. Un élément de carré nul dans $A_S^1(E)$ est alors une structure de S-algèbre au sens de [16].

II.d. L'algèbre de Lie graduée $\Lambda^*(E') \otimes S^*(E)$

On considère le produit tensoriel de l'algèbre extérieure sur le dual de E par l'algèbre symétrique sur E ; on obtient une algèbre associative pour le produit tensoriel. On peut faire opérer cette algèbre sur elle-même par dérivations:

$$\Lambda^*(E') \otimes S^*(E) \hookrightarrow Der(\Lambda^*(E') \otimes S^*(E))$$

$$(\alpha \otimes X_1 \vee \ldots \vee X_p).(\beta \otimes S) = \sum_{i=1}^{p} \alpha_{\wedge} i_{X_i} \beta \otimes X_1 \vee \ldots \vee \hat{X}_i \ldots \vee X_p \vee S$$

(ici $\vee$ désigne le produit symétrique).

On a alors pour le crochet gradué une formule analogue à $(++)$. Pour

$$\alpha \otimes X_p \vee \ldots \vee X_p \in \Lambda^{a+1}(E') \otimes S^{p+1}(E)$$

et

$$\beta \otimes Y_0 \vee \ldots \vee Y_q \in \Lambda^{b+1}(E') \otimes S^{q+1}(E)$$

alors

$$[\alpha \otimes X_0 \vee \ldots \vee X_p, \beta \otimes Y_0 \vee \ldots \vee Y_q] =$$

$$\sum_{i=0}^{q} \beta \wedge i Y_i \alpha \otimes Y_0 \ldots \vee \hat{Y}_i \ldots \vee Y_q \vee X_0 \vee \ldots \vee X_p +$$

$$(-1)^{ab+1} \sum_{i=0}^{p} \alpha_{\wedge} i_{X_i} \beta \otimes X_0 \vee \ldots \vee \hat{X}_i \ldots \vee X_p \vee Y_0 \ldots \vee Y_q$$

Le crochet est dans $\Lambda^{a+b+1}(E') \otimes S^{p+q+1}(E)$.

On graduera un élément de $\Lambda^{a+1}(E') \otimes S^{p+1}(E)$ par $(2-a)$. L'algèbre $A(E) \hookrightarrow \Lambda^* E' \otimes S^*(E)$ est alors une sous-algèbre de Lie graduée.

Un élément c de $\Lambda^2 E' \otimes S^2 E$ est une application bilinéaire antisymétrique sur E à valeurs dans les 2-tenseurs symétriques sur E. Supposons maintenant E de dimension finie et soit $X_i, i = 1 \ldots n$ une base m; c est alors déterminée par $c(X_i, X_j) = P^{ij}$ où $P_{ij} \in S^2 E$ est un polynôme quadratique homogène $P_{ij} = \sum_{k,\ell=1}^{n} P_{ij}^{k\ell} X_k X_\ell$. Les scalaires (P_{ij}^{k1}) sont donc symétriques en les indices (k, ℓ) et antisymétriques en les indices (i, j).

Si on note R l'anneau des polynômes à n indéterminées, l'élément c permet de définir une application bilinéaire antisymétrique, homogène en degré, de R à valeurs dans lui même.

$$R \times R \xrightarrow{\{.,.\}} R$$

$$\{f, g\} = \sum_{1 \leq i < j \leq n} P_{ij} \frac{\partial f}{\partial X_i} \frac{\partial g}{\partial X_j}.$$

Cette application vérifie la règle de Leibnitz:

$$\{fg, h\} = f\{g, h\} + g\{f, h\}.$$

Il est facile de voir qu'elle vérifie l'identité de Jacobi si et seulement si $[c, c] = 0$ dans $\Lambda^3(E') \otimes S^3(E)$. Les structures de Poisson quadratiques, dont l'étude a été abordée par Sklyanin ([**36**]) pour classifier des solutions de l'équation de Yang-Baxter, sont donc associées à des objets de carré nul dans $\Lambda^*(E') \otimes S^*(E)$. Le lecteur trouvera quelques éléments sur la classification cohomologique et la rigidité de ces structures dans l'article à paraître ([**33**]).

Remarques: (1) $\Lambda^*(E') \otimes S^*(E)$ est à la fois une algèbre associative graduée et une algèbre de Lie graduée. Les axiomes de l'algèbre de Poisson graduée sont aisément vérifiés.

(2) On peut s'intéresser aux algèbres de Poisson homogènes de degré p. Il faut alors étudier des éléments de $\Lambda^2(E') \otimes S^p(E)$ de carré nul pour le crochet précédent. Le cas $p = 1$ nous ramène à $A(E)$: les structures de Poisson linéaires s'identifient aux structures d'algèbre de Lie, par le résultat bien connu de Kirillov.

(3) La cohomologie associée s'identifie à la cohomologie de Poisson algébrique ([**26**], [**17**]).

II.e. L'algèbre de Poisson $P_\omega(E)$ et ses algèbres associées

Les résultats de ce paragraphe ont été annoncés dans [**25**]. L'idée essentielle est d'étendre les constructions de Richardson et Nijenhuis au cas d'un espace vectoriel muni d'une forme bilinéaire. Soit $\xi : E \times E \to k$

une forme bilinéaire non dégénérée. On peut l'étendre facilement en une application bilinéaire:

$$\omega : A(E) \times A(E) \to \Lambda^*(E')$$

$$\omega(\alpha \otimes X, \beta \otimes Y) = \xi(X, Y)\alpha \wedge \beta$$

Un élément A de $A(E)$ sera dans $A_\omega(E)$ s'il laisse la forme ω invariante dans le sens suivant:

Pour tous B et C dans $A(E)$, on a:

$$i(A).\omega(B, C) = \omega([A, B], C) + (-1)^{ab+1}\omega(B, [A, C])$$

On peut alors montrer par un calcul direct la:

Proposition: *$A_\omega(E) \hookrightarrow A(E)$ est une sous-algèbre de Lie graduée. Si ω est antisymétrique, (E, ω) est alors un espace symplectique et on montre aisément que $A_\omega^k(E) = 0$ pour $k \geq 1, A_\omega^{(-1)}(E) = E$ et $A_\omega^0(E) = sp(E, \omega)$. (L'argument est analogue à celui de la nullité des prolongements d'ordre supérieur pour une algèbre de Lie compacte).*

On supposera donc que ω est symétrique non dégénérée et nous allons expliciter la structure de $A_\omega^*(E)$ dans ce cas.

Notons ω' la forme quadratique induite sur l'espace dual E' via l'isomorphisme naturel entre E et E' induit par ω: si $\alpha \in E'$ on trouve $x_\alpha \in E$ unique tel que $\alpha(x) = \omega(x_\alpha, x)$ quel que soit x dans E, et on pose $\omega(x_\alpha, x_\beta) = \omega'(\alpha, \beta)$. Nous allons associer à (E, ω) une algèbre de Poisson graduée $P_\omega(E)$. Posons $P_\omega^k(E) = \Lambda^{k+2}(E')$ et $P_\omega^k(E) = 0$ si $k < -2$.

Le produit extérieur fait naturellement de $P_\omega^k(E)$ une algèbre associative.

On définit le crochet sur $P_\omega^{-1}(E) = E'$ par $\{\alpha, \beta\} = \omega'(\alpha, \beta) \in k = P_\omega^{-2}(E)$. Le crochet se définit ensuite par induction sur $P_\omega^*(E)$ tout entier en imposant la règle de Leibnitz graduée:

$$\{\alpha, \beta \wedge \gamma\} = \{\alpha, \beta\} \wedge \gamma + (-1)^{ab}\beta \wedge \{\alpha, \gamma\}.$$

Cette algèbre a été introduite pour la première fois, sous une forme légèrement différente par Kostant et Sternberg ([**21**]). On remarquera la fonctorialité de la construction $(E, \omega) \to P_\omega(E)$ ce qui n'était pas le cas pour les algèbres de Lie graduées vues précédemment. On peut montrer également que le crochet de Lie de $P_\omega(E)$ est un crochet gradué de dérivations: un élément α de $P_\omega^{-1}(E)$ opère sur $\Lambda^*(E')$ comme dérivation de degré (-1) par la formule:

$$i(\alpha).(\alpha_1 \wedge \ldots \wedge \alpha_p) = \sum_{i=1}^{p}(-1)^{i-1}\omega'(\alpha, \alpha_i)\alpha_1 \wedge \ldots \wedge \widehat{\alpha_i} \wedge \ldots \wedge \alpha_p$$

que l'on peut étendre aisément à tous les éléments de $P_\omega^*(E)$.

Nous allons maintenant construire un homomorphisme d'algèbres de Lie graduées:

$$\phi_\omega^* : P_\omega^*(E) \to A_\omega^*(E)$$

Heuristiquement parlant, ϕ_ω consiste à descendre un indice à l'aide de ω pour passer des tenseurs $(0, p+1)$ aux tenseurs $(1, p)$. Si E est de dimension finie, on a une formule explicite à l'aide d'une base (e_i) de E pour laquelle $\omega_{ij} = \omega(e_i, e_j)$

$$\phi_\omega(\alpha) = \Sigma \omega_{ij}^{-1} i(e_i)\alpha \otimes e_j$$

On voit alors immédiatement que le diagramme ci-dessous est commutatif:

$$
\begin{array}{ccc}
P_\omega^*(E) & \xrightarrow{\phi_\omega} & A_\omega^* \\
\searrow i & & \nearrow i \\
& Der\Lambda^*(E) &
\end{array}
$$

Par conséquent, ϕ_ω est bien un homomorphisme d'algèbres de Lie graduées.

Nous allons maintenant nous restreindre à un cas particulier: si F est un espace vectoriel quelconque, posons $E = F \oplus F$ et ω la forme bilinéaire symétrique non dégénérée donnée par:

$$\omega(x + \xi, y + \eta) = \eta(x) + \xi(y)$$

Posons $\mathcal{P}^*(F) = P_\omega^*(F \oplus F')$. Ces espaces peuvent se décomposer suivant les tenseurs de type "pur":

$$\mathcal{P}^k(F) = \Lambda^{k+2}(F \oplus F') = \bigoplus_{p+q=k} \Lambda^{p+1}(F) \otimes \Lambda^{q+1}(F')$$

et on vérifie facilement que le crochet est bigradué pour cette décomposition.

On a la sous-algèbre $\mathcal{P}_0(F) = \oplus \mathcal{P}_0^k(F)$ avec

$$\mathcal{P}_0^k(F) = \bigoplus_{\substack{p+q=k \\ p \geq 0 \, q \geq 0}} \Lambda^{p+1}(F) \otimes \Lambda^{q+1}(F).$$

Les inclusions naturelles $A^*(F) \to \mathcal{P}^*(F)$ et $A^*(F') \to \mathcal{P}^*(F)$ sont des morphismes d'algèbres de Lie graduées. Les objets de carré nul dans $\mathcal{P}_0^*(F)$ sont les structures de bigèbre de Lie sur F et feront l'objet d'une étude détaillée dans le chapitre suivant.

On va maintenant considérer une autre sous-algèbre de $\mathcal{P}^*(F)$.

Posons $\tilde{A}(F) = \bigoplus\limits_{p \geq 0,\, q = 0 \ ou\ -1} \Lambda^{p+1}(F') \otimes \Lambda^{q+1}(F)$. On voit immédiatement qu'il s'agit bien d'une sous-algèbre, le crochet étant bigradué pour
la bigraduation. Cette algèbre $\tilde{A}(F)$ s'identifie au produit semi-direct de
l'algèbre de Richardson-Nijenhuis $A(F)$ par $\Lambda^*(F')$ sur lequel elle opère par
dérivations; soit $\tilde{A}(F) = A(F) \rtimes \Lambda^*(F')$ (la construction du produit semi-
direct d'une algèbre de Lie par une représentation s'étend sans difficultés
au cas gradué). Supposons maintenant que F soit un espace symplectique
muni de la 2-forme $\sigma \in \Lambda^2(F)$ et posons $A_\sigma(F) = \{c \in A(F)/[c,\sigma] = 0\}$, $A_\sigma(F) \to A(F)$ est clairement une sous-algèbre de Lie graduée. On
explicite facilement la condition $[c,\sigma] = 0$: si $c \in A^1(F)$ alors $[c,\sigma] \in \Lambda^3(F^*)$ et on a

$$[c,\sigma](x_1, x_2, x_3) = \sum_{i,jk} \varepsilon_{ijk}\sigma(c(x_i, x_j), x_k)$$

si $[c,c] = 0$ alors σ est un 2 cocycle pour la cohomologie de (F,c) à coefficients scalaires. Un élément $c \in A_\sigma^1(F)$ est de carré nul si il définit sur
F une structure d'algèbre de Lie telle que la forme symplectique soit invariante (*). Ces structures ont été étudiées et classifiées par A. Médina et
P. Revoy ([**29**]).

II.f. L'algèbre de Schouten et ses généralisations

Dans cet exemple et ceux qui vont suivre nous considérons un espace
différentiel gradué et ses endomorphismes. La différentielle va jouer un rôle
essentiel dans la construction du crochet gradué, par une opération du type
"dérivée de Lie".

Soit V une variété différentiable et $(\Omega^*(V), d)$ son complexe de De
Rham. Soit $\Omega_*(V)$ l'espace des tenseurs antisymétriques contravariants
sur V. Le produit intérieur définit une opération:

$$\Omega_k(V) \xrightarrow{\ \ \ i\ \ \ } g\ell^{-k}(\Omega^*(V))$$

qui est un homomorphisme d'algèbres associatives graduées. On a ensuite
l'opération de dérivée de Lie.

$$\Omega_k(V) \xrightarrow{\ \ \ L\ \ \ } g\ell^{1-k}(\Omega^*(V))$$

avec $L_X = [d, i_X]$ le crochet gradué de $g\ell^*(\Omega^*(V))$ soit $L_X = d \circ i_X + (-1)^{k+1} i_X \circ d$. Le crochet de Schouten des tenseurs X et Y, noté $[\![\, X, Y\,]\!]$,
est alors défini de façon unique par la formule $L_{[\![\, X, Y\,]\!]} = [L_X, L_Y]$.

(*) Attention: ne pas confondre avec le $A_\omega(E)$ défini ci-dessus.

Ce crochet fait de $\Omega_*(V)$ une algèbre de Lie graduée avec un décalage d'indices: $X \in \Omega_p(V)$ est de degré $(p-1)$. Une autre définition de ce crochet est la suivante: soit N l'anneau des fonctions différentiables sur V, un p tenseur sur V induit alors un opérateur multidifférentiel sur N: $X \in \Omega_p(V)$ définit $\tilde{X} \in A^{p-1}(N)$ par la formule

$$\tilde{X}(f_1, \ldots, f_p) = <X, df_1 \wedge \ldots \wedge df_p> \ . \tag{$*$}$$

On a ainsi un plongement $\Omega_*(V) \to A^*(N)$ qui fait de $\Omega_*(V)$ une sous-algèbre de Lie graduée de l'algèbre de Richardson-Nijenhuis $A^*(N)$: un calcul explicite en terme de coordonnées permet de montrer que l'on a bien la même structure d'algèbre de Lie graduée. Ce point de vue a été développé dans [22]. Ici, on a encore une structure associative qui fait de $\Omega_*(V)$ une algèbre de Poisson graduée.

Un élément de carré nul dans $\Omega_2(V)$ est une structure de Poisson sur la variété V, la formule ($*$) définissant alors le crochet de Poisson sur l'anneau N. Cette utilisation du crochet de Schouten pour la géométrie des structures de Poisson est due à Tulczyjew [39] et à Lichnerowicz [27].

La cohomologie associée est alors la Λ-cohomologie de la structure de Poisson, qui avait été définie par Lichnerowicz [26] et Brylinski [6]. Dans le cas où la structure de Poisson est de rang maximal, on a une structure symplectique et la descente des indices induit un morphisme de complexes:

$$
\begin{array}{ccccc}
\longrightarrow & \Omega_p(V) & \xrightarrow{\partial_\wedge} & \Omega_{p+1}(V) & \longrightarrow \ldots \\
& \uparrow \Lambda^* & & \uparrow \Lambda^* & \\
\longrightarrow & \Omega^p(V) & \xrightarrow{d} & \Omega^{p+1}(V) & \longrightarrow \ldots
\end{array}
$$

et on a $H_\Lambda^p(\Omega_*(V)) = H_{dR}^{p+1}(V)$. On retrouve ainsi la cohomologie de De Rham de la variété V. Ce résultat est dû à Lichnerowicz [27].

Dans le cas général, le calcul de la Λ-cohomologie, important pour l'étude des déformations de la structure de Poisson, est un problème ouvert et difficile, analogue au calcul des cohomologies feuilletées. Voir par exemple les travaux de Dominique Melotte [30].

L'algèbre construite dans le paragraphe II.d peut se réaliser comme sous-algèbre graduée de celle de Schouten. Si E est de dimension finie n, avec la base $(e_i)i = 1, \ldots n$, on a un morphisme naturel:

$$\Lambda^p(E') \otimes S^*(E) \xrightarrow{\quad I \quad} \Omega_p(E)$$

défini par $I(c) = \displaystyle\sum_{1 \leq i_1, < \ldots < i_p \leq n} c(e_{i_1}, \ldots, e_{i_p})e_1 \wedge \ldots \wedge e_p$. L'image de I se compose alors des tenseurs à coefficients polynomiaux.

Nous pouvons maintenant donner quelques variantes et conséquences de la construction de Schouten ; la définition montre clairement la naturalité du crochet: si φ est un difféomorphisme de V, on a $[\varphi^*(X), \varphi^*(Y)] = \varphi^*([X,Y])$. Par conséquent si G est un groupe de Lie d'algèbre $\mathbb{G}$, les tenseurs invariants par les translations (à gauche ou à droite) forment une sous-algèbre de Lie graduée de $\Omega_*(G)$. Ces tenseurs s'identifient à l'algèbre extérieure de $\mathbb{G}$, notée $\Lambda^*(\mathbb{G})$. On peut décrire simplement cette structure d'algèbre de Lie graduée avec l'axiome d'algèbres de Poisson.

$$[X_\wedge Y, Z] = X_\wedge[Y,Z] + (-1)^{ZY}[X,Z]_\wedge Y$$

et $[X,Y]$ est le crochet dans $\mathbb{G}$ si X et Y sont des 1-tenseurs.

Nous appellerons *crochet de Schouten algébrique* le crochet de Lie gradué ainsi obtenu. (cf. Par exemple [10] et [20]).

Cette algèbre de Lie graduée est en relations avec celle définie au II.d. de la façon suivante: $\Lambda^1(E') \otimes S^1 E$ s'identifie à $\mathbb{G} = g\ell(E)$. L'identification $I_1 : I\mathbb{G} \to \Lambda^1(E') \otimes S^1 E$ va se prolonger multiplicativement en $I_p : \Lambda^p(\mathbb{G}) \to \Lambda^p(E') \otimes S^p(E)$. Explicitement:

$$I_p(X_1 \wedge \ldots \wedge X_p)(a_1, \ldots, a_p) = \sum_{\sigma \in \Sigma_p} \varepsilon(\sigma) X_1(a_{\sigma(1)}) \vee \ldots \vee X_p(a_{\sigma(p)})$$

pour $X_i \in \mathbb{G}$ et $a_i \in E$. I_* est donc un homomorphisme d'algèbres associatives qui est un homomorphisme d'algèbre de Lie en degré 0 ; c'est donc un homomorphisme d'algèbre de Lie graduées, car nous avons ici deux algèbres de Poisson. Ce point de vue est utile dans la classification des structures de Poisson quadratiques (cf. [33]).

Un élément $r \in \Lambda^2(\mathbb{G})$ de carré nul est une solution de l'équation de Yang-Baxter classique sur $\mathbb{G}$. Les applications des méthodes d'algèbres de Lie graduées aux équations de Yang Baxter seront développées dans le chapitre IV de ce travail.

Une généralisation intéressante du crochet de Schouten peut se construire à partir des (R, A) algèbres de Sridharan, étudiées par Hübschmann [17]. Si R est un anneau commutatif et A une R-algèbre commutative, une (R, A) algèbre est la donnée d'une algèbre de Lie L sur R avec une action $L \to Der(A)$ et d'une structure de A-module sur L, soit $A \times L \to L$ qui vérifient:

$$(a\alpha).b = a\alpha(b)$$

$$[\alpha, a\beta] = a[\alpha, \beta] + \alpha(a)\beta$$

Pour α et β dans L, a et b dans A.

Le lecteur averti aura reconnu l'axiomatisation du cas où $A = C^\infty(V)$ et $L = \mathfrak{a}(V)$ l'algèbre des champs de vecteurs tangents à la variété V. On sait associer à la $(R.A)$ algèbre de Lie L la généralisation du complexe de De Rham ; on note D_A l'espace des différentielles de Kähler, cet espace représente le foncteur dérivations: pour tout A-module M, on a $Hom_A(D_A, M) = Der(A, M)$. Le complexe de De Rham généralisé est alors l'algèbre des formes A-multilinéaires alternées sur D_A, c'est-à-dire $\Lambda(A) = (\Lambda_A(D_A), d)$.

Grâce à l'identification $L \to Der(A) = Hom_A(D_A, A) = \Lambda^1(A)$ on a un produit intérieur $L \xrightarrow{\ i\ } g\ell^{-1}(\Lambda(A))$ et de même pour les tenseurs antisymétriques $\Lambda_k^A(L) \xrightarrow{\ i\ } g\ell^{-k}(\Lambda(A))$. On a ainsi une généralisation canonique du crochet de Schouten sur $\Lambda_*^A(L)$, que l'on peut aussi réaliser comme restriction du crochet de Richardson-Nijenhuis sur A: on a une inclusion $\Lambda_*^A(L) \to A^*(A)$. Un p tenseur $\Lambda \in \Lambda_p^A(L)$ induit une application p-linéaire de A dans A, notée $\tilde{\Lambda}$, par la formule:

$$\tilde{\Lambda}(a_1, \ldots, a_R) = <\Lambda, da_1 \wedge \ldots \wedge da_p>$$

(on a $da_i \in D_A$ et $da_1 \wedge \ldots \wedge da_p \in \Lambda_p(D_A)$).

Un exemple géométrique de cette construction est fourni par les structures de Poisson; si V est une variété de Poisson avec le 2-tenseur $\Lambda \in \Omega_2(V)$, on peut construire une structure de (R, N) algèbre sur l'espace $L = \Omega^1(V)$ des 1 formes différentielles sur $V.L$ est naturellement un N-module, et la formule suivante définit sur L un crochet de Lie:

$$\{\alpha, \beta\} = L_{\Lambda \# \alpha}\beta - L_{\Lambda \# \beta}\alpha - d(<\Lambda, \alpha \wedge \beta>)$$

où $\Lambda^\# : \Omega^1(V) \to a(V)$ est défini par $\Lambda^\#(\alpha) = i(\alpha).\Lambda$
L opère sur N par dérivations grâce à la formule:

$$\alpha.f = <\Lambda, \alpha \wedge df>$$

La construction précédente donne alors une structure d'algèbre de Lie graduée sur $\Lambda_*^N(L) = \Omega^*(V)$ les formes différentielles sur V.

Cette structure a été définie et étudiée indépendamment par divers auteurs: De Wilde et Lecomte [22], J.L. Koszul [20], Coste, Dazord, Weinstein, Magri [8]. Notons que dans le cas symplectique, l'application $\Lambda^\#$ étant un isomorphisme, cette algèbre est isomorphe à l'algèbre de Schouten.

II.g. L'algèbre de Fröhlicher-Nijenhuis et ses généralisations

Cette algèbre apparait dans l'étude des dérivations du complexe de De Rham $(\Omega^*(V), d)$. Considérons l'espace des formes différentielles à valeurs vectorielles sur V et notons cet espace $\Omega^*(V, TV)$. Ces formes opèrent naturellement sur les formes scalaires:

$$\Omega^*(V, TV) \xrightarrow{\ i\ } Der(\Omega^*(V))$$

Sur les formes décomposables du type $\alpha \otimes X \in \Omega^*(V, TV)$ avec $\alpha \in \Omega^*(V)$ et $X \in \mathfrak{a}(V)$, on a:

$$i(\alpha \otimes X).\omega = \alpha_\wedge i(X).\omega.$$

Le crochet gradué ainsi obtenu n'est autre que le crochet de Richardson-Nijenhuis fibré au-dessus de V: on a en effet $\Omega^p(V, TV)_x = A^{p-1}(T_x V)$ pour tout $x \in V$. On peut utiliser comme précédemment la différentielle de De Rham pour définir une dérivée de Lie: si $\theta \in \Omega^*(V, TV)$ avec $a = \delta^\circ(\theta)$ on pose $L_\theta = [d, i_\theta] = d \circ i_\theta + (-1)^a i_\theta \circ d$.

Le calcul montre alors que le crochet dans $Der^*(\Omega^*(V))$ de L_θ et L_η est encore une dérivation du type L, ce qui définit le crochet de Fröhlicher-Nijenhuis par la formule:

$$L_{[\![\theta, \eta]\!]} = [L_\theta, L_\eta].$$

Pour les formes décomposées, on a la formule suivante, due à P. Michor:

$$[\![\alpha \otimes X, \beta \otimes Y]\!] = \alpha_\wedge \beta \otimes [X, Y] + \alpha_\wedge L_X \beta \otimes Y - L_Y \alpha_\wedge \beta \otimes X +$$
$$(-1)^{\delta^\circ \alpha}(d\alpha_\wedge i_X \beta \otimes Y + i_Y \alpha_\wedge d\beta \otimes X)$$

Le résultat de Fröhlicher et Nijenhuis [11] montre que l'algèbre des dérivations du complexe de De Rham s'identifie à deux copies de $\Omega^*(V, TV)$ munies des crochets de Richardson-Nijenhuis et de Fröhlicher-Nijenhuis respectivement.

Un élément de degré 1 de cette algèbre, $R \in \Omega^1(V, TV)$ s'interprète géométriquement comme un champ d'endomorphismes du fibré tangent à V ; son carré gradué $[\![R, R]\!] \in \Omega^2(V, TV)$ est connu sous le nom de Torsion de Nijenhuis. Si R est une structure presque complexe, alors $[\![R, R]\!]$ est l'obstruction à l'intégrabilité de cette structure (résultat de Newlander et Nirenberg). Voir par exemple [19] pour des développement récents de cette théorie.

De même que le crochet de Schouten, celui de Fröhlicher est naturel pour l'action des difféomorphismes de la variété: soit $[\![\varphi^*(\theta), \varphi^*(\eta)]\!] = \varphi^*([\![\theta, \eta]\!])$ si $\varphi \in Diff(V)$.

Par conséquent pour un groupe de Lie G d'algèbre $\mathbb{G}$, on a un crochet induit sur l'espace des formes différentielles invariantes à valeurs vectorielles, c'est-à-dire l'espace des formes multilinéaires alternées sur $\mathbb{G}$ à valeurs dans $\mathbb{G}$. Si $R \in \Lambda^1(\mathbb{G}, \mathbb{G})$ son carré gradué $[\![R, R]\!] \in \Lambda^2(\mathbb{G}, \mathbb{G})$ est défini par la formule: $2[\![R, R]\!](X, Y) = -(R \circ R)([X, Y]) - [R(X), R(Y)] + R([X, R(Y)]) + R([R(X), Y])$. Si $R \circ R = Id$, alors $[\![R, R]\!] = 0$ implique que:

$$-[RX, RY] + R([X, RY]) + R(RX, Y]) = [X, Y].$$

C'est l'équation de *Yang Baxter classique modifiée* (cf [35], [18]). Rappelons que cette formule implique que $[X, Y]_R = \frac{1}{2}([X, RY] + [RX, Y])$ est un autre crochet de Lie sur $\mathbb{G}$, et si l'on note $\mathbb{G}_R$ l'algèbre de Lie ainsi obtenue, on a des homomorphismes d'algèbres de Lie

$$(R \pm Id) : \mathbb{G}_R \to \mathbb{G}.$$

Ce point de vue sera développé dans le chapitre V. Remarquons que si l'on prend $R = Id$, on a trivialement $[\![\, Id, Id \,]\!] = 0$. Un calcul fastidieux mais direct, montre alors que: $\partial_{Id} : \Lambda^p(\mathbb{G}, \mathbb{G}) \to \Lambda^{p+1}(\mathbb{G}, \mathbb{G})$ est la différentielle du complexe de Chevalley-Eilenberg de $\mathbb{G}$ à valeurs dans la représentation adjointe et par conséquent:

$$H^p_{Id}(\Lambda^*(\mathbb{G}, \mathbb{G})) = H^p_{Lie}(\mathbb{G}, \mathbb{G}).$$

Le crochet de Fröhlicher-Nijenhuis se généralise au cas des (R, A) algèbres: si L est une (R, A) algèbre, l'analogue des formes à valeurs vectorielles est l'espace des applications A-multilinéaires alternées de L dans lui-même, noté $\Lambda^*_A(L, L)$. La construction précédente s'étend sans difficulté et la formule de P.Michor reste valable.

Les algèbres de Lie graduées de Schouten et de Fröhlicher-Nijenhuis contiennent comme sous-algèbre de degré 0, l'algèbre de Lie $\mathfrak{a}(V)$ des champs tangents à la variété ; elles constituent donc deux manières distinctes d'étendre $\mathfrak{a}(V)$ en une algèbre de Lie graduée. Le problème de construire une "grande" algèbre de Lie graduée englobant comme sous-algèbres celles de Schouten et de Fröhlicher-Nijenhuis a motivé un travail récent de A.M. Vinogradov ([40]) systématisant la construction de la dérivée de Lie dans un cadre très général. Ces résultats seront exposés dans le chapitre V.

III. Bigèbres de Lie et groupes quantiques

Les résultats de cette partie ont été obtenus en collaboration entre P. Lecomte et l'auteur.

III.a. Caractérisation des structures de bigèbre de Lie

E étant un espace vectoriel de dimension finie, nous allons étudier les objets de carré nul dans l'algèbre $\mathcal{P}^*_0(E)$.

On peut décrire le crochet de façon plus explicite en utilisant une base $(e_i)i = 1 \ldots n$ de E et sa base duale $(\varepsilon^i)i = 1 \ldots n$ de E'. On notera les tenseurs sur E par des lettres latines et les tenseurs sur E' par des lettres grecques.

Si $X \in \Lambda^*(E)$ et $\chi \in \Lambda^*(E')$ on pose

$$\mathcal{K}(X\chi) = \sum_{i=1}^{n} i(\varepsilon^i) X . i(e_i) \chi$$

On peut alors exprimer le crochet à l'aide de cet opérateur $\mathcal{K}$:

Si $\alpha \otimes A \in \Lambda^{p+1}(E') \otimes \Lambda^{q+1}(E), \beta \otimes B \in \Lambda^{r+1}(E') \otimes \Lambda^{s+1}(E)$, alors

$$\{\alpha \otimes A, \beta \otimes B\} = (-1)^{q+1}\alpha\mathcal{K}(A,\beta)B + (-1)^{(p+q)(r+s)+1}(-1)^{s+1}\beta\mathcal{K}(B\alpha)A.$$

Grâce à ses formules, on peut vérifier explicitement que:

$A^*(E) \hookrightarrow \mathcal{P}_0^*(E)$ et $A^*(E') \hookrightarrow \mathcal{P}_0^*(E)$ sont des inclusions d'algèbres de Lie graduées.

Un élément $\mathcal{C} \in \mathcal{P}_0^1(E)$ se décompose en $\mathcal{C} = c + \gamma$ avec $c \in \Lambda^2(E') \otimes E$ et $\gamma \in \Lambda^2(E) \otimes E' = A^1(E')$.

Le crochet $[\mathcal{C}, \mathcal{C}] \in \mathcal{P}_0^2(E)$ se décompose suivant la bigraduation en trois composantes:

$[c, c] \in \Lambda^3(E') \otimes E = A^2(E)$

$[\gamma, \gamma] \in \Lambda^3(E) \otimes E' = A^2(E')$

$[c, \gamma] \in \Lambda^2(E) \otimes \Lambda^2(E')$.

Par conséquent $[\mathcal{C}, \mathcal{C}] = 0$ implique $[c, c] = 0, [\gamma, \gamma] = 0, [c, \gamma] = 0$. Donc c définit une structure d'algèbre de Lie sur E, γ définit une structure d'algèbre de Lie sur E' ; la condition $[c, \gamma] = 0$ peut s'écrire $\partial_c(\gamma) = 0$ ou $\partial_\gamma(c) = 0$. Elle signifie que γ est un 1-cocycle pour la cohomologie de l'algèbre de Lie (E, c), ou que c est un 1-cocycle pour la cohomologie de l'algèbre de Lie (E', γ).

Nous avons donc obtenu la notion de bigèbre de Lie, due à Drinfeld [10] ; notre formalisme permet de voir immédiatement l'aspect auto-dual de cette notion: (E, c, γ) est une structure de bigèbre de Lie sur E, si et seulement si (E', γ, c) est une structure de bigèbre de Lie sur E'. Nous avons montré la:

Proposition: *La correspondance $(E, c, \gamma) \to \mathcal{C} = c + \gamma$ est une bijection entre les structures de bigèbre de Lie sur E et les éléments de carré nul dans $\mathcal{P}_0^1(E)$.*

On peut utiliser maintenant le morphisme d'algèbres de Lie graduées: $\Phi_\omega : \mathcal{P}_0^*(E) \to A^*(E \otimes E')$ construit au chapitre précédent. Si on pose $\Phi_\omega(\mathcal{C}) = \mathcal{C}'$, on a $\Phi_\omega([\mathcal{C}, \mathcal{C}]) = [\mathcal{C}, \mathcal{C}]$. Donc $[\mathcal{C}, \mathcal{C}] = 0$ le morphisme Φ_ω étant injectif. Par conséquent $\mathcal{C} = c + \gamma$ définit une structure de bigèbre sur E si et seulement si $\mathcal{C}' = \Phi_\omega(\mathcal{C})$ définit une structure d'algèbre de Lie sur $E \oplus E'$.

La structure d'algèbre de Lie obtenue sur $E \oplus E'$, est le double de la structure de bigèbre, définie par Drinfeld [10] ; on retrouve ici de façon très simple et rapide, la caractérisation des bigèbres de Lie en terme de leurs doubles, donnée dans [10] partie 3.

III.b. Cohomologie des bigèbres de Lie

Si $\mathcal{C}$ est une structure de bigèbre sur E, les algèbres de Lie (E, c), (E', γ) et $(E \otimes E', \Phi_\omega(\mathcal{C}))$ seront notées L, L' et $D(L)$ respectivement. Le complexe $(\mathcal{P}_0^*(E), \partial_\mathcal{C})$ s'identifie alors à celui défini par Drinfeld dans [**10**] §9. On peut donc écrire avec les notations de [**10**]: $Der^p(L) = H_\mathcal{C}^p(\mathcal{P}_0^*(E))$. Les déformations de la bigèbre L dépendent donc de $Der^1(L)$. et les obstructions à ces déformations sont donc dans $Der^2(L)$. D'autre part, la cohomologie graduée de $\mathcal{P}_0^*(E)$ contient les déformations universelles des structures de bigèbre sur E.

Avant de donner des exemples de calculs explicites de ces cohomologies, rappelons leur rôle dans la construction des *groupes quantiques* associés à la bigèbre.

Si L est une bigèbre de Lie, L' la structure d'algèbre de Lie duale, et $U(L)$ l'algèbre enveloppante universelle de L, on a:

Définition: Un groupe quantique associé à L est la donnée sur $U(L) \otimes_k k[[h]] = U(L)[[h]]$ d'une multiplication: $\mu_h : U(L) \otimes U(L) \to U_h(L)$ qui définit sur $U_h(L)$ une structure d'algèbre de Hopf non cocommutative telle que
(1) la structure induite sur $U_h(L) \otimes_{k[[h]]} k = U(L)$ est la structure d'algèbre enveloppante universelle sur $U(L)$.
(2) La comultiplication $\Delta_h : U(L) \to U_h(l) \otimes U_h(L)$ vérifie $\Delta_h(X) = \Delta(X) + h\gamma(X) + h\varepsilon(h)$ pour tous les $X \in L$, Δ étant la comultiplication de $U(L)$.

Rappelons ici, pour la commodité du lecteur, les axiomes d'algèbres de Hopf. Voir [**1**] pour les détails.

Une algèbre de Hopf est la donnée sur un k-espace vectoriel H d'une multiplication $\mu : H \otimes H \to H$, d'une comultiplication $\Delta : H \to H \otimes H$, d'une unité $\varepsilon : k \to H$, d'une co-unité $\eta : H \to k$ d'une application antipodale $S : H \to H$ telles que:
(1) μ est une multiplication associative d'unité ε
(2) Δ est coassociative de counité η
(3) μ et Δ sont compatibles en le sens suivant: le diagramme ci-dessous est commutatif:

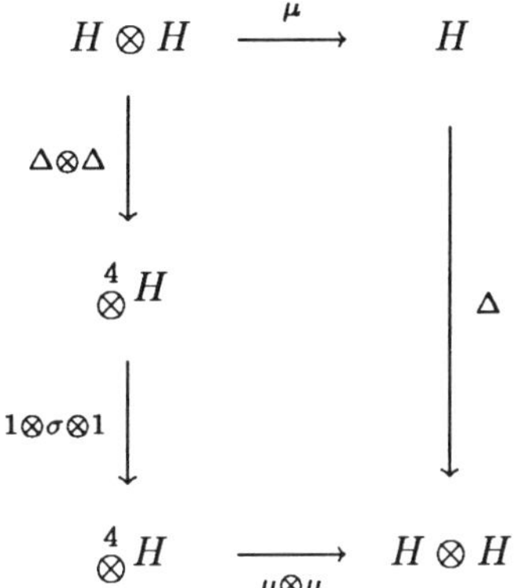

avec $\sigma(x \otimes y) = y \otimes x$

(4) S vérifie $\mu_0(S \otimes 1) \circ \delta = \mu \circ (1 \otimes S) \circ \delta = \varepsilon \circ \eta$.

On appellera quantification de la bigèbre L la donnée d'un groupe quantique associé à L.

D'après Drinfeld [10] les espaces de cohomologie $Der^i(L)$ pour $i = 1, 2$ interviennent dans la quantification de la façon suivante:

Supposons que l'on ait deux quantifications (μ_h, δ_h) et (μ'_h, δ'_h) telles que les multiplications coincident à l'ordre $\leq n$: $\delta_h = \delta'_h + h^{n+1}\delta_{n+1} + h^{n+1}\varepsilon(h)$. On peut maintenant écrire les axiomes d'algèbres de Hopf:

- L'équation d'associativité pour μ'_h nous donne, si on identifie le coefficient de h^n, que μ_n est déterminée par ses valeurs sur les éléments primitifs de $U(L)$, c'est-à-dire les éléments de l'algèbre de Lie L, et qu'elle est un cocycle pour la cohomologie de L, et donc $\mu_n \in Z^2(L, U(L))$.

- L'équation de coassociativité pour Δ'_h montre que si on identifie le coefficient de h^{n+1}, l'application Δ_{n+1} restreinte aux primitifs nous donne une application $\Delta_{n+1} : L \to L \wedge L$. Le coefficient de h^{n+2} montre ensuite que Δ_{n+1} est un cocycle pour la cohomologie de L soit $[\Delta_{n+1}, \gamma] = 0$.

- Enfin l'équation de compatibilité entre Δ_h et μ_h calculé à l'ordre n montre que

$$\Delta\mu_n = (\mu_n \otimes \mu + \mu \otimes \mu_n) \circ \tau \circ (\Delta \otimes \Delta)$$

Soit, pour X et Y dans $L \subset U(L)$

$$\Delta(\mu_n(X, Y)) = \mu_n(X, Y) \otimes 1 + 1 \otimes \mu_n(X, Y).$$

Par conséquent, μ_n est à valeurs dans les primitifs et $\mu_n : L \wedge L \to L$ est un 2-cocycle. On a donc $[\mu_n, c] = 0$.

La même équation à l'ordre $(n + 1)$ nous donne:

$$\Delta_{n+1}\mu + \Delta_1\mu_n =$$
$$(\mu_n \otimes \mu + \mu \otimes \mu_n) \circ \tau \circ (\Delta_1 \otimes \Delta + \Delta \otimes \Delta_1) +$$
$$(\mu \otimes \mu) \circ \tau \circ (\Delta_{n+1} \otimes \Delta + \Delta \otimes \Delta_{n+1})$$

C'est une égalité entre applications de $U(L) \otimes U(L)$ dans lui même. Si on calcule sur les primitifs, on voit que, sachant que $\Delta_1 = \gamma$.

$$(\mu_n \otimes \mu + \mu \otimes \mu_n) \circ \tau \circ (\Delta_1 \otimes \Delta + \Delta \otimes \Delta_1) - \Delta_1 \mu_n = [\mu_n, \gamma].$$

Et de même:

$$(\mu \otimes \mu) \circ \tau \circ (\Delta_{n+1} \otimes \Delta + \Delta \otimes \Delta_{n+1} - \Delta_{n+1} \circ \mu = [\Delta_{n+1}, c]$$

après antisymétrisation.

En résumé μ_n et δ_{n+1} sont déterminés par leurs valeurs sur les éléments de L, ils définissent $\mu_n : L \wedge L \to L$ et $\Delta_{n+1} : L \to L \wedge L$ et on a $[\mu_n, c] = 0, [\Delta_{n+1}, \gamma] = 0, [\mu_n, \gamma] + [\Delta_{n+1}, c] = 0$.

Soit $\partial_C(\mu_n + \Delta_{n+1}) = 0$ et la différence entre les deux quantifications nous donne un élément $[\mu_n + \Delta_{n+1}] \in Der^{(1)}(L)$.

Si cet élément est trivial en cohomologie on a $\tau : L \to L$ tel que $[\tau, c] = \mu_n$ et $[\tau, \gamma] = \Delta_{n+1}$; l'endomorphisme défini sur les primitifs par la formule $X \to X + h^n \tau(X)$ échange les structures (μ'_h, Δ'_h) et (μ_n, Δ_h) à l'ordre $(n+1)$.

On voit donc comment les éléments non triviaux de $Der^{(1)}(L)$ permettent de perturber terme à terme les éléments d'une quantification de la bigèbre de Lie L (voir [9]).

De même, les éléments de $Der^{(2)}(L)$ sont des obstructions à la construction par étapes d'une quantification. Pour simplifier les formules, nous décrivons seulement la première étape:

Soient μ_1 et Δ_2 tels que $\partial_C(\mu_1 + \Delta_2) = 0$ et posons $\tilde{\mu}_h = \mu + h\mu_1, \tilde{\Delta}_h = \Delta + h\Delta_1 + h^2\Delta_2$.

Cherchons maintenant des prolongements μ_2 et Δ_3 tels que si l'on pose $\mu_h = \tilde{\mu}_h + h^2\mu_2 \Delta_h = \tilde{\Delta}_h + h^3\Delta_3$, on ait μ_h associative à l'ordre 2, Δ_h associative à l'ordre 4, et la relation de compatibilité vérifiée à l'ordre 3.

Des calculs fastidieux mais élémentaires permettent alors de voir que l'équation d'associativité implique que μ_2 est déterminée par ses valeurs sur les primitifs. Compte tenu de $[\gamma, \mu_1] + [c, \Delta_2] = 0$ l'équation de compatibilité à l'ordre 2 montre que $\mu_2 : L \wedge L \to L$ et l'identification du terme en h^2 dans l'équation d'associativité montre enfin que $2[c, \mu_2] = [\mu_1, \mu_1]$ (α).

Pour l'équation de coassociativité, on peut de même se ramener à un calcul sur les primitifs: l'associativité à l'ordre 3 implique que, compte tenu de $[\gamma, \delta_2] = 0$ on a $\delta_3 : L \to L \wedge L$ et l'associativité à l'ordre 4 implique:
$2[\gamma, \Delta_3] = [\Delta_2, \Delta_2]$ (β).

On peut enfin expliciter l'équation de compatibilité à l'ordre 3 toujours en calculant sur les primitifs, et on obtient:

$$[\gamma, \mu_2] + [c, \Delta_3] = [\mu_1, \Delta_2] \qquad\qquad (\gamma)$$

Ces trois équations (α, β, γ) peuvent s'écrire:

$$\partial_C[\mu_2 + \Delta_3) = \frac{1}{2}[\mu_1 + \Delta_2, \mu_1 + \Delta_2]$$

$\mu_2 + \Delta_2$ étant dans $Der^{(1)}(L)$ on voit que la classe de cohomologie de $[\mu_1 + \Delta_2, \mu_1 + \Delta_2]$ dans $Der^{(2)}(L)$ est une obstruction à l'existence de $\mu_2 + \Delta_3$, et donc à la construction de l'étape suivante de la quantification.

L'analyse complète des quantifications d'une bigèbre de Lie L nécessite donc le calcul de $Der^{(i)}(L)$, $i = 1, 2$ et de l'application

$$Der^{(1)}(L) \to Der^{(2)}(L)$$
$$c \to [c, c]$$

III.c. Calculs pour le cas semi-simple

Si l'on considère la bigraduation de $\mathcal{P}_0(E)$, l'opérateur $\partial_{\mathcal{C}}$ fait de $\mathcal{P}_0(E)$ un complexe bigradué, et se décompose suivant $\partial_{\mathcal{C}} = \partial_c + \partial_\gamma$ avec ∂_c le bidegré $(1,0)$ et ∂_γ de bidegré $(0,1)$. On peut ainsi ramener le calcul de $Der^*(L)$ à des calculs cohomologiques pour L et L' respectivement.

En degré 0, la cohomologie se calcule facilement et on trouve:

$$\mathcal{P}_0^0(E) = E' \otimes E = Hom(E, E) = Hom(E', E')$$

$$\mathcal{P}_0^1(E) = E' \wedge E' \otimes E + E' \otimes E \wedge E.$$

Pour $\ell \in Hom(E, E), \partial_c \ell = \partial_c \ell + \partial_\gamma \ell$ avec $\partial_c \ell \in E' \wedge E' \otimes E$ et $\partial_\gamma \ell \in E' \otimes E \wedge E$. Donc $\partial_c \ell = 0$ si et seulement si $\partial_c \ell = 0$ et $\partial_\gamma \ell = 0$. On retrouve ainsi le résultat de Drinfeld [18].

Proposition: $Der^0(L)$ *est l'espace des applications linéaires de* E *dans* E *qui sont des dérivations de* L *et de* L'.

Pour le calcul des autres espaces nous supposerons désormais que $L = (E, c)$ est une structure d'algèbre de Lie semi-simple de dimension finie.

Considérons la suite spectrale associée à la seconde filtration sur $\mathcal{P}_0(E)$: on a alors

$$E_0^{q,p} = \Lambda^{p+1}(E^*) \otimes \Lambda^{q+1}(E) \text{ et } d_0^{q,p} = \partial_c : E_0^{q,p} \to E_0^{q,p+1}.$$

Le complexe $(E_0^{q,p}, d_0^{q,p})$ s'identifie donc au complexe de Chevalley-Eilenberg de l'algèbre de Lie L à coefficients dans les $(q + 1)$ tenseurs alternés sur L, décalé d'une unité:
donc $E_1^{q,0} = H^{p+1}(L, \Lambda^{q+1}(L))$ si $p > 0$
$\qquad E_1^{q,0} = Z^1(L, \Lambda^{q+1}(L)).$

Le théorème bien connu, de Whitehead, sur les algèbres de Lie semi-simples nous donne alors, si $p > 0$:

$$E_0^{q,p} = H^{p+1}(L; k) \otimes Inv_L(\Lambda^{q+1}(L)) = Inv_L\Lambda^{p+1}(L') \otimes Inv_L\Lambda^{q+1}(L).$$

Etudions maintenant les termes en $E_2^{q,0}$: on doit calculer la cohomologie du complexe ci-dessous:

$$Z^1(L, L) \xrightarrow{\ \partial_\gamma\ } Z^1(L, \Lambda^2 L) \to \ldots$$

On a pour toute algèbre de Lie semi simple L et toute représentation M : $H^1(L, M) = 0$ et $H^0(L, M) = Inv_L(M)$; soit pour chaque q une suite:

$$0 \to Inv_L\Lambda^{q+1}(L) \to \Lambda^{q+1}(L) \xrightarrow{\ \partial_c\ } Z^1(L, \Lambda^{q+1}(L)) \to 0$$

et par conséquent une suite exacte de complexes:

$$
\begin{array}{ccccccc}
0 & & 0 & & 0 & & \\
\uparrow & & \uparrow & & \uparrow & & \\
Z^1(L,L) & \xrightarrow{\partial_\gamma} & Z^1(L,\Lambda^2 L) & \cdots \longrightarrow & Z^1(L,\Lambda^{q+1}(L)) & \longrightarrow & \cdots \\
\uparrow {\scriptstyle \partial_c} & & \uparrow {\scriptstyle \partial_c} & & \uparrow {\scriptstyle \partial_c} & & \\
L & \xrightarrow{\partial_\gamma} & \Lambda^2 L & \cdots \longrightarrow & \Lambda^{q+1}(L) & \longrightarrow & \cdots \\
\uparrow & & \uparrow & & \uparrow & & \\
Inv_L(L) & \longrightarrow & Inv_L(\Lambda^2 L) & \cdots \longrightarrow & Inv_L\Lambda^{q+1}(L) & \longrightarrow & \cdots \\
\uparrow & & \uparrow & & \uparrow & & \\
0 & & 0 & & 0 & &
\end{array}
$$

La différentielle du dernier complexe est clairement triviale. Notons Λ_+^* le complexe du milieu: si on identifie l'algèbre extérieure sur L aux cochaînes multilinéaires alternées sur L', ce complexe est celui qui donne la cohomologie de L' à coefficients scalaires, tronquée en degré zéro.

D'où $H^p(\Lambda_+^*(L)) = H^p(L';k)$ si $p > 1$

$H^1(\Lambda_+^*(L)) = Z^1(L';k)$.

On peut maintenant étudier la suite exacte longue associée à cette suite exacte courte de complexes. Notons $I_p : Inv_L\Lambda^p(L) \to H^p(L';k)$, l'application induite par l'inclusion ; la suite exacte longue s'écrit:

$$\xrightarrow{I_p} H^p(\Lambda_+^*(L)) \to E_2^{p-1,0} \to Inv_L\Lambda^{p+1} \xrightarrow{I_{p+1}} H^{p+1}(\Lambda_+^*(L)) \to E_2^{p,0} \to \dots$$

On en déduit $E_2^{p,0} = Ker I_{p+2} \oplus Coker I_{p+1}$.

Dans les bas degrés, on trouve: $Der^0(L) = E_2^{0,0} = Z^1(L';k)$.

Sachant qu'un élément x de $Z^1(L';k)$ détermine un élément de $Der^0(L)$ représenté par $\partial_c(x) : L \to L$, on retrouve bien le calcul de $Der^0(L)$ fait précédemment.

On a $Inv_L\Lambda^1(L) = 0$, $Inv_L\Lambda^2(L) = 0$, $Inv_L\Lambda^3(L) = k$.

D'où la suite exacte courte:

$$0 \to H^2(L';k) \to E_2^{1,0} \to Inv_L\Lambda^3(L) \xrightarrow{I_3} H^3(L';k) \to E_2^{2,0} \to 0$$

On en déduit:
$$Der^{(1)}(L) = H^2(L';k) \otimes Ker I_3$$
$$Der^{(2)}(L) = Coker I_3.$$

Le calcul complet de $Der^*(L)$ repose donc sur le calcul de la cohomologie $H^*(L';k)$ et de l'identification des invariants de L dans celle-ci. Nous avons ainsi montré le:

Théorème: (c'est la prp. 3.2. de [**23**]) *La cohomologie $Der^*(L)$ de la bigèbre de Lie L est l'aboutissement d'une suite spectrale dégénérant au terme E_2 et vérifiant:*

$$E_2^{p,q} = Inv_L\Lambda_{p+1}(L') \otimes Inv_L\Lambda^{q+1}(L) \text{ si } p > 0$$

$$E_2^{0,q} = Ker I_{q+2} \otimes Coker I_{q+1}$$

où I_q désigne l'application naturelle

$$I_q : Inv_L\Lambda^q(L) \to H^q(L';k).$$

Nous allons maintenant expliciter le crochet gradué induit par $\mathcal{P}_0(E)$ sur sa cohomologie $Der^*(L)$ pour les termes du type $E_2^{0,q}$. Des cochaînes

$X \in \Lambda^{a+1}(L)$ et $Y \in \Lambda^{b+1}(L)$ déterminent des classes de $Der^a(L)$ et $Der^b(L)$ représentées respectivement par $\partial_c(X) \in Z^1(L, \Lambda^a(L))$ et $\partial_c(Y) \in Z^1(L, \Lambda^b(L))$. On peut alors montrer (cf. partie (4)) que l'on a:

$$[\partial_c(X), \partial_c(Y)] = \partial_c([\![\,X, Y\,]\!])$$

où

$$[\![\,X, Y\,]\!] \in \Lambda^{a+b+1}(L)$$

désigne le crochet de Schouten de X et de Y. Le crochet gradué des éléments de $Der^*(L)$ dépend donc du crochet de schouten sur $\Lambda^*(L)$.

Nous pouvons maintenant détailler davantage le calcul de $Der^i(L)$ pour $i = 1, 2$. De la relation $[c, \gamma] = 0$ on déduit que le crochet γ ou comme application de L dans $L \wedge L$ est un 1-cocycle de L. La cohomologie de L étant triviale, γ est un cobord (on dit que l'on a une bigèbre de Lie cobord [10]) donc il existe $u \in L \wedge L$ tel que $\gamma = \partial_c(u) = [c, u]$. On déduit alors de $[\gamma, \gamma] = 0$ que le crochet de Schouten $[\![\,u, u\,]\!]$ est un invariant (c'est l'équation de Yang-Baxter généralisée. cf. partie IV). On voit également que $\partial_\gamma(u) = [\![\,u, u\,]\!]$ dans $\Lambda^2(L)$.

Si $[\![\,u, u\,]\!] = 0$ alors $\partial_\gamma(u) = 0$ dans $\Lambda^3(L)$ et u détermine alors une classe de cohomologie de $H^2(L'; k)$ donc un élément de $Der^{(1)}(L)$.

Si $[\![\,u, u\,]\!] \in Inv_L\Lambda^3(L)$ alors $[\![\,u, u\,]\!] = \partial_\gamma(u)$ est nul dans $H^3(L'; k)$ et détermine donc un élément de $Ker I_3$.

Dans tous les cas, u détermine un élément non trivial de $Der^{(1)}(L)$ représenté par $\gamma = \partial_c(u)$. Le crochet gradué de cet élément avec lui-même dans $Der^3(L)$ est évidemment nul ; nous allons donner maintenant des exemples de calculs explicites.

(a) $\mathbf{L} = s\ell(\mathbf{2}, \mathbb{R})$

Considérons la base (X, Y, H) avec les crochets
$[X, Y] = H, [H, X] = 2X, [H, Y] = -2Y$.
Si on prend $u = aX \wedge Y + bX \wedge H + cY \wedge H$, le crochet de Schouten vérifie $[\![\,u, u\,]\!] = 2(a^2 - 4bc)X \wedge Y \wedge H$. Si on note x, y, h la base duale de (X, Y, H), le crochet de Lie sur L' s'écrit alors:

$$\begin{cases} \{x, h\} &=& 2bh & + & ax \\ \{y, h\} &=& -2ch & + & ay \\ \{x, y\} &=& 2cx & + & 2by \end{cases}$$

On voit facilement que cette algèbre de Lie est résoluble, et que son algèbre dérivée est engendrée par $ax + 2bh$ et $ay - 2ch$. On en déduit que $H^1(L'; \mathbb{R}) = \mathbb{R}$ et $H^2(L'; \mathbb{R}) = H^3(L'; \mathbb{R}) = 0$; d'où $Der^{(0)}(L) = \mathbb{R}$ et $I_3 : Inv_L\Lambda^3(L) \rightarrow H^3(L')$ vérifier $Ker I_3 = Inv_L\Lambda^3(L) = \mathbb{R}$. Donc

$Der^{(1)}(L) = \mathbb{R}$ et on peut prendre l'éléement $u \in \Lambda^2(L)$ comme représentant de la classe de cohomologie. On a également $Der^{(2)}(L) = Der^{(3)}(L) = 0$ et $Der^{(5)}(L) = Inv_L\Lambda^3(L) \otimes Inv_L\Lambda^3(L^1) = \mathbb{R}$. L'espace $Der^{(0)}(L)$ est engendré par la dérivation intérieure donnée par le crochet par $aH - 2bX + 2cY$.

On peut alors vérifier facilement que l'algèbre de Lie graduée $Der^{(*)}(L)$ est abélienne. On déduit donc de $Der^{(1)}(L) = \mathbb{R}$ et $Der^{(2)}(L) = 0$ qu'il existe une et une seule quantification de la bigèbre L (à isomorphisme près). Pour $a = 1$ et $b = c = 0$ on obtient le groupe quantique $U_h(sl(2))$ de Drinfeld [**10**].

(b) Algèbres simples de dimension finie, calcul sur un exemple

Soit L une algèbre simple de dimension finie, A un système de racines et $(X_\alpha, X_{-\alpha}, H_\alpha)_{\alpha \in A}$ une base de Cartan.

Posons $R = \sum_{\alpha \in A} X_\alpha \wedge X_{-\alpha}$. On peut montrer que $[R, R]$ est un multiple de l'invariant canonique de $Inv_L\Lambda^3(L)$. Donc $\gamma = \partial_c(R)$ détermine une structure de bigèbre de Lie sur L. En considérant pour chaque racine positive α, l'ensemble des racines β telles que $\alpha + \beta$ soit une racine, on montre que: $\gamma(X_\alpha) = X_\alpha \wedge H_\alpha, \gamma(X_{-\alpha}) = X_{-\alpha} \wedge H_{-\alpha}$ et γ est nulle sur l'algèbre de Cartan. Dans la base duale $(x_\alpha, x_\alpha, h_\alpha)$ le crochet de L' va s'écrire $\{x_\alpha, h_\alpha\} = x_\alpha, \{x_{-\alpha}, h_{-\alpha}\} = x_{-\alpha}$, $\alpha \in A$, tous les autres crochets étant triviaux.

Notons J l'algèbre des commutateurs de L' ; cette algèbre est engendrée par les $x_\alpha, x_{-\alpha}$ et est abélienne, on a donc la suite exacte $0 \to J \to L' \xrightarrow{\pi} I \to 0$ où I et J sont abéliennes. Remarquons que J s'identifie à l'annulateur de la sous-algèbre de Cartan $H \to L$ et on peut donc identifier I au dual de H.

La suite spectrale de Hochschild-Serre de l'extension ci-dessus permet alors de calculer la cohomologie de L' à coefficients scalaires. I et J étant abéliennes, $E_2^{p,q} = H^p(I, H^q(J)) = H^p(I, \Lambda^q(J'))$. Pour $q \neq 0$, l'algèbre abélienne I opère non trivialement sur $H^p(I, \Lambda^q(J'))$ alors qu'elle opère trivialement sur $H^*(L'))$. Les classes de $E_2^{p,q}$ pour $q \neq 0$ ne donnent pas de classes de cohomologie dans L' d'où $H^*(I) = \Lambda^*(I') \xrightarrow{\pi^*} H^*(L')$ est un isomorphisme. On peut donc identifier $H^*(L')$ à l'algèbre extérieure de l'algèbre de Cartan $\Lambda^*(H)$.

Il est alors clair que $I_p : Inv_L\Lambda^p(L) \to H^p(L') = \Lambda^p(H)$ est l'application nulle et donc:

Proposition: *Soit L une algèbre de Lie semi-simple de sous-algèbre de Cartan $H \subset L$ munie de la structure de bigèbre définie par*

$$R = \sum_{\alpha \in A} X_\alpha \wedge X_{-\alpha} \text{ on a alors}$$

$$Der^{(p+1)}(L) = Inv_L \Lambda^{p+1} L \oplus \Lambda^p(H)$$

On voit de plus que l'algèbre graduée $Der^*(L)$ est abélienne: le crochet gradué sur les générateurs provenant de $\Lambda^*(H)$ et $Inv_L \Lambda^*(L)$ est induit par le crochet de Schouten de $\Lambda^*(L)$ comme nous l'avons vu précédemment ; ce crochet est donc clairement nul. La théorie d'obstruction à la quantification décrite plus haut est donc complètement triviale dans ce cas, et par conséquent **toutes les classes de** $Der^{(1)}(L) = \Lambda^2(H) \oplus \mathbb{R}$ (ici $\mathbb{R}$ représente le générateur canonique de $Inv_L \Lambda^3(L)$) **peuvent être utilisées pour construire des quantifications de la bigèbre.**

Donc si $\Lambda^2(H) \neq 0$, c'est-à-dire si $L \neq SL(2, \mathbb{R})$ la dimension de $Der^{(1)}(L)$ est plus grande que un et il n'y a pas de quantification "canonique". Notamment le groupe quantique $U_h(SL(n))$ de Drinfeld n'est pas la seule quantification possible pour la bigèbre associée à $SL(n)$. Rappelons que ce groupe quantique est défini par les formules:

$$\Delta_h(X_{\pm\alpha}) = X_{\pm\alpha} \otimes \exp(\frac{hH_\alpha}{4}) + \exp(-\frac{hH_\alpha}{4}) \otimes X_{\pm\alpha}$$
$$\Delta_h(H_\alpha) = H_\alpha \otimes 1 + 1 \otimes H_\alpha \text{ (cf. [10] p. 807 ex. 6.2).}$$

Un terme de $\Lambda^2(H)$ peut servir à modifier la comultiplication à chaque étape de façon à obtenir une quantification non équivalente à la précédente. Ceci contredit une conjecture de L. Takhtadjan [38], formulée dans le preprint "Quantum groups and integrable models": "... there should be a cohomology describing these deformations and having the property that its H^2 group is one dimensional".

Remarquons cependant que Drinfeld dans [10] donne une condition supplémentaire pour assurer l'unicité de cette quantification: le groupe quantique $U_h(SL(n))$ donne une quantification unique parmi celles qui respectent la sous algèbre de Cartan.

Notons enfin que la quantification Δ_h est donnée par une suite de termes homogènes pour le degré de h. On obtient en développant:

$$\Delta_h(X_\alpha) = \Delta(X_\alpha) + \sum_{n=1}^{\infty} h^n \Delta_n(X_\alpha)$$

dans

$$\Delta_n(X_\alpha) = X_\alpha \otimes \frac{H_\alpha^n}{n!} + (-1)^n \frac{H_\alpha^n}{n!} \otimes X_\alpha.$$

L'introduction d'une éventuelle perturbation au degré $n(n \geq 2)$ ajouterait à Δ_n un terme de degré $< n$ en les X_i, H_i. On peut donc considérer qu'en un sens approprié, le terme symbolique (si on pense aux éléments de $U_h(SL(n))$ comme des opérateurs différentiels) est défini de façon unique.

Ce phénomène devrait s'interpréter dans le cadre du travail de R.Berger ([5]): à un groupe quantique $U_h(\mathbb{G})$ quantifiant l'algèbre enveloppante $U(\mathbb{G})$, il associe un diagramme d'algèbres de Hopf filtrées

$$
\begin{array}{ccc}
U_h(\mathbb{G}) & \longrightarrow & S_h(\mathbb{G}) \\
\downarrow & & \downarrow \\
U(\mathbb{G}) & \longrightarrow & S(\mathbb{G})
\end{array}
$$

Chaque flèche horizontale est la projection d'une algèbre filtrée sur son algèbre graduée associée ; l'algèbre $U(\mathbb{G})$ est cocommutative et non commutative. Les remarques précédentes montrent alors que la quantification est bien définie de façon canonique au niveau de $S_h(\mathbb{G})$.

(c) Calculs dans le cas où L est l'algèbre de Heisenberg

Soit $L = H$ l'algèbre de Heisenberg en dimension 3 engendrée par les générateurs X, Y, Z avec comme crochet non trivial $[X, Y] = Z$. Les structures de bigèbres de Lie associées à H ont été classifiées indépendamment par M. Cahen et C. Ohn [7] et Sczymczak et Zakrzewski [37] d'autre part. Nous appliquerons ici les calculs cohomologiques à l'étude des déformations des bigèbres associées à H, vers les bigèbres de $SL(2)$. Il existe une déformation bien connue de H vers $SL(2)$; notre but est d'essayer de déformer simultanément les structures de bigèbre. Nous allons établir le résultat suivant:

Théorème: *Il existe à isomorphisme près, une seule structure de bigèbre de Lie sur H qui se déforme vers une structure de bigèbre de $SL(2)$.*

Dans [7] M. Cahen et C. Ohn donnent une classification complète des structures de bigèbre sur H et décrivent la géométrie des structures de Poisson associées sur le groupe de Heisenberg. Nous allons donner brièvement leurs résultats:

L'application $\gamma : H \to \Lambda^2(H)$ est représentée par une matrice $U \in M_3(\mathbb{R})$ dans les bases (X, Y, Z) et $(Y \wedge Z, Z \wedge X, X \wedge Y)$ respectivement. On exprime ensuite les conditions $[\gamma, \gamma] = 0, [\gamma, c] = 0$ et on effectue la classification modulo les automorphismes de l'algèbre H.

La condition de cocycle $\gamma(Z) = X.\gamma(Y) - Y.\gamma(X)$ implique que

$$
U = \begin{pmatrix} & & v_1 \\ & T & v_2 \\ -v_1 & -v_2 & 0 \end{pmatrix}
$$

avec $T \in M_2(\mathbb{R})$. La condition de Jacobi donne alors

$$(t_{12} + t_{21})v_1 - 2t_{11}v_2 = 0 \qquad - 2t_{22}v_1 + ('t_{12} + t_{21})v_2 = 0.$$

Les automorphismes de H sont déterminés par $A \in GL(2, \mathbb{R})$ et $a \in \mathbb{R}^2$. Ils opèrent suivant:

$$\begin{bmatrix} X \\ Y \\ Z \end{bmatrix} \longrightarrow \begin{bmatrix} A & a \\ 0 & dtA \end{bmatrix} \begin{bmatrix} X \\ Y \\ Z \end{bmatrix}$$

Modulo ces automorphismes, on peut se ramener si $v \neq 0$ à $v = (1,0)$ et finalement à

$$U = \begin{bmatrix} \lambda & 0 & 1 \\ 0 & 0 & 0 \\ -1 & 0 & 0 \end{bmatrix}$$

Si $v = 0$. la structure est déterminée entièrement par T matrice d'une forme bilinéaire (elle se transforme suivant $T \to AT^t A$) que l'on peut mettre sous forme canonique, soit

$$\begin{bmatrix} \varepsilon & \lambda \\ -\lambda & \eta \end{bmatrix} \begin{bmatrix} \varepsilon & 1 \\ -1 & 0 \end{bmatrix} \begin{bmatrix} \varepsilon & 0 \\ 0 & 0 \end{bmatrix} \begin{bmatrix} 0 & 1 \\ -1 & 0 \end{bmatrix}$$

avec ε et η égaux à ± 1.

On obtient finalement cinq structures de bigèbres distinctes modulo automorphismes que nous noterons en utilisant la base duale (x, y, z).

Type I: $x \otimes (\lambda Y \wedge Z + X \wedge Y) + z \otimes (-Y \wedge Z)$
Type II: $x \otimes (\varepsilon Y \wedge Z - \lambda X \wedge Z) = y \otimes (\eta X \wedge Z - \lambda Y \wedge Z)$
Type III: $x \otimes (\varepsilon Y \wedge Z - X \wedge Z) + y \otimes -Y \wedge Z$
Type IV: $x \otimes \varepsilon Y \wedge Z$
Type V: $x \otimes -X \wedge Z + y \otimes -Y \wedge Z.$

Le type II doit être subdivisé suivant les signes de ε et η si on s'intéresse au type géométrique des feuilles. Le type V est une bigèbre de Lie cobord.

Nous allons maintenant calculer la cohomologie $Der^*(H)$ suivant les divers types. On doit calculer la cohomologie du bicomplexe suivant:

$$
\begin{array}{ccccc}
\Lambda^1(H,H) & \xrightarrow{\ \partial_c\ } & \Lambda^2(H,H) & \xrightarrow{\ \partial_c\ } & \Lambda^3(H,H) \\
\ \downarrow{\scriptstyle\partial_\gamma} & & \ \downarrow{\scriptstyle\partial_\gamma} & & \ \downarrow{\scriptstyle\partial_\gamma} \\
\Lambda^1(H,\Lambda^2(H)) & \xrightarrow{\ \partial_c\ } & \Lambda^2(H,\Lambda^2(H)) & \xrightarrow{\ \partial_c\ } & \Lambda^3(H,\Lambda^2(H)) \\
\ \downarrow{\scriptstyle\partial_\gamma} & & \ \downarrow{\scriptstyle\partial_\gamma} & & \ \downarrow{\scriptstyle\partial_\gamma} \\
\Lambda^1(H,\Lambda^3(H)) & \xrightarrow{\ \partial_c\ } & \Lambda^2(H,\Lambda^3(H)) & \xrightarrow{\ \partial_c\ } & \Lambda^3(H,\Lambda^3(H))
\end{array}
$$

En calculant d'abord la cohomologie par rapport à la différentielle ∂_c on obtient:

$$
\begin{array}{ccc}
Z^1(H,H) & H^2(H,H) & H^3(H,H) \\
\ \downarrow{\scriptstyle\partial_\gamma} & \ \downarrow{\scriptstyle\partial_\gamma} & \ \downarrow{\scriptstyle\partial_\gamma} \\
Z^1(H,\Lambda^2(H)) & H^2(H,\Lambda^2(H)) & H^3(H,\Lambda^2(H)) \\
\ \downarrow{\scriptstyle\partial_\gamma} & \ \downarrow{\scriptstyle\partial_\gamma} & \ \downarrow{\scriptstyle\partial_\gamma} \\
Z^1(H,\Lambda^3(H)) & H^2(H,\Lambda^3(H)) & H^3(H,\Lambda^3(H))
\end{array}
$$

$H^2(H,\Lambda^2H)$ est engendré par
$$
\begin{bmatrix}
x\wedge z\otimes AX\wedge Z & + & BY\wedge Z \\
y\wedge z\otimes CX\wedge Z & + & DY\wedge Z \\
x\wedge y\otimes \varepsilon X\wedge Y &&
\end{bmatrix}
$$

$H^2(H,H)$ est engendré par
$$
\begin{bmatrix}
x\wedge y\otimes aX & + & bY \\
x\wedge z\otimes \varphi X & + & \psi Y - bZ \\
y\wedge z\otimes \xi X & - & \varphi Y + aZ
\end{bmatrix}
$$

Nous pouvons maintenant calculer $\partial_\gamma : H^2(H,H) \to H^2(H,\Lambda^2 H)$.

$H^2(H,\Lambda^2H)$ est engendré par
$$
\begin{bmatrix}
x\wedge z\otimes AX\wedge Z & + & BY\wedge Z \\
y\wedge z\otimes CX\wedge Z & + & DY\wedge Z \\
x\wedge y\otimes \varepsilon X\wedge Y &&
\end{bmatrix}
$$

Type I: pour $\tilde{c} \in H^2(H, H)$ comme ci-dessus, on a:

$$\partial_\gamma(\tilde{c}) = \begin{bmatrix} x \wedge y \otimes \alpha\lambda Z \wedge Y & + & \lambda\zeta X \wedge Y \\ x \wedge z \otimes \lambda\zeta Z \wedge X & - & \zeta\lambda\varphi Z \wedge Y \\ y \wedge z \otimes -\lambda\zeta Z \wedge Y \end{bmatrix}$$

donc $\partial_\gamma(\tilde{c}) = 0$ en cohomologie si et seulement si $\zeta = \varphi = 0$. Dans ce cas $E_2^{0,1} = Ker\partial_\gamma$ est engendré par

$$\begin{bmatrix} x \wedge y \otimes aX & + & bY \\ x \wedge z \otimes \psi Y & - & bZ \\ y \wedge z \otimes aZ \end{bmatrix}$$

Type II à V: $\partial_\gamma(\tilde{c})$ est donné par:

$$\begin{bmatrix} y \wedge z \otimes 2\eta\varphi Z \wedge X + (\eta\psi + \varepsilon\zeta)Z \wedge Y \\ w \wedge z \otimes -(\varepsilon\zeta + \eta\psi)Z \wedge X + 2\varepsilon\varphi Z \wedge Y \\ x \wedge y \otimes (\lambda - \eta b)Z \wedge X + (b\lambda + a\varepsilon)Z \wedge Y + (\varepsilon\zeta - \eta\psi)X \wedge Y \end{bmatrix}$$

$\partial_\gamma(\tilde{c}) = 0$ en cohomologie implique donc que:

$$\begin{bmatrix} \eta\varphi = 0 & \varepsilon\varphi = 0 \\ \varepsilon\zeta = 0 & \eta\psi = 0 \end{bmatrix}$$

Pour le type II on doit donc avoir $\varphi = \psi = \zeta = 0$ donc $E_2^{0,1} = Ker\partial_\gamma$ est engendré par:

$$\begin{bmatrix} x \wedge y \otimes aX & + & bY \\ x \wedge z \otimes & - & bZ \\ y \wedge z \otimes aZ \end{bmatrix}$$

Pour les type III et IV il suffit d'avoir $\varphi = \zeta = 0$ et donc:

$$E_2^{0,1} = Ker\partial_\gamma \text{ est engendré par } \begin{bmatrix} x \wedge y \otimes aX & + & bY \\ x \wedge z \otimes \psi Y & - & bZ \\ y \wedge z \otimes aZ \end{bmatrix}$$

Par conséquent, pour les types de I à IV la déformation de H vers $SL(2)$ engendrée par le cocycle $c = x \wedge z \otimes X - y \wedge z \otimes Y$ vérifie $[c, \gamma] \neq 0$ en cohomologie et donc la déformation ne s'étend pas en une déformation de bigèbre. Calculons maintenant en détails le cas du type V.

Pour le type V on a $\partial_\gamma = 0$ et donc $E_2^{0,1} = H^2(H, H)$. Le calcul montre en effet que $[c, \gamma]$ est donné par $x \wedge y \otimes \alpha Z \wedge X + \beta Z \wedge Y$ qui est nul en cohomologie. Pour les autres termes de la cohomologie, on calcule aisément que:

$Z^1(H, H)$ est engendré par les termes du type$M = \begin{bmatrix} x \otimes aX + \alpha Y + AZ \\ y \otimes \beta X + bY + BZ \\ z \otimes (a + b)Z \end{bmatrix}$

$$Z^1(H, \Lambda^2 H) \text{ est engendré par } M' = \begin{bmatrix} x \otimes AX \wedge Z + BY \wedge Z + aX \wedge Y \\ y \otimes CX \wedge Z + DY \wedge Z + bX \wedge Y \\ z \otimes bX \wedge Z + aZ \wedge Y \end{bmatrix}$$

L'élément $\partial_\gamma(M) \in Z^1(H, \Lambda^2 H)$ est donné par la formule

$$\partial_\gamma(M) = x \otimes (a+b)Z \wedge X + y \otimes (a+b)Z \wedge Y$$

D'où $Der^{(0)}(L) = Ker\partial_\gamma \subset Z^1(H, H)$ est engendré par les $x \otimes aX + \alpha Y + AZ$ et $y \otimes \beta X - aY + BZ$ et donc de dimension 5.

L'image $\partial_\gamma(M)$ dans $Z^1(H, \Lambda^3 H)$ est engendrée par les $x \otimes bZ \wedge X \wedge Y + y \otimes aZ \wedge X \wedge Y$; donc $\partial_\gamma : Z^1(H, \Lambda^2 H) \to Z^1(H, \Lambda^3 H)$ est surjective et $E_2^{2,0} = 0$. On en déduit que $E_2^{2,0} = Ker\partial_\gamma \subset Z^1(H, \Lambda^2 H)$ est de dimension 4 et par conséquent $DimDer^{(1)}(L) = 9$.

Calculons maintenant $Der^{(2)}(L)$: $H^3(H, H)$ est engendré par les $x \wedge Y \wedge z \otimes aX + bY$ et $H^3(H, \Lambda^2 H)$ est engendré par $x \wedge y \wedge z \otimes X \wedge Y$. On voit alors facilement que $\partial_\gamma : H^3(H, H) \to H^3(H, \Lambda^2 H)$ est nulle. Donc $E_2^{0,2} = H^3(H, H)$ est de dimension 2.

Enfin comme $\partial_\gamma : H^2(H, \Lambda^2 H) \to H^2(H, \Lambda^3 H)$ est nul, on a $E_2^{1,1} = Ker\partial_\gamma : H^2(H, \Lambda^2 H)$. On voit là encore que ∂_γ est nul et donc $E_2^{1,1} = H^2(H, \lambda^2 H)$ est de dimension 5. Au total, on obtient $DimDer^{(2)}(L) = 7$.

On peut enfin calculer le crochet

$$\mathcal{T} \to [\mathcal{T}, \mathcal{T}]$$

$$Der^{(1)}(L) \to Der^{(2)}(L)$$

Soit $\mathcal{T} = \bar{C} + \bar{\gamma}$ avec $\bar{C} \in H^2(H, H)$ et $\bar{\gamma} \in E_2^{1,0}$. On a $\overline{[\gamma, \gamma]} = 0$ car l'espace correspondant est nul. $[\overline{C}, \overline{C}] = 0$ comme le montre le calcul (toute déformation infinitésimale de H est une déformation vraie). $[\overline{C}, \overline{\gamma}]$ est donné par la formule suivante:

Pour $\bar{\gamma} = x \otimes (AX \wedge Z + BY \wedge Z) + y \otimes (CX \wedge Z - AY \wedge Z)$

$\bar{C} = w \wedge y \otimes (aX + bY) + x \wedge z \otimes (\varphi X + \psi Y - bZ) + y \wedge z \otimes (\xi X - \varphi Y + aZ)$.

On obtient

$$[\bar{C}, \bar{\gamma}] = \begin{bmatrix} x \wedge z \otimes (B\xi - C\psi)Z \wedge X + 2(A\psi - B\varphi)Z \wedge Y \\ y \wedge z \otimes (C\varphi - A\xi)Z \wedge X + (C\psi - B\xi)Z \wedge Y \\ x \wedge y \otimes [C\psi - B\xi - 2A\varphi]X \wedge Y \end{bmatrix}$$

On remarquera que les termes a et b n'interviennent pas dans cette formule. On trouve que $[\bar{C}, \bar{\gamma}]$ est nul en cohomologie si et seulement si il est identiquement nul et donc si $C\psi = B\xi$, $B\varphi = A\psi$, $C\varphi = A\xi$, $A\varphi = 0$.

Si on s'intéresse plus particulièrement à la déformation aboutissant vers $SL(2)$, soit $\bar{C} = x \wedge z \otimes X - y \wedge z \otimes Y$, on a alors:

$$[\bar{C}, \bar{\gamma}] = x \wedge z - 2BZ \wedge Y) + 2Cy \wedge z \otimes Z \wedge X - 2Ax \wedge y \otimes X \wedge Y.$$

408   Claude Roger

D'où $\overline{[C,\gamma]} = 0 \Rightarrow \bar{\gamma} = 0$. Par conséquent $(C_1 = C + t\bar{C}, \gamma_t = \gamma)$ constitue une déformation vraie de bigèbre de Lie.

Nous avons donc montré que la bigèbre cobord, induite par le tenseur $X \wedge Y$ sur $SL(2,\mathbb{R})$ est obtenue par déformation de la bigèbre de type V de H, que cette déformation est unique, et que seule la bigèbre de type V admet une déformation de ce type.

Remarque: La structure de type V admet comme crochet de Lie-Poisson associé sur le groupe de Heisenberg, le crochet suivant:

$$\{f, g\} = \frac{\partial f}{\partial z}(x\frac{\partial g}{\partial x} + y\frac{\partial g}{\partial y}) - \frac{\partial g}{\partial z}(x\frac{\partial f}{\partial x} + y\frac{\partial f}{\partial y}).$$

Les feuilles symplectiques dans le groupe de Heisenberg paramétré par $(x, y, z) \in \mathbb{R}^2 \times \mathbb{R}_+^*$ ont pour équation $x = \lambda y [(x, y) \neq (0, 0)]$ (cf. [7]). Dans les coordonnées (x, z) le tenseur symplectique s'écrit $\Lambda = 2x\partial_x \wedge \partial_z$

On peut quantifier la structure de bigèbre de type V sur H par des formules analogues à celles qui définissent le groupe quantique $U_h(SL(2))$ de Drinfeld, soit:

$$\begin{bmatrix} \Delta_h(X) &=& X \otimes \exp(\frac{hZ}{4}) &+& \exp(-\frac{hZ}{4}) \otimes X \text{ idem pour } Y \\ \Delta_h(Z) &=& Z \otimes 1 + 1 \otimes Z \end{bmatrix}$$

alors la multiplication induite sur les fonctions différentiables sur les feuilles symplectiques n'est autre que le star-produit de moyal associé à la structure symplectique ci-dessus. Voir par exemple [22] pour les généralités sur les star-produits. On voit ainsi de façon évidente la grande similitude entre les formules de Drinfeld et les star-produits de Moyal.

IV. Applications à l'équation de Yang-Baxter quantique

L'équation de Yang-Baxter quantique s'est introduite comme équation de compatibilité dans les modèles de physique statistique et de physique des particules dont nous ne parlerons pas ici (voir par ex. [4] et [36]). Si E est un espace vectoriel, on considère des opérateurs $R : E \otimes E \to E \otimes E$ auxquels on associe R^{12}, R^{13} et R^{23} opérateurs de $E \otimes E \otimes E$ dans lui-même, obtenus en intercalant l'identité avec R. On dira que l'équation de Yang-Baxter quantique est vérifiée si on a:

$$R^{12}R^{13}R^{23} = R^{23}R^{13}R^{12} (Y.B.Q.)$$

On dit alors que R est une R-matrice quantique.

On posera $\sigma : \otimes^2 E \to \otimes^2 E$ $\sigma(x \otimes y) = y \otimes x$ et $R^{21} = \sigma \circ R^{12}$. Une R-matrice quantique sera dite unitaire si $R^{12}R^{21} = Id(\otimes^2 E)$. Pour

$L \to End(E)$ une sous-algèbre de Lie, on pourra chercher des solutions de $Y.B.Q.$ dans $U(L) \otimes U(L)$ (cf. [9], [10]).

Le formalisme de la R-matrice quantique peut servir à décrire l'obstruction à la cocommutativité d'une algèbre de Hopf. Plus précisément, soit A une algèbre de Hopf, Δ sa comultiplication et $\Delta' = \sigma \circ \Delta$ sa comultiplication opposée ; s'il existe $R \in A \otimes A$ inversible telle que $R\Delta(x)R^{-1} = \Delta'(x)$ quel que soit $x \in A$, alors R est une matrice quantique unitaire. Ce point de vue a été utilisé notamment par M. Rosso pour la construction de groupes quantiques ([34]).

L'application identité de $E \otimes E$ dans lui-même est bien évidemment une R-matrice que nous noterons $\mathbb{1}$. Elle correspond aux algèbres de Hopf commutatives. On peut maintenant obtenir la limite semi-classique de l'équation $(Y.B.Q.)$ en cherchant les solutions sous la forme $\mathbb{1} + hr + h\varepsilon(h)$ avec $r \in End(\otimes^2 E)$, puis en écrivant un développement limité à l'ordre 2. La première relation non triviale s'obtient en considérant le terme d'ordre 2 et on trouve en antisymétrisant:

$$\left[\begin{matrix} [r^{12}, r^{13}] & + & [r^{12}, \quad r^{23}] + [r^{13}, r^{23}] = 0 \\ r^{12} + r^{21} & = & 0 \end{matrix} \right. \qquad (Y.B.Q.)$$

Cette équation de Yang-Baxter classique a un sens pour toute algèbre de Lie $L \subset End(E)$. On a alors $r \in \Lambda^2 L$ car la seconde condition implique l'antisymétrie, et on reconnaît facilement dans l'équation ci-dessus l'expression du crochet de Schouten $[\![r, r]\!]$ dans $\Lambda^3 L$.

La démarche suivie ici est inverse des démarches habituelles dans la mathématisation de la mécanique quantique: ici il ne s'agit pas pour l'instant de "quantifier" un système "classique" mais au contraire de passer à la limite semi-classique à partir d'un système quantique donné.

L'analogie semi-classique de l'application aux algèbres de Hopf indiquée ci-dessus est la relation avec les bigèbres de Lie ; on rappelle ici (cf. [18]) que pour $r \in \Lambda^2(L)$ le cobord $\gamma = \partial_c(r) : L \to \Lambda^2(L)$ définit une structure de bigèbre de Lie si et seulement si le crochet de Schouten $[\![r, r]\!]$ est un invariant de $\Lambda^3(L)$. (c'est l'équation de Yang-Baxter "modifiée" qui a été étudiée en détail dans [18]).

Nous allons maintenant interpréter les équations $Y.B.C.$ et $Y.B.Q.$ dans notre formalisme d'algèbres de Lie graduées et d'objets de carré nul, ce qui nous permetrra de proposer un formalisme général pour la quantification algébrique. On vient de voir que $Y.B.C.$ n'était autre que la nullité d'un crochet de Schouten. Nous allons montrer un résultat analogue au niveau quantique.

Proposition: *Soit H une algèbre de Hopf cocommutative de comultiplication Δ et U^* l'algèbre de Lie graduée définie en 2b.(*). Soit $F \in U^1$*

(*) Nous noterons son crochet $[\ ,\ ]$

un élément inversible et posons $R = [\sigma(F)]^{-1}F$. *Alors* $[F,F] = 0$ *si et seulement si* R *vérifie Y.B.Q.*

Démonstration: Nous allons définir une comultiplication modifiée $\Delta_F : H \to H \otimes H$ en posant $\Delta_F(x) = F^{-1}\Delta(x)F$ quel que soit $x \in H$.

On vérifie aisément que Δ_F est coassociative si et seulement si $[F,F] = 0$ (on utilise ici la formulation du crochet de Grabowski à l'aide de la comultiplication).

Si on veut maintenant étudier l'obstruction à la cocommutativité de la comultiplication Δ_F, on a

$$\Delta'_F(x) = \sigma(\Delta_F(x)) = \sigma(F)^{-1}\sigma(\Delta(x))\sigma(F) =$$

$$\sigma(F)^{-1}\Delta(x)\sigma(F) = \sigma(F)^{-1}F\Delta_F(x)F^{-1}\sigma(F)$$

Donc $R = \sigma(F)^{-1}F$ est l'obstruction à la cocommutativité et vérifie l'équation de Yang-Baxter quantique d'après ce qui précède.

Remarque: Cette proposition n'est autre que le théorème 2 de Drinfeld ([9]) que nous avons reformulé en terme d'algèbre de Hopf, ce qui rend les choses un peu plus naturelles.

Nous allons maintenant exprimer ce dernier résultat en terme d'algèbres de Lie graduées: soit toujours H une algèbre de Hopf cocommutative, on va lui associer une algèbre de Lie graduée notée $M'(H)$ qui est une version duale de $M(H)$ définie en [22].

$M'^p(H)$ est l'espace des applications linéaires $H \to \bigotimes^{p+1} H$ à partir de $A : H \to \bigotimes^{a+1} H$ on définit $A^i_b : \bigotimes^{b+1} H \to \bigotimes^{ab+1} H$ par $A^i_b(x_0 \otimes \ldots \otimes x_b) = x_0 \otimes \ldots \otimes x_{i-1} \otimes A(x_i) \otimes x_{i+1} \otimes \ldots \otimes x_b$ pour i variant de 0 à b.

Pour $B \in M'^{b+1}(H)$ on a $i(A).B \in M'^{a+b+1}(H)$ défini par

$$i(A).B = \sum_{i=0}^{b}(-1)^i A^i_b \circ B.$$

Comme dans la partie II, on pose $[A,B] = i(B).A + (-1)^{ab+1}i(B).A$. Les objets de carré nul dans $M'(H)$ sont alors les comultiplications coassociatives sur H. H étant une algèbre de Hopf, les itérés successifs de la comultiplication Δ sont bien définis et on a $\Delta^k \in M'^k(h)$. On considère alors les applications: $J_k : U^k \to M'^k(H)$ définies par $J_k(T)(a) = T^{-1}\Delta^k(a)T$.

Proposition: J_* *est un morphisme d'algèbres de Lie graduées.*

Démonstration: On peut établir la formule pour le produit intérieur (pour ne pas alourdir les notations, on notera i les produits intérieurs des algèbres U^* et $M'^*(H)$). On a alors:

$$J_{a+b}(i(A).B) = i(J_a(A))J_b(B).$$

Cette formule se vérifie directement en utilisant la description de la structure d'algèbre de Lie graduée sur U^* en terme du coproduit. On avait (cf. Grabowski [15]):

$$i(A).B = \sum_{i=0}(-i)_{i_a}[Id_i \otimes \Delta^k \otimes Id_{b-1}](B).\underbrace{1 \otimes \ldots \otimes 1}_{i} \otimes A \otimes \underbrace{1 \otimes \ldots \otimes 1}_{b-i}$$

Cette formule permet de conclure en utilisant les axiomes d'algèbre de Hopf. Le calcul est laborieux mais sans difficultés majeures.

L'analogue classique de cette proposition va se situer au niveau des bigèbres de Lie.

Si L est une algèbre de Lie de dimension finie, considérons l'algèbre $A^*(L') \hookrightarrow \mathcal{P}_0^\omega(L \oplus L')$. C'est l'algèbre des tenseurs sur L de type $(1,p)$, alternés en les p derniers termes. Nous avons d'autre part étudié dans la partie II l'algèbre de Schouten $\Lambda * (L)$. Posons $j* : \Lambda * (L) \to A^*(L')$ l'application définie par $j * (r) = (-1)^{|r|}\partial_c(r)$ où ∂_c est l'application bord de la cohomologie de L. On a alors la:

Proposition: *$j*$ est un homomorphisme d'algèbre de Lie graduées.*

Démonstration: Nous allons nous placer dans l'algèbre de Lie graduée $P_*^\omega(L \oplus L')$. Si $r \in \Lambda * (L)$, on a $j * (r) = \partial_c(r) = [c, r]$ pour le crochet de $P_*^\omega(L \oplus L')$. Donc:

$[j * (r_1), j * (r_1)] = [[c; r_1], [c, r_2]] = [c, [r_1, [c, r_2]]]$ en utilisant l'identité de Jacobi graduée. Soit $[j*(r_1), j*(r_2)] = j*([r_1, [c, r_2]])$. La proposition se ramène alors à montrer que le crochet de Schouten sur $\Lambda * (L)$ peut s'écrire en terme du crochet gradué de $P_*^\omega(L \otimes L')$ sous la forme $[\![r_1, r_2]\!] = [r_1, [c, r_2]]$.

Ce résultat se montre facilement en remarquant que $[r_1[c, r_2]] = [\![r_1, r_2]\!]$ si r_1 et r_2 sont dans L, puis en montrant que ce crochet est un crochet de Poisson gradué pour la multiplication dans l'algèbre extérieure $\Lambda * (L)$, et donc identique au crochet de Schouten, par unicité (le lecteur sceptique pourra utiliser une base de L et les formules de la partie II pour calculer le crochet en coordonnées explicites).

On peut maintenant résumer nos résultats sous forme de diagramme. On notera H l'algèbre enveloppante de l'algèbre de Lie L.

<table>
<tr><td>Yang-Baxter
quantique</td><td>$U^* \xrightarrow{\ J^*\ } M'^*(H)$
crochet de Grabowski</td></tr>
<tr><td>Yang-Baxter
classique</td><td>$\Lambda^*(L) \xrightarrow{\ j^*\ } A^*(L') \hookrightarrow P^\omega(L \oplus L')$
crochet de Schouten</td></tr>
</table>

On a deux morphismes d'algèbres de Lie graduées. Les objets de carré nul dans les algèbres de gauche sont des solutions des équations de Yang-Baxter quantique et de Yang-Baxter classique respectivement ; pour les algèbres de droite, ce sont des structures de coalgèbre ou de bigèbre de Lie respectivement.

La flèche verticale de gauche est l'application qui à une algèbre de Lie graduée et à objet de carré nul associe son espace des cohomologie qui est une nouvelle algèbre de Lie graduée (cf; partie I). En effet, si on prend $\mathbb{1} = 1 \otimes 1 \in U^1$ on a $[\mathbb{1}, \mathbb{1}] = 0$ et $H^*_{\mathbb{1}}(U^*) = \Lambda'_*(L)$ avec le crochet de Schouten: c'est une version du fameux théorème de Hochschild, Kostant, et Rosenberg (voir Grabowski [15]). La flèche verticale de droite symbolise une limite semi classique: à un groupe quantique on associe sa partie infinitésimale qui est une bigèbre de Lie. On peut conjecturer également que la comultiplication de H, soit $\Delta \in M'^1(H)$, qui est de carré nul vérifie $H^*_{\Delta}(M'(H)) = A^*(L')$ avec sa structure d'algèbre de Lie graduée.

Ce diagramme suggère ce que pourrait être une *formalisation de la quantification algébrique*: on a une algèbre de Lie graduée L^* avec un élément $C \in L^1$ tel que $[C, C] = 0$. Une *quantification de* (L^*, C) serait la donnée d'une algèbre de Lie graduée $\mathcal{L}^*$ avec un élément canonique $I \in \mathcal{L}^1$ et $\mathcal{T} \in \mathcal{L}^1$ vérifiant:

(1) $H^*_I(\mathcal{L}^*) = L^*$

(2) $[\mathcal{T}, \mathcal{T}] = 0$

(3) $\mathcal{T} = I + \varepsilon C + \ldots \varepsilon$ étant un infiniment petit en un sens approprié.

Remarquons que le passage de YBC à YBQ répond exactement à ces conditions $-\mathcal{L}^*$ n'est pas nécessairement définie de façon unique: on peut chercher des solutions de Yang-Baxter quantique dans d'autres algèbres que l'enveloppante universelle.

La construction d'un star produit associé à une variété de Poisson en est également un exemple: soit $(V, \{,\})$ la variété de Poisson et $N = C^\infty(V)$; posons $\mathcal{L}^* = M^*(N)$ et $I = \mu$ la multiplication de N. alors d'après le théorème de Hochschild Kostant et Rosenberg déjà cité $H^*_I(\mathcal{L}^*) = L^* = \Omega * (V)$ l'algèbre des tenseurs contravariants antisymétriques munie du crochet de Schouten. Un élément $C \in L^1$ de carré nul est une structure de Poisson et l'élément $\mathcal{T}$ est bien un star-produit associé à celle-ci. Ici l'algèbre $\mathcal{L}^*$ est fixée a priori.

Le point de vue exposé ici est implicite dans le travail de J. Grabowski ([15] 6) où des résultats sur U^* sont utilisés pour construire des star-produits. Sa construction est la suivante: si l'algèbre de Lie L opère sur une variété V, on a un morphisme d'algèbres de Lie $L \to \mathfrak{a}(V)$ qui s'étend en un morphisme d'algèbres de Schouten $\Lambda_*(L) \xrightarrow{\;g\;} \Omega_*(V)$.

Au niveau quantique, on utilise le fait que les éléments de l'algèbre enveloppante se réalisent comme des opérateurs différentiels et on a $U(L) \xrightarrow{\;G\;} M^0(N)$ qui est un morphisme d'algèbres associatives.

Cette application se prolonge multiplicativement en $U^k \xrightarrow{\;G^k\;} M^k(N)$ par $G^k(X_0 \otimes \ldots \otimes X_k) = G(X_0) \ldots G(X_k)$ avec la multiplication induite par celle de N. Il montre alors que G^* est un morphisme d'algèbres de Lie graduées et obtient ainsi des exemples de star-produits.

V. Un crochet de Fröhlicher-Nijenhuis algébrique. Applications

Les crochets de Lie gradués de Schouten et de Fröhlicher-Nijenhuis font appel à la différentielle extérieure pour construire une dérivée de Lie, qui permet ensuite de définir le crochet $[X, Y]$ vérifiant la formule $L_{[X,Y]} = [L_X, L_Y]$ (cf. II.f, II.g ci-dessus). Le crochet de Schouten est défini sur les champs de tenseurs de type $(p, 0)$ alternés et celui de Fröhlicher-Nijenhuis sur les champs de type $(1, p)$ alternés en les p derniers termes ; ils constituent tous les deux des prolongements à une algèbre de Lie graduée du crochet des champs de vecteurs.

Le problème de la construction d'un crochet gradué sur les champs de tenseurs (p, q) alternés en les variables covariantes et contravariantes, qui serait une extension des crochets de Fröhlicher-Nijenhuis et de Schouten a suscité beaucoup de travaux, sans véritable succès avant une publication récente de A.M. Vinogradov [40].

Nous donnons ici les grandes lignes des constructions de Vinogradov, valables dans un cadre très général, puis nous donnerons notre version purement algébrique du crochet de Fröhlicher-Nijenhuis, qui devrait fournir un formalisme de quantification pour certaines versions de l'équation de Yang-Baxter.

a) Les travaux de A.N. Vinogradov

Soit (E^*, d) un espace différentiel d étant une différentielle de degré 1, et soit $L^* = g\ell(E^*)$ l'algèbre de Lie graduée des endomorphismes gradués de l'espace E^*. Si $A \in L^a(I.e.x \in E^k \Rightarrow A(x) \in E^{k+a}$ et $B \in L^b$, on a le crochet de L^*, $[A, B] = A \circ B + (-1)^{ab+1} B \circ A$. La présence de la différentielle d permet de définir pour chaque élément de L^* sa "liefication" (*) $L_A = [A, d]$.

On définit alors un nouveau crochet $[\![\, A, B\,]\!]$ par la formule

$$[\![\, A, B\,]\!] = \frac{1}{2}[L_A; B] + (-1)^{b+1}\frac{1}{2}[A, L_B]$$

(remarquons que degré $[\![\, A, B\,]\!]$ = degré A + degréB + 1) on montre

---

(*) Traduction littérale du néologisme de Vinogradov ("liévizatsia")

facilement $[\![A, B]\!] = (1)^{(a+1)(b+1)+1}[\![B, A]\!]$. $[\![\,,\,]\!]$ est une application antisymétrique graduée en graduant L^a par $a+1$.

Un calcul de commutateurs gradués montre ensuite que

$$L_{[\![A,B]\!]} = [L_A, L_B].$$

Malheureusement, $[\![\,,\,]\!]$ ne vérifie pas en général l'identité de Jacobi graduée.

Vinogradov définit une "anomalie"

$$Flor(A, B, C) = \sum_{(cycl)} (-1)^{a(c+1)+b+1}[A, [\![B, C]\!]]$$

qui vérifie

$$L_{Flor(A,B,C)} = \sum_{(cycl)} (-1)^{(a+1)(c+1)}[\![A, [\![B, C]\!]]\!]$$

L'identité de Jacobi graduée pour $[\![\,,\,]\!]$ est vérifiée modulo les éléments du type L_X mais en général $L_{Flor(A,B,C)}$ est non nul. Les crochets de Schouten et de Fröhlicher-Nijenhuis sont bien construits suivant ce modèle (cf. en particulier la prop. 4.2 de la partie précédente). En appliquant ce formalisme à $(E^*, d) = (\Omega^*(V), d)$ le complexe de De Rham d'une variété, on a $L^* = SDiff^*(V)$, l'algèbre graduée des opérateurs "superdifférentiels" selon Vinogradov (en imposant certaines conditions de continuité sur les opérateurs des formes dans elles-mêmes, cf. [40]). Si on pose $\mathcal{N}^*(V) = SDiff^*(V)_{/L(SDiff^*(V))}$ les formules ci-dessus montrent que $[\![\,,\,]\!]$ est un crochet de Lie gradué sur $\mathcal{N}^*(V)$ et que $\Omega_*(V) \hookrightarrow \mathcal{N}^*(V)$ et $\Omega_*(V, TV) \hookrightarrow \mathcal{N}^*(V)$ sont des sous-algèbres de Lie graduées ; ceci répond bien au problème de l'unification des deux crochets, quoi que de façon peu satisfaisante car on ne décrit pas explicitemment la structure de $\mathcal{N}^*(V)$.

b) La construction du crochet de Fröhlicher algébrique

On suppose donnée une algèbre de Lie graduée L^* dont le crochet noté Δ est défini à l'aide de "produit intérieur". On a si $A \in L^a, B \in L^b, i(A).B \in L^{a+b}$ et le crochet est défini par $A\Delta B = i(B).A + (-1)^{ab+1}i(A).B$. L'identité de Jacobi pour Δ résulte alors de la formule:

(i) $i(i(C).B).A - i(C).i(B).A = (-1)^{bc}[i(i(B).C)A - i(B)i(C)A]$ et on a de plus:

(ii) $i(B\Delta C)A = i(C)i(B)A + (1)^{bc+1}i(B)i(C)A.$

Voir [22] pour plus de détails sur le formalisme algébrique adapté. Les algèbres de Lie graduées de ce type comprennent $M^*(E), A^*(E)$ et U^* (cf. la partie II).

Supposons maintenant donné un élément $d \in L^1$ de carré nul. Une construction analogue à celle de Vinogradov va nous permettre de définir le crochet de Fröhlicher algébrique.

Si $A \in L^a$, on définit la dérivée de Lie par rapport à A par la formule de Cartan: $L_A(B) = i(A).(B\Delta d) + (-1)^{a+1}(i(A)B\Delta d)$.

On peut maintenant remplacer Δ par son expression en terme de produits intérieurs ; en utilisant la formule (ii):

$$L_A(B) = (-1)^{a+1}i(d).(i(A).B)+(-1)^b i(i(A).B)d+$$
$$i(A)(i(d).B) + (-1)^{b+1}i(A).i(B).d$$

(iii) $L_A(B) = (-1)^{a+1}i(A\Delta d)B + (-1)^d[i(i(A).B)d - i(A).i(B)d]$.

Dans le cas particulier où $L^* = M^*(E)$ (resp. $A^*(E)$) et où d est donc une structure associative sur E (resp. une structure de Lie sur E), L^* est munie d'une structure *associative* graduée, la multiplication sur E induisant une multiplication des cochaînes, que nous appellons le cup produit, noté $\cup$. Par exemple, dans le cas associatif, si on note la multiplication par $\mu \in M^1(E)$, on a:

$$(C_1 \cup C_2)(X_1, \ldots, X_{p+a}) = \mu(C_1(C_1(X_1, \ldots, X_p), C_2(X_{p+1}, \ldots, X_{p+q}))$$

Pour U^* on retrouve le produit tensoriel naturel.

Un calcul élémentaire montre alors que:

$$(-1)^b[i(i(A).B)d - i(A).i(B).d] = A \cup B + (-1)^{ab+a+b}B \cup A.$$

Nous utiliserons cette notation dans le cas général, et nous aurons donc:

(iv) $L_A(B) = A \cup B + (-1)^{(a+1)(b+1)+1}B \cup A + (-1)^{a+1}i(A\Delta d)B$.

Nous pouvons définir maintenant le crochet de Fröhlicher algébrique: on définit l'algèbre de Lie graduée $\mathbb{L}^*$ par $\mathbb{L}^a = L^{a+1}$ (à partir de maintenant, tous les degrés sont supposés décalés d'une unité).

On pose:

$$[\![A, B]\!] = \frac{1}{2}(L_A B + (-1)^{ab+1}L_B A) = A \cup B + (-1)^{ab+1}B \cup A$$
$$+(-1)^a \frac{1}{2}[i(A\Delta d)B + (-1)^{(a+1)(b+1)}i(B\Delta d).A]$$

Théorème: *Le crochet $[\![\, , \,]\!]$ fait de $\mathbb{L}^*$ une algèbre de Lie graduée et on a* $[L_A, L_B] = L_{[\![A,B]\!]}$.

Démonstration: Elle se fait par un calcul direct. On écrit explicitement l'identité de Jacobi $\sum_{(cycl)} (-1)^{ac}[\![A[\![B]\!]C]\!]$, on remplace le crochet par sa valeur donnée par la formule ci-dessus, et on regroupe les termes suivant le nombre de d qu'ils contiennent. Les formules (i) et (ii) ci-dessus permettent de conclure: une étape essentielle pour memner à bien les calculs est l'établissement de la formule:

$$i(A\Delta d)(B\Delta d) = (A\Delta d)\cup B + (-1)^{ab+b+1}B\cup(A\Delta d) + (-1)^a[i(A\Delta d)B]\Delta d.$$

Dans le cas particulier où $L* = A^*(E)$ et $d \in A^1(E)$ est une structure d'algèbre de Lie, on retrouve bien la formule du crochet de Fröhlicher algébrique (cf. partie II.h) et la formule de P.W. Michor.

On devrait pouvoir établir une formule identique à celle de Vinogradov en considérant $L^* \hookrightarrow g\ell(L^*, d)$ par son action adjointe, puis démontrer que dans notre cas l'anomalie $Flor(A, B, C)$ s'annule.

c) Application: Une version de la R-matrice quantique

Rappelons tout d'abord l'approche de Semenov-Tian-Shansky du problème de la R-matrice classique (voir [18], [35] pour les détails).

Si L est une algèbre de Lie, une R-matrice classique est une application linéaire $R : L \to L$ telle que l'application bilinéaire de L dans lui-même définie par $[X, Y]_R = \frac{1}{2}([X, RY] + [RX, Y])$ soit un crochet de Lie. On notera L_R l'algèbre de Lie d'espace sous-jacent L munie du crochet $[,]_R$. La condition nécessaire et suffisante pour que R soit une R-matrice s'écrit alors:

$$\sum_{(X,Y,Z)} [Z, [RX, RY] - R([X, Y]_R)] = 0.$$

En particulier si R vérifie l'équation de Yang-Baxter:

$$(Y.B.) \qquad [RX, RY] = R([X, Y]_R).$$

On dit alors que R est une R-matrice classique.
De même si R vérifie l'équation de Yang-Baxter modifiée

$$(Y.B.M.) \qquad [RX, RY] - R([X, Y]_R) + [X, Y] = 0.$$

Dans ce cas $R \pm Id : L_R \to L$ est un morphisme d'algèbres de Lie (voir [35] 2 et 3 pour les détails concernant ces résultats).

Si L admet une forme bilinéaire invariante $K : L \times L \to k$ (par exemple si L est semi-simple classique ou une algèbre de Kac-Moody) les deux approches de l'équation de Yang-Baxter sont équivalentes. Plus précisément si R est antisymétrique (i.e. $K(RX, Y) + K(X, RY) = 0$) on lui fait correspondre $r \in \Lambda_2(L)$ par l'isomorphisme $L \to L^*$ induit par K: alors R vérifie $Y.B.$ si et seulement si r vérifie $Y.B.$ (c'est la prop.2 de [35]). On peut voir également que R vérifie $Y.B.M.$ si et seulement si le crochet de

Schouten de r par lui-même vérifie $[\![\, r, r \,]\!] = I$, où I est l'invariant de $\Lambda^3(L)$ correspondant à $(X, Y, Z) \to \sum_{(cycl)} K([X, Y], Z)$ (voir [18] pour diverses considérations sur les différentes versions de l'équation de Yang-Baxter).

Dans le cas où L n'admet pas de forme bilinéaire invariante, on n'a plus de moyen naturel de comparer les deux approches ; d'après [35] l'étude des $R : L \to L$ serait plus naturelle pour les algèbres de Lie de dimension infinie quelconques.

Nous avons vu dans la partie II.h que l'équation de Yang-Baxter modiée pouvait s'exprimer au moyen du crochet de Fröhlicher sur $\Lambda^*(L, L)$: si $R : L \to L$ vérifie $R \circ R = Id$ alors l'équation de Yang-Baxter modifiée est équivalente à $[\![\, R, R \,]\!] = 0$.

Nous allons utiliser la version associative du crochet de Fröhlicher algébrique vue précédemment afin de décrire un candidat pour la quantification de cette R-matrice.

Pour toute algèbre associative A, on sait construire un crochet de Fröhlicher algébrique sur $M^*(A)$. Soit $\omega \in M^0(A)$ (donc de degré 1 pour le crochet de Fröhlicher), nous allons expliciter $[\![\, \omega, \omega \,]\!]$.

On remarque qu'ici $\omega \Delta d$ s'identifie au bord de Hochschild de ω vue comme cochaîne sur l'algèbre A, à valeurs dans elles-même. On a:

$$[\![\, \omega, \omega \,]\!](a, b) = \omega(a)\omega(b) + \omega^2(ab) - \omega(a\omega(b)) - \omega(\omega(a)b).$$

On peut alors définir une autre multiplication sur A: pour a et b dans A, on pose $a * b = \omega(a)b + a\omega(b) - \omega(ab)$. On a alors la

Proposition: *1) Si $[\![\, \omega, \omega \,]\!] = 0$ l'application $(a, b) \to a * b$ définit une multiplication associative sur A. On note A_ω l'algèbre obtenue.*

2) Sous les mêmes hypothèses $\omega : A_\omega \to A$ est un homomorphisme d'algèbres associatives.

Démonstration: La multiplication $*$ vue comme élément de $M^1(A)$ s'écrit $\omega \Delta d$ et son associativité équivaut à la nullité de $(\omega \Delta d)\Delta(\omega \Delta d)$. Or, les formules du crochet de Fröhlicher algébrique nous montrent que:

$$(\omega \Delta d)\Delta(\omega \Delta d) = [\![\, \omega, \omega \,]\!]\Delta d, \text{ d'où le résultat.}$$

La propriété 2) se vérifie alors immédiatement. Remarquons que la condition de normalisation $\omega(1) = 1$ assure que A_ω est unitaire.

Grâce à ce résultat nous pouvons proposer une définition de la R-matrice quantique correspondant à la R-matrice classique vue comme application de L dans L.

Pour $A = U(L)$ une R-matrice quantique sera une application $\omega \to A$ vérifiant $[\![\, \omega, \omega \,]\!] = 0$. Une quantification de $R : L \to L$ vérifiant $(Y.B.M.)$ et

$R \circ R = Id$ est la donnée d'une R-matrice quantique $\omega : U(L)[h] \to U(L)[h]$ telle que pour $X \in L$ on ait $\omega(X) = X + hR(X) + h\varepsilon(h)$.

On aurait ainsi deux structures associatives distinctes sur la même algèbre $U(L)[h]$.

Appendice: Les structures de bigèbres de Lie et les R-matrices associées à l'algèbre de Virasoro

Rappelons brièvement la définition de celle-ci: soit $\mathcal{A}(S^1)$ l'algèbre de Lie des champs de vecteurs tangents au cercle, et les champs $e_k = i \exp(ik\theta)\frac{\partial}{\partial\theta}$ pour $k \in \mathbb{Z}$. Soit $\mathcal{L} \subset \mathcal{A}(S^1)$ la sous-algèbre partout dense engendrée par les sommes finies des e_k. On calcule facilement les crochets:

$$[e_k, e_\ell] = (\ell - k)e_{k+\ell}.$$

Un résultat bien connu de Gelfand et Fuks montre que la cohomologie scalaire vérifie $H^2(\mathcal{L}; \mathbb{C}) = H^2(\mathfrak{a}(S^1), \mathbb{C}) = \mathbb{C}$ et cette cohomologie est engendrée par le 2-cocycle:

$$\omega(e_k, e_\ell) = 2(k^3 - k)\delta_k^{-\ell}.$$

L'algèbre de Virasoro, notée Vir est l'extension centrale (universelle) de $\mathcal{L}$.

Vir est donc engendrée par la base $(e_k, k \in \mathbb{Z}; z)$ avec les crochets ;

$$[z, e_k] = 0 \qquad [e_k, e_\ell] = (\ell - k)e_{k+\ell} + 2(k^3 - k)_{-k}^{\ell}z.$$

D'où la suite exacte:

$$0 \to \mathbb{C} \to Vir \to \mathcal{L} \to 0$$

Considérons l'élément de $\Lambda_2\mathcal{L}$ suivant:

$$A_k = a_k e_k \wedge e_{-k} + b_k e_0 \wedge e_k + c_k e_0 \wedge e_{-k} \text{ pour un } k \text{ entier} > 0$$

on calcule aisément $[\![A_k, A_k]\!] = -4k(a_k^2 + b_k c_k)e_0 \wedge e_k \wedge e_{-k}$ dans $\Lambda_3\mathcal{L}$ donc si $a_k^2 + b_k c_k = 0$, A_k est une solution de Yang-Baxter classique sur $\mathcal{L}$. Mais par contre le même calcul dans $\Lambda_3 Vir$ montre qu'on ne peut pas annuler les termes contenant z si A_k est non nul ; on n'obtient pas de solutions de $Y.B.C.$ sur Virasoro par cette méthode.

Comme nous l'avons vu dans la dernière partie, on peut obtenir des exemples de R-matrice classique au sens de Semenov-Tian-Chansky de la façon suivante: si l'algèbre L contient deux sous-algèbres de Lie L' et L'' telles que L soit la somme directe de L' et de L'' comme espace vectoriel, on peut prendre $R : L \to L$ définie par $R_{|L'} = Id_{L'} R_{|L''} = -IdL''$; alors R vérifie l'équation de Yang-Baxter modifiée comme on le voit presque immédiatement.

Pour l'algèbre $\mathcal{L}$, on peut considérer deux décompositions distinctes. Soit $W_1 \hookrightarrow \mathcal{L}$ la sous-algèbre engendrée par les e_k pour $k \geq -1$, $L_a \hookrightarrow \mathcal{L}$ la sous-algèbre engendrée par les e_k pour $k \geq a(a \geq 0)$.

W_1 s'identifie à l'algèbre des champs de vecteurs formels à une indéterminée, L_a s'identifiant alors à la sous-algèbre fermée des champs dont le jet d'ordre $\leq a$ est nul.

Remarquons que L_a s'identifie à la sous-algèbre engendrée par les e_k pour $k \leq -a$.

On a alors deux décompositions distinctes:

$$\mathcal{L} = W_1 + L_2 \quad \text{et} \quad \mathcal{L} = L_0 + L_1.$$

Il existe des décompositions analogues pour Vir (le cocyle engendrant l'extension se trivialise quand on le restreint à W_1 ou L_a) qui fournissent donc des exemples de R-matrice classique au sens de Semenov-Tian-Shansky.

REFERENCES

[1] E. Abe, "Hopf algebras", Cambridge University Press.

[2] V. Aldaya & J. Navarro Salas, *Quantization on the Virasoro group*, Communications in Mathematical Physics **126** (1990), 575–595.

[3] R. Aminou, *Bigèbres de Lie, structures de Poisson et variantes de l'équation de Yang-Baxter,*. Publ. IRMA Lille, vol. **9** n3 (1987).

[4] R.J. Baxter, "Exactly Solved Models in Statistical Mechanics", Academic Press, 1989.

[5] R. Berger, *Algebraic quantization*, Letters in Mathematical Physics **17** (1989), 275–283.

[6] J.L. Brylinski, *A differential complex for Poisson manifolds*, Journal of Differential Geometry **288** (1988), 93–114.

[7] M. Cahen & C. Ohn, *Bialgebra structures on the Heisenberg algebra*, Travaux mathématiques de l'ULB (1989, vol. 2, p. 183).

[8] A. Coste, P. Dazord, A. Weinstein, *Groupoïdes symplectiques*, Publ. du Dép. de Math. de l'Université de Lyon 2.A (1987) 1-62.

[9] V.G. Drinfeld, *On constant quasi classical solutions of the Yang-Baxter quantum equation*, Sov. math. Doklady **28** (1983) 667–671.

[10] V.G. Drinfeld, *Quantum groups*, in "Actes du Congrès International des Mathématiciens", Berkeley, 1986.

[11] A. Fröhlicher & A. Nijenhuis, *Theory of vector valued differential forms*, Indagationes Mathematicae **18** (1956), 338–359.

[12] M. Gerstenhaber *The cohomology structure of an associative ring*, Annals of Math. **78** (1963), 267–288.

[13] M. Gerstenhaber *On the deformations of rings and algebras*, Annals of Math. **79** (1964), 59–103.

[14] M. Gerstenhaber & S.D. Schack, *Algebraic cohomology and deformation theory*, in "Deformation theory of algebras and structures", NATO-ASI Series C. **297**, Kluwer Academic Publishers, 1988.

[15] J. Grabowski, *Abstract Jacobi and Poisson structures. Quantization and star-products*, Preprint.

[16] D. Gurevich, *Quantum Y.B. equation and a generalization of Formal Lie theory*, Rep. Dept. Math. Univ. Stockholm **24** (1986).

[17] J. Huebschmann, *Poisson cohomology and quantization*, J. Reine Angew. Math. **408** (1990), 57–113.

[18] Y. Kosmann-Schwarzbach, *Equations de Yang-Baxter et structures de Poisson*, Journées Relativistes de Chambéry (15.05.87).

[19] Y. Kosmann-Schwarzbach & F. Magri, *Poisson-Nijenhuis structures*, Preprint.

[20] J.L. Koszul, *Crochet de Schouten-Nijenhuis et cohomologie*, Astérisque hors série (1985), 251–271.

[21] B. Kostant & S. Sternberg, *Symplectic reduction, BRS cohomology, and infinite dimensional Clifford algebras*, Annals of Physics **176** (1987), 49–113.

[22] P. Lecomte & M. De Wilde, *Formal deformations of the Poisson-Lie algebra of a symplectic manifold and star products. Existence, equivalence, derivations*, in "Deformation theory of algebras and structures", NATO-ASI Series C **297**, Kluwer Academic Publishers.

[23] P. Lecomte & C. Roger, *Rigidity of the current Lie algebra of complex simple type*, Bull. Lond. Math. Soc. (2) **37** (1988), 232–240.

[24] P. Lecomte, *Applications of the cohomology of graded Lie algebras to Formal deformations of Lie algebra*, Letters in Mathematical Physics **13** (1987), 157–166.

[25] P. Lecomte & C. Roger, *Modules et cohomologie des bigèbres de Lie*, C.R.A.S. Paris **310** Série I (1990), 405–410 .

[26] A. Lichnerowicz, *Cohomologie 1-différentiable et déformation de l'algèbre de Lie dynamique d'une variété canonique*, C.R.A.S. Paris **280** Série I (1975), 1217–1220.

[27] A. Lichnerowicz, *Les variétés de Poisson et leurs algèbres de Lie associées*, Journal of Diff. Geom. **12** (1977), 253–300.

[28] S.L. Loukyanov, *Quantization of the Gelfand-Dikii brackets*, Funct. Anal. et Appl. **22** (1989), 255–262.

[29] A. Medina & P. Revoy, *Groupes de Lie à structure symplectique invariante*. Preprint.

[30] D. Melotte, *Cohomologie de Chevalley associée aux variétés de Poisson*, Bull. Soc. Royale des Sciences de Liège **LVIII** (1989), 319–413.

[31] A. Nijenhuis, *Jacobi type identities for bilinear differential concommitants of certain tensor Fields I*, Indagationes Mathematicae **17** (1955), 390–403.

[32] A. Nijenhuis & R. Richardson, *Deformations of Lie algebra structures*, J. Math. Mech. **171** (1967), 89–105.

[33] C. Roger, A. Tihami, M. Elgaliou, *Une cohomologie pour les algèbres de Lie de Poisson homogènes* A paraître aux publ. du Dép. de Mathématiques de Lyon.

[34] M. Rosso, *An analogue of the P.B.W. theorem and the universal R-matrix for $U_h(SL(N))$*, Comm. Math. Phys. **124** (1989), 307–318.

[35] M.A. Semenov Tian-Shansky, *What is a classical R-matrix ?*, Funct.

Anal. et Appl. **17** (1983), 17–35.

[36] A.K. Sklyanin, *On some algebraic structures connected with the Yang-Baxter equation*, Funct. Anal. et Appl. **16** (1982), 27–39.

[37] I. Szymczak & S. Zakrzewski, *Quantum deformations of the Heisenberg group obtained by geometric quantization*, Preprint.

[38] L.A. Takhtadjan, *Quantum groups and integrable models*, Preprint, LOMI, Leningrad.

[39] W.M. Tulczyjew, *The graded Lie algebra of multivector Fields and the generalized Lie derivative of forms*, Bull. Acad. Pol. Sci. Ser. Math. Astr. Phys. **22** (1974), 937–942.

[40] A.M. Vinogradov, *Unifications des crochets de Schouten et de Fröhlicher, cohomologies et opérateurs super différentiels (en russe)*, Math. Zametki **47** Vol. 6 (1990).

U.R.A. CNRS n 746 - Laboratoire de Géométrie et Analyse
Université Claude Bernard - Lyon 1
43, boulevard du 11 novembre 1918
69622 Villeurbanne Cedex   -   France

An Analogue of B.G.G. Resolution for the Quantum $SL(N)$-Group

MARC ROSSO

Dedicated to Jean-Marie Souriau

Introduction

Quantized enveloping algebras associated with any simple Lie algebra $\mathcal{G}$ were introduced by Drinfeld and Jimbo ([3],[6]). For $\mathcal{G}$ of type A_N, which is the case we shall be interested in, the definition is as follows:

Let $t \in \mathbb{C}^*$. $U_t\mathcal{G}$ is the unital algebra generated by:

$$E_i, F_i, k_i^{\pm 1} \qquad 1 \leq i \leq N$$

with relations:

$$k_i k_i^{-1} = k_i^{-1} k_i = 1; \quad k_i k_j = k_j k_i$$

$$k_i E_j k_i^{-1} = t^{(\alpha_i, \alpha_j)} E_j; \quad k_i F_j k_i^{-1} = t^{-(\alpha_i, \alpha_j)} F_j$$

$$E_i F_j - t^{-2(\alpha_i, \alpha_j)} F_j E_i = \delta_{ij} \frac{k_i^4 - 1}{t^4 - 1}$$

and, for $|i - j| = 1$:

$$E_i^2 E_j - (t^2 + t^{-2}) E_i E_j E_i + E_j E_i^2 = 0$$

$$F_i^2 F_j - (t^2 + t^{-2}) F_i F_j F_i + F_j F_i^2 = 0$$

where $(\alpha_i)_{1 \leq i \leq N}$ is a basis of simple roots in the root system, and $(\ ,\ )$ is the inner product:

$$(\alpha_i, \alpha_i) = 2$$

$$(\alpha_i, \alpha_{i\pm 1}) = -1$$

$$(\alpha_i, \alpha_j) = 0 \qquad |i - j| \geq 2.$$

(These generators are related to Jimbo's e_i and f_i by: $E_i = e_i k_i$, $F_i = f_i k_i$.) The coproduct is given by:

$$\Delta(k_i) = k_i \otimes k_i$$

$$\Delta(E_i) = E_i \otimes 1 + k_i^2 \otimes E_i$$

$$\Delta(F_i) = F_i \otimes 1 + k_i^2 \otimes F_i.$$

S will denote the antipode, ε the augmentation. Let $\mathbb{C}[T]$ be the subalgebra generated by $k_i^{\pm 1}$'s

$U_t n_+$ the subalgebra generated by 1 and the E_i's

$U_t n_-$ the subalgebra generated by 1 and the F_i's.

Then, we have a triangular decomposition:

$$U_t \mathcal{G} = U_t n_- \otimes \mathbb{C}[T] \otimes U_t n_+.$$

We shall denote by $U_t b_+$ the subalgebra generated by the E_i's and $k_i^{\pm 1}$'s. Giving E_i (resp. F_i) the degree α_i (resp. $-\alpha_i$), and $k_i^{\pm 1}$ the degree 0, we see that $U_t \mathcal{G}$ is Q-graded, where Q is the root lattice.

In the classical theory, the (co)homology of the Lie algebra n_- (with obvious notations) has played an important role, for example in the Borel–Weil–Bott–Kostant theorem or in the construction of the Bernstein–Gelfand–Gelfand resolution. In these questions, one uses the fact that the Lie algebra homology with values in a $\mathcal{G}$-module V may be computed using the Koszul complex $\Lambda \mathcal{G} \otimes V$. Our aim here is to give an analogous framework for the quantum group $U_t n_-$, in the A_N-case.

In fact, as is well known, $H_*(\mathcal{G}, V)$ is given by the Tor functor for the associative algebra $U\mathcal{G}$: $H_*(\mathcal{G}, V) = \mathrm{Tor}_*^{U\mathcal{G}}(V, \mathbb{C})$. The interest is that $\mathrm{Tor}_*^A(R, L)$ is defined for any associative augmented algebra A and right (resp. left) A-module R (resp. L). It can be computed as the homology of the so-called Bar construction, which is defined as follows: let $\varepsilon : A \to \mathbb{C}$ the augmentation, and $I = \ker \varepsilon$. Put $\quad B_n(R, A, L) = R \otimes I^{\otimes n} \otimes L$
$$\partial : B_n \to B_{n-1}$$

$$\partial(r, a_1, \ldots, a_n, l) = (ra_1, a_2, \ldots, a_n, l)$$

$$+ \sum_{i=1}^{n-1} (-1)^i (r, \ldots, a_i a_{i+1}, \ldots, a_n, l)$$

$$+ (-1)^n (a, a_1, \ldots, a_{n-1}, a_n l).$$

Then $\mathrm{Tor}_*^A(R, L) = H_*(B(R, A, L))$.

Dually, one also defines Ext functors as the cohomology of the dual complex:

$$\mathrm{Ext}_A^*(R, L^*) = H_*(B(R, A, L)^*).$$

Now, in the case of $U_t n_-$, although there is no Lie algebra around, $U_t n_-$ is an augmented associative algebra and we can consider $\mathrm{Tor}_*^{U_t n_-}$. The problem is that the Bar construction is quite big and not very easy to handle. We shall see that $U_t n_-$ has a special structure (it is what Priddy calls a Koszul algebra) and thanks to this, one can construct a small subcomplex of the Bar complex to compute $\mathrm{Tor}_*^{U_t n^-}$ (Koszul resolutions). Furthermore, $U_t b_+$ acts on this complex, so that we have an action of $U_t \mathcal{G}$ on the complex giving the resolution of $\mathbb{C}$.

This allows to follow closely the original paper of Bernstein, Gelfand and Gelfand [2] to give an analog of the Borel–Weil–Bott–Kostant theorem and of the BGG resolution (thanks to some properties of the center of $U_t \mathcal{G}$). Then, following ideas of Zelevinski [11] and Akin [1] for the classical case, we will use this BGG resolution to give an analog of Frobenius character formula for Hecke algebras of type A_N.

I. Koszul algebras and Koszul resolutions [7] [8]

1. **Definition 1.** *A pre-Koszul algebra is a unital algebra A, with an augmentation $\varepsilon : A \to \mathbb{C}$, and which has a presentation of the following form:*
   - generators: $(a^i)_{i \in I}$, $\varepsilon(a^i) = 0$, I a countable totally ordered set
   - relations: $\sum_i f_i a^i + \sum_{j,k} f_{j,k} a^j a^k$
A is said to be homogeneous if $f_i = 0$.

One sees that A is filtered, with $F_p A$ the set of monomials of length smaller than p and one can consider the associated graded space $\tilde{A} = \oplus \tilde{A}_p$, with $\tilde{A}_p = F_p A / F_{p-1} A$. $\tilde{A}$ is said to be the pre-Koszul homogeneous associated algebra. It is a quadratic algebra.

2. Let B be a basis of A consisting of 1, the a^i's and certain monomials $a^{i_1} \ldots a^{i_n}$. A subset S of $\cup_{n=1}^\infty I^n$ is called a *labeling set* for B if, for each $a \in B \backslash \{1\}$, there is a unique $(i_1, \ldots, i_n) \in S$ such that $a = a^{i_1} \ldots a^{i_n}$. The pair (B, S) is called a *labeled basis* for A. Let's note $B^* = \{1, \alpha(i), \ldots, \alpha(i_1, \ldots, i_n), \ldots\}$ the dual basis.

3. Let A a homogeneous pre-Koszul Algebra. $\mathrm{Ext}_A^*(\mathbb{C}, \mathbb{C})$ may be computed by the Bar complex and has a natural algebra structure. One has $\mathrm{Ext}_A^1(\mathbb{C}, \mathbb{C}) = \{\varphi \in A^* / \forall x, y \in A \backslash \{1\}, \varphi(xy) = 0\}$. (i.e. linear forms which are 0 on decomposable elements.)

Now, the a^i's form a basis of the indecomposables of A, so the dual $\alpha(i)$'s define cocycles. Let α_i the cohomology class of $\alpha(i)$. We have $\mathrm{Ext}_A^1(\mathbb{C}, \mathbb{C}) = \mathrm{Vect}(\alpha_i)$.

Definition 2. *A is said to be a homogeneous Koszul algebra if $\mathrm{Ext}_A^*(\mathbb{C}, \mathbb{C})$ is generated, as an algebra, by the α_i's.*

Definition 3. *A pre-Koszul algebra A is said to be Koszul if its associated graded $\tilde{A}$ is a homogeneous Koszul algebra.*

4. Let A be a homogeneous Koszul algebra with generators $(a^i)_{i \in I}$, and

(B, S) a labeled basis. Then each monomial $a^k a^l$ has a unique expression:

$$a^k a^l = \sum_{(i,j) \in S} f_{ij}^{kl} a^i a^j.$$

Furthermore, one has a very explicit description of the cohomology algebra $\mathrm{Ext}_A^*(\mathbb{C}, \mathbb{C})$ in term of those constants f_{ij}^{kl}.

Theorem 4. $\mathrm{Ext}_A^*(\mathbb{C}, \mathbb{C})$ *is the unital algebra generated by the* α_i*'s,* $i \in I$*, with the relations:*
   For $(i, j) \in S$

$$\alpha_i \alpha_j + \sum_{(k,l) \notin S} f_{ij}^{kl} \alpha_k \alpha_l = 0.$$

5. Let A be a Koszul (not necessarily homogeneous) algebra with generators $(a_i)_{i \in I}$. Let L and R be respectively left and right A-modules. Then, there exists a small subcomplex of the bar complex $B_*(R, A, L)$, which is called the Koszul complex $K_*(R, A, L)$ and whose homology is $\mathrm{Tor}_*^A(R, L)$. Let $\bar{K}_*(A) = K_*(\mathbb{C}, A, \mathbb{C})$. Then $K_*(R, A, L) = R \otimes \bar{K}_*(A) \otimes L$ as a linear space, and $K_*(A, A, L)$ is a resolution of L by free left A-modules.

The complex $\bar{K}_*(A)$ is defined as follows: Put $J = \tilde{A}_1$ (i.e. the linear space generated by the images of the a_i's in $\tilde{A}$), and for $1 \leq p \leq n - 1$, define linear maps:

$$\partial_p : J^{\otimes n} \to \tilde{A}^{\otimes(n-1)}$$

$$(x_1, \ldots, x_n) \mapsto (x_1, \ldots, x_p x_{p+1}, \ldots, x_n).$$

Then $\bar{K}_p(A) = \cap_{1 \leq p \leq n-1} \ker \partial_p$.

The differential ∂ in $\bar{K}_*(A)$ is obtained from a presentation of A; let (B, S) a labeled basis for A, and let us write the relations in A as:

$$a^k a^l = \sum_{(i,j) \in S} f_{ij}^{kl} a^i a^j + \sum_m f_m^{kl} a^m.$$

One gets a basis $\tilde{B}$ for $\tilde{A}$ from B by replacing the letter a by the letter b and the relations are:

$$b^k b^l = \sum_{(i,j) \in S} f_{ij}^{kl} b^i b^j.$$

Now, let $\sum c_{i_1 \ldots i_n}(b^{i_1}, \ldots, b^{i_n}) \in \bar{K}_n(A)$. Then

$$\partial\left(\sum c_{i_1 \ldots i_n}(b^{i_1}, \ldots, b^{i_n})\right) =$$

$$\sum c_{i_1 \ldots i_n}\left(\sum_{l=1}^{n-1}(-1)^l(b^{i_1}, \ldots, \sum f_i^{b_{i_l} b_{i_{l+1}}} b^i, \ldots)\right).$$

Let R be a right A-module. The differential in the complex $R \otimes \bar{K}(A)$ is such that $R \otimes \bar{K}(A)$ is a subcomplex of the Bar complex, i.e.:

$$d(r \otimes \sum c_{i_1 \ldots i_n}(b^{i_1}, \ldots, b^{i_n})) =$$

$$\sum_m r a_m \otimes \sum_{\substack{(c_{i_1}, \ldots, i_n) \\ i_1 = m}} c_{i_1 \ldots i_n}(b^{i_2}, \ldots, b^{i_n}) + r \otimes \partial(\sum c_{i_1 \ldots i_n}(b^{i_1}, \ldots, b^{i_n})).$$

The dual complex $\bar{K}^* = (\bar{K}_*)^*$ is easier to describe. Let $(\beta(i_1, \ldots, i_n))$ the dual basis $(\tilde{B})^*$ and β_i the cohomology class of $\beta(i)$ in $\mathrm{Ext}^1_{\bar{A}}(\mathbb{C}, \mathbb{C})$.

Theorem 5. *The complex $\bar{K}^*(A)$ is the differential graded algebra generated by the β_i, $i \in I$, with, for each $(i, j) \in S$, the relation:*

$$\beta_i \beta_j + \sum_{(k,l) \notin S} f_{ij}^{kl} \beta_k \beta_l = 0.$$

The differential δ is defined by:

$$\delta \beta_m = \sum_{(k,l) \notin S} f_m^{kl} \beta_k \beta_l.$$

6. In order to use this machinery, one needs a simple way to decide whether a given augmented algebra is a Koszul algebra. Priddy gave a sufficient condition, which allows to do this only from the presentation of the algebra and without any preliminary cohomological computations. We give it now.

Let A be a homogeneous pre-Koszul algebra, $(a^i)_{i \in I}$ the generators, (B, S) a labeled basis. As I is totally ordered, so is $\cup_{n=1}^{\infty} I^n$: order first by length, and then by lexicographic ordering. Now $S \subset \cup_{n=1}^{\infty} I^n$ inherits an ordering.

Definition 6. *One says that (B, S) is a P.B.W. basis if:*
1. *$(i_1, \ldots, i_k)$ and $(j_1, \ldots, j_l) \in S \Leftrightarrow$ (either $(i_1, \ldots, i_k j_1, \ldots, j_l) \in S$ or the label of each monomial appearing in a reduced expression of $a^{i_1} \ldots a^{i_k} a^{j_1} \ldots a^{j_l}$ is strictly bigger than $(i_1, \ldots, i_k, j_1, \ldots, j_l)$).*
2. *For $k > 2 : (i_1 \ldots, i_k) \in S \Leftrightarrow \forall j \in \{1, \ldots, k-1\}(i_1, \ldots, i_j) \in S$ and $(i_{j+1}, \ldots, i_k) \in S$*

Theorem 7. *If A is a pre-Koszul homogeneous Koszul algebra which has a P.B.W. basis, then A is a Koszul homogeneous algebra.*

II. Koszul resolutions for $U_t n_-$

Our first task is to show that $U_t n_-$ is a Koszul algebra, that is, to give a presentation with relations which are at most quadratic. In order to do that, it is useful to introduce new generators, which are analogues of root vectors, for each negative root. This was already done in [9] for the case of $U_t n_+$, so we just have to adapt those computations to $U_t n_-$.

Let R_+ the set of positive roots of $\mathcal{G}$. Each $\alpha \in R_+$ is of the form $\alpha = \alpha_i + \alpha_{i+1} + \cdots + \alpha_j$, which we depict as

$$\underset{\alpha_i}{\bullet\!\!-\!\!\!-\!\!\!-\!\!\bullet} \cdots \bullet\!\!-\!\!\!-\!\!\!\underset{\alpha_j}{-\!\!\bullet} \ .$$

Recall that we have an adjoint representation: $\mathrm{ad} : U_t\mathcal{G} \to \mathrm{End}(U_t\mathcal{G})$ defined by: $\mathrm{ad} = (L \otimes R)(\mathrm{Id}\otimes S)\Delta$, where L (resp. R) is the left (resp. right) regular representation.

We define F_α by induction on the length of α : let $\alpha' = \alpha_{i+1} + \cdots + \alpha_j$ and put $F_\alpha = \mathrm{ad}\, F_i(F_{\alpha'})$. The commutation relations for the F_α's are easily derived from [9] and are summarized as follows, according to the relative positions of α and β.

$$F_\beta F_\alpha - t^{2(\alpha,\beta)} F_\alpha F_\beta = F_{\alpha+\beta} \times (-t^{-2})$$

$$F_\beta F_\alpha = F_\alpha F_\beta$$

$$F_\beta F_\alpha - t^{2(\alpha,\beta)} F_\alpha F_\beta = 0$$

$$F_\beta F_\alpha - t^{2(\alpha,\beta)} F_\alpha F_\beta = 0$$

$$F_\beta F_\alpha - t^{2(\alpha,\beta)} F_\alpha F_\beta = (t^2 - t^{-2}) F_\gamma F_{\gamma'}$$

$$F_\beta F_\alpha = F_\alpha F_\beta.$$

Our index set I will be R_+, with the following total ordering (we write the roots in decreasing order) $\alpha_1, \alpha_1 + \alpha_2, \alpha_1 + \alpha_2 + \alpha_3, \ldots, \alpha_1 + \cdots + \alpha_N, \alpha_2, \alpha_2 + \alpha_3, \ldots, \alpha_2 + \cdots + \alpha_N, \ldots, \alpha_N$ which we shall also denote $\alpha(1) > \alpha(2) > \cdots > \alpha(n)$. From [9], one easily deduces that the $F_{\alpha(1)}^{m_1} \ldots F_{\alpha(n)}^{m_n} (m_1, \ldots, m_n) \in \mathbb{N}^n$ form a basis for $U_t n_-$.

All this implies immediately that the associated graded to $U_t n_-$ is a homogeneous pre-Koszul algebra with a P.B.W. basis so that $U_t n_-$ is a

Koszul algebra. Furthermore, from the presentation given above, one immediately deduces the complex $(\bar{K}^*(U_t n_-), \delta)$: it is the differential graded algebra generated by elements ξ_α, $\alpha \in R_+$, of degree -1, with the following relations $\xi_\alpha^2 = 0$, for $\alpha > \beta$ $\xi_\alpha \xi_\beta + t^{2(\alpha,\beta)} \xi_\beta \xi_\alpha = 0$ in all cases except when

$$\xi_\alpha \xi_\beta + \xi_\beta \xi_\alpha + (t^2 - t^{-2}) \xi_{\gamma_1} \xi_{\gamma_2} = 0.$$

The differential:

$$\delta(\xi_\alpha) = \sum_{\substack{\gamma_1 + \gamma_2 = \alpha \\ \gamma_1 > \gamma_2}} \xi_{\gamma_1} \cdot \xi_{\gamma_2}.$$

The relations allow to rewrite each monomial in the ξ_α's as a linear combination of monomials in which ξ_α appears at most once and in which the indices α are in decreasing order.

Giving to ξ_α the degree $-\alpha \in Q$ (root lattice), we see that as Q-graded vector space, $\bar{K}^*(U_t n_-)$ is isomorphic to the exterior algebra $\Lambda^*(n_-^*)$, that is to the usual Koszul complex for the trivial module $\mathbb{C}$. So, the general theory gives us finite length complexes to compute the (co)homology of $U_t n_-$ in various $U_t n_-$-modules.

In order to apply the BGG strategy, we need a resolution of $\mathbb{C}$ by $U_t \mathcal{G}$-modules.

III. $U_t\mathcal{G}$-module structure on $U_t n_- \otimes \bar{K}_*$

We first define an action of $U_t b_+$ on $\bar{K}_*$. Then the induced module $U_t \mathcal{G} \otimes_{U_t b_+} \bar{K}_*$, which is isomorphic to $U_t n_- \otimes \bar{K}_*$ as $U_t n_-$-modules, will carry a natural action of $U_t \mathcal{G}$ by left multiplication.

1. Recall that we have an adjoint representation: $\mathrm{ad} : U_t \mathcal{G} \to \mathrm{End}(U_t \mathcal{G})$. Then $\mathrm{ad}\, E_i$ has the following "twisted derivation" property: for $\xi \in U_t \mathcal{G}$, Q-homogeneous of degree β,

$$\mathrm{ad}\, E_i(\xi \cdot \eta) = \mathrm{ad}\, E_i(\xi) \cdot \eta + t^{2(\alpha_i, \beta)} \xi \, \mathrm{ad}\, E_i(\eta).$$

We have:

$$\mathrm{ad}\, E_i(F_\alpha) = \begin{cases} F_{\alpha - \alpha_i}, & \text{if } \alpha = \alpha_i + \dots, \\ -t^2 F_{\alpha - \alpha_i} k_i^4, & \text{if } \alpha = \dots + \alpha_i \\ \frac{k_i^4 - 1}{t^4 - 1} & \text{if } \alpha = \alpha_i \\ 0 & \text{other cases.} \end{cases}$$

Recall that J is the degree 1 subspace of the associated graded to $U_t n_-$. We define the action of E_i in $J^{\otimes n}$ by induction on n. Call f_α the image of F_α in J.

$$\text{On } J : \rho(E_i)(f_\alpha) = \begin{cases} f_{\alpha-\alpha_i} & \text{if } \alpha = \alpha_i + \dots \\ -t^2 f_{\alpha-\alpha_i} & \text{if } \alpha = \dots + \alpha_i \\ 0 & \text{if not.} \end{cases} \quad \text{Suppose it is defined}$$

on $J^{\otimes(n-1)}$. Then, we put

$$\rho(E_i)(f_\alpha, f_{\alpha'}, \dots) = \left\{ \begin{array}{l} (f_{\alpha-\alpha_i}, f_{\alpha'}, \dots) \\ -t^2 \cdot t^{-4(\alpha_i, \alpha'+\dots)}(f_{\alpha-\alpha_i}, f_{\alpha'}, \dots) \\ 0 \end{array} \right\}$$

$$+ \, t^{-2(\alpha_i, \alpha)} f_\alpha \otimes \rho(E_i)(f_{\alpha'}, \dots)$$

according to the same three possibilities for f_α as above. Put also

$$\rho(k_i)(f_\alpha, f_{\alpha'}, \dots) = t^{-(\alpha_i, \alpha+\alpha'+\dots)}(f_\alpha, f_{\alpha'}, \dots) \, .$$

Then one checks:

Proposition 8. *These formulas define a representation of $U_t b_+$ in each $J^{\otimes n}$, and the subspace $\bar{K}_n \subset J^{\otimes n}$ is stable by this action.*

2. Consider now $U_t \mathcal{G} \otimes \bar{K}_*$. $U_t b_+$ acts on $U_t \mathcal{G}$ on the left by $R \circ S$, where R is the right regular representation and S the antipode. So, we have a representation $\tilde{\rho} = [(R \circ S) \otimes \rho]\Delta$ of $U_t b_+$ on $U_t \mathcal{G} \otimes \bar{K}_*$, and the induced module $U_t \mathcal{G} \otimes_{U_t b_+} \bar{K}_*$ is nothing but the space of coinvariants under this action.

On $U_t \mathcal{G} \otimes \bar{K}_*$, consider the differential given by the same formula as in §1, for $R \otimes \bar{K}_*(A)$. It clearly commutes with the left action of $U_t \mathcal{G}$ and with $\tilde{\rho}(\mathbb{C}[T])$. Let $\overline{U_t b_+}$ be the kernel of the augmentation.

Lemma 9. *The commutator of the differential with $\tilde{\rho}(E_i)$ is in $\tilde{\rho}(\overline{U_t b_+})$.*

Corollary 10. *As the space of coinvariants is the quotient of $U_t \mathcal{G} \otimes \bar{K}_*$ by the image $\tilde{\rho}(\overline{U_t b_+})(U_t \mathcal{G} \otimes \bar{K}_*)$, the differential on $U_t \mathcal{G} \otimes \bar{K}_*$ descends on a differential on the space of coinvariants.*

Furthermore, in the isomorphism with $U_t n_- \otimes \bar{K}_*$, this differential goes to the one given by the Koszul resolution.

IV. The BGG resolution

Let us first recall some notations. Let $\mathcal{H}^*$ the complex linear space generated by the α_i's, and $\mathcal{H}$ its dual. Define $H_i \in \mathcal{H}$ by: $\alpha_j(H_i) =$

(α_i, α_j) $\forall j$. Then, each $\lambda \in \mathcal{H}^*$ defines a character

$$e^\lambda : \mathbb{C}[T] \to \mathbb{C}^*$$

$$e^\lambda(k_i) = t^{\lambda(H_i)}.$$

Let $\mathbb{C}_\lambda$ be the $U_t b_+$-module $\mathbb{C}$ where $\mathbb{C}[T]$ acts via e^λ and the E_i's by 0. Using the triangular decomposition, the Verma module $M(\lambda)$ is defined by the usual induction method:

$$M(\lambda) = U_t \mathcal{G} \otimes_{U_t b_+} \mathbb{C}_\lambda.$$

It has a unique irreducible quotient, which we denote by $V(\lambda)$, which is finite dimensional if and only if λ is a dominant weight.

Let Z be the center of $U_t \mathcal{G}$ and $\chi_\lambda : Z \to \mathbb{C}$ the central character associated with $M(\lambda)$. It is defined by:

$$\chi_\lambda(z) = e^\lambda[\varepsilon_- \otimes \mathrm{id} \otimes \varepsilon_+(z)],$$

where $\varepsilon_\pm : U_t n_\pm \to \mathbb{C}$ is the augmentation, and where we use the triangular decomposition. We have proved in [10]:

Theorem 11. *Let λ, μ be weights. Then $\chi_\lambda = \chi_\mu \Leftrightarrow e^{\lambda+\rho} \in W(e^{\mu+\rho})$ where W is the Weyl group and ρ half the sum of the positive roots.*

Using this, one can deduce the basic facts about inclusions of Verma modules, following step by step BGG in [2], and get:

Theorem 12. *For each dominant weight λ, one has an exact sequence of $U_t \mathcal{G}$-modules*

$$0 \leftarrow V(\lambda) \leftarrow C_0 \leftarrow C_1 \leftarrow \cdots \leftarrow C_s \leftarrow 0$$

where $C_k = \oplus_{l(w)=k} M(w(\lambda + \rho) - \rho)$ is a free $U_t n_-$-module.

Following Zelevinski [11], one can obtain other useful exact sequences as follows: If V is a finite dimensional $U_t \mathcal{G}$-module, it is also a $U_t n_-$-module and:

$$0 \leftarrow V \otimes V(\lambda) \leftarrow V \otimes C_0 \leftarrow \cdots \leftarrow V \otimes C_s \leftarrow 0$$

is a projective resolution of $V \otimes V(\lambda)$ (tensor products of $U_t n_-$-modules).

Considering the spaces of coinvariants under $U_t n_-$ (recall that if L is a $U_t n_-$-module, the coinvariants are $L/\overline{U_t n_-}(L)$ where $\overline{U_t n_-}$ is the kernel of the augmentation in $U_t n_-$) and then their weight subspaces of a given weight μ, one gets another exact sequence.

For L a $U_t n_-$-module, call $\phi_{V,\mu}(L) = (L/\overline{U_t n_-}(L))(\mu)$. Then:

$$\phi_{V,\mu}(M(\lambda)) = V_{\mu-\lambda}, \text{ the subspace of weight } \mu - \lambda \text{ in } V$$

$$\phi_{V,\mu}(V(\lambda)) = V_{\mu-\lambda,\lambda} = \{v \in V_{\mu-\lambda}; E_i^{\lambda(H_i^V)+1} v = 0\},$$

where H_i^V is the coroot vector corresponding to α_i.

So, we have

Theorem 13. *If V is a finite dimensional $U_t\mathcal{G}$-module and μ, ν two dominant weights, then one has an exact sequence*

$$0 \leftarrow V_{\mu-\nu,\nu} \leftarrow C_0' \leftarrow C_1' \leftarrow \cdots \leftarrow C_s' \leftarrow 0$$

where $C_k' = \oplus_{l(w)=k} V_{\mu+\rho-w(\nu+\rho)}$.

In particular: for $\nu = 0$, $V_{\mu,0}$ is the space of highest weight vectors of weight μ in V.

V. Application to Hecke algebras

Take $V = (\mathbb{C}^{N+1})^{\otimes p}$, the p-th tensor power of the fundamental representation. Jimbo [6] has shown that the image of the R-matrix of $U_t sl(N+1)$ in the fundamental representation allows to define a representation of the Hecke algebra $\mathcal{H}_p(t^4)$ in V, and that the images of $\mathcal{H}_p(t^4)$ and $U_t\mathcal{G}l(N+1)$ in $\text{End}(V)$ are exactly commutant of each other.

So, one has a decomposition $V = \oplus_\mu [V(\mu) \otimes R_\mu]$, where μ describes the set of weights $(\mu_1, \ldots, \mu_{N+1})$ such that $\mu_1 + \cdots + \mu_{N+1} = p$ and R_μ is the irreducible $\mathcal{H}_p(t^4)$-module corresponding to the partition μ. $\mathcal{H}_p(t^4)$ acts on each term of the resolution given by Theorem 13 and the maps are morphisms of $\mathcal{H}_p(t^4)$-modules. Now, $V_{\mu,0} \simeq R_\mu$ and so we have a resolution of this irreducible $\mathcal{H}_p(t^4)$-module. Let's identify in terms of $\mathcal{H}_p(t^4)$ the different terms of resolution.

Recall that for each $w \in S_p$ the symmetric group on p elements, there is a basis element T_w in $\mathcal{H}_p(t^4)$ such that if $w = s_{i_1} \ldots s_{i_r}$ is a reduced expression of w in the elementary transpositions $T_w = T_{s_{i_1}} \ldots T_{s_{i_r}}$. Note $T_i = T_{s_i}$. For $\lambda = (\lambda_1, \lambda_2, \ldots) \in \mathbb{N}^{N+1}$, with $\sum \lambda_i = p$, let $\mathcal{H}_\lambda$ the subalgebra of $\mathcal{H}_p(t^4)$ generated by

$$T_1, \ldots, T_{\lambda_1-1}, T_{\lambda_1+1}, \ldots, T_{\lambda_1+\lambda_2-1},$$

$$T_{\lambda_1+\lambda_2+1}, \ldots, T_{\lambda_1+\cdots+\lambda_{1N+1}-1}.$$

Now, with $V = (\mathbb{C}^{N+1})^{\otimes p}$, the weight space V_λ is isomorphic, as a $\mathcal{H}_p(t^4)$-module, to the module induced from trivial representation of $\mathcal{H}_\lambda$. So, equating the Euler characteristic of the resolution of R_μ to 0, we see that the trace of any element of $\mathcal{H}_p(t^4)$ in R_μ is given by an alternating sum of products of traces in such induced representations (Generalization of the Frobenius character formula for the symmetric groups). The interest of such a formula is that one has simple matrix representations for the images of the T_i's in those induced representations (which can be found for example in Hoefschmidt [5]).

REFERENCES

1. K. Akin, *On complexes relating the Jacobi-Trudi Identify with the Bernstein-Gelfand-Gelfand Resolution*, J. of Alg. **117** (1988), 494–503.
2. I. N. Bernstein, I. M. Gelfand and S. I. Gelfand, *Differential operators on the base affine space and a study of $\mathcal{G}$-modules, part* II: *The resolution of a finite dimensional $\mathcal{G}$-module*, in "Lie Groups and Their Representations", (I. M. Gelfand, Ed.), Wiley, New York (1975).
3. V. G. Drinfeld, *Quantum Groups*, Proceedings of the I.C.M., Berkeley, 1986.
4. B. L. Feigin and B. L. Tsygan, *Cyclic homology of algebras with quadratic relations, universal enveloping algebras and group algebras*, Lect. Notes in Math. Vol. 1289, 210–240.
5. Hoefschmidt, *Representations of Hecke algebras of finte groups with* BN *pairs of classical type*, Univ. of British Columbia Thesis (1974).
6. Jimbo, *A q-difference analogue of $U(\mathcal{G})$ and the Yang–Baxter equation*, Lett. Math. Phys. **10** (1985), 63–69.
7. I. Y. Manin, *Some remarks on Koszul algebras and quantum groups*, Ann. Inst. Fourier **38** (1988), 191–206.
8. S. B. Priddy, *Koszul resolutions*, Trans. A.M.S. **152** (1970), 39–60.
9. M. Rosso, *An analogue of* P.B.W. *theorem and the universal R-matrix for $U_h sl(N+1)$*, Comm. Math. Phys. **124** (1989), 307–318.
10. ______, *Analogues de la forme de Killing et du Théorème d'Harish-Chandra pour les groupes quantiques*, to appear in Ann. Sc. de l'Ecole Normale Supérieure (1990).
11. A. V. Zelevinski, *Resolvents, dual pairs and character formulas*, Funct. Anal. and Applic. **21** (1987), 152–154.

Miller Fellow   CNRS, Centre de Mathématiques
Department of Mathematics   Ecole Polytechnique
UC Berkeley   and   F-91128 Palaiseau Cedex
Berkeley, CA 94720   France
USA

Hamiltonian Constraints,
the Jacobi Principle and Gravity

JĘDRZEJ ŚNIATYCKI

Dedicated to Jean-Marie Souriau

Abstract

The inverse Legendre transformation, applied to the Hamiltonian constraints in general relativity, leads to a homogeneous Lagrangian L defined on an open domain $\mathcal{D}$ of the space of Riemannian metrics on the Cauchy surface. Curves of metrics on the Cauchy surface with tangent vectors in $\mathcal{D}$ give rise to solutions of Einstein equations if and only if they are critical points of L and satisfy the supermomentum constraints.

For homogeneous cosmological models the the Lagrangian L corresponds to the Jacobi metric, and solutions of Einstein equations are given by time-like geodesics. In some cosmological models the prolongation of the Jacobi metric allows for an extension across the singularity of evolutions leading to the gravitational collapse.

1. Hamiltonian constraints in general relativity

In the Arnowitt-Deser-Misner formulation of general relativity the Einstein equations are presented as a constrained Hamiltonian system on the cotangent bundle of a space $\mathcal{M}$ of Riemannian metrics on a typical Cauchy surface M, [1]. It is assumed that M is a smooth oriented compact manifold, possibly with non-empty boundary ∂M, and $\mathcal{M}$ is the space of Riemannian metrics of Sobolev class H^k, for sufficiently high k, which satisfy appropriate boundary conditions if $\partial M \neq \emptyset$, [2]. The L^2-cotangent bundle of $\mathcal{M}$ is the space $T^*\mathcal{M}$ consisting of symmetric contravariant 2-tensor densities. For each $g \in \mathcal{M}$, $\dot{g} \in T_g\mathcal{M}$ and $\pi \in T_g^*\mathcal{M}$, the evaluation of π on $\dot{g}$ is given by integration,

$$\langle \pi | \dot{g} \rangle = \int_M \pi \dot{g} \, . \tag{1}$$

Let $\mathcal{H} : T^*\mathcal{M} \to \mathbf{R}$ be defined by

$$\mathcal{H}(g, \pi) = (\det\ g)^{-1/2} [tr(\pi^2) - \frac{1}{2}(tr\ \pi)^2] - (\det\ g)^{1/2} R[g] \, , \tag{2}$$

where $R[g]$ denotes the scalar curvature of the metric g, and the trace operator tr is defined in terms of g, and let $\mathcal{J}$ be the covariant divergence of π with respect to the Levi-Civita connection of g,

$$\mathcal{J}[g,\pi] = Div\ \pi\ .\tag{3}$$

The constraint equations are

$$\mathcal{H} = 0\ ,\tag{4}$$

and

$$\mathcal{J} = 0\ ;\tag{5}$$

they are called the superhamiltonian and the supermomentum constraints, respectively. The Hamiltonian of the theory is

$$H = \int_M N\mathcal{H}\ ,\tag{6}$$

where N is an arbitrary nowhere vanishing function called lapse. In the general ADM formalism one uses the integrand of the form $(N\mathcal{H} + N^i\mathcal{J}_i)$, where $\mathcal{J}_i$ are components of $\mathcal{J}$ and (N^i) is called a shift vector. Here we choose vanishing shift which corresponds to the study of evolution in the space-time in the direction normal to instantaneous Cauchy surfaces.

Let $(a,b) \to \mathcal{M} : t \mapsto g(t)$ be a curve in $\mathcal{M}$. It gives rise to a metric $\mathbf{g}$ on $(a,b) \times M$ such that

$$\mathbf{g}_{(t,m)} = -N^2(dt)^2 + g(t)(m)\tag{7}$$

for every $(t,m) \in (a,b) \times M$. The metric $\mathbf{g}$ satisfies vacuum Einstein equations if and only if the curve $g(t)$ is a projection to $\mathcal{M}$ of a solution of the constrained Hamiltonian system described above.

The supermomentum constraints $\mathcal{J} = 0$ are quite well understood. The Hamiltonian system under consideration is invariant under the group of diffeomorphisms of M (preserving the boundary conditions) and the supermomentum constraints are the consequence of this invariance. If ξ is a vector field on M generating a one parameter group $\exp(\xi)$ of diffeomorphisms of M, then

$$J_\xi = \int_M \mathcal{J}\,\xi\tag{8}$$

is the momentum corresponding to the Hamiltonian action of $\exp(\xi)$ on T^*M. Thus, the supermomentum constraints give the vanishing of the equivariant momentum map corresponding to the Hamiltonian action on $T^*\mathcal{M}$ of the group of diffeomorphisms of the typical Cauchy surface M.

The superhamiltonian constraints $\mathcal{H} = 0$ are the consequence of the invariance of the Einstein theory under diffeomorphisms which deform

Cauchy surfaces. These diffeomorphisms do not act in $T^*\mathcal{M}$ and are much more difficult to describe in the Hamiltonian formalism, see [3,4]. The aim of this paper is to discuss the consequences of the superhamiltonian constraints in terms of the structure of $T^*\mathcal{M}$. In order to get some insight into the general setting we start analysis of Hamiltonian constraints for finite dimensional configuration spaces.

2. Homogeneous Lagrangians and Hamiltonian constraints

Let Q be a manifold and L a homogeneous Lagrangian of degree 1 smooth on a conical domain D in TQ. That is, for every $u \in D$ and every $\lambda > 0$, $\lambda u \in D$ and

$$L(\lambda u) = \lambda L(u) \; . \tag{9}$$

For example, if g is a Riemannian metric on Q, then $L(u) = \sqrt{g(u, u)}$ is a homogeneous Lagrangian of degree 1 smooth on $D = \{u \in TQ | u \neq 0\}$. If g is not positive definite then $D = \{u \in TQ | g(u, u) > 0\}$. Since L is homogeneous of degree 1, it follows that

$$\frac{d}{d\lambda} L(\lambda u)|_{\lambda=1} - L(u) = 0 \; , \tag{10}$$

which implies that the Hamiltonian H associated to L vanishes identically. Thus, the Legendre transformation corresponding to L maps D onto a codimension 1 submanifold P of T^*Q given by a Hamiltonian constraint $H = 0$.

Conversely, given a sufficiently regular codimension 1 submanifold P of T^*Q, the inverse Legendre transformation, adapted to the presence of constraints, gives a homogeneous Lagrangian L of degree 1 such that P is the image of the Legendre transformation corresponding to L, and the solutions of the constraint Hamiltonian equations and the Euler-Lagrange equations have the same projections on Q, $c.f.$ ref. [5]. The geometric interpretation of the inverse Legendre transformation in the presence of constraints given below is due to W.M. Tulczyjew (unpublished). Let θ be the canonical 1-form of T^*Q, associating to each $w \in T_p T^*Q$ the value of p on the projection of w to TQ,

$$\theta(w) = \langle p | T\pi_Q(w) \rangle \; , \tag{11}$$

where $\pi_Q : T^*Q \to Q$ denotes the cotangent bundle projection, and $\iota_P : P \to T^*Q$ be the inclusion map. The pull back of $d\theta$ to P has 1-dimensional kernel $K \subset TP$. Let D be the set of non-zero vectors in the projection of K to TQ,

$$D = T(\pi_Q \circ \iota_P)(K - 0) \; . \tag{12}$$

Given $u \in D$, choose $w \in T(\pi_Q \circ \iota_P)^{-1}(u)$ and set

$$L(u) = \theta(w) \,. \tag{13}$$

If D is open in TQ and projects onto Q, and L is well defined by Eq. (13), then L is the required homogeneous Lagrangian of degree 1, for details see ref. [6].

Introducing local coordinates (q^i) in Q, and the corresponding coordinates $(q^i, \dot{q}^i)$ and (q^i, p_i) in TQ and T^*Q, respectively, and expressing P as the zero level of a Hamiltonian function $H(q^i, p_i)$, we see that $(q^i, \dot{q}^i) \in D$ if and only if

$$\dot{q}^i = N \frac{\partial H}{\partial p_i} \tag{14}$$

for some $(q^i, p_i) \in H^{-1}(0)$ and some Lagrange multiplier N. The presence of the Lagrange multiplier N in Eq. (14) is the only modification of the inverse Legendre transformation needed to handle the Hamiltonian constraint $H = 0$. The Lagrangian L is given by

$$L(q^i, \dot{q}^i) = p_i \dot{q}^i \,, \tag{15}$$

where the summation convention is assumed and $p_i's$ are solutions of Eq. (14).

3. The Jacobi principle

As an example of the construction given above consider the energy level of a Hamiltonian

$$H = \frac{1}{2} p^2 + V(q) \tag{16}$$

corresponding to energy E. In this case

$$P = H^{-1}(E) \,, \tag{17}$$

and Eq. (14) yields

$$\dot{q}^i = N p_i \,. \tag{18}$$

Substituting this expression into the equation $H = E$ we obtain

$$N^2 = 2(E - V)/\dot{q}^2 \,, \tag{19}$$

and

$$L(q^i, \dot{q}^i) = \sqrt{2[E - V(q)]\dot{q}^2} \,. \tag{20}$$

The variational principle with the Lagrangian given by Eq. (20) is the classical Jacobi principle, cf. ref. [5]. The corresponding Lagrange-Euler equations are equations of geodesics of the classical Jacobi metric

$$g_Q = 2[E - V(q)]dq^2 \,. \tag{21}$$

Relativistic dynamics of massive particles provides also an example of the Jacobi principle, [6]. In this case the homogeneous Lagrangian corresponds to the standard Minkowski metric multiplied by the mass parameter.

Another example of the Jacobi principle is given by the propagation of weak waves in elastic solids. In this case the Hamiltonian, which is constrained to vanish, is a polynomial in momenta of degree 6, and the corresponding homogeneous Lagrangian gives a Finslerian metric on the configuration space. The action integral along a path equals to the time elapsed, which implies that elastic waves satisfy the Fermat principle, [7].

4. Back to gravity

Hamiltonian constraints in gravity are not a single constraint, as in the examples above, but an infinite system labelled by points in the Cauchy surface M. Nevertheless, one can follow the above construction and obtain an equivalent homogeneous Lagrangian.

Given $\dot{g} \in T_g \mathcal{M}$, the solution $\pi \in T_g^* \mathcal{M}$ of

$$\dot{g} = N \frac{\delta \mathcal{H}}{\delta \pi} \tag{22}$$

and the constraint equation (4) is

$$\pi = (2N)^{-1} (\det\ g)^{1/2} [\dot{g} - g\ tr\dot{g}]\ , \tag{23}$$

where

$$N = \pm \sqrt{\frac{tr(\dot{g}^2) - (tr\dot{g})^2}{4R[g]}}\ . \tag{24}$$

Hence, the corresponding homogeneous Lagrangian, defined up to a sign, is given by

$$L[g, \dot{g}] = \int_M \sqrt{R[g] \det(g) [tr(\dot{g}^2) - (tr\dot{g})^2]} \tag{25}$$

It is smooth in the domain

$$\mathcal{D} = \{(g, \dot{g}) \in T\mathcal{M} | R[g] \det(g) [tr(\dot{g}^2) - (tr\dot{g})^2] > 0\}. \tag{26}$$

Theorem. *Let $(a,b) \to \mathcal{M} : t \mapsto g(t)$ be a curve in $\mathcal{M}$ contained in $\mathcal{D}$ satisfying the supermomentum constraints, Eq. (5), and let N be given by Eq. (24). The metric $\mathbf{g}$ on $(a,b) \times M$, given by Eq. (7), satisfies the vacuum Einstein equations if and only if $t \to g(t)$ satisfies the Euler-Lagrange equations for the Lagrangian L given by Eq. (25).*

Proof. The supermomentum constraints are assumed. The construction of L ensures that its critical points correspond to solutions of the constraint Hamiltonian system with the Hamiltonian given by Eq. (6) and the constraints $\mathcal{H} = 0$.

It should be noted that, because of the integration over M, the Lagrangian (23) does not correspond to a Riemannian metric on $\mathcal{M}$. It could be considered as a generalization of a Finslerian metric on $\mathcal{M}$, which is singular when $R[g] = 0$.

There are several Riemannian or pseudo-Riemannian metrics on $\mathcal{M}$. The most convenient one is the DeWitt metric $\mathcal{G}$ defined by

$$\mathcal{G}(h, k) = \int_M [h \cdot k - tr(h)tr(k)]vol[g] \qquad (27)$$

for every $h, k \in T_g\mathcal{M}$, where $vol[g] = \sqrt{\det(g)}d_3x$ is the volume form defined by g and the orientation of M. Einstein equations give rise to Lagrangian system on $\mathcal{M}$ with the kinetic energy term given by the DeWitt metric and the potential energy term given by

$$V[g] = \int_M N \ R[g]vol[g] \ . \qquad (28)$$

However, in this case the Hamiltonian constraints (4) have to be imposed in addition to the Euler-Lagrange equations, [9,10]. In ref. [9], DeWitt imposed a normalization of N by requiring $V[g] = 1$. This allows for a correspondence between solutions of the Einstein equations and geodesics of the DeWitt metric, which depends on the choice of N satisfying the normalization condition.

5. A cosmological application

In cosmological models, in which M is a homogenous space and one considers only invariant metrics on M, the integration over M is trivial and the homogeneous Lagrangian L corresponds to an indefinite Riemannian metric; in analogy with the classical case I shall refer to this metric as the Jacobi metric. I should like to illustrate its application to a Robertson-Walker universe with homogeneous massles scalar field. The terminology and notation adopted here follows ref. [11].

Consider a curve of metrics on M of the form

$$g(t) = \frac{a(t)^2}{[1 + \frac{k}{4}(x^2 + y^2 + z^2)]}[dx^2 + dy^2 + dz^2] \qquad (29)$$

on M, where $a(t)$ is interpreted as the "radius" of the universe and k is a parameter taking values 1, 0, and -1. The choice of k determines the sign of the intrinsic curvature of $g(t)$. Assume that the space M is filled with a homogeneous massless scalar field ϕ. This model leads to a constrained Hamiltonian system with the configuration space Q parametrized by the variables (a, ϕ), $a > 0$, and the Hamiltonian constraint

$$H \equiv \frac{1}{24a}\pi_a^2 - \frac{1}{a^3}\pi_\phi^2 + 6ka = 0 \ , \tag{30}$$

where (π_a, π_ϕ) are the canonical momenta.

Given $(\dot{a}, \dot{\phi}, a, \phi)$, the inverse Legendre transformation yields

$$\pi_a = N^{-1}12a\dot{a} \ , \ \pi_\phi = -N^{-1}a^3\dot{\phi} \ , \tag{31}$$

$$N^2 = \frac{a^2\dot{\phi}^2 - 12\dot{a}^2}{12ka} \ , \tag{32}$$

and the corresponding homogeneous Lagrangian

$$L = \sqrt{12k(a^4\dot{\phi}^2 - 12a^2\dot{a}^2)} \ . \tag{33}$$

The variational principle with the Lagrangian (33) yields equations of the geodesics of the Jacobi metric

$$g_Q = 12k(a^4 d\phi^2 - 12a^2 da^2) \ , \tag{34}$$

with unit tangent vectors $\dot{q} = (\dot{a}, \dot{\phi}, a, \phi)$,

$$g_Q(\dot{q}, \dot{q}) = 1 \ . \tag{35}$$

For $k = 0$, the Jacobi metric vanishes identically and our method does not apply. Let g_{Q+} and g_{Q-} be the metric on Q corresponding to $k = 1$ and $k = -1$, respectively. Clearly,

$$g_{Q+} = -g_{Q-} \ . \tag{36}$$

Consider $W = \mathbf{R}^2$ with coordinates (u, v) and the Lorentz metric

$$g_W = du^2 - dv^2 \ , \tag{37}$$

and let

$$W^+ = \{(u, v) \in W | u^2 - v^2 > 0\} \tag{38}$$

be the set of "time-like" vectors in W. Similarly, we denote by W^- the set of "space-like" vectors in W,

$$W^- = \{(u, v) \in W | u^2 - v^2 < 0\} \ , \tag{39}$$

and by g_{W+} and g_{W-} the pull-backs of g_W to W^+ and W^-, respectively. The mapping

$$u = 6a^2 sinh(\phi/\sqrt{3}) \, , \, v = 6a^2 cosh(\phi/\sqrt{3}) \tag{40}$$

gives an isometry of (Q, g_{Q+}) onto (W^+, g_{W+}). Similarly, the mapping

$$u = 6a^2 cosh(\phi/\sqrt{3}) \, , \, v = 6a^2 sinh(\phi/\sqrt{3}) \tag{41}$$

gives an isometry of (Q, g_{Q-}) onto (W^-, g_{W-}). Moreover, the singular boundary of Q, given by $a = 0$, corresponds under both mappings to the "light-cone"

$$\Gamma = \{(u, v) \in W | u^2 - v^2 = 0\} \tag{42}$$

The evolution of our constrained Hamiltonian system is given by the geodesics of the metric g_Q with the velocities constrained by the condition (33). It is convenient to study their behaviour in terms of the appropriate embbedding of Q into W. The geodesics of g_W satisfying the condition (33) are "time-like" lines in W. The geodesics originating at a point $w \in W^+$ intersects Γ in two points, the point of intersection in the "past" of w corresponds to the "big bang" and the point in the "future" of w corresponds to the "big crunch". On the other hand, the points of intersection with Γ of the geodesics originating at a point $w \in W^-$ are both either in the "past" of w or both in the "future" of w. Thus, for $k = 1$, we have two singularities: "big bang" and "big crunch", while, for $k = -1$, there is only one singularity, either "big bang" or "big crunch", in agreement with the standard results.

It is interesting that in our picture the cases with $k = 1$ and $k = -1$ are open submanifolds of the 2-dimensional Minkowski space W which allows for continuation of the geodesics through the gravitational collapse. This suggests that the gravitational collapse need not be a genuine singularity of the dynamical system under consideration, but it may be only the breakdown of our description of the system.

It may be that this possibility of continuation of geodesics across the singularity appears only in the special case presented here. However, similar pattern seems to appear also in some other cosmological models, [12].

Acknowledgments

I should like to express my gratitude to Larry Bates, Mark Gotay, Jim Isenberg, Hélène Kerbrat-Lunc and Yvan Kerbrat for stimulating discussions. In particular, I should like to thank Jorma Louko for pointing out an error in the argument given in the preliminary draft.

REFERENCES

[1] R. Arnowitt, S. Deser, and C. Misner, *The dynamics of General Relativity*, in "Gravitation, an Introduction to Current Research", pp. 227–265, L. Witten (ed.), John Wiley, New York, 1962.

[2] E. Binz and J. Śniatycki, *Conservation Laws in Space-times with Boundaries*, Class. Quantum Grav. **3** (1986), 1191–1197.

[3] E. Binz, J. Śniatycki and H. Fischer, "Geometry of Classical Fields", North Holland, Amsterdam, 1988.

[4] M. Gotay, J. Isenberg, J. Marsden, and R. Montgomery, with collaboration of J. Śniatycki and Ph. Yasskin, "Momentum Maps and Classical Relativistic Fields", in preparation.

[5] J.L. Synge, *Classical Dynamics*, in "Handbuch der Physik", vol. 3, part 1, pp. 1–225, Springer Verlag, Berlin, 1960.

[6] J. Śniatycki, *Lagrangian systems, canonical systems and Legendre transformation*, in "Proceedings of 13th Biennial Seminar of Canadian Mathematical Congress", vol. 2., pp. 125–143, 1971.

[7] M. Epstein and J. Śniatycki, *Fermat Principle in Elastodynamics*, to appear in Journal of Elasticity.

[9] C. W. Misner, *Minisuperspace*, in "Magic Without Magic: John Archibald Wheeler", J. Klauder (ed.), W.H. Freeman, San Francisco.

[9] B.S. DeWitt, *Spacetime as a Sheaf of Geodesics in Superspace*, in "Relativity", pp 359–374, C. Carmeli, S. Fickler and L. Witten (eds.), Plenum Press, New York, 1970.

[10] A. Fischer and J. Marsden, *The Einstein's equations of evolution - a geometric approach*, J. Math. Phys. **13** (1972), 546–568.

[11] M. Gotay and J. Isenberg, *Geometric Quantization and Gravitational Collapse*, Phys. Rev. D **22** (1980), 235–248.

[12] Y. Kerbrat, H. Kerbrat-Lunc and J. Śniatycki, *Geometry of minisuperspace in examples*, submitted for publication.

Department of Mathematics and Statistics
The University of Calgary
Calgary, AB T2N 1N4
Canada

Orbites coadjointes et caractères

MICHÈLE VERGNE

Soit M une variété hamiltonienne symplectique préquantifiable. Nous avions proposé dans [V] une formule conjecturale pour le caractère de la représentation correspondant par la quantification géométrique de Kostant-Souriau à la variété M. Nous expliquons ici quel sens donner à cette formule pour les représentations associées à une orbite coadjointe fermée d'un groupe de lie semi-simple.

Si M est une variété, on note $\mathcal{A}(M)$ l'algèbre des formes différentielles sur M. Nous suivrons les notations de [BGV] (chapitres 1,7,8) pour la cohomologie équivariante.

Soit G un groupe réel semi-simple connexe de centre fini. Soit $\mathfrak{g}$ son algèbre de Lie. Soit $\lambda \in \mathfrak{g}^*$ un élément du dual de l'espace vectoriel $\mathfrak{g}$. Soit M l'orbite de λ. C'est une variété hamiltonienne, munie d'une forme symplectique équivariant canonique σ_M. Par définition, pour $X \in \mathfrak{g}$,

$$\sigma_M(X) = J(X) + \sigma.$$

où σ est la forme symplectique canonique de l'orbite coadjointe M et $J(X)$ la restriction à $M \subset \mathfrak{g}^*$ de la forme linéaire $J(X)(f) = (X, f)$, $f \in \mathfrak{g}^*$. Dans la suite, M sera toujours munie de l'orientation définie par sa structure symplectique canonique.

Nous supposons que l'orbite de λ est fermée. Le stabilisateur H de λ dans G est alors un sous-groupe réductif H de G. Il existe donc une décomposition H-invariante

$$\mathfrak{g} = \mathfrak{h} \oplus \mathfrak{r}.$$

Cette décomposition détermine une connexion G-invariante θ sur l'espace fibré principal $G \to G/H$ et donc sur tous les fibrés associés. Notons $pr_{\mathfrak{h}}$ la projection de $\mathfrak{g}$ sur $\mathfrak{h}$. Le moment ([BGV] chapitre 7 ou [V]) de la connexion θ est donné par

$$\mu(X)(g) = pr_{\mathfrak{h}}(g^{-1}X), \quad \text{pour } X \in \mathfrak{g},\ g \in G.$$

Notons $\Omega \in \mathcal{A}^2(G, \mathfrak{h})$ la courbure de la connexion θ. Soit

$$\Omega(X) = \Omega + \mu(X)$$

la courbure équivariante.

Si a est une fonction analytique sur $\mathfrak{h}$, et si $\alpha \in \mathcal{A}(G) \otimes \mathfrak{h}$, alors $a(\alpha) \in \mathcal{A}(G)$. Si a est une fonction analytique H-invariante sur $\mathfrak{h}$, la fonction sur $\mathfrak{g}$ à valeurs dans $\mathcal{A}(G)$ définie par $X \cdot a(\Omega(X))$ passe au quotient en une forme différentielle équivariante fermée sur $M = G/H$.

Soit $j_{\mathfrak{g}}(X)$ la fonction G-invariante sur $\mathfrak{g}$ définie par

$$j_{\mathfrak{g}}(X) = \frac{(\sin h(\mathrm{ad}(X/2))}{\mathrm{ad}(X/2)}).$$

Définition 1. Soit

$$J(X, M) = \det_{\mathfrak{g}/\mathfrak{h}}\left(\frac{\sin h(\Omega(X)/2)}{\Omega(X)/2}\right) \in \mathcal{A}(M)$$

et soit

$$J(X, M^{\perp}) = \det_{\mathfrak{h}}\left(\frac{\sin h(\Omega(X)/2)}{\Omega(X)/2}\right) \in \mathcal{A}(M).$$

Les classes de cohomologie équivariante déterminées par $J(X, M)$ et $J(X, M^{\perp})$ v'érifient la relation

$$J(X, M)J(X, M^{\perp}) \cong j_{\mathfrak{g}}(X).$$

La forme $J^{-1/2}(X, M)$ est la version équivariante du genre $\hat{A}(M)$. En particulier si $\mathfrak{g}$ est l'algèbre de Lie d'un groupe compact, si λ est un paramètre G-admissible et si D_{λ} est l'opérateur de Dirac tordu sur M associé à λ (voir [BGV], chapitre 8), alors on peut exprimer l'indice équivariant de D_{λ} par

$$indice(D_{\lambda})(\exp X) = (2i\pi)^{-\dim M/2} \int_{M} e^{i\sigma_{M}(X)} J(X, M)^{-1/2}$$

$$= (2i\pi)^{-\dim M/2} j_{\mathfrak{g}}(X)^{-1/2} \int_{M} e^{i\sigma_{M}(X)} J(X, M^{\perp})^{1/2}.$$

Dicté par la théorie de l'indice équivariant d'Atiyah-Segal-Singer, il est donc important de considérer la fonction sur $\mathfrak{g}$ donnée par

$$\Theta_{M}(X) = (2i\pi)^{-\dim M/2} j_{\mathfrak{g}}(X)^{-1/2} \int_{M} e^{i\sigma_{M}(X)} J(X, M^{\perp})^{1/2}.$$

Nous expliquons maintenant quel est le sens à donner à cette intégrale lorsque M n'est pas compacte.

Lemme 2. *Soit a une fonction polynomiale H-invariant sur $\mathfrak{h}$. Si ϕ est une densité C^∞ à support compact sur $\mathfrak{g}$, la forme différentielle sur M définie par*

$$\int_{\mathfrak{g}} e^{i\sigma_M(X)} a(\Omega(X)) d\phi(X)$$

est intégrable sur M.

On peut alors définir la fonction généralisée $D(M,a)$ sur $\mathfrak{g}$ par

$$\int_{\mathfrak{g}} D(M,a)(X) d\phi(X) = (2i\pi)^{-\dim M/2} \int_M \left(\int_{\mathfrak{g}} e^{i\sigma_M(X)} a(\Omega(X)) d\phi(X) \right).$$

Si a est identiquement égale à 1, $D(M,1)$ est la transformée de Fourier de l'orbite M munie de sa mesure canonique.

Proposition 3. *Soit a_n une suite de fonctions polynomials H-invariante sur $\mathfrak{h}$ et convergeant uniformément sur tout compact vers une fonction analytique a, alors la suite de fonctions généralisées G-invariantes $D(M,a_n)$ converge vers une fonction généralisée G-invariante $D(M,a)$ ne dépendent que de a et non de la suite a_n choisie.*

Moralement, on a $D(M,a)(X) = (2i\pi)^{-\dim M/2} \int_M e^{i\sigma_M(X)} a(\Omega(X))$.

Soit $a = j_{\mathfrak{h}}^{1/2}$. C'est une fonction analytique H-invariante sur $\mathfrak{h}$. On peut donc définir $D(M, j_{\mathfrak{h}}^{1/2})$ et ceci est le sens à donner à $\int_M e^{i\sigma_M(X)} J(X, M^\perp)^{1/2}$ lorsque M n'est pas compacte.

Supposons pour simplifier que λ soit un paramètre G-admissible elliptique. Il existe alors une réprésentation unitaire irréductible T_λ par le foncteur de Zuckermann. On a alors le

Théorème 4. *Dans un voisinage de 0 dans $\mathfrak{g}$, on a l'égalité de fonctions généralisées suivante :*

$$Tr T_\lambda(\exp X) = j_{\mathfrak{g}}(X)^{-1/2} D(M, j_{\mathfrak{h}}^{1/2})$$

Lorsque M est une orbite elliptique de dimension maximum, alors $\mathfrak{h}$ est une algèbre de Cartan et $j_{\mathfrak{h}}^{1/2}$ est identiquement égale à 1. Dans ce cas, la représentation T_λ est une représentation de la série discrète. La formule précédente pour la trace de T_λ est la formule conjecturée par Kirillov [K] et démontrée par Rossmann [R].

REFERENCES

[BGV] N. Berline, E. Getzler and M. Vergne, *Heat kernels and Dirac operators.* Springer-Verlag, Berlin-Heidelberg-New York, 1991.

[K] A. A. Kirillov, *Method of orbits in the theory of unitary representations of Lie groups.* Funks. Anal. i Prilozhen. **2**(1968), 1, 96–98.

[R] W. Rossmann, *Kirillov's character formula for reductive groups.* Invent. Math. **48** (1978), 207–220.

[V] M. Vergne, *Formule de Kirillov et indice de l'opérateur de Dirac.* Proc. Intl. Congress Mathematicians, 1984, Varsovie PWN-Polish Scientific Publishers. North Holland, Amsterdam, Nw York, Oxford, 1984.

École Normale Supérieure
45 rue d'Ulm
75005 Paris, France

Noncommutative Geometry and Geometric Quantization

ALAN WEINSTEIN*

Dedicated to Jean-Marie Souriau

1 Introduction

This article is intended to explain a program initiated separately by the author, M. Karasev [12], and S. Zakrzewski [42], in which the geometric quantization of objects called *symplectic groupoids* is suggested as a means for relating the symplectic (and Poisson) manifolds of classical mechanics to noncommutative algebras. These algebras, of fundamental importance in quantum mechanics, have also become of increasing interest because of their applications to geometry [6] and to certain Hopf algebras called quantum groups [8].

It is a great pleasure to include this article in the proceedings of a conference dedicated to Jean-Marie Souriau, whose work and personal influence have done so much to unite geometry (and geometers) and quantization. I have tried to preserve here the informal nature of my talk at the meeting.

2 The quantization problem

In classical mechanics, the observables are functions on a space P, which is usually either the phase space or the space of motions of a physical system. Although one can formulate all of mechanics in terms of the algebra $C^\infty(P)$ of functions on P and its additional Poisson bracket structure, it often helps to appeal directly to the geometry of the points in the space P itself.

In quantum mechanics, the algebra $\mathcal{A}$ of observables is noncommutative, and the objects which play the role of points, namely the irreducible representations of $\mathcal{A}$, may be quite scarce. For instance, there is only one such representation for the algebra of compact operators on Hilbert space $L^2(\mathbb{R}^n)$, while the corresponding classical algebra of functions admits all the elements of $\mathbb{R}^{2n}$ as points.

*Research partially supported by NSF Grant DMS-87-01609 and DOE contract AT 381-ER 12097.

The correspondence referred to in the preceding paragraph is, usually based on the supposition that the algebra $\mathcal{A}$ belongs to a family $\mathcal{A}_\hbar$ of algebras, all with a common underlying vector space, but with a variable multiplication $*_\hbar$ such that as $\hbar \to 0$, the algebra approaches a commutative algebra $\mathcal{A}_0$, which may be interpreted as the algebra of functions on a classical phase space. In addition, the properties of $\mathcal{A}_\hbar$ for $\hbar$ small are supposed to be controlled by $\mathcal{A}_0$ with its Poisson bracket $\{f,g\} \stackrel{\text{def}}{=} \lim_{\hbar\to 0}(f *_\hbar g - g *_\hbar f)/i\hbar.$ [1]

The theory (or program) of quantization turns the correspondence principle around and seeks to construct $\mathcal{A}_\hbar$ (and its representations) from a *Poisson algebra* consisting of a commutative algebra $\mathcal{A}_0$ and a Lie algebra structure $\{\ ,\ \}$ on $\mathcal{A}_0$ which satisfies the compatibility condition $\{fg,h\} = \{f,h\}g + f\{g,h\}$. In *deformation quantization* [1], the product $*_\hbar$ is constructed as a formal power series in $\hbar$, and convergence questions are generally ignored; furthermore, deformation quantization is essentially an algebraic (cohomological) procedure, although affine connections play a role in the case where $\mathcal{A}_0$ is the algebra of C^∞ functions on a manifold.

Unlike deformation quantization, the technique of symplectic groupoids is a geometric procedure which attempts to construct the algebra $\mathcal{A}_\hbar$ for a particular value of $\hbar$ (effectively, $\hbar = 1$).

Although the most common phase spaces for classical mechanics are symplectic manifolds, on which the Poisson bracket has the local normal form $\{f,g\} = \sum_{i=1}^n [\frac{\partial f}{\partial q_i}\frac{\partial g}{\partial p_i} - \frac{\partial g}{\partial q_i}\frac{\partial f}{\partial p_i}]$ in coordinates $q_1,\ldots,q_n,p_1,\ldots,p_n$, our approach, like that of deformation quantization will apply as well to more general Poisson structures on manifolds. (A Poisson structure on P is a bracket on $C^\infty(P)$ which makes it into a Poisson algebra.) This generality is useful for several reasons, including:

- dealing directly with reduced versions of systems with symmetry;

- quantizing Poisson Lie groups to produce quantum groups (Hopf algebras);

- dealing with the classical limit $\hbar \to 0$ in "classical terms".

The last point above deserves the following further explanation, which is inspired by Souriau's insistence in [28] on the role of units in applications of geometry to mechanics.

Let ω be the symplectic structure on a manifold S which represents the phase space of a physical system. The value of ω on a pair of tangent vectors is *not* a real number but rather an element of a 1-dimensional

---

[1]The idea that commutator brackets approach Poisson brackets in the limit of large quantum numbers seems to go back to Dirac, Kramers, and Pauli, (see Chapter IV of [17], especially p. 140), but I have found it difficult to find a source for the corresponding statement in the limit as $\hbar \to 0$. The earliest statement of this kind which I have found is in [2].

vector space $\mathbf{A}$ whose elements represent "quantities of action" (energy $\times$ time). Thus, the deRham cohomology class $[\omega]$ belongs to $H^2(S, \mathbf{A})$, so the prequantizability of S (i.e. the integrality of $[\omega]$) can be determined only after one chooses a basis of $\mathbf{A}$ ("unit of action") which gives an isomorphism of $\mathbf{A}$ with $\mathbb{R}$.

Although nature presents us with Planck's constant $\hbar$ as a distinguished basis of $\mathbf{A}$, the correspondence principle suggests that we pretend that all units of action might be admissible. A convenient way to implement this idea is to replace S by the product $S \times \mathbf{A}$, with a family $\tilde{\omega}$ of symplectic structures along the fibres of the projection $\pi_{\mathbf{A}}$ from $S \times \mathbf{A}$ to $\mathbf{A}$. Following convention, we denote by $\hbar$ the general element of $\mathbf{A}$, rather than the particular value found in nature. [2]

For $\hbar \in \mathbf{A}$, we denote by $i_\hbar : S \to S \times \mathbf{A}$ the natural embedding onto $\pi_{\mathbf{A}}^{-1}(\hbar)$. The family $\tilde{\omega}$ is defined by setting, for tangent vectors v and w in a fibre of TS, $\tilde{\omega}(T_{i_\hbar} v, T_{i_\hbar} w) = \hbar^{-1}\omega(v, w)$. Although this form blows up when $\hbar = 0$, the corresponding Poisson structure is everywhere defined. If we denote by $\{\ ,\ \}_\omega$ the Poisson bracket on S associated with ω, then the bracket $\{\ ,\ \}$ on $P \times \mathbf{A}$ is determined by the equation $\{f, g\} \circ i_\hbar = \hbar \{f \circ i_\hbar, g \circ i_\hbar\}_\omega$. (The "infinite" 2-form corresponds to the zero Poisson bracket.) Note that, if f and g are real-valued functions on $S \times \mathbf{A}$, then $f \circ i_\hbar$ and $g \circ i_\hbar$ are real valued functions on S, but their bracket $\{f \circ i_\hbar, g \circ i_\hbar\}_\omega$ is a function on S with values in $\mathbf{A}^*$, so that the "product" $\hbar \{f \circ i_\hbar, g \circ i_\hbar\}_\omega$ is again real valued.

We call $(S \times \mathbf{A}, \{\ ,\ \})$ a *Heisenberg-Poisson* manifold since, if $S = \mathbb{R}^{2n}$ with canonical coordinates $q_1, \ldots, q_n, p_1, \ldots, p_n$, then the Poisson bracket relations on $\mathbb{R}^{2n} \times \mathbf{A}$ are

$$\{q_i, p_j\} = \hbar \delta_{ij}, \{q_i, q_j\} = \{p_i, p_j\} = \{\hbar, q_i\} = \{\hbar, p_i\} = 0,$$

which are just the bracket relations for the dual of the $2n + 1$-dimensional Heisenberg Lie algebra.

3   Classical models of algebras

There are two fundamental ideas behind the method of symplectic groupoids. The first is that, using geometric quantization, one may attempt to construct directly a noncommutative algebra of observables, rather than a Hilbert space on which it operates. The second is that, between the geometric structures of classical mechanics and the algebraic structures of quantum mechanics there are objects which are still geometric but which incorporate some quantum properties. (The (pre)quantum manifolds of

---

[2]This is not as strange as it may seem at first; in essence, we are letting $\hbar$ be a "dynamical variable" whose constancy in time will be a consequence of the equations of motion.

[28] illustrate this second idea. Although, as principal $U(1)$ bundles with connection, they remain quite geometrical, they are sufficient to explain such typically quantum phenomena as the Bohm-Aharanov effect and the quantization of electric charge in the presence of a magnetic monopole [9].

We know that, when the geometric quantization machine runs smoothly, a vector space V is produced from a symplectic manifold S, and lagrangian submanifolds of S correspond to elements or classes of elements in V. If we wish V to be an associative *-algebra $\mathcal{A}$ with unit element, like the algebras of quantum mechanics, then the underlying symplectic manifold Γ, must also carry additional structures, which can be described as lagrangian submanifolds in the following way.

The multiplication in $\mathcal{A}$, a bilinear map from $\mathcal{A} \times \mathcal{A}$ to $\mathcal{A}$, can also be considered as an element μ of a tensor product $\mathcal{A}^* \otimes \mathcal{A}^* \otimes \mathcal{A}$, where $\otimes$ signifies a suitable completion of the algebraic tensor product when $\mathcal{A}$ is infinite dimensional. If $\mathcal{A}$ is the quantization of Γ, then $\mathcal{A}^*$ becomes in a natural way the quantization of $\overline{\Gamma}$ (Γ with its symplectic structure multiplied by -1), and $\mathcal{A}^* \otimes \mathcal{A}^* \otimes \mathcal{A}$ is the quantization of $\overline{\Gamma} \times \overline{\Gamma} \times \Gamma$. The algebra multiplication μ should be represented, then, by a lagrangian submanifold $M \subset \overline{\Gamma} \times \overline{\Gamma} \times \Gamma$. In a similar way, the unit element of $\mathcal{A}$ should be represented by a lagrangian submanifold Γ_0 of Γ itself, and the star operation in $\mathcal{A}$ by a lagrangian submanifold I in $\Gamma \times \Gamma$. (The star operation in $\mathcal{A}$ is conjugate linear, hence it is linear from $\mathcal{A}^*$ to $\mathcal{A}$ and is thus an element of $\mathcal{A} \otimes \mathcal{A}$.)

The axioms relating multiplication, unit element, and star in $\mathcal{A}$ translate into properties of M, Γ_0, and I with respect to the calculus of lagrangian submanifolds, using the composition of relations. An attempt to write down such properties may be found in [32], under the name "*-algebra in the symplectic category," but the correct version of these properties was found only recently by Karasev [12], the author, [34], and Zakrzewski [42] and may be expressed concisely by the statement "Γ is a symplectic groupoid." More precisely:

- M is the graph $\{(x, y, xy)\}$ of a partially defined multiplication whose domain is a subset Γ_2 of $\Gamma \times \Gamma$;

- there are submersions $\alpha : \Gamma \to \Gamma_0$ and $\beta : \Gamma \to \Gamma_0$ such that, for each $x \in \Gamma$, $\alpha(x)x = x = x\beta(x)$; furthermore, (x, y) belongs to Γ_2 if and only if $\beta(x) = \alpha(y)$;

- for each $x \in \Gamma$, $xI(x) = \alpha(x)$ and $I(x)x = \beta(x)$ (we write x^{-1} for $I(x)$);

- when xy and yz are defined, $(xy)z$ and $x(yz)$ are both defined and equal. [3]

---

[3]Algebraically, a groupoid is a small category Γ in which all morphisms are invertible. Our assumption that M is lagrangian is what makes the groupoid sym-

Many algebraic properties of groupoids follow simply from the axioms above. For instance, $\alpha(xy) = \alpha(x)$ and $\beta(xy) = \beta(y)$, α and β are retractions from Γ to Γ_0, $\alpha(x^{-1}) = \beta(x)$, etc. There is also a very important geometric consequence of the axioms and the fact that M is lagrangian: there is a unique Poisson structure on Γ_0 for which α is a Poisson map. In addition, β is an anti-Poisson map (β^* reverses the sign of Poisson brackets), and the subalgebras $\alpha^* C^\infty(\Gamma_0)$ and $\beta^* C^\infty(\Gamma_0)$ commute with one another in the Lie algebra $C^\infty(\Gamma)$. Hence, $\Gamma_0 \leftarrow \Gamma \rightarrow \overline{\Gamma_0}$ is a symplectic dual pair in the sense that $\ker T\alpha$ and $\ker T\beta$ are the symplectic orthogonal complements to one another in $T\Gamma$.

A lagrangian submanifold of Γ for which the restrictions of α and β are diffeomorphisms onto Γ_0 will be called *unitary*. The unitary lagrangian submanifolds form a group $U(\Gamma)$ under the multiplication of subsets induced from the product on Γ. (Under quantization, these manifolds correspond to unitary elements of $\mathcal{A}$.) The Lie algebra of $U(\Gamma)$ may be identified in a natural way with the space $Z^1(\Gamma_0)$ of closed 1-forms on Γ_0, with the bracket induced from the Poisson bracket on $C^\infty(\Gamma_0)$. Thus, the symplectic groupoid represents a kind of "geometric integration" of the Poisson manifold Γ_0, and so it is natural to pose the following problem, which is also the first step in our quantization procedure.

Integration problem for Poisson manifolds. Given a Poisson manifold P, find a symplectic groupoid Γ and a Poisson isomorphism of Γ_0 with P.

If we can solve the integration problem for P, we may identify P with Γ_0 so that the structure maps α and β go from Γ directly to P. We then refer to Γ as a *symplectic groupoid over P* and say that P is *integrable*.

Example 3.1 If X is any Poisson manifold with the zero Poisson structure, T^*X with its standard symplectic structure is a symplectic groupoid over X. The maps α and β are both equal to the cotangent bundle projection, and the groupoid product is addition in the fibres.

Example 3.2 If $\mathfrak{g}$ is any Lie algebra, then the dual space $\mathfrak{g}^*$ with its Lie-Poisson structure is integrable. For any Lie group G whose Lie algebra is $\mathfrak{g}$, a symplectic groupoid over $\mathfrak{g}^*$ is given by T^*G with its cotangent symplectic structure but with a product different from the one in Example 3.1. The maps α and β for this new structure are given by the right and left translation of cotangent vectors to the fibre at the identity element of G.

Thus, the integrability of all Lie-Poisson spaces $\mathfrak{g}^*$ is a consequence of the existence of a Lie group with a given Lie algebra, and the integration problem for Poisson manifolds may be considered as a generalization of the

---

plectic. It then follows that the manifold Γ_0 of identities and the graph I of inversion must be lagrangian as well.

one for finite-dimensional Lie algebras. We also see from this example that the Poisson manifold $\mathfrak{g}^*$ may be integrated in many ways, but the groupoid arising from the simply connected G is distinguished in that it is α-*simply connected*; i.e. the fibres of α (and hence of β) are simply connected.

Example 3.3 If P is any symplectic manifold, then $P \times \overline{P}$ with the multiplication $(p, q)(q, r) = (p, r)$ is a symplectic groupoid over P. Its unitary lagrangian submanifolds are the graphs of symplectomorphisms of P. By analogy with the case of Example 3.2, one may also construct an α-simply connected symplectic groupoid over P by taking the *fundamental groupoid* $\pi(P)$ consisting of homotopy classes of paths in P with fixed endpoints. It is a covering space of the components of $P \times \overline{P}$ containing the diagonal, and so it inherits a symplectic structure by pullback.

Example 3.4 If S is a "physical" symplectic manifold, whose symplectic form ω is **A**-valued, then the Heisenberg-Poisson manifold $S \times \mathbf{A}$ is integrable if and only if the set of periods for the pullback of ω to the universal covering $\tilde{S}$ is a discrete (hence zero or cyclic) subgroup of **A** (see [37]). In this case, an α-simply connected Γ is constructed by using a prequantizations of $\pi(S)$ together with a blowing-up construction at $\hbar = 0$.

4   Prequantization of the symplectic groupoid

Once we have constructed a symplectic groupoid Γ over a Poisson manifold P, the next step toward building an algebra is to prequantize Γ in a way which is compatible with its groupoid structure. This problem has been investigated in general, with the following result (Theorems 3.1 and 3.2 in [40]).

Theorem 4.1 *Let Γ be a symplectic groupoid for which α is a locally trivial fibration with connected and simply connected fibres. If the cohomology class of the symplectic structure on Γ is integral, then there is a unique prequantization of Γ (i.e. a principal $U(1)$ bundle Q with a connection θ whose curvature is ω) for which the induced flat connection over the lagrangian submanifold Γ_0 has trivial holonomy.*

For any parallel section Q_0 of Q over Γ_0, there is a unique groupoid structure on Q, with Q_0 as the set of units, such that Q is a central extension of Γ by $U(1)$ and the graph of multiplication in Q is an integral manifold of the form $(-\theta, -\theta, \theta)$ on $Q \times Q \times Q$.

The groupoid extension Q corresponds to a 2-cocycle on the groupoid Γ, from which one can construct a twisted groupoid algebra, following the methods described in [22]. Actually, one must alter the construction in

[22] slightly, since the bundle Q is nontrivial, so that the cocycle is not continuous. More concretely, if E is the complex vector bundle over Γ associated to the principal bundle Q, then there is for each $(x, y) \in \Gamma_2$ an isomorphism from $E_x \otimes E_y$ to E_{xy}, which induces by integration an associative multiplication on the space of smooth, compactly supported sections of the line bundle $E \otimes \Omega^{\frac{1}{2}}(\ker T\alpha) \otimes \Omega^{\frac{1}{2}}(\ker T\beta)$. ($\Omega^{\frac{1}{2}}$ signifies half-densities.)

In a sense, the results in [40] serve to connect the global, continuous theory in [22] with the infinitesimal, smooth version of our theory obtained earlier in [10].

We are still short of our goal, since the associative algebra just constructed has the "size" of the space of functions on Γ rather than on the Poisson manifold Γ_0. This is, of course, the usual problem associated with prequantization, and to correct it we must use a polarization.

5   Polarization of the symplectic groupoid

As is common in geometric quantization, this is the hard part. First of all, we must find a polarization F of Γ which is nice enough so that the space V of F-parallel sections of E (or its tensor product with some natural line bundle) can be constructed. Next, we must "quantize" the lagrangian submanifold $M \subset \overline{\Gamma} \times \overline{\Gamma} \times \Gamma$ to get a bilinear multiplication on E. Finally, we must deduce the associativity and other properties of this multiplication from those of the groupoid product.

The last step in the list above is especially difficult–it requires a "precise" quantization of lagrangian submanifolds rather than an asymptotically correct one. This difficulty is of the same nature as that of quantizing symplectomorphisms to produce unitary transformations in a functorial way: although it is impossible in general, it can be done effectively in special situations, usually in the presence of a large symmetry group, or when the intersection of the lagrangian submanifolds with the leaves of the polarization have constant dimension.

The rest of this paper is devoted to a series of examples, together with problems which arise from them.

Example 5.1 In Example 3.1, give T^*X the usual prequantization by the trivial bundle, and the polarization by the fibres. The space V is then $C^\infty(X, \mathbb{C})$ with the operations of pointwise multiplication and complex conjugation, and the constant function 1 as unit element.

Problem 5.2 An interesting technical point arises here. The twisted groupoid algebra of T^*X consists of densities along the fibres, with the operation of convolution. If we try to restrict this convolution to F-parallel densities, the result is always infinite. Is there a natural way to "renormalize" the groupoid algebra multiplication to obtain the pointwise multiplication of functions on X? This problem is similar to that of constructing the inner product on half-densities on X from that on sections of the prequantum line bundle, and its solution should probably involve tensoring E with another line bundle.

Example 5.3 In Example 3.2, give T^*G the same prequantization and polarization as in Example 5.1. The space V is still $C^\infty(G, \mathbb{C})$, but the new groupoid structure quantizes to the operation of convolution, with the antiinvolution $f(x) \mapsto \overline{f(x^{-1})}$, and the delta-function at the group identity as unit element.

Problem 5.4 The alert reader may have noticed a flaw in the preceding example. The natural domain of the convolution operation on a group consists not of functions but of densities. It appears then that the construction of V itself, and not just the multiplication, should depend upon the groupoid structure on Γ. What is an invariant way to describe the construction of V? Here again, tensoring with an appropriate line bundle should play a role.

Example 5.5 In Example 3.3, suppose that the symplectic manifold P is equipped with a prequantization. This induces in a natural way a prequantization of $P \times \overline{P}$, following the prescriptions on pp. 326-328 of [28]. This prequantization has the same properties as the ones produced by Theorem 4.1; the groupoid structure is just that of the *gauge groupoid* [7] of the original prequantization of P.

Suppose now that P is also equipped with a polarization F_0, so that the manifold P is "quantized" to a complex vector space H. Then, if we give $\Gamma = P \times \overline{P}$ the polarization F which is the product of two copies of F_0, the resulting vector space V is the space of linear endomorphisms of H (modulo topological technicalities which we shall ignore here).

If all goes well, the diagonal in Γ should quantize to the identity endomorphism, and the graph of multiplication in Γ should quantize to the composition of endomorphisms. (The second problem can be reduced to the first if one replaces P by P^3.)

This seems to work out correctly for both real and complex polarizations P_0. In the case of a complex polarization, a good way to quantize a lagrangian submanifold L (carrying a density) seems to be to form the superposition of the coherent states (see [21] or Rawnsley's paper in this volume) corresponding to a parallel unitary section over L of the prequantum line bundle.

Example 5.6 If we take $P = \mathbb{R}^{2n}$ with its standard symplectic structure, there is a polarization of $P \times \overline{P}$ which does not depend on any polarization of P but only on the affine structure. In fact, the polarization comes from the isomorphism of $\Gamma = \mathbb{R}^{2n} \times \overline{\mathbb{R}^{2n}}$ with $T^*\mathbb{R}^{2n}$ (see [31])

$$(x, y) \mapsto (\frac{1}{2}(x + y), \tilde{\omega}(y - x)),$$

where $\tilde{\omega}$ is the map from $\mathbb{R}^{2n}$ to $\mathbb{R}^{2n\,*}$ associated with the symplectic structure. Using this polarization to quantize Γ, we get as V the space of smooth functions on the diagonal, which we identify with $\mathbb{R}^{2n}$ itself, and the multiplication on V turns out to be the so-called *Moyal product* (see [1]). If one uses the Blattner-Kostant-Sternberg pairing to go from this space to the one obtained from two copies of an affine real polarization of $\mathbb{R}^{2n}$ itself, the result is the isomorphism which takes each function h on phase space to the function (actually, the half-density) on $\mathbb{R}^n \times \mathbb{R}^n$ which is the Schwartz kernel of the operator whose Weyl symbol (see [30]) is h.

There does not seem to be a detailed exposition of the preceding example in print, but it is very close to the next one.

Example 5.7 The quantization procedure which we are describing in this paper is completely carried out in [38] when P is a torus $\mathbb{T}$ with a translation-invariant Poisson structure. The symplectic groupoid for $\mathbb{T}$ always turns out to be symplectomorphic to $T^*\mathbb{T}$ (with a groupoid structure depending on the Poisson structure, of course), so that the vector space V may be identified with $C^\infty(\mathbb{T})$. When the Poisson structure on $\mathbb{T}$ is zero, the product on V is of course the pointwise multiplication, otherwise one gets a noncommutative multiplication on $C^\infty(\mathbb{T})$ which is precisely that of (the functions on) a *noncommutative torus*, one of the basic examples of noncommutative geometry [6]. (The relation between Poisson tori and noncommutative tori was studied from the point of view of deformation quantization in [23].)

6 Double symplectic groupoids and quantum groups

Although the two multiplications associated with the two groupoid structures on T^*G described in Examples 5.1 and 5.3 live on different spaces, it is possible to relate them more closely by dualizing one of them, say convolution. In this way, we obtain the *coproduct* Δ on $C^\infty(G)$ defined as a map from $C^\infty(G)$ to $C^\infty(G) \otimes C^\infty(G) = C^\infty(G \times G)$ by the formula $(\Delta f)(x, y) = f(xy)$. The coproduct satisfies a coassociative law and is related to pointwise multiplication by the simple identity $\Delta(fg) = \Delta(f)\Delta(g)$, making $C^\infty(G)$ into a *Hopf algebra*.

Of course, one can also leave the convolution operation alone and dualize the pointwise multiplication to get the coproduct which takes a density on G and pushes it forward by the diagonal map to produce a (distributional) density on $G \times G$.

The compatibility of multiplication and coproduct on $C^\infty(G)$ is reflected at the geometric level by the fact that T^*G is a *double groupoid* with respect to its two groupoid structures over G and $\mathfrak{g}^*$. This means that each of the structure maps for one groupoid structure is a groupoid morphism for the other. It is natural to try to model more general Hopf algebras by looking at more general *symplectic double groupoids*.

Remark 6.1 S. Zakrzewski has pointed out to me that not every symplectic double groupoid can be expected to model a Hopf algebra. For instance, if G is any Lie groupoid, T^*G is a symplectic double groupoid, but the corresponding quantum object $C^\infty(G)$ is generally not a Hopf algebra, but rather only a *Hopf algebroid* in the sense of [20]. (See the comments on p. 116 of [3] as well.) For example, if M is any manifold, it carries the trivial groupoid structure, and then the double groupoid structure on T^*M consists of the same groupoid structure used twice. (The fact that this is a double groupoid reflects the commutativity of addition in the fibres.) The "coproduct" on $C^\infty(M)$ wants to map a function on M to the corresponding multiple of the delta function along the diagonal in $M \times M$, but this lies in the space of generalized densities on $M \times M$ rather than (generalized) functions. The correct coproduct (coproductoid?) for the Hopf algebroid structure maps $C^\infty(M)$ to the quotient algebra of $C^\infty(M \times M)$ consisting of the functions on the diagonal.

Zakrzewski further points out that, for a symplectic groupoid to model a Hopf algebra, it is natural that the second geometric operation should be a coproduct rather than a product. The compatibility conditions here are somewhat different then for products; details may be found in [42].

It turns out that a symplectic double groupoid always arises when one integrates a *Poisson Lie group*, i.e. a Lie group G carrying a Poisson structure which is *multiplicative* in the sense that the multiplication map from $G \times G$ to G is a Poisson map. More precisely [14]:

Theorem 6.2 *Let G be any Poisson Lie group. Then G is integrable as a Poisson manifold, and the group structure on G lifts to give a second groupoid structure on the symplectic groupoid Γ of G, making Γ into a double groupoid. The space of identities for the second groupoid structure is another Poisson group G^* which is the dual of G in the sense of Drinfel'd [8].*

Remark 6.3 The cautions of Remark 6.1 do not appear to pose any difficulties in the case of the symplectic double groupoids arising from Theorem 6.2.

The symplectic double groupoid of a Poisson Lie group is very closely related to the *double group* [8] whose Lie algebra as a vector space is a direct sum of subalgebras isomorphic to $\mathfrak{g}$ and $\mathfrak{g}^*$ and which carries an *ad*-invariant inner product for which the summands are isotropic. In fact, for many Poisson Lie groups (called *complete*) the symplectic groupoid as a symplectic manifold is isomorphic [14] [15] to the double group with the symplectic structure studied in [26].

The Hopf algebras obtained by quantizing (by any means available) Poisson Lie groups are called (algebras of functions on) *quantum groups.* It would be very interesting to be able to construct quantum groups from the symplectic double groupoids of Theorem 6.2. Leaving aside for the moment the relatively simple prequantization problem, we shall describe here an idea due to Szymczak and Zakrzewski [29] (also see [18]) for polarizing the symplectic groupoids of certain Poisson Lie groups.

At any point p of a Poisson manifold P where the Poisson tensor is zero, there is an induced *linearization* [33], which is a Lie-Poisson structure on $T_p P$, dual to a Lie algebra structure on $T_p^* P$. The Poisson structure is said to be (locally or globally) linearizable at p if there is a (local or global) Poisson isomorphism from $T_p P$ to P which maps 0 to p and whose derivative at 0 is the identity. If H is a Poisson Lie group, its Poisson tensor vanishes at the identity element, so there is a Lie algebra structure on $\mathfrak{h}^*$ which is just the Lie algebra of the dual group H^*. If H is the dual of G, then the dual of H is G, so that the linearized structure on $\mathfrak{h}$ is just the Lie-Poisson structure of $\mathfrak{g}^*$.

Suppose now that the dual of the simply connnected Poisson Lie group G is globally linearizable. This means that there is a Poisson isomorphism from the Lie-Poisson space $\mathfrak{g}^*$ to G^*. We know from Example 3.2 that T^*G is a symplectic groupoid over $\mathfrak{g}^*$, so it must be a symplectic groupoid over G^* as well. On the other hand, we know from Theorem 6.2 that the symplectic groupoid of G^* has a second groupoid structure which makes it into a symplectic groupoid over G. We conclude:

Proposition 6.4 *Suppose that the dual of the Poisson Lie group G is globally linearizable. Then the cotangent bundle T^*G has a structure of symplectic double groupoid over G and G^*, with the "second" groupoid structure being that of T^*G over $\mathfrak{g}^*$. Unless G has the zero Poisson structure, the groupoid structure of T^*G over G is not given by addition in the fibres; instead, it is obtained by lifting from $\mathfrak{g}^*$ the noncommutative group structure induced by the Poisson isomorphism with G^*.*

The application of the preceding proposition to quantization is immediate. If G^* is globally linearizable, then we give its symplectic groupoid T^*G the usual prequantization and polarization so that, by Example 5.3, we obtain the vector space $C^\infty(G)$ with its usual convolution (or coproduct) operation. On the other hand, as long as the Poisson structure on G is not zero, the groupoid structure on T^*G over G is not addition in the fibres, and so the *multiplication* on $C^\infty(G)$ is not pointwise multiplication, but is instead a noncommutative, "deformed" multiplication.

Computing the new multiplication on $C^\infty(G)$ is generally a difficult problem. It seems to require having an explicit linearization of G^*, together with explicit calculation of the new groupoid structure on T^*G.

We are currently trying to carry this out for the case of the Poisson Lie group structures on compact semisimple Lie groups described in [15]. According to the theorem of Conn [4], the Poisson structure on the dual of a semisimple Poisson Lie group is always locally linearizable (since it is analytic). In addition, for the structures of [15], Ginzburg, Lu, and Ratiu ([13], plus work in progress) have found a diffeomorphism ϕ from $\mathfrak{g}^*$ to G^* which takes symplectic leaves to symplectic leaves, preserving the cohomology classes of their symplectic structures. This is very strong evidence for global linearizability since, by the theorem of Moser [19], it is possible to deform ϕ leafwise to a family of symplectic diffeomorphisms, while Conn's theorem says that, near the identity, the leafwise maps can be fit together smoothly.

In the case of $SU(2)$, the linearization has been found explicitly by P. Xu (unpublished), but even in this simple case it has not yet been possible to finish the construction of the deformed multiplication. We hope, of course, that it will agree in some sense with the quantum $SU(2)$ of [8] and [41].

One case in which the entire procedure has been carried out is that of Heisenberg groups. Szymczak and Zakrzewski [29] classify all the multiplicative Poisson structures on the simply connected Heisenberg groups G. They show that each such structure is linearizable; in fact, the exponential map is a Poisson isomorphism from $\mathfrak{g}$ to G. Next, they observe that, for a large class of the multiplicative structures, the dual group G^* is the semidirect product of the positive real numbers acting linearly on the $2n$-dimensional hyperplane $W \subset \mathfrak{g}^*$ annihilated by the center of G. When this action preserves the natural symplectic structure on W, the dual group G^* is linearizable as well, this time not by the exponential map but by a modification known as "coordinates of the second kind." Let us call this situation the *bilinearizable* case.

It turns out that every bilinearizable Heisenberg group G is complete, from which it follows without too much difficulty that its symplectic double groupoid can be identified as a symplectic manifold with the symplectic vector space $\mathfrak{g} \times \mathfrak{g}^*$. It is convenient to use a different polarization for each of the two groupoid structures, so that each one quantizes to a group convolution. As we explained above, we get the space $C^\infty(G)$ with its usual coproduct. The "deformed multiplication" on $C^\infty(G)$, on the other hand, is realized as the convolution on densities on G^*. To put both of these operations on the same space, one uses the Fourier transform to identify densities on $\mathfrak{g}^*$ with smooth functions on $\mathfrak{g}$, together with the exponential map for G and the coordinates of the second kind for G^*. The resulting Hopf algebra structures seem to resemble very closely the ones discovered by Kac and Palyutkin [11], which were recently analyzed by Rieffel [25] in

458 A. Weinstein

terms of Poisson Lie groups and their duals; we do not know whether they are identical.

Example 6.5 As a final example, we look at the case of simply connected abelian Poisson Lie groups. These are just the Lie-Poisson spaces $\mathfrak{g}^*$ considered in Example 3.2. We have already seen that T^*G is the symplectic double groupoid for this example, just as it is for the group G with zero Poisson structure. We have described its quantization at the beginning of this section, but now we must consider the convolution of densities on G as the deformed multiplication on $\mathfrak{g}^*$, while the deformed coproduct is now the push-forward along the diagonal in $G \times G$. To relate these operations to the original multiplication and coproduct on $\mathfrak{g}^*$, it is necessary to identify functions on $\mathfrak{g}^*$ with densities on G. This can be done, via the Fourier transform to densities on $\mathfrak{g}$, if we have a global diffeomorphism of G with $\mathfrak{g}$, i.e. if $\mathfrak{g}$ with the zero Poisson structure is globally linearizable. The exponential map is such a diffeomorphism when G is nilpotent, and this case has been found [24] to be completely in order, even from the analytic point of view. When G is exponential solvable but not nilpotent, the Schwartz functions on $\mathfrak{g}$ are no longer closed under convolution, so problems begin to appear. Finally, when G is not diffeomorphic to euclidean space, the relation between the convolution on G and the Poisson bracket on $\mathfrak{g}^*$ must be localized at the identity of G. To accomplish this, one may restrict attention to distributional densities supported at the identity, which correspond to the universal enveloping algebra $U(\mathfrak{g})$, a quantization of the Poisson algebra of polynomials on $\mathfrak{g}^*$. Alternatively, one may follow the suggestion in [24] and transfer to G functions on larger and larger neighborhoods of $0 \in \mathfrak{g}^*$ as $\hbar \to 0$.

7 Conclusion

Examination of the examples above, and some others not described here, suggests a conjecture relating Poisson manifolds and their quantizations.

Definition 7.1 *A Poisson manifold P is of* **exponential type** *if its α-simply connected symplectic groupoid Γ exists and admits a diffeomorphism to T^*P which is symplectic and is $\mathbb{Z}_2$-equivariant with respect to inversion in Γ and multiplication by -1 in T^*P.*

We conjecture that the property of exponential type for a Poisson manifold P is equivalent to the existence of a good quantization of the Poisson algebra $C^\infty(P)$ by a family (not just formal) of deformed multiplications on the same space. Of course, the notion of good quantization must be made more precise, and the condition of exponential type may need to be refined. A good test case may be the compact Riemann surfaces, considered as symplectic manifolds. They are of exponential type [27], but the work in [30] suggests that the quantization of these manifolds by multiplications on a fixed vector space may be impossible.

REFERENCES

[1] Bayen, F., Flato, M., Fronsdal, C., Lichnerowicz, A., and Sternheimer, D., *Deformation theory and quantization, I and II*, Ann. Phys. **111**, (1977), 61–151.

[2] Bohm, D., "Quantum Theory", Prentice-Hall,New York, 1951.

[3] Brown, R., *From groups to groupoids: a brief survey*, Bull. London Math. Soc. **19** (1987), 113–134.

[4] Conn, J.F., *Normal forms for analytic Poisson structures*, Annals of Math. **119** (1984), 576–601.

[5] Conn, J.F., *Normal forms for smooth Poisson structures*, Annals of Math. **121** (1985), 565–593.

[6] Connes, A., "Géométrie non Commutative," InterEditions, Paris, 1990.

[7] Coste, A., Dazord, P. et Weinstein, A., *Groupoïdes symplectiques*, Publications du Département de Mathématiques, Université Claude Bernard-Lyon I **2A** (1987), 1–62.

[8] Drinfel'd, V. G., *Quantum groups*, Proc. ICM, Berkeley, 1986, vol.1, 789–820.

[9] Horváthy, P.A., *Classical action, the Wu-Yang phase factor, and pre-quantization*, Lecture Notes in Math. **836** (1980), 67–70.

[10] Huebschmann, J., *Poisson cohomology and quantization*, J. Reine Angew. Math. **408** (1990), 57–113.

[11] Kac, G.I. and Palyutkin, V.G., *An example of a ring group generated by Lie groups*, Ukrain. Math. J. **16** (1964), 99–105.

[12] Karasev, M.V., *Analogues of objects of Lie group theory for nonlinear Poisson brackets*, Math. USSR Izvestiya **28** (1987), 497–527.

[13] Lu, J.-H., and Ratiu, T., *On the nonlinear convexity theorem of Kostant*, preprint (1990).

[14] Lu, J.-H. and Weinstein, A., *Groupoïdes symplectiques doubles des groupes de Lie-Poisson*, C. R. Acad, Sci. Paris **309** (1989), 951–954.

[15] Lu, J.-H. and Weinstein, A., *Poisson Lie groups, dressing transformations, and the Bruhat decomposition*, J. Diff. Geom. **31** (1990), 501–526.

[16] Mackenzie, K., "Lie Groupoids and Lie Algebroids in Differential Geometry," LMS Lecture Notes Series, **124**, Cambridge Univ. Press, 1987.

[17] Mehra, J. and Rechenberg, H., "The Historical Development of Quantum Theory;" v. 4, Springer-Verlag, New York, 1982.

[18] Mikami, K., *Symplectic double groupoids over Poisson* $(ax+b)$*-groups*, Trans. Amer. Math. Soc. (to appear).

[19] Moser, J., *On the volume elements on a manifold*, Trans. Amer. Math. Soc. **120** (1965), 280–296.

[20] Ravenel, D.C. "Complex cobordism and stable homotopy groups of spheres," Academic Press, Orlando, 1986.

[21] Rawnsley, J., *Coherent states and Kahler manifolds*, Quart. J. Math. **28** (1977), 403–415.

[22] Renault, J., *A groupoid approach to C^* algebras*, Lecture Notes in Math. **793** (1980).

[23] Rieffel, M.A., *Deformation quantization of Heisenberg manifolds*, Commun. Math. Phys. **122** (1989), 531–562.

[24] Rieffel, M.A., *Lie group convolution algebras as deformation quantizations of linear Poisson structures*, Amer. J. Math. **112** (1990), 657–686.

[25] Rieffel, M.A., *Some solvable quantum groups*, in "Proceedings of the Rumanian conference OAET II," Sept. 1989 (to appear).

[26] Semenov-Tian-Shansky, M. A., *Dressing transformations and Poisson Lie group actions*, Publ. RIMS, Kyoto University **21** (1985), 1237–1260.

[27] Sikorav, J.-C., *Problèmes d'intersections et de points fixes en géométrie hamiltonienne*, Comment. Math. Helvetici **62** (1987), 62–73.

[28] Souriau, J.M., "Structure des systèmes dynamiques," Dunod, Paris, 1970.

[29] Szymczak, I. and Zakrzewski, S., *Quantum deformations of the Heisenberg group obtained by geometric quantization*, preprint, Warsaw University, 1990.

[30] Unterberger, A., *Quantification et analyse pseudo-différentielle*, Ann. Sci. Éc. Norm. Sup. **21** (1988), 133–158.

[31] Weinstein, A., *Symplectic manifolds and their lagrangian submanifolds*, Advances in Math. **6** (1971), 329–346.

[32] Weinstein, A., *The symplectic "category"*, Lect. Notes Math. **905** (1982), 45–50.

[33] Weinstein, *A., The local structure of Poisson manifolds*, J. Diff. Geom. **18** (1983), 523–557.

[34] Weinstein, A.,*Symplectic groupoids and Poisson manifolds*, Bull. Amer. Math. Soc. **16**, (1987), 101–104.

[35] Weinstein, A., *Coisotropic calculus and Poisson groupoids*, J. Math. Soc. Japan **40** (1988), 705–727.

[36] Weinstein, A., *Some remarks on dressing transformations*, J. Fac. Sci. Univ. Tokyo. Sect. 1A, Math. **36** (1988), 163–167.

[37] Weinstein, A., *Blowing up realizations of Heisenberg-Poisson manifolds*, Bull. Sc. Math. **113** (1989), 381–406.

[38] Weinstein, A., *Symplectic groupoids, geometric quantization, and irrational rotation algebras*, in "Symplectic geometry, groupoids, and integrable systems," Seminaire sud-Rhodanien à Berkeley (1989), Springer-MSRI Series (to appear in 1991).

[39] Weinstein, A., *Affine Poisson structures*, Intern. J. Math. **1** (1990), 343–360.

[40] Weinstein, A. and Xu, P., *Extensions of symplectic groupoids and quantization*, J. Reine. Angew. Math. (to appear).

[41] Woronowicz, S. L., *Twisted SU(2) group. An example of a noncommutative differential calculus*, Publ. RIMS, Kyoto University **23** (1987), 117–181.

[42] Zakrzewski, S., *Quantum and classical pseudogroups, I and II*, Comm. Math. Phys. **134** (1990), 347–370, 371–395.

Department of Mathematics
University of California
Berkeley, CA 94720
USA
(alanw@math.berkeley.edu)

Reduction Procedures for Poisson Manifolds

DON C. WILBOUR and JUDITH M. ARMS

To J.-M. Souriau,
whose ideas have provided the foundation for all our work

Abstract.

A reduction procedure gives an answer to the following question. Given a Poisson algebra of functions on a manifold M and a subset V of M, how may one construct in a natural way a Poisson algebra on some quotient space of V? If M is a symplectic manifold, V is a submanifold, and the kernel of the symplectic form is fibrating, then there is a well-known geometric reduction procedure which yields a symplectic quotient of V. On the other hand, if V is the zero set of an equivariant momentum map, then an algebraic reduction procedure of Śniatycki and Weinstein yields a reduced Poisson algebra which is often (but not always) a function algebra on a quotient of V. (It may contain nilpotents, for instance.) Consideration of these and other methods of reduction raises the following questions. 1) How may these procedures be generalized? 2) How do different reductions compare: when are they the same, how do they differ? Substantial progress has been made recently on these questions. Several approaches now are available for the case of singular V, and some comparison results have been proved. This paper reviews previous work and presents a "generalized Dirac reduction" that applies to an arbitrary subset of a Poisson manifold. The Dirac reduction has both algebraic and geometric characterizations, facilitating new comparison results.

I. Introduction

In classical mechanics the concept of reduction arose in the study of systems with first integrals in involution, that is, conserved quantities that Poisson commute. The reduced space is a quotient of a level set, call it V, of the first integrals, and the Poisson structure and dynamics descend in a natural way to the reduced space. In the modern formulation of this problem, due largely to Professor Souriau [**So**], V is a level set of an equivariant momentum map for the canonical action of a group on a symplectic manifold M, and we quotient by the action. In this form reduction has played a role in the study of completely integrable systems, representation theory, and geometric invariant theory. (See e.g. [**KKS**] and [**Kr**].) One also would like to consider cases where V arises from physical constraints on configuration or phase space. The basic question is the following. Given a Poisson algebra of functions on a manifold M and a subset V of M, how may one construct in a natural way a Poisson algebra on some quotient space of V?

About twenty years ago, Meyer [**Me**] and Marsden and Weinstein [**MaW74**] formulated a reduction procedure that applies when V is the level set of a regular value of the momentum and the group action is sufficiently nice. In recent years several reduction procedures have been articulated in order to extend the reduction process to various singular situations. Consideration of these methods raises the following questions. 1) How generally do they apply, and how may they be further generalized? 2) How do different reductions compare: when are they the same, how do they differ?

This paper reviews various reduction procedures and discusses recent progress in answering the questions posed above. In particular we will present a "generalized Dirac reduction" that has both algebraic and geometric characterizations, and thus facilitates comparisons. For simplicity, we will restrict attention to the finite dimensional case; but many of our constructions and results may be extended, if properly formulated, to the infinite dimensional case. (Cf. [**Wi**].)

II. Basic Reduction Procedures

The reduction procedures in the literature place varying emphasis on geometry, algebra, and, in the case that V is defined by a momentum map, the group action. We consider first a basic geometric approach which concentrates on construction of a reduced space, and leaves the reduced algebra to follow as a consequence. Although the best known version of this reduction, due to Meyer, Marsden, and Weinstein, usually is described with reference to the group action, in fact the construction is essentially geometric. In contrast the second approach considered below, algebraic reduction, ignores geometry (i.e. the reduced space) completely, and uses the group action only as represented in the Poisson algebra by the components of the momentum map. Recently two additional procedures have been proposed which have both geometric and algebraic aspects, and produce both a reduced space and a reduced algebra. In this section we briefly describe these reductions and some of their advantages and disadvantages.

The basic geometric reduction constructs the reduced space as a symplectic quotient of the constraint set V; the reduced Poisson algebra then arises in the usual way as the algebra of smooth functions on this reduced space. This construction appears in many places, for instance in Abraham and Marsden [**AbM**], who credit it to E. Cartan. We will formulate it as follows.

Theorem 2.1 *Let V be a differentiable manifold with a closed form ω. Suppose that*

$$\ker \omega = \{X \in TV \mid i(X)\omega = 0\}$$

has constant dimension and thus defines a distribution on V. Then this distribution is integrable. Suppose further that the foliation defined by this distribution is fibrating; i.e. there exists a manifold W and a submersion $\pi : V \to W$ such that the leaves of the foliation defined by this distribution are the preimages of points in W. Then there exists a unique symplectic form $\widetilde{\omega}$ on W with $\omega = \pi^*\widetilde{\omega}$.

In particular if V is a submanifold of a symplectic manifold (M, ω), we obtain a symplectic form $\widetilde{\omega}$ with $\omega|_V = \pi^*\widetilde{\omega}$. We call W the (geometrically) reduced space; the (geometrically) reduced algebra $C^\infty(W)$ we shall denote by $\mathcal{A}_G$.

Recall that when a group G acts symplectically on M, a momentum map is a mapping

$$J : M \to \mathfrak{g}^*$$

such that for each $\xi \in \mathfrak{g}$, the Hamiltonian vector field for the function $J^\xi(x) = \langle J(x), \xi \rangle$ is ξ_M, the corresponding infinitesimal generator of the group action on M. We shall always require momentum maps to be equivariant with respect to the coadjoint action on the dual Lie algebra $\mathfrak{g}^*$.

If μ is a (weakly) regular value of J, then $V = J^{-1}(\mu)$ is a manifold. Let G_μ be the isotropy group of $\mu \in \mathfrak{g}^*$. From the definition of J one easily computes that $\ker \omega$ at a point p is equal to the tangent space to the G_μ–orbit of p, that is to $T(G_\mu \cdot p)$. If G_μ is connected and acts properly and freely on V, then the theorem above gives us a reduced space

$$W = \frac{V}{G_\mu}.$$

The Poisson bracket on W is induced by that on M. If f and g are functions on W, they give rise to G_μ-invariant functions on V; it is easily shown that these may be extended to functions on M whose Poisson bracket restricts and projects to the bracket of f and g on W. We will designate this reduction as the MMW reduction because it was first noted by Meyer [**Me**] and Marsden and Weinstein [**MaW74**] (all influenced by work of Smale [**Sm**]).

(Remarks: This geometric reduction procedure still works, with minor modifications, even if we require of G_μ only that it act properly on V with equidimensional orbits. If G_μ is not connected, W will be a k-fold cover of V/G_μ, where k is at most the number of components of G_μ. For simplicity in presenting results, we will always hypothesize that G_μ is connected; but with minor modifications to account for this k-fold cover, our results still hold if G_μ is not connected. If the G_μ-action on V is not free but only locally free, i.e., has discrete isotropy, then W is not necessarily a manifold. In this case, or even if we merely require the orbits $G_\mu \cdot p$ to have the same dimension, one can easily show using arguments as in [**We**] that W is a symplectic orbifold.)

The basic geometric approach has several limitations. As presented above, it applies to symplectic but not more general Poisson manifolds. The constraint set V must be a manifold, whereas in many important applications it is not. (See for example [**AMM**].) There are restrictions on the kernel of $\omega|_V$. In the momentum map case, these become restrictions on the group action. Also this procedure may ignore information about the momentum map J. For example, on a two dimensional phase space with global coordinates (x, y), the two real line actions generated by $J = y$ and $J = y^2$ have the same zero momentum set. Thus the corresponding geometric reduced spaces are the same, and the reduction process loses the information that J is quadratic in the latter case.

Śniatycki and Weinstein [**SnW**] proposed a more algebraic reduction procedure which avoids these problems. Let $I(J)$ be the (commutative) ideal of functions generated by the components of the momentum map J; thus each function in $I(J)$ is a sum of terms with components of J as factors. Let $N(I(J))$ indicate the Poisson normalizer of this ideal; that is,

$$N(I(J)) = \{f \in C^\infty(M)| \ \{f, I(J)\} \subseteq I(J)\}.$$

The equivariance of J implies that $I(J)$ is closed under Poisson bracket, and therefore $I(J)$ is a Poisson ideal in $N(I(J))$; that is, $\{I(J), N(I(J))\} \subseteq I(J)$ as well as $I(J) \cdot N(I(J)) \subseteq I(J)$. Thus the Poisson algebra structure on $C^\infty(M)$ descends to one on

$$\mathcal{A}_A = \frac{N(I(J))}{I(J)}.$$

Śniatycki and Weinstein [**SnW**] show that if G is connected, zero is a regular value of the momentum map, and $V = J^{-1}(0)$, this reduced algebra may be identified with the Poisson algebra of the reduced space V/G.

This algebraic reduction has several advantages over the basic geometric reduction discussed above. It does distinguish the exact form of the momentum map (e.g. in the y^2 example above). Given an equivariant momentum map, the reduced algebra $\mathcal{A}_A$ always exists. Momentum maps can be defined for Poisson manifolds which are not symplectic, so the algebraic reduction procedure extends to this more general case. (In the definition of momentum map above, require that the group G act by Poisson maps, i.e. maps that preserve the Poisson bracket. Hamiltonian vector fields and equivariance can be defined as before; or the latter may be rephrased by requiring J to be a Poisson map with the (appropriately signed) Lie-Poisson structure on $\mathfrak{g}^*$. See [**MaW83**] for more discussion.)

One apparent disadvantage is the restriction to the zero value of momentum. This restriction is easily overcome by a standard construction which reduces the case of a nonzero value to the zero value case; cf. [**AGJ**]. Section IV below gives another method for dealing with nonzero values.

Although in regular cases $\mathcal{A}_A$ is equal to $C^\infty(V/G)$, in some cases $\mathcal{A}_A$ is not the function algebra of any space. A simple example of this phenomenon is the circle action on $\mathbf{C}^2$ (with the standard symplectic structure) generated by $J = |z_1|^2 + |z_2|^2$. Define f_1, f_2, and f_3 by

$$f_1 = |z_1|^2 - |z_2|^2$$

$$f_2 + i f_2 = 2 z_1 z_2.$$

The reduced algebra $\mathcal{A}_A$ is generated by (the equivalence classes of) f_1, f_2, and f_3, but

$$f_1^2 + f_2^2 + f_3^2 = 0 \quad \text{in } \mathcal{A}_A \text{ (i.e., modulo } I(J)).$$

As f_1, f_2, and f_3 are nonzero elements of $\mathcal{A}_A$, an algebra over the reals, $\mathcal{A}_A$ cannot be the function algebra of any space. In such cases by the algebraic reduction procedure we have a reduced algebra but no reduced space in the usual sense. In some examples, "real nilpotent" elements like the f_i above make the geometric quantization of $\mathcal{A}_A$ agree with the result of quantizing first, then imposing symmetry constraints. (Cf [**SnW**].) From a classical point of view, however, these nilpotents seem problematic. (For more discussion of this and other examples with such "real nilpotents," see [**AGJ**].)

A third approach which we will call "reduction by invariants" (introduced as "universal reduction" by Arms, Cushman, and Gotay [**ACG**]) falls between the basic geometric and algebraic procedures. Let $V = J^{-1}(\mu)$, a level set of an equivariant momentum map on a symplectic, or more generally a Poisson, manifold and let $I(V)$ be the ideal of smooth functions that vanish on V. The set of G-invariant functions on M, $C^\infty(M)^G$, inherits a Poisson algebra structure which descends to

$$\mathcal{A}_I = \frac{C^\infty(M)^G}{I(V) \cap C^\infty(M)^G}.$$

(See [**ACG**] for proof.) In this quotient, the denominator is the intersection of the numerator with $I(V)$, so elements of $\mathcal{A}_I$ can be distinguished by their values on V. Thus, unlike $\mathcal{A}_A$, the invariantly reduced algebra $\mathcal{A}_I$ is a function algebra on a quotient space of V. Let $\pi : M \to M/G$ be projection onto the orbit space of M. The invariantly reduced space is $\pi(V)$. This space can be identified with V/G_μ, the reduced space in the geometric MMW procedure. (Recall that for simplicity we assume G_μ is connected.) An element of $\mathcal{A}_I$ can be identified with a function on the reduced space $\pi(V)$ because by G-invariance any representative of that element in $C^\infty(M)^G$ induces a function on V/G_μ; this function is uniquely determined by the original element of $\mathcal{A}_I$. Thus with only the requirement that V be the level set of a momentum map, invariant reduction always produces both a reduced space and a reduced Poisson algebra of functions on that space.

However, $\mathcal{A}_I$ can be strictly smaller than the reduced algebra obtained by the geometric MMW procedure in some cases where both apply, even though both have the same reduced space. For example, suppose that the coadjoint orbit $G \cdot \mu$ is open. Then $\pi(V)$ is also open (because $\pi^{-1}(\pi(V)) = J^{-1}(G \cdot \mu)$), and a G_μ-invariant function on V may have no G-invariant extension to M. Such an example is described in [**ACG**].

The basic geometric reduction procedure is more general than the algebraic or invariant approaches in that no momentum map is supposed. At the same time it is much more limited because M must be symplectic, V must be smooth, and $\omega|_V$ must be constant rank. In part these restrictions occur because the reduced space must be nice enough (i.e. a manifold or orbifold) to define, intrinsically, a function algebra. One way to circumvent this problem is illustrated in the reduction by invariants: define the reduced algebra directly from $C^\infty(M)$ and independently of the reduced space.

This approach was used by Arms, Gotay, and Jennings [**AGJ**] to extend the geometric approach to some cases where V is singular. We recall the ideas of this procedure only briefly here and refer to [**AGJ**] for details. A tangent space TV_p is defined at each singular point p of V. (This tangent space must be a vector space at each point, but may vary in dimension from point to point.) Define two points in V to be equivalent if they are connected by a curve whose tangent is everywhere in the kernel of ω restricted to TV, and let the reduced space be the quotient of V by this equivalence relation. The reduced algebra $\mathcal{A}_G$ consists of all functions on V which have smooth extensions to M and have zero derivative in the directions of $\ker \omega$. Existence of a Poisson algebra structure on $\mathcal{A}_G$ is not automatic, but for example it suffices to know that V is the level set of the momentum for a compact group action. In this case (if the group is connected) the resulting reduced space and algebra are the same as for reduction by invariants. On the other hand if V has high order singularities then there may be no natural Poisson algebra structure on $\mathcal{A}_G$. We use the symbol $\mathcal{A}_G$ because this reduced algebra coincides with the geometric reduced algebra defined earlier when both apply. (Actually, there is one exception but it rarely occurs in examples of interest. If V is not closed, $C^\infty(V)$ may contain functions which do not extend even continuously off V. Then the reduced algebra of [**AGJ**] may be strictly smaller than the reduced algebra $C^\infty(W)$ for the basic geometric reduction.) In particular if the MMW reduction applies, it will coincide with the present geometric reduction. The examples discussed earlier show that $\mathcal{A}_G$ and $\mathcal{A}_A$ need not coincide, even if the group G is compact. (See [**AGJ**] for proofs of these results and additional examples.)

Thus we have three basic philosophies of reduction: geometric, algebraic, and invariant. The algebraic and invariant approaches have the same domain of applicability (level sets of equivariant momentum), but this neither contains nor is contained in the domain of the geometric approach. If the group is compact, we have noted that the geometric and invariant reductions coincide. More generally, comparisons have been difficult to make. Combining the previous examples shows that the three reductions all can be different, even when V is a manifold! Somewhat surprisingly, the introduction of yet another method of reduction in the next section helps simplify the picture.

III. The Generalized Dirac Reduction

In this section we propose a new reduction procedure which we call the generalized Dirac reduction. This reduction procedure applies in all cases: no group action or momentum is required for the construction, and there are no restrictions on the set V. Moreover, there are two ways to construct the Dirac reduced algebra $\mathcal{A}_D$, one algebraic in flavor and the other more geometric. Using both these constructions will facilitate comparisons in Section IV to the reduction procedures of Section II. The geometric approach to the (generalized) Dirac reduction is a special case of a general geometric reduction discussed in [**Wi**] and outlined briefly at the end of this section.

Dirac's basic prescription for reduction was the following. Given two functions on a constraint set V, extend these functions arbitrarily off V to functions f and g in $C^\infty(M)$, compute $\{f,g\}$, and restrict the result to V. In general $\{f,g\}|_V$ depends on the values of f and g off V. The largest class of functions for which $\{f,g\}|_V$ depends only on $f|_V$ and $g|_V$ is $N(I(V))$, the Poisson normalizer of the ideal $I(V)$ of functions that vanish on V. Routine computations show that $N(I(V))$ is a Poisson algebra with $I(V) \cap N(I(V))$ as a Poisson ideal. We define the Dirac reduced algebra to be

$$\mathcal{A}_D = \frac{N(I(V))}{I(V) \cap N(I(V))}.$$

(For the reader familiar with the terminology of first and second class functions and constraints, this definition gives $\mathcal{A}_D$ as the quotient of the first class functions by the first class constraints.) As in the case of the invariantly reduced algebra $\mathcal{A}_I$, the elements of the Dirac reduced algebra $\mathcal{A}_D$ can be distinguished by their values on V. Identifying those points of V which are not separated by any element of $\mathcal{A}_D$ produces the reduced space.

Recall that on a Poisson manifold, a function f defines a derivation $\{\cdot, f\}$ and hence a vector field, the Hamiltonian vector field of f, which we will denote by X_f. Let

$$EV = \{X \in TM \mid X = X_f(p) \quad \text{where} \quad f \in I(V) \quad \text{and} \quad p \in V\}.$$

When V is a manifold and M is symplectic, EV is the symplectic orthogonal complement of TV, i.e.

$$EV = TV^{\perp} = \{X \in TM \mid \omega(X, TV) = 0\}.$$

By definition h is in $N(I(V))$ if and only if for every $f \in I(V)$, we have $\{h, f\} = dh(X_f) \in I(V)$. The latter holds exactly when $dh(EV) = 0$, so

$$N(I(V)) = \{h \in C^{\infty}(M) \mid dh(EV) = 0\}.$$

Thus we have a more geometric characterization of $N(I(V))$.

This geometric characterization generalizes to produce still more reduction procedures. For simplicity in outlining of these procedures, we restrict for the moment to the symplectic case. (Cf. [**Wi**] for details and the Poisson, presymplectic, and infinite dimensional cases.) For any subset V, let

$$ZV = \{X \in TM \mid df(X) = 0 \quad \text{for every } f \in I(V)\}.$$

Note that $ZV \subset TM|_V$. Even though V may not be a manifold, we may choose as a "tangent space" TV any subset of ZV which is a vector space at each point of V. (If $TV = ZV$, then $TV^{\perp} = EV$ as defined above.)

Let

$$\begin{aligned}
F(TV) &= \{f \in C^{\infty}(M) \mid X_f(p) \in TV \quad \text{for every } p \in V\} \\
&= \{f \in C^{\infty}(M) \mid df(TV^{\perp}) = 0\}, \quad \text{and} \\
F(TV + TV^{\perp}) &= \{f \in C^{\infty}(M) \mid X_f(p) \in TV + TV^{\perp} \quad \text{for every } p \in V\} \\
&= \{f \in C^{\infty}(M) \mid df(TV \cap TV^{\perp}) = 0\}.
\end{aligned}$$

Then one can define a bracket "pointwise" on the quotient (commutative) algebras

$$\frac{F(TV)}{I(V) \cap F(TV)} \quad \text{and} \quad \frac{F(TV + TV^{\perp})}{I(V) \cap F(TV + TV^{\perp})}.$$

That is, given two functions f and g representing classes in one of these quotients, one can use Hamiltonian vector fields projected to TV to define the value of a new function $\{f, g\}$ at each point of V. This value will be well-defined but one must check that $\{f, g\}$ defines an element in the quotient and that the Jacobi identity holds. If so, we obtain a new reduced algebra—possibly several, using either $F(TV)$ or $F(TV+TV^{\perp})$ and making various choices for TV. A reduced algebra defines a corresponding reduced space as above (by identifying points of V not separated by functions in the algebra). For $F(ZV) = N(I(V))$, we obtain the Dirac reduced algebra $\mathcal{A}_D$. With a different choice of TV, $F(TV + TV^\perp)$ gives the geometric reduction of [**AGJ**].

Thus potentially we have whole families of generalized geometric reductions. In practice it is difficult to prove that one actually obtains a Poisson algebra for a particular choice of TV except in simple cases, when the resulting algebra usually coincides with $\mathcal{A}_D$. There are examples, however, where some choices yield strictly larger reduced algebras. Again, see [**Wi**] for details.

IV. Comparison Results

In many cases of interest, the Dirac and geometric reductions coincide. In particular when V is the level set of a momentum map and the MMW reduction applies, the Dirac and MMW reductions agree with each other and with the algebraic reduction. With mild additional assumptions they also agree with reduction by invariants. More generally $\mathcal{A}_A$ projects onto a subalgebra of $\mathcal{A}_D$, and this subalgebra contains $\mathcal{A}_I$. In this section we discuss the relation between the geometric and Dirac reductions, propose a new way to generalize the algebraic reduction to the $\mu \neq 0$ case, and prove the comparison results claimed above. The Dirac reduction also will agree with the Poisson reduction of Marsden and Ratiu [**MaR**] in many cases, including most of their examples; for this comparison, please see [**Wi**]. (The procedure in [**MaR**] includes a choice of a bundle on V, so this Poisson reduction is actually a family of reductions. Some coincide with $\mathcal{A}_D$, some with other of the generalized geometric reductions outlined in Section III.)

When the basic geometric reduction procedure (Theorem 2.1) applies, V is a submanifold. If V is not closed, then an extension problem similar to that mentioned near the end of section II may occur. That is, $\mathcal{A}_G$ may contain functions which do not extend smoothly off V and therefore are not in $\mathcal{A}_D$. Excepting this case, however, the Dirac reduction agrees with the basic geometric reduction.

Proposition 4.1 *Suppose V is closed and the geometric reduction procedure of Theorem 2.1 applies; i.e. the hypotheses of Theorem 2.1 hold. Then*

$$\mathcal{A}_G = \mathcal{A}_D.$$

Proof. As V is a submanifold, $\ker \omega = TV \cap TV^\perp$. Recall from Theorem 2.1 that we have a projection $\pi : V \to W$ with $\ker \omega$ tangent to the fibers of π. Suppose that $f \in N(I(V))$; thus $f|_V$ can be identified with an element of $\mathcal{A}_D$. Then $df(TV^\perp) = 0$, so f is constant along the fibers of π and determines a smooth function $\tilde{f}$ on W such that $f|_V = \tilde{f} \circ \pi$. Define a map φ from $\mathcal{A}_D$ to $\mathcal{A}_G$ by setting

$$\varphi(f|_V) = \tilde{f}.$$

If $f \in I(V)$, then $\tilde{f} \equiv 0$, and vice versa. Therefore φ is well defined and also one to one. (The latter depends on the fact that φ clearly preserves the commutative algebra structure.)

To see that φ is surjective, suppose that $\tilde{f} \in C^\infty(W)$ and let $g = \tilde{f} \circ \pi$. Then g is a smooth function on V with $dg(\ker \omega) = 0$. Because V is a closed submanifold, locally we can extend g smoothly off V so that $dg(TV^\perp) = 0$. By a partition of unity argument we obtain $f \in N(I(V))$ with $f|_V = g$, and thus $\varphi(f|_V) = \tilde{f}$.

It remains to show that φ preserves the Poisson bracket. First note that $f \in N(I(V))$ implies that X_f is tangent to V. From this one easily computes that $\pi_* X_f = X_{\tilde{f}}$, using $\pi^* d\tilde{f} = i^* df$ and $\pi^* \tilde{\omega} = i^* \omega$, where $i : V \to M$ is inclusion. Then on V we have

$$\{f, g\} = (i^*\omega)(X_f, X_g) = \pi^*\tilde{\omega}(X_f, X_g) = \tilde{\omega}(X_{\tilde{f}}, X_{\tilde{g}}) = \{\tilde{f}, \tilde{g}\}_W,$$

where $\{\ ,\ \}_W$ indicates the Poisson bracket on W. ∎

With minor changes this proof also shows that the geometric reduction defined using $F(TZ + TZ^\perp)$ at the end of Section III will also agree with $\mathcal{A}_G$ and $\mathcal{A}_D$.

The relationship between the Dirac reduction and the geometric reduction of [**AGJ**] is less simple. Examples exist in which $\mathcal{A}_G$ is strictly larger than $\mathcal{A}_D$. However, under the conditions shown in [**AGJ**] to suffice for the geometric reduction procedure to work, it is straightforward to show that $\mathcal{A}_G = \mathcal{A}_D$. Even when the reduced algebras coincide, the reduced spaces may not. In the geometrically reduced space there may be points that are not separated by functions in the reduced algebra. Typically the resulting space is not Hausdorff; see e.g. [**AGJ, Ex. 3.4**]. For more discussion of these ideas, see [**Wi**].

Now suppose that $V = J^{-1}(\mu)$. In order to include the case $\mu \neq 0$ neatly into our comparison results, we generalize the definition of the algebraic reduction as follows. Let $K = J - \mu$, so $V = K^{-1}(0)$, and let $I(K)$ be the ideal generated by the components of K. Routine calculations show that $N(I(K))$ is a Poisson algebra with $I(K) \cap (N(I(K))$ as a Poisson ideal, so we define

$$\mathcal{A}_A = \frac{N(I(K))}{I(K) \cap N(I(K))}.$$

When $\mu = 0$ this agrees with the previous definition. (It is an open question whether this method for handling the nonzero case agrees with the standard construction mentioned in Section II above.)

As $I(K) \subset I(V)$, there is a natural projection of $\mathcal{A}_A$ onto

$$\mathcal{B} = \frac{N(I(K))}{I(V) \cap N(I(K))}.$$

This quotient $\mathcal{B}$ plays a central role in comparing Dirac, algebraic, and invariant reductions. To see this, we shall use the following fact.

Proposition 4.2 $N(I(K)) \subset N(I(V))$.

Proof. Let $f \in N(I(K))$, and let $\varphi(t)$ be an integral curve of the Hamiltonian vector field X_f of f with $a \leq t \leq b$ and $\varphi(0) \in V$. Then $\{f, K\} \subset I(K)$, which implies that $d/dt(K \circ \varphi)$ is a sum of terms each of which has a component of K as a factor. Let $M(t)$ be the maximum of the absolute value of the components of $K(\varphi(t))$. If $M(t)$ is not identically zero, then one can show that it is bounded below by an exponential function on any finite interval. But $M(0) = 0$, so $M(t) \equiv 0$ and so $K(\varphi(t)) \equiv 0$. Thus $\varphi(t)$ must lie in V, and if $g \in I(V)$, then $\{g, f\} = X_f(g) = 0$ on V. In other words, $\{I(V), N(I(K)\} \subset I(V)$. ∎

Corollary 4.3 *The natural projection $\mathcal{A}_A \to \mathcal{B}$ is a Poisson algebra homomorphism. In particular, $\mathcal{B}$ is a Poisson algebra.*

Proof. This is a routine calculation; Proposition 4.2 is useful in showing that $I(V) \cap N(I(K))$ is a Poisson ideal in $N(I(K))$. ∎

Now we can prove our main comparison result.

Theorem 4.4 *The Poisson algebra $\mathcal{B}$ has a natural inclusion into the Dirac reduced algebra $\mathcal{A}_D$, and contains a Poisson subalgebra naturally isomorphic to the invariantly reduced algebra $\mathcal{A}_I$. Thus we have*

$$\begin{array}{ccccc} & & \mathcal{A}_A & & \\ & & \downarrow & & \\ \mathcal{A}_I & \hookrightarrow & \mathcal{B} & \hookrightarrow & \mathcal{A}_D. \end{array}$$

Proof. First we show the inclusion $\mathcal{A}_I \subset \mathcal{B}$. Invariant functions Poisson commute with components of J, so $\{C^\infty(M)^G, K\} = \{C^\infty(M)^G, J\} = 0$. Thus $C^\infty(M)^G \subset N(I(K))$, and we have a natural map from $\mathcal{A}_I$ to $\mathcal{B}$ whose image is all sets of the form $f + I(V) \cap N(I(K))$, where $f \in C^\infty(M)^G$. Suppose f and g are two functions in $C^\infty(M)^G$ that represent the same class in $\mathcal{B}$; thus $f - g \in I(V) \cap N(I(K))$. To show injectivity, it suffices to observe that $f - g \in I(V)^G = I(V) \cap C^\infty(M)^G$. Similar arguments, starting from Proposition 4.2, give the injection of $\mathcal{B}$ into $\mathcal{A}_D$. Corollary 4.3

gives the projection of $\mathcal{A}_A$ onto $\mathcal{B}$. All these maps preserve Poisson algebra structure because in each case it is descended from that on $N(I(V))$. ∎

If $\mathcal{A}_A$ differs from $\mathcal{B}$, then $\mathcal{A}_A$ is not a function algebra on V in a natural way. Let f be a function on M which represents a nonzero element of $\mathcal{A}_A$ and projects to zero in $\mathcal{B}$. Then f is identically zero on V. Thus $\mathcal{B}$ gives us a function algebra on V which is arguably the best such function algebra for approximating $\mathcal{A}_A$.

The example from [**ACG**] cited in Section II shows that the inclusion of $\mathcal{A}_I$ into $\mathcal{B}$ may be proper. It is possible for $\mathcal{B}$ to be a proper subalgebra of $\mathcal{A}_D$ but known examples are rather artificial. For instance, let J be a real valued function that vanishes exactly on the closed unit disk V in $\mathbf{R}^2$. Then the function x is not in $N(I(J))$ because $\{x, J\} = \partial J/\partial y$ is not in $I(J)$. This can be seen by observing that the ratio $(\partial J/\partial y)/J$ blows up on the boundary of V. But $N(I(V)) = C^\infty(\mathbf{R}^2)$ because df vanishes on V for each function f in $I(V)$. Therefore x represents a (nonzero) class in $\mathcal{A}_D$ which is not in $\mathcal{B}$.

Clearly the condition $I(K) = I(V)$ suffices to make $\mathcal{A}_A = \mathcal{B} = \mathcal{A}_D$. When J is the momentum for the action of a compact, connected group, the results of [**AGJ**] give weaker conditions sufficient for $\mathcal{A}_A = \mathcal{B}$, and show that in general for such groups $\mathcal{A}_I = \mathcal{B} = \mathcal{A}_D$.

As an application of these remarks, consider the homogeneous Yang-Mills fields studied by Gotay [**G**]. In order to construct a Yang-Mills field theory with gauge group G, there must exist an adjoint-action invariant inner product on the Lie algebra $\mathfrak{g}$ of G. We will assume that this inner product is positive definite. Gotay considered Yang-Mills fields on a trivial G-bundle over a flat spacetime $T^n \times \mathbf{R}$. With the assumption of spatial homogeneity, he identified the configuration space with

$$(\mathbf{R}^n)^* \otimes \mathfrak{g} \approx \bigoplus_{i=1}^{n} \mathfrak{g},$$

where gauge transformations are given by the adjoint action on the second factor. Using the inner product we may identify $\mathfrak{g}$ and $\mathfrak{g}^*$; then the phase space M may be identified with the direct sum of $2n$ copies of $\mathfrak{g}$. The induced action of G on M is the direct sum of $2n$ copies of the adjoint action, and is therefore orthogonal. If H is the image of G in the orthogonal group of the vector space M, then the action of G factors through the action of H, which is compact. For each reduction procedure, we will obtain the same results whether we use G or H, because the nonzero components of the momentum map and therefore the constraint set V are the same for H as for G. Applying the discussions after Proposition 4.1 and Theorem 4.4 to the action of H, we see that the Dirac reduction, the geometric reduction of [**AGJ**], and the reduction by invariants all produce the same reduced space and the same reduced algebra. If we specialize to the cases

$G = SO(3)$ or $SU(2)$, Gotay [**G**] shows that the algebraic reduction also yields the same reduced algebra. Thus in these examples, all these methods of reduction are equivalent.

Our final results show that if the MMW reduction procedure applies, then there is essential agreement among the reduction procedures.

Theorem 4.5 *Suppose the conditions necessary for the reduction of Meyer and Marsden-Weinstein hold. That is, suppose that M is a symplectic manifold, that J is an equivariant momentum map, that μ is a weakly regular value of J (so that $V = J^{-1}(\mu)$ is a manifold with $TV = \ker dJ|_V$), that G_μ acts properly on V, and that $\ker \omega$ has constant dimension. Then the MMW, algebraic, and Dirac reductions all agree (up to finite covers if G_μ fails to be connected). If in addition the coadjoint orbit of μ is closed and G acts properly on M, the reduction by invariants also agrees. Thus for weakly regular values of the momentum for a compact group action, all these reductions procedures agree.*

Proof. Weak regularity implies that $I(V) = I(K)$, so by Theorem 4.4 and the remarks immediately following, the algebraic and Dirac reductions agree. If the action of G_μ on V is free, then the orbit space of V is a manifold and the agreement with the geometric MMW reduction is immediate from Proposition 4.1. For the more general case, the orbit space is an orbifold. The local orbifold structure is given by S/Γ, where S is a local slice for the G_μ action and Γ is a finite subgroup of G_μ. (Cf. [**We**].) Smooth functions on the orbifold are exactly those that can be lifted to invariant functions on V. As $\ker \omega$ is the tangent space to the G_μ-orbits, the argument now proceeds as in the proof of Proposition 4.1. Theorem 3 of [**ACG**] gives the conclusion about reduction by invariants. ∎

Theorem 4.5 and the remarks and examples preceding it suggest that in most applications of interest, the MMW, invariant, algebraic, and Dirac reduction procedures are essentially in agreement. When these reductions differ, the Dirac reduction has the following advantages. By Theorem 4.5, among these reductions only the Dirac reduction is truly a generalization of the MMW reduction: it is the only one which both agrees with the MMW reduction whenever the latter applies and also always produces an underlying reduced space. (The algebra $\mathcal{B}$ also satisfies these conditions, but uses a mixture of ideals for its definition which has no particular motivation. Thus this reduction does not seem to be as natural as the Dirac reduction.) Furthermore, by Proposition 4.1 and Theorem 4.4 the Dirac reduction is useful for comparisons among the various reductions even if it is not the preferred reduction in a particular application. Finally, of all the reductions considered above, only the Dirac reduction procedure is truly universal in that it can be applied to any constraint set, no matter how the constraint set is defined.

REFERENCES

[AbM] R. Abraham & J. E. Marsden, "Foundations of Mechanics", Benjamin/Cummings, Reading MA (1978), pp. 298, 371.

[ACG] J. M. Arms, R. H. Cushman, & M. J. Gotay, *A Universal Reduction Procedure for Hamiltonian Group Actions*, in "Proceedings of the 1989 MSRI Workshop on Hamiltonian Systems," Springer (to appear 1991).

[AGJ] J. M. Arms, M. J. Gotay, & G. Jennings, *Geometric and Algebraic Reduction for Singular Momentum Mappings*, Adv. in Math. **74** (1990), 43–103.

[ArMM] J. M. Arms, J. E. Marsden, & V. Moncrief, *Symmetry and Bifurcations of Momentum Mappings*, Commun. Math. Phys. **78** (1981), 455-478.

[G] M. J. Gotay, *Reduction of Homogeneous Yang-Mills Fields*, J. Geom. Phys. **6** (1989), 349–365.

[KKS] D. Kazhdan, B. Kostant, & S. Sternberg, *Hamiltonian Group Actions and Dynamical Systems of Calogero Type*, Comm. Pure Appl. Math. **31** (1978), 481–507.

[Kr] F. Kirwan, "Cohomology of Quotients in Symplectic and Algebraic Geometry," Princeton University Press, Princeton (1984); *Partial Desingularisations of Quotients of Nonsingular Varieties and Their Betti Numbers*, Ann. of Math. **122** (1985), 41–85.

[MaR] J. E. Marsden & T. Ratiu, *Reduction of Poisson Manifolds*, Lett. Math. Phys. **11** (1986), 161–170.

[MaW74] J. E. Marsden & A. Weinstein, *Reduction of Symplectic Manifolds with Symmetry*, Rep. Math. Phys. **5** (1974), 121–130.

[MaW83] Jerrold E. Marsden & Alan Weinstein, *Coadjoint Orbits, Vortices, and Clebsch Variables for Incompressible Fluids*, Physica D **7** (1983), 305–323.

[Me] K. R. Meyer, *Symmetrics and Integrals in Mechanics*, in "Dynamical Systems" (Proc. Sympos. Univ. Bahia, Salvador, 1971), ed. M. M. Pexioto, Academic Press, New York (1973), 259–272.

[Sm] S. Smale, *Topology and Mechanics I, II*, Inv. Math. **10** (1970), 305–331, and **11** (1970), 45–64.

[SnW] J. Śniatycki and A. Weinstein, *Reduction and Quantization for Singular Momentum Mappings*, Lett. Math. Phys. **7** (1983), 155–161.

[So] J.-M. Souriau, "Structure des Systèmes Dynamiques," Dunod, Paris (1970).

[We] A. Weinstein, *Symplectic V-Manifolds, Periodic Orbits of Hamiltonian Systemes, and the Volume of Certain Riemannian Manifolds*, Commun. Pure Appl. Math. **30** (1977), 265–271.

[Wi] Don C. Wilbour, "Ph.D. thesis", University of Washington (in preparation).

Departement of Mathematics;   University of Washington
Seattle, WA 98195;   U.S.A.

List of Participants

Arms Judith, University of Washington
Audin Michèle, IRMA Strasbourg
Barthelemy Benoît, Marseille
Benenti Sergio, Università di Torino
Berenguier Guy, Marseille
Berger Roland, St Etienne
Bigot Georges, Marseille
Blattner Robert, UCLA Los Angeles
Bongaarts P., Instituut Lorentz, Leiden
Bordemann Martin, Freiburg
Borghero Francesco, Università di Cagliari
Boualem Hassan, Montpellier
Boucetta Mohamed, Montpellier
Boutat Driss, Lyon
Brada Claude, Avignon
Breuneval Jacques, Marseille
Burdet Guy, CPT-CNRS Marseille
Carinena José, Zaragoza
Chastenet De Gery Jerôme, CNAM Paris
Combescure Jacques, Montpellier
Coste Antoine, CPT-CNRS Marseille
Cushman Richard, Utrecht

Dardie Jean-Michel, Montpellier
Dazord Pierre, Lyon
Desolneux Nicole, Lyon
Donato Paul, Marseille
Dufour Jean-Paul, Montpellier
Duval Christian, Marseille
Elgradechi Mohamed Amine, Paris VII
Elhadad Jimmy, Marseille
Emmrich Claudio, Freiburg
Ewen Holger, Karlsruhe
Fliche Henri-Hugues, Marseille
Garcia Pérez Pedro Luis, Salamanca
Gotay Mark, Annapolis
Gourevitch Dmitri, Moscow
Gracia Xavier, Barcelona
Guieu Laurent, Marseille
Habch Bouchaib, Montpellier
Helmstetter Jacques, Grenoble
Holm Darryl, Los Alamos
Huebschmann Johannes, Lille
Iglesias Patrick, CPT-CNRS Marseille
Itzykson Claude, CEA Saclay

Kellendonk Johannes, Bonn
Koch Martin, Freiburg
Kosmann-Schwarzbach Yvette, Lille
Kostant Bertram, MIT Cambridge
LaFontaine Jacques, Montpellier
Lalonde François, Montrèal
Lee Ho Joong, Nantes
Lerman Eugene, Utrecht
Liardet Pierre, Marseille
Libermann Paulette, Paris
Lichnerowicz André, Collège de France
Maillot Henry, Lyon I
Manchon Dominique, Nancy
Marle Charles-Michel, Paris VI
Marsden Jerry, Berkeley
Medina Alberto, Avignon
Meinrenhen Eckhard, Freiburg
Molinier Jean-Christophe, Montpellier
Molino Pierre, Montpellier
Montgomery Richard, Berkeley
Morvan Jean-Marie, Lyon
Muñoz Lecanda M.C., Barcelona
Muñoz Masqué Jaime, Madrid
Nguiffo Boyom Michel, Montpellier
Niglio Louis, Avignon
Orro Patrice, Chambéry
Ouadfel Ali, Montpellier
Ouzilou René, Lyon
Patissier Guy, Lyon

Perrin Martine, CPT-CNRS Marseille
Pichon G., Paris-Nord Villetaneuse
Pradines Jean, Toulouse
Ratiu Tudor, Santa Cruz
Raugel Geneviève, Orsay
Rawnsley John, Warwick
Reyman Alexei, Steklov Inst. Leningrad
Roger Claude, Lyon
Roman-Roy N., Barcelona
Rosso Marc, Polytechnique Palaiseau
Roubtsov Vladimir, Moscow
Scheffold Bernard, Freiburg
Simms David, Dublin
Sjamaar Reyer, Utrecht
Śniatycki Jędrzej, Calgary
Souriau Jean-Marie, Marseille
Sow Mohamadou, Lyon
Sternberg Shlomo, Harvard
Taquet Bernadette, IRMA Strasbourg
Toure Alfred, Montpellier
Trotman David, Marseille
Tsenkos Ngae, Metz
Tuynman Gijs, Marseille
Vallée Claude, Poitiers
Vergne Michèle, Ecole Normale, Paris
Weinstein Alan, Berkeley
Zakrzewski, Stanislaw, Warsaw
Ziegler François, Marseille

Progress in Mathematics

Edited by:

J. Oesterlé
Département de Mathématiques
Université de Paris VI
4, Place Jussieu
75230 Paris Cedex 05, France

A. Weinstein
Department of Mathematics
University of California
Berkeley, CA 94720
U.S.A.

Progress in Mathematics is a series of books intended for professional mathematicians and scientists, encompassing all areas of pure mathematics. This distinguished series, which began in 1979, includes authored monographs and edited collections of papers on important research developments as well as expositions of particular subject areas.

We encourage preparation of manuscripts in some form of TeX for delivery in camera-ready copy which leads to rapid publication, or in electronic form for interfacing with laser printers or typesetters.

Proposals should be sent directly to the editors or to: Birkhäuser Boston, 675 Massachusetts Avenue, Cambridge, MA 02139, U. S. A.

A complete list of titles in this series is available from the publisher.